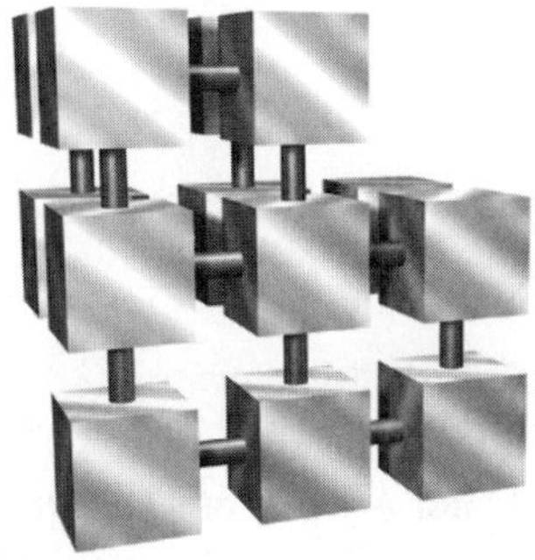

Modeling and Control of Dynamic Systems

Narciso F. Macia
George J. Thaler

THOMSON
DELMAR LEARNING™

Australia Canada Mexico Singapore Spain United Kingdom United States

THOMSON

DELMAR LEARNING

Modeling and Control of Dynamic Systems

Narciso Macia

Vice President, Technology and Trades SBU:
Alar Elken

Editorial Director:
Sandy Clark

Senior Acquisitions Editor:
Steve Helba

Senior Development Editor:
Michelle Ruelos Cannistraci

Marketing Director:
Dave Garza

Channel Manager:
Dennis Williams

Marketing Coordinator:
Casey Bruno

Production Director:
Mary Ellen Black

Production Manager:
Andrew Crouth

Production Editor:
Stacy Masucci

Editorial Assistant:
Dawn Daugherty

Printed in the United States of America
1 2 3 4 5 XX 06 05 04

For more information contact Thomson Delmar Learning
Executive Woods
5 Maxwell Drive, PO Box 8007,
Clifton Park, NY 12065-8007
Or find us on the World Wide Web at
www.delmarlearning.com

Library of Congress Cataloging-in-Publication Data:
Macia, Narciso.
Modeling and control of dynamic systems / Narciso Macia.
p. cm.
Includes bibliographical references and index.
ISBN: 1-4018-4760-9
1. Automatic control—Mathematical models.
2. Feedback control systems—Mathematical models.
I. Title.
TJ213+

2004058056

NOTICE TO THE READER

Publisher does not warrant or guarantee any of the products described herein or perform any independent analysis in connection with any of the product information contained herein. Publisher does not assume, and expressly disclaims, any obligation to obtain and include information other than that provided to it by the manufacturer.

The reader is expressly warned to consider and adopt all safety precautions that might be indicated by the activities herein and to avoid all potential hazards. By following the instructions contained herein, the reader willingly assumes all risks in connection with such instructions.

The publisher makes no representation or warranties of any kind, including but not limited to, the warranties of fitness for particular purpose or merchantability, nor are any such representations implied with respect to the material set forth herein, and the publisher takes no responsibility with respect to such material. The publisher shall not be liable for any special, consequential, or exemplary damages resulting, in whole or part, from the readers' use of, or reliance upon, this material.

This book is dedicated to the late George J. Thaler, who originally was going to be the first author of this book, and who exhibited commendable self-control in his life: by keeping the proper balance in sharing his life among God, family, physical activity (tennis) and his academic profession. Without his contribution this book would not exist.

And to my wonderful family, Donna, Maria, Graciela, Rebeca, Melinda and Cisito who have supported me whole heartily in this effort.

Contents

5 Stability and Dynamic Response

6 Time Domain Performance Characteristics

7 Root Locus Analysis

9 Introduction to State Space Methods

10 Design of Control Systems

Preface

Introduction

This book is intended as a first text in dynamic systems and their control, appropriate for undergraduate engineering technology and engineering students. It can also be used as a graduate text if additional assignments are given to graduate students.

What type of student should take such a dynamic system course? In my opinion all engineering technology and engineering students, regardless of their area of specialization, because a good grasp of dynamic behavior will open doors to a better understanding of real-world problems. Critical elements in the formulation of a solution to problems are often overlooked if only the static or steady-state conditions are considered. This limited perspective produces solutions that are short lived or do not solve the problem for its entire range of operation. Every effort has been made in writing this book to equip the reader with the tools necessary for assessing the issues associated with the dynamic aspects of the problem. Needless to say, we have also encouraged order-of-magnitude analysis, based on static conditions to 'bracket' the solution.

Today's students find curriculums that are over packed with subjects. Such curriculum takes students on an endless introduction of many subject areas and often they fail to grasp the beautiful harmony that exists among the various disciplines. I am a strong believer that studying dynamic systems has the potential of galvanizing the underlying concepts found in the many subjects taken by students, and allows them to apply these concepts to the real world. It also helps students transform a bunch of apparently disconnected ideas and principles into a beautifully interrelated framework. We have made every effort to assist the student in this endeavor. It is a great experience to see students discover this harmony in the behavior of dynamic systems, and help them become more confident in the subjects they have already learned.

This book uses simple, every-day examples that most students have previously experienced, as vehicles for teaching dynamic system principles. It provides a step-by-step explanation of how to obtain a "model schematic" for the physical system, a graphical representation that captures the essence of the dynamic system under question. It shows how to transform this model schematic into a mathematical model (differential equation or transfer function), which then can be simulated (used to obtain a time response) with MALTAB/SIMULINK.

The book immerses students in one of the most powerful and easy-to-use software environments ever developed: MATLAB/SIMULINK. The SIMULINK environment provides a similar environment to that used by engineering students 25 years ago: solving differential equations and dynamic systems with an analog computer. However, today's students, by using MATLAB/SIMULINK, possess a much richer set of displaying and troubleshooting tools than their predecessors did. In addition, today's students have at their disposal a large number of nonlinear blocks that permit the evaluation of their effect on the response. Similarly, students can easily see the effect of changing a time constant or a gain.

Implementation

In Chapter 11, Implementing the Controls Scheme with Hardware: PLCs, students are exposed to the reality that there are many ways of realizing the control task. For instance, the engineer could use a mainframe computer, a micro-controller, or a PLC (Programmable Logic Controller) to mention just a few possibilities. It offers a taste of the process that an engineer has to go through in making this selection, by bringing into the picture issues such as cost, development time, computational power, size, and so on. The book assumes that, for the most part, the PLC will be the most likely vehicle for the reader to implement his control scheme. For this reason, it offers an introduction of the discrete-event controlling capability of the PLC and then goes into detail of how to use the PLC for implementing a closed-loop PID (Proportional + Integral + Derivative) loop.

DC Solenoid

One of the instruments I use to facilitate a deepening sense of inquiry is a series of exercises and experiments that focus one specific piece of hardware: a DC-solenoid, position control system. This allows students to develop depth with at least one dynamic system and its controller. This goal is embodied in Chapter 13 and also the accompanying Lab Manual. It is my intent that at the end of the semester students will feel very comfortable with the drivers, sensors, and hardware, and with the underlying control methods used in this particular control system's characterization, analysis, and design.

Idea Portfolio

In addition, I use a suggestion made by a former graduate student, Sapto Susilo, to increase the interest and perceived relevance of dynamic systems and control. Each student picks an idea of their own, and relates it to all the tools and methods presented in the control class. The intention here is that this idea, which is of particular interest to the student, might serve as an effective motivator for him or her to invest additional time exploring this field of study. The implementation of this suggestion is what I call the Idea Portfolio. It represents a series of questions at the end of each chapter that will hopefully motivate students to engage in an investigative process where they produce sketches, estimate model parameters, and perform simulations as they investigate their ideas from the perspective of a controls engineer. Appendix G contains additional suggestions and provides a useful form that can guide students through this creative process.

Implementation Platforms

One goal of this class is to help students gain an appreciation for the various platforms for performing the control action. There are many ways of performing the control action, as discussed in Chapter 11. However, after reflection and experimention regarding which piece of hardware would be the most practical as a teaching tool, I chose the Programmable Logic Controller (PLC). This takes into consideration the fact that I have students in the dynamic systems and control class that come from electronics, computing, mechanical, and manufacturing backgrounds. The PLC is a controller platform that all engineering students can use, regardless of whether they are familiar with electronics or programming. The PLC is also the backbone of the manufacturing operation and is used in the testing of many systems due to the ease in which it can be programmed.

Supplements

Lab Manual includes over 20 hands-on projects and experiments.

(ISBN#: 1401885977)

Instructor's CD-ROM includes solutions to textbook and lab manual.

(ISBN#: 140189254X)

About the Authors

This book integrates dynamic systems and their control, two fields that are closely intertwined. I have been involved (and fascinated) with dynamic systems since my sophomore year as a mechanical engineering student. I have nurtured this interest on a consistent basis as a graduate student, an engineer employed in the industry, and as a university professor. I have also been involved with the analysis, design, and implementation of controllers ranging from fluidic circuits to Programmable Logic Controllers (PLC). In the course of this journey, I have come to appreciate the strengths and weaknesses of these platforms. My late co-author, the distinguished professor George J. Thaler, dedicated his professional life to the development and teaching of control theory. He authored and co-authored numerous books and articles on the subject. His contribution regarding how to design a control system will prove extremely valuable for any control engineer. Without any doubt, this book would not have been possible without his contribution.

Acknowledgments

I am extremely grateful to those individuals who encouraged me in the pursuit of writing a book. I also would like to acknowledge those teachers and professors who believed in me or imparted to me the knowledge that has allowed me to engage in this monumental (for me) writing task. Among them are Linda Darnell, Charles C. Blackwell, Robert L. Woods, David Hullender, Robert White, Walter Higgings and William J. Dorson.

The team at Delmar has been extremely supportive and patient: Michelle Ruelos Cannistraci has provided a steady and politely firm hand trying to keep this book on schedule. It has also been a pleasure working with her collaborators Linda DeMasi, Kathleen Vonk, Jon Duff, and Stacy Masucci. Other individuals have been a source of encouragement in this pursuit: Mimi Villaca, Charles Okonkwo, Antonio Rodriguez, and Jose Manuel Campoy. I would like to convey my gratitude to the many reviewers who reviewed the manuscript. Their suggestions have made this book several orders of magnitude more clear. Thank you to the following reviewers:

Sohail Anwar, Penn State University

Ray Bachnak, Texas A & M University

Harold Broberg, Indiana Purdue University Indianapolis

Warren Hill, Weber State University

Kathy Ossman, University of Cincinnati

Chia-Chi Tsui, DeVry University

I am also extremely grateful to the many students who corrected errors, suggested control problems, or expressed a better way to convey a principle. They also gave me the opportunity to collaborate on their thesis or applied projects. We both learned, who learned more I am not sure.

I am extremely eager to improve this effort and welcome any suggestions or corrections that you might have.

Narciso F. Macia
Arizona State University Polytechnic
macia@asu.edu

1

Introduction

Learning Objectives

In this chapter, you will learn how to:

- Recognize dynamic systems
- Characterize closed-loop systems
- Differentiate between open- and closed-loop systems
- Identify uses of open and closed-loop systems
- Be aware of unstable feedback systems
- Name the tests a control engineer performs to identify models

1.1 Introduction

This book is concerned with systems. A system is a device, or a collection of devices, for accomplishing some task. This definition implies that a system must have at least one input and one output and that there is a cause-and-effect relationship between the two (i.e., a change in the input causes a change in the output).

In this book, we are concerned with dynamic systems—systems for which the input causes the output to vary with time. For example, let the dynamic system be a simple integrator, and let the input be a positive constant. For this type of input, the output of the integrator increases linearly with time (e.g., the output is a ramp). Dynamic systems are encountered in every engineering discipline, and it might even be argued that all systems are dynamic systems. We will deal primarily with electrical, fluid, mechanical, and thermal systems, but the skills, tools, and approaches you will develop can be used with other types of systems, including chemical, economic, and population dynamic systems.

Most of this text is devoted to one class of dynamic systems: closed-loop feedback controls. Such systems are used throughout our modern culture. Applications include:

- Speed and fuel controls in vehicles
- Electronic circuits
- Ship and airplane controls
- Satellite tracking
- Fluid and chemical process controls
- Temperature controls (e.g., home heating)
- Medical instruments
- Computer hardware and software

New ideas requiring feedback controls arise daily.

The dynamics of any physical system are governed by the laws of nature. To understand the behavior of a system, you must determine which laws apply. At this point, you typically draw a simplified model schematic, which captures the relevant characteristics. You then use this model schematic to obtain a mathematical model of the system. One common form of mathematical model is a differential equation or a set of them. Once these equations have been obtained, alternate methods for solving them are available. Some methods are analytical, and others are numerical. The latter generally use computer solutions.

1.2 Models

To design, analyze, or modify a dynamic system, you must understand how the system works. A primary objective of this text is to help you learn a procedure that will enable you to determine the characteristics of a dynamic system. Another objective of this text is to help you develop the ability to analyze the performance of dynamic systems, which requires the use of computers. It is too time-consuming and expensive to develop a system using trial and error with a prototype model. Today, the hardware prototype is needed only after extensive computer studies of the system model have been performed.

Because computer studies (analysis and design) are the intended uses of a mathematical model, the model itself should be in a form that can be entered easily into the computer. Two common forms are the transfer function block diagram form and the state-space form. The choice of model form may depend on user preference, on software, on the nature of the dynamic system, on the type of behavior that is expected, or on the intended use of the system.

Simulation is the process of entering the mathematical model into the computer and applying chosen signals that help analyze the system. The computer programs determine the resulting output based on the chosen inputs. The most commonly used model frequently consists of block diagrams and transfer functions. The individual transfer functions represent sensors, the dynamic system under investigation or the controller itself. Commercially available simulation programs, such as MATLAB and its toolboxes (e.g., SIMULINK), enable the engineer to readily enter a model into the computer. This reduces the burden of having to write detailed software and enables the engineer to concentrate on analyzing or designing the problem.

1.3 Open-Loop Systems

Consider a gas furnace that is used to heat a house. If you have no thermostat and no suitable temperature measurement, you can calibrate the valve that is used to adjust the gas flow. On a given day, you can estimate room temperature in the morning and set the gas valve to a calibrated value intended to keep the house comfortably warm all day. This is an open-loop system and is shown in Figure 1–1. The input is the valve setting. The output is the temperature in the house. For some weather conditions, the house temperature remains acceptable. Notice, however, that if external weather conditions are colder than expected, the house will be colder than desired, because the valve setting, and hence the gas flow are constants.

In an open-loop system, note that the signal flow is directly from input to output. If the output is not at an acceptable value, there is no automatic correction available; you may only make a correction manually by adjusting the input valve setting. Open-loop systems, if their performance is adequate, tend to be simpler and less expensive than feedback controls and usually require less maintenance. They are acceptable for many applications.

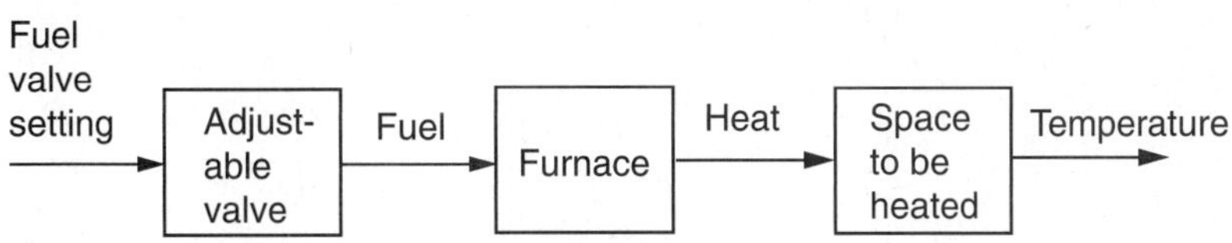

Figure 1–1 House heating—open loop

1.4 Feedback Controls

A feedback or closed-loop control may have a block diagram similar to an open-loop system, but it operates according to the following simple principle: The signal to be measured (and controlled) is the output signal, and it is compared with the input signal by subtraction. Most feedback controls are duplicators in the sense that they try to produce an output that duplicates the input, especially when the system is in steady state. The difference between the input and the output is called the "error signal."

In normal practice, the error signal is used to reduce the difference between input and output. Consider once again the open-loop gas furnace, and assume that a thermostat is available. The thermostat contains a temperature-measuring device (usually a bimetallic strip), which is calibrated for the desired range of room temperatures and can be set at a chosen value. When the temperature is at the chosen value, the gas valve allows a corresponding flow of gas. When the room temperature increases more than a specified amount, the thermostat reduces the valve opening. The process of measuring the room temperature and using this measurement to modify the input gas flow makes the system a feedback control system (closed-loop system). It automatically corrects for changes in the output (room temperature). This process is shown in Figure 1–2.

The advantage of automatic correction is one obvious reason for using feedback control. Other reasons arise with other systems. The major drawback is cost, which, in the case of the gas furnace, comes from the thermostat.

Sample Feedback Controls

Consider the following examples, and reflect on how to transform them into closed-loop (feedback) systems. Think about the benefits and disadvantages of both open-loop and closed-loop systems.

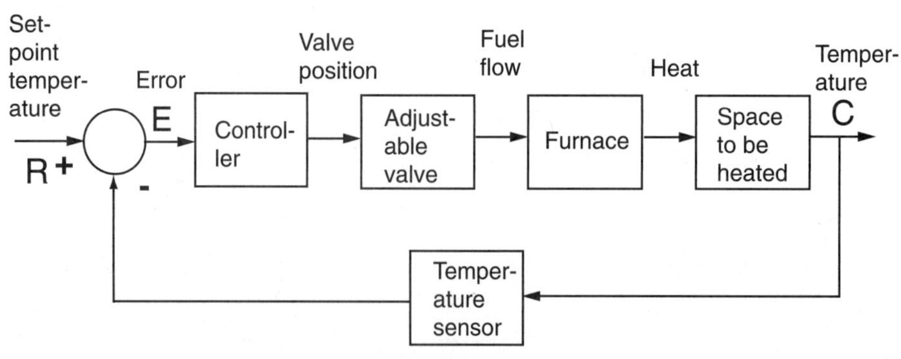

Figure 1–2 House heating with feedback (closed loop)

Conveyor Belts Envision a system that moves boxes from one level to the next, like the system that carries luggage in an airport. The system moves along a 30° incline. The driver for this conveyor belt is a motor connected to a constant 12 V source. What happens to the speed of the conveyor belt when the weight of the boxes being moved increases?

Level Controllers Consider a tank whose level must be kept constant. This tank receives water from three pipes with valves and fixed orifices at the end. The valves turn on and off randomly, producing a varying inlet flow rate. The tank has a modulating valve at the bottom that can be manually or automatically adjusted to control the water leaving the tank. Figure 1–3 shows the system.

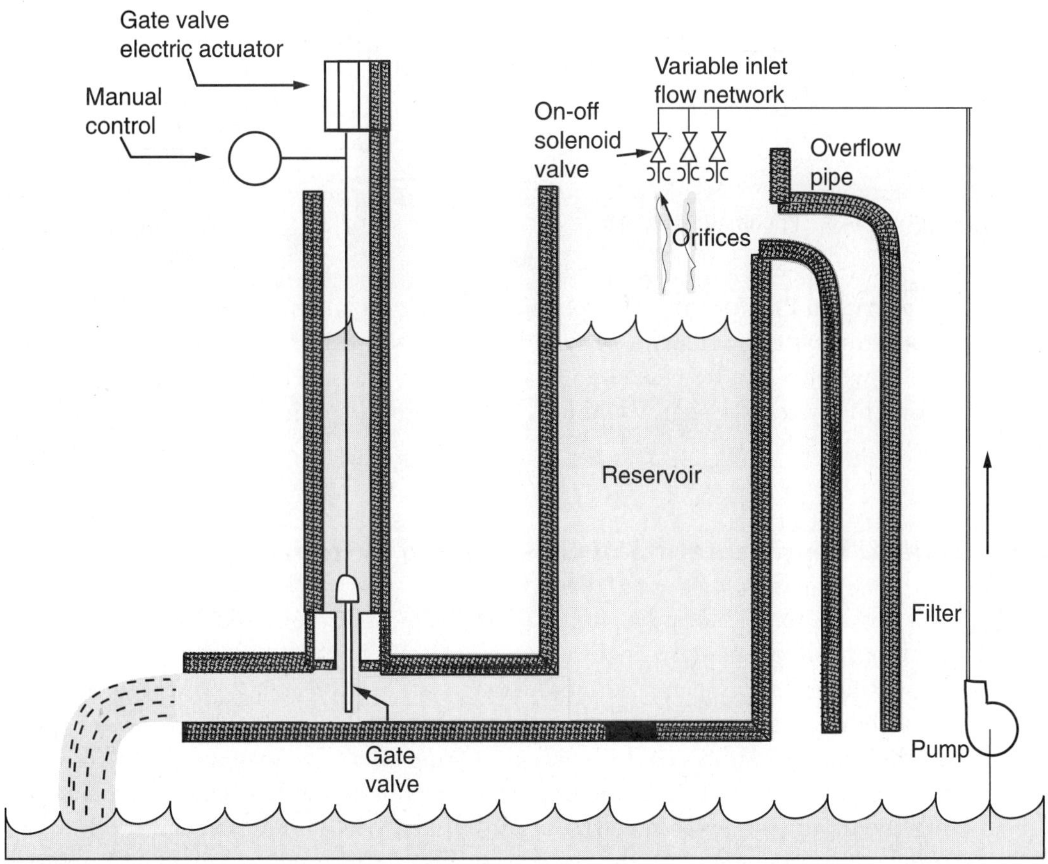

Figure 1–3 Level control system

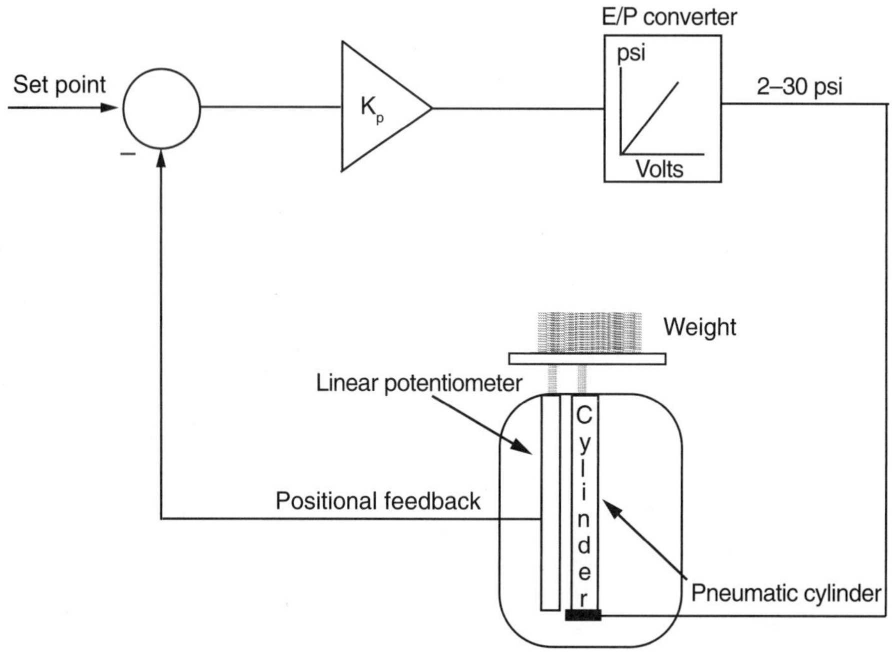

Figure 1–4 Pneumatic platform

Pneumatic Platforms The pneumatic platform is similar to a pneumatic elevator. In this system, a pneumatic cylinder supports a platform on which boxes of different weights can be placed and moved from one level to another. Figure 1–4 shows the system.[1] What would likely happen to the position of the cylinder if the weight of the box was to increase?

1.5 General Representation of Closed-Loop Systems

It is possible to use block diagrams to describe most of the closed-loop systems covered in this book, namely single-input, single-output (SISO) systems. The following example describes another closed-loop system. Consider a potter's wheel with an automatic speed control. This system is designed to maintain a constant speed on the wheel, regardless of the braking/resisting effect the system experiences when the potter squeezes the clay. Figure 1–5 shows this system's block diagram, including the input (reference speed), output (wheel speed), and disturbance (braking torque the potter applies to the clay piece). This representation, paired with transfer functions (covered in the next chapter), provides a unified framework with which control engineers, regardless of their areas of specialization, communicate.

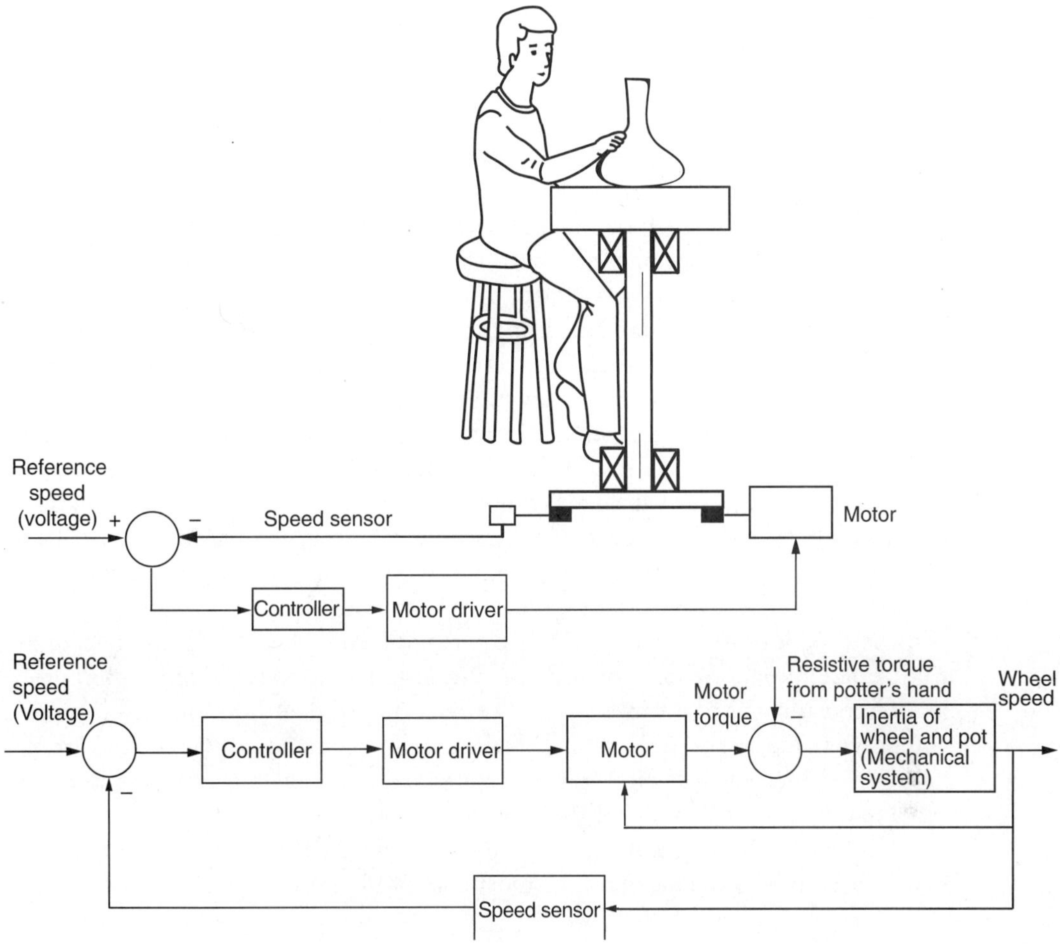

Figure 1–5 Schematic and block diagram of a potter turning a piece on an electric wheel

1.6 System Analysis from Models

A device or a system may need more than one model. For example, architects planning a house may start with sketches of the outside structure to convey the expected appearance. They may even prepare a three-dimensional model, which is static rather than dynamic. While initial models may not be to scale, precise parameter values (e.g., lengths, wall locations) will be needed before construction can begin.

Similarly, the model of a dynamic system may start as a sketch or a diagram in which each block represents a component or a piece of hardware. Some blocks may represent static devices, but the most significant components will be dynamic.

For each block, a mathematical description is needed if the dynamic behavior of the system is to be analyzed. The equations describing the dynamic parts are usually ordinary differential equations derived using the basic laws of nature (e.g., Newton's laws, Kirchhoff's laws). Often these differential equations are transformed into transfer functions to simplify their solutions. Note that accurate values for the parameters are needed if the analysis is to be trusted.

The blocks in the block diagram model may be represented by equations, suitable for computer implementation. Initially, long-hand solution of the equations may be desirable, but simulation to aid system analysis often requires multiple solutions of the equations for different initial conditions, different kinds of test signals, and different parameter values (to check tolerances). The most commonly used block diagram models use transfer functions to describe the hardware in the blocks.

To obtain meaningful simulation results, the model itself must be an accurate description of the system. If the hardware exists and the basic equations are derived, then parameter values for components can be obtained from calculations, dimensions, and handbook data. Finally, if test data from the hardware is available, the computer model is tested under identical conditions and the results compared. If there are substantial differences, the computer model is considered to be in error, and the model must be improved. If the results from the model and system are comparable, the model may be used in system analysis. Note that in design problems, frequently the hardware has not been selected, and an "ideal model" is used (a computer model, of course). This "ideal model" is considered the equivalent of a hardware model, and its performance is compared to that of the proposed system to validate the computer model of the proposed system.

The nature of the test signals used in model validation and model studies are varied, as seen in later chapters. The most commonly used test signals are:

Sinusoidal input—A characteristic of a linear system is that when it is excited with a sinusoidal input, it produces a sinusoidal output with the same frequency. To evaluate the system, the system can be excited with a sinusoidal input of a specific frequency. The output signal is then compared to the input using magnitude and phase data. When the process is repeated for many frequencies it is called a frequency response test. The result can be graphed using a special chart called a Bode Plot.

Step input—If the input is suddenly changed from one value to another (as quickly as possible), the output can be evaluated with respect to its total change and the speed of the transition.

Ramp input—The input can be a signal that increases linearly with respect to time. Of course, this process can only be maintained for a finite time, because eventually the physical system will reach a limit.

Impulse input—An impulse is a very short input relative to the dynamics of a system. For instance, a kick to a floating log and one hammer strike on a nail are impulse inputs.

Step change in a parameter value—A step change in a parameter value occurs when a parameter in the system (e.g., friction coefficient, weight) changes suddenly. The output is then analyzed with respect to the modified parameter.

To perform such tests and analyze the results, an engineer must know what to expect and how to interpret the results. The following chapters provide the needed background.

1.7 Stable and Unstable Systems

Even though closed-loop systems can provide much better control than open-loop systems (they are more accurate, faster, etc.), they are more difficult to implement. It is also much easier for a closed-loop system to become unstable when the loop is closed.

The concept of instability might be new, because almost all the systems you see are stable. During early development, however, these systems were likely unstable. Some engineers worked diligently to render them stable.

When trying to understand stability, it helps to separate the instability of the dynamic system (the plant) from the instability of the complete feedback system (e.g., a closed-loop system). It is possible to have a stable feedback system even though the plant (the physical system being controlled) is inherently unstable.

Sample Stable and Unstable Systems

Following are examples of stable, conditionally stable, unstable, and almost-unstable systems for plants (or physical systems) and complete feedback systems.

Plants/Physical Systems

- Ball on top of a peak (unstable)
- Inverted pendulum (a stick attached to a bearing at the bottom) (unstable)
- The process of heating water with a heating element (conditionally stable, depending on applied voltage) (Too much heat dissipation causes a vapor film to form around the heating element, which impedes further heat dissipation. This in turn causes the heating element's temperature to increase and subsequently burn up.)
- Lightbulb with a carbon filament (stable) (Carbon's resistance increases with increasing temperature until a stable temperature is reached.)

- Ball suspended by an upward air stream (stable) (Bernoulli's Law maintains the ball in the center of the air stream.)

Complete Feedback (Closed-Loop) Systems

- Braking a bicycle using the front brake (almost unstable) (The center of gravity tries to get in front of the point of restraint, the front wheel.)
- Tail dragger airplane, as it slows after landing (almost unstable) (Same as previous example. A large portion of the weight (and the center of gravity) is behind the two front wheels. As these wheels brake, there is a tendency for the center of gravity (and tail) to continue moving. This effect can easily make the airplane rotate.)
- Balancing an upside-down broom on the palm of your hand (stable, but difficult)
- Bicycle (stable) (Even though the center of gravity is above the point of support, where the wheels touch the ground, the cyclist and the gyroscopic effect of the wheels make it stable.)
- Riding a bicycle by holding the handlebars with arms crossed (unstable for most people)
- Sound system or public address (PA) system with too much feedback (unstable)
- Two people in heated argument (unstable)
- Cancer, which involves uncontrollable cell growth (unstable without chemical intervention)
- Trying to slowly stop a vehicle when sitting on a slippery seat (unstable) (As the vehicle decelerates, the driver slips forward, applying more pressure on the brakes and increasing the rate of deceleration.)
- Telephone call center that is experiencing an overload (unstable) (people call, they wait for a while, hang up, and then call again)
- Stopping an 18-wheeler (semitruck) on a:
    - Slippery surface (almost unstable)
    - Dry surface (stable)
- Trying to land an airplane when the trim is not adjusted (almost unstable)
- Balance/vestibular abnormality

Balance/vestibular abnormality is a balance disorder that keeps individuals from being able to walk. To walk, the brain receives three types of inputs:

1. **Vestibular**—The sensors associated with vestibular input are in the inner ear. They act like the gyroscope in an aircraft, and they are very effective in responding to quick changes in position.
2. **Propioreceptor**—The sensors associated with propioreceptor input are in the joints of the body. They sense the amount of force the joint must transmit.
3. **Visual**—The image received by the eyes and transmitted to the brain through the optical nerve.

> Normal individuals receive signals from the three preceding sensors and arrive at proper muscle coordination/command so as to walk properly (stable system). Individuals with balance/vestibular abnormalities, on the other hand, cannot integrate the signals and cannot walk without falling (unstable system). Interestingly, when such individuals are blindfolded, they can walk without falling, and the system is stable with respect to walking.

As you can see, any complicated closed-loop system can be unstable. Consequently, there is a need to predict whether the system will be stable or not. This is important, because an unstable system can damage itself, the surrounding structure, and/or personnel.

It is also important to determine how close a system is to instability, which can be done by performing a process similar to that which mechanical engineers use to determine the "factor of safety" in a mechanical part under stress. A large factor of safety allows for worse material properties than assumed, greater loading than assumed, and small errors/ignored effects in the calculation process. In control systems, some parameters and, consequently, the entire model, may change in an undesirable direction in response to temperature, wear and tear, or manufacturing tolerances.

1.8 The Control Engineer's Role

A control engineer is involved in three types of assignments:

1. **Modeling and simulation of a dynamic system**—The modeling and simulation task seeks to determine the behavior of a dynamic system, such as the level in a reservoir after a 100-year storm. In another example, when someone bungee jumps, this task could determine the maximum tension on the cord and how far the cord would stretch. Notice that these examples are not feedback control systems.
2. **Converting an open-loop system to a closed-loop (feedback) system**—The engineer must carry out all the steps to convert an open-loop system to a closed-loop one. These steps include determining adequate specifications, selecting sensors, selecting modulating components, designing the control law, and selecting and programming the appropriate controller.
3. **Improving the performance/behavior of a closed-loop system**—Here the feedback system already exists but its performance must improve. Typical examples of improvement include:
    - Improving accuracy
    - Improving response time
    - Ensuring that the system remains stable over time

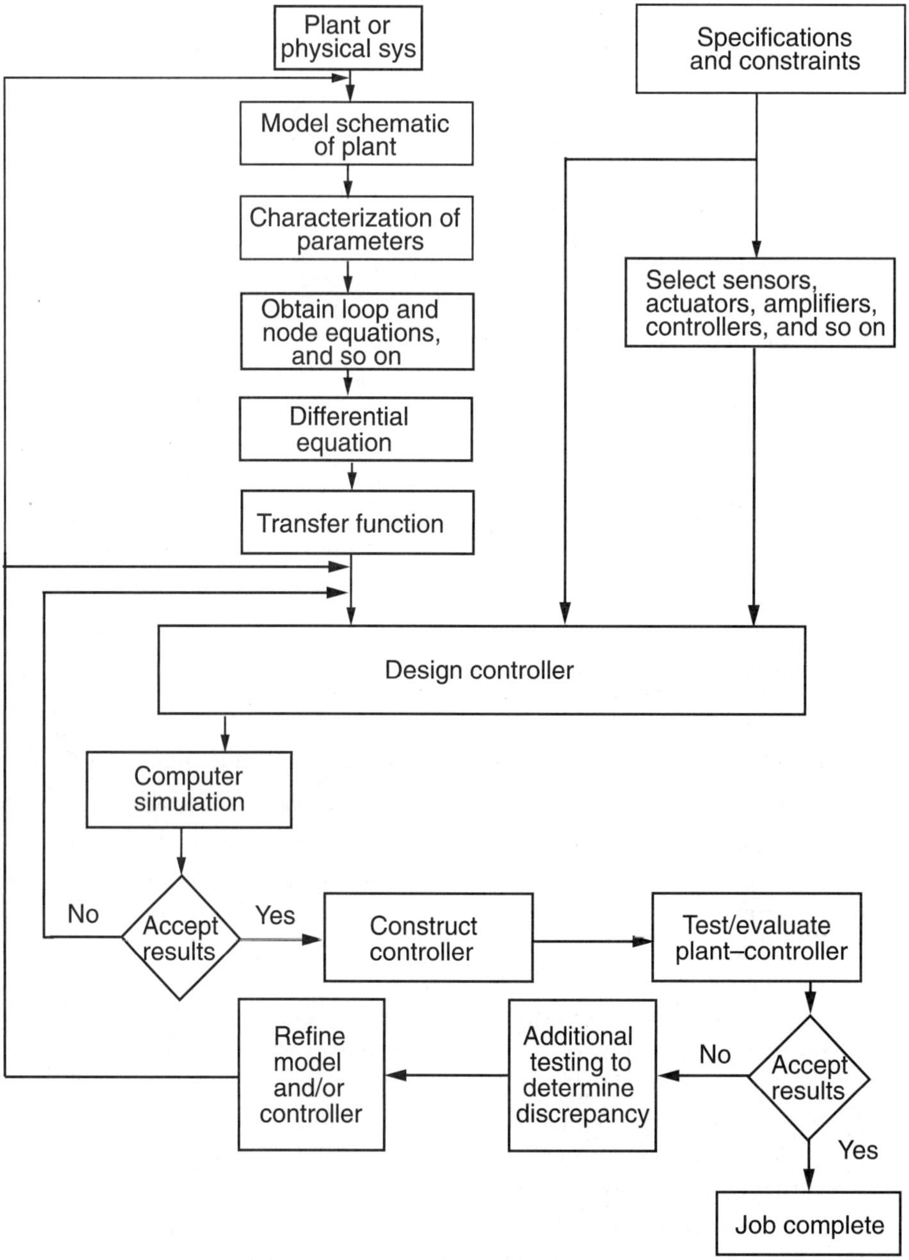

Figure 1–6 Overview of the design process control engineers use

Tasks 2 and 3 typically performed by a control engineer, can be represented in the block diagram in Figure 1–6.

The purpose of this text is to give you the methodology and tools you need to carry out such tasks.

Summary

1. Closed-loop systems have significant advantages over open-loop systems.
2. Block diagrams are good for representing the interconnections of dynamic system components.
3. Some systems are inherently stable, while others are unstable. Appropriate feedback can make unstable systems stable.
4. A control engineer improves overall system performance by designing an appropriate control system.

Idea Portfolio

For an existing system

1. What dynamic system interests you? How is this system controlled? Which device performs the control action? Is it analog or digital? How does it accomplish the control action? What are the consequences of not controlling the system within the limits? Which situations keep the controller from doing what it is supposed to do?
2. If you can access the control system of interest, perform a series of experiments to familiarize yourself with it. Of course, only experiment when safe for you and the equipment. Typical experiments could be:
   - Turn the system on and observe its output.
   - Did the output oscillate before achieving its final value?
   - Was there some noise in the output or input?
   - When you disconnected the sensor that measures the output variable, what happened to the system?
3. Create a block diagram of the control system that includes:
   - Input
   - Prefilter controller
   - Plant (dynamic system that must be controlled)
   - Output
   - Sensor (device that converts the output signal to a signal that can be measured and used to perform the control function)

4. What kind of controller is used?
5. Can the control strategy be modified?
6. Which changes in the control strategy would increase the overall value of the system?
7. Is it possible to instrument the system so that you could monitor its performance and input/output relationship?

For a new system and/or controller (hardware does not exist or is inaccessible)

8. What is the ideal way to control the system?
9. Which control scheme would you use?
10. Would it use an open-loop or a closed-loop controller?
11. What are the benefits of increasing the accuracy of the controlled variable?
12. Can the system tolerate a large discrepancy between the command (input) and the output?

Problems

1. Provide ten examples of open-loop systems.
2. For each of the following fields, provide one example of a closed-control system. Use a block diagram to explain it.
   - Manufacturing operation (e.g., machining process)
   - Medical
   - Economics
   - Home application
   - Aerospace
   - Automotive
   - Human relations
   - Communication systems
   - Water distribution
   - Chemical processing
   - Cooking
   - Transplantation
   - National park management
   - Employment behavior in a manufacturing plant
3. Give an example of a system that is operated in the open-loop mode, and that should be converted to a closed-loop control system.
4. List two main advantages of an open-loop system.

5. List two main disadvantages of an open-loop system.
6. List two main advantages of a closed-loop control system.
7. List two main disadvantages of a closed-loop control system.
8. A runner is running on a motorized treadmill, moving the belt at a constant speed. Is this an open- or a closed-loop system? What is the output (i.e., the controlled variable)? What is the desired condition of the controlled variable? Produce a block diagram of this system.
9. In each of the following scenarios, indicate the type of problem. You need not come up with a solution, just one of the three following answers for each of the following scenarios.

   DYN SYS For dynamic system analysis

   OL to CL For conversion from an open-loop system to a closed-loop one

   CL IMPR For improvement to a closed-loop system

   - A patient is on 100 percent oxygen, then suddenly removed from it and allowed to breath room air. How long will it take for the patient's blood oxygenation level to achieve a new value?
   - A house thermostat allows too much variation in the house's temperature. You are asked to reduce this variation.
   - You want to improve/enhance your drip irrigation system so that it delivers the exact amount of water every day.
   - You like to listen to the radio in your car, but you find yourself continually adjusting the volume. When you are at an intersection, the radio seems too quiet. Solve the problem.
   - When a box with some protective equipment, surrounded by protective foam, falls from a 5-foot shelf, what is the maximum acceleration (g's) the delicate equipment experiences?
   - When you place a room-temperature (80° F) soda in a refrigerator (32° F), how long does the can take to cool to 50° F?
   - As a space capsule reenters the earth's atmosphere, it produces extremely high surface temperatures. What would be the increase in the capsule's inside temperature?
10. Indicate three types of forcing functions that are used to investigate the behavior of dynamic systems.
11. Provide as many examples as you can of open-loop systems in your home.
12. Provide as many examples as you can of closed-loop systems in your home.
13. Make a schematic of the cruise control in your car. Indicate any of its special features.

14. Provide two examples of inherently unstable systems that, with control, are made stable.
15. Explain why a ping-pong ball suspended in the air is a stable system.
16. Sketch the process control engineers typically follow in doing their jobs.
17. How does the job of control engineers differ when those engineers work for small companies and large companies?
18. Suppose your task is to create a controller for a new, automatic, water-conserving, antiburning valve and faucet that dispenses water into a bathtub. This valve is to be used at home, where there are many children. Upon pressing a button, the water comes out of the nozzle at 80° F (+/− 3° F). The water comes out for 3 minutes and then automatically shuts off.
    - List all the components needed for this task.
    - Make a block diagram of the control system.
    - Identify the type of control system you would use.
    - Describe how the system fails when the electricity goes off.
    - Assume you are the control engineer in charge, and make a diagram of how you would tackle this job.
19. What is a model?
20. Provide three examples of physical systems and their corresponding models.

2

Mathematical Background for Dynamic Systems

Learning Objectives

In this chapter, you will learn how to:

- Identify the various methods of capturing a physical system's dynamic characteristics.
- Apply the Laplace Transformation to the function of time.
- Use the Laplace method for solving a differential equation.
- Represent dynamic subsystems using block diagrams.
- Confirm the dimensional homogeneity of a differential equation using dimensional analysis.
- Use MATLAB/SIMULINK to solve differential equations.

2.1 Historical Perspective

During the nineteenth century, scientists used operator methods to solve equations dealing with dynamic methods. In these methods, which were developed, applied, and promoted by English electrical engineer Oliver Heaviside (1850–1925), operators were manipulated according to algebraic rules. These methods solved dynamic problems accurately, which inspired more scientists and engineers to use them.

Despite its success, this not-so-rigorous manipulation concerned some contemporary mathematicians, who argued that the manipulation should be more rigorous. Some of these mathematicians rejuvenated the work of Pierre Simons

Laplace (1749–1827), who had developed a similar procedure but in a more organized and mathematically rigorous manner. Apparently, Laplace's work had been somewhat ignored, because in those days there was little demand for the numerical solution of specific problems.[1] Interestingly, Gottfried Wilhelm Leibnitz (1646–1716) and Leonhard Euler (1707–83), who preceded Laplace, had used similar methods to solve differential equations.[2,3]

A major portion of this chapter is dedicated to the Laplace method, because it is one of the principal tools of dynamic system analysis. The Laplace method is not only useful for solving differential equations, it is useful for evaluating integrals and solving integral equations.

2.2 Introduction

You are about to acquire a new language and a new set of tools. You have probably dealt with problems that involve static relationships. By static relationships we mean relationships that do not change as a function of time. Static relationships are described by means of algebraic equations. Now you deal with components that have dynamic behaviors. For these components, algebraic tools do not suffice. These components require calculus, the mathematics of change, which includes differential and integral calculus and differential equations.

A differential equation is one of several methods for describing the dynamic characteristics of systems and components. It describes, in mathematical terms, the relationship between a system's input and output. A differential equation for a dynamic system depends on the components that are part of the dynamic system and the manner in which the components are connected. The Laplace transformation provides an algebraic approach to dealing with dynamic systems.

We will begin with some mathematical definitions and fundamentals, then we will apply the Laplace approach to analyzing a dynamic problem, namely a ball dropped from the top of a building.

2.3 Calculus

As a prerequisite for this book, you should have taken a course in differential and integral calculus and know basic calculus definitions and relationships. Some who learn calculus focus only on definitions and mechanics and fail to appreciate its ability to effectively describe things that are changing. This book uses calculus principles in several applied physical problems to emphasize calculus's importance and use.

Example 2.1

Consider a car trip in which you travel at a decreasing speed as you move from your favorite spot outside the city to your home. Your speed is indicated by the graph in Figure 2–1. What is the total distance traveled?

Solution We can solve this problem using calculus, because:

$$d(t) = \int_0^t s(\tau)d\tau \qquad \text{Eq. 2.1}$$

where:

$d(t)$ = distance traveled in t hours (in miles)

$s(t)$ = car speed as a function of time (in miles per hour)

Remember that when you take the integral of a function, it is equivalent to finding the area under the curve. Thus, the area under the speed versus time curve is:

$$d = \left(60\frac{\text{miles}}{\text{hr}}\right)(1\text{hr}) + \left(40\frac{\text{miles}}{\text{hr}}\right)(2\text{hr}) + \left(20\frac{\text{miles}}{\text{hr}}\right)(1\text{hr})$$
$$= 160 \text{ miles}$$

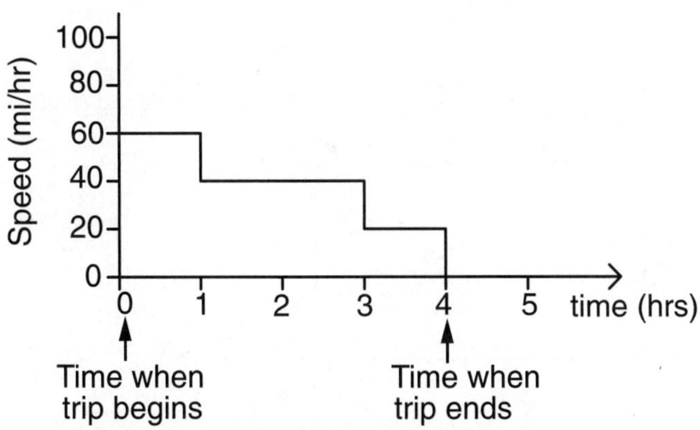

Figure 2–1 Car speed history for Example 2.1

Example 2.2

Consider an empty water tank of cross-sectional area A_{tank}, as shown in Figure 2–2. Obtain an expression for the flow out of the tank as a function of the water level h.

Solution Assuming that the tank has a constant cross-sectional area A_{tank}, the water volume in the tank is:

$$V_{water} = A_{tank}h \quad \text{Eq. 2.2}$$

Substituting into the previous equation:

$$Q_{in} - Q_{out} = \frac{d}{dt}(V_{water}) = \frac{d}{dt}(A_{tank}h)$$

but because

$$Q_{in} = 0, \text{ then}$$

$$Q_{out} = -A_{tank}\frac{dh}{dt}$$

Thus, if we were to plot the water level as a function of time, the slope of the curve times $-A_{tank}$ yields the outgoing flow.

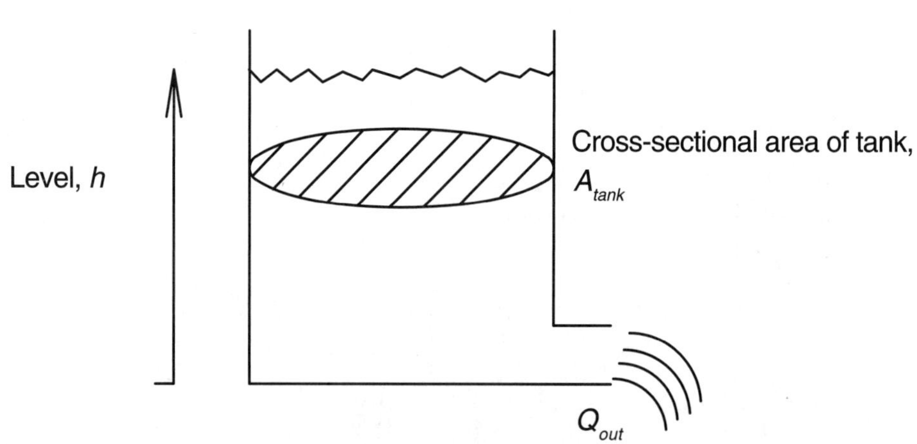

Figure 2–2 Emptying tank of Example 2.2

Example 2.3

Consider a similar tank problem in which the tank is being filled by a flow Q_{in}, and in which Q_{in} is not constant, as shown in Figure 2–3. What is the tank level h for a particular Q_{in} time history?

Solution We can use the previous equation:

$$Q_{in} - Q_{out} = -A_{tank}\frac{dh}{dt} \quad \textbf{Eq. 2.3}$$

In this case, $Q_{out} = 0$. Solving for $\frac{dh}{dt}$, we obtain:

$$\frac{dh}{dt} = \frac{1}{A_{tank}} Q_{in}$$

Integrating, we obtain:

$$h(t^*) = \frac{+1}{A_{tank}} \int_0^{t^*} Q_{in} dt + h_o$$

where t^* is the time at which the level is desired and h_o is the initial level of the tank when the measurement process began ($t = 0$).

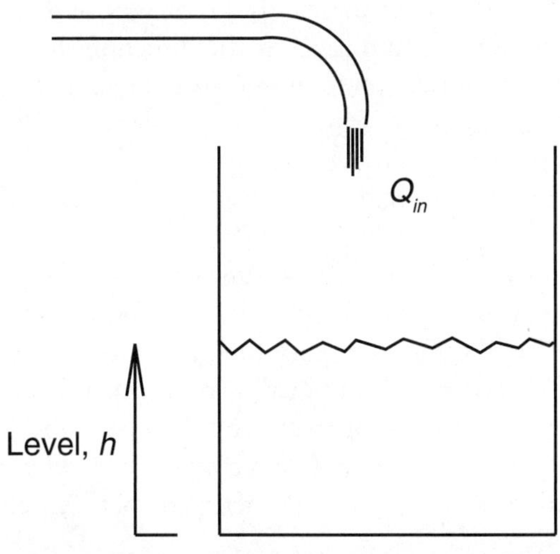

Figure 2–3 Filling tank problem of Example 2.3

2.4 Differential Equations

As mentioned in the beginning of this chapter, a differential equation is one of several methods for describing the behavior of a system with dynamic characteristics. Typically, a differential equation relates a system's input and output. To introduce differential equations, consider a racquetball that is released from the top of a building. Modeling is not an exact science. Instead, it combines *a priori* knowledge of system/component behavior (knowledge one has before encountering the problem), various physical laws, and algebraic and dynamic relationships. Sometimes, this process is called the "Application of First Principles." It often is an iterative process: One proposes a model, evaluates its behavior through simulation techniques, compares it to the real system, and then changes the model to make it behave more like the real system.

Let us return to the racquetball problem. One of the decisions we must make is the dependent variable to be used in the differential equation. We say "dependent" variable, because the value of the variable at any point in time depends on the initial value and a history of the external (independent) input(s) to the system. Our choice of variable depends primarily on our interest, and often we may choose from several variables. In this problem, we will choose displacement, but we could have also chosen the ball's velocity.

When solving this kind of problem, it is helpful to draw a schematic that emphasizes the relevant variables and objects. Figure 2–4a shows the ball at its initial position ($t = 0$) and a symbol that is an indicator for y, the position of the ball at any value of t. This little symbol is a tell-tale (after the piece of yarn that sailors hang from their sails to tell the direction of the wind). Here, we use this symbol to show the reference position where $y = 0$ and to show the direction that produces a positive y. It is not essential that the direction is chosen correctly (i.e., the direction in which the real system will move initially). However, it is easier to work with if we choose the direction in which the ball will move initially.

Figure 2–4b is a closeup of the ball, or a free-body diagram indicating active forces. First, we have the weight of the ball, $W = mg$, where m is the mass of the ball and g is the acceleration of gravity. In creating the free-body diagram, we assume that the ball is in motion, so we include the air-friction forces acting on the ball. Notice that these friction forces oppose the ball's motion. We can also assume that the magnitude is proportional to the ball's speed, even though in reality the force is proportional to the square of the speed. We will simplify things and assume a linear relationship between friction force and speed, where the friction force is equal to the product of velocity and the friction coefficient b. That is,

$$\text{friction} = b\frac{dy}{dt}.$$

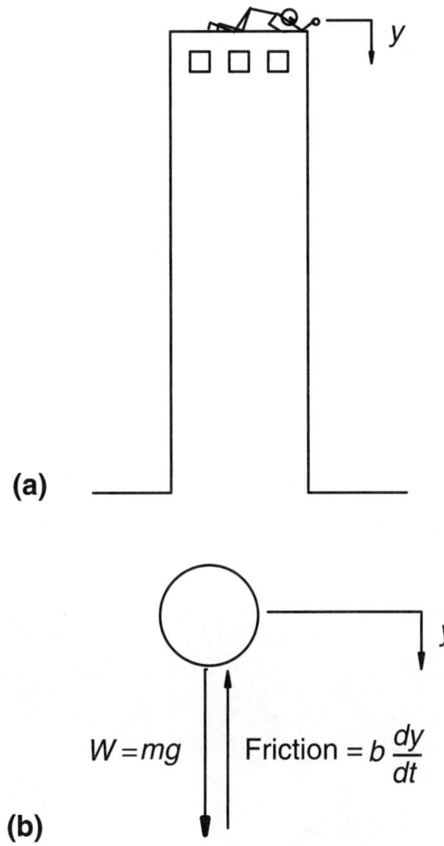

Figure 2–4 Releasing of a racketball (a) schematic and (b) free-body diagram

Now that all the forces acting on the ball have been accounted for and are shown in the free-body diagram, we apply Newton's second law:

$$\sum forces = ma \qquad \textbf{Eq. 2.4}$$

Because our goal is to obtain a differential equation in terms of y, we will substitute $\frac{d^2y}{dt^2}$ for the ball's acceleration, a. Next, we include in the equation all the applicable forces. Notice that in creating Figure 2–4b, we indicated the direction and the magnitude of the forces. Any force in the direction of the y

tell-tale will be considered positive; any forces opposite to the y tell-tale will be considered negative. Gathering the forces while applying the previous guidelines results in:

$$W - \text{friction} = m\frac{d^2y}{dt^2}$$

or **Eq. 2.5**

$$mg - b\frac{dy}{dt} = m\frac{d^2y}{dt^2}$$

One typically shows a differential equation in the following form:

functions of the dependent variable = functions of the independent variables

Therefore, we rearrange the previous equation to the form:

$$m\frac{d^2y}{dt^2} + b\frac{dy}{dt} = mg$$ **Eq. 2.6**

This is the resulting differential equation describing the ball. You will also see the same differential equation presented using the shorthand notation developed by Newton for the time derivative, namely:

$$\dot{y} = \frac{dy}{dt}$$

and **Eq. 2.7**

$$\ddot{y} = \frac{d^2y}{dt^2}$$

Thus, the differential equation would be presented as:

$$m\ddot{y} + b\dot{y} = mg$$ **Eq. 2.8**

Before we attempt to solve the differential equation, we should check it. Remember that, when you are in industry, there is no such thing as partial credit, nor can you go to the back of the book to confirm your answer. For these reasons, we will present some techniques that are helpful when checking the derived differential equation. Instead of waiting until the end of the derivation process to begin checking, perform several checks throughout the process. For example, make sure that the differential equation is dimensionally homogeneous, as demonstrated in the next section.

2.5 Dimensional Homogeneity

The units of variables and coefficients in a differential equation can be checked on two levels. First, in general terms, are the units of variables and coefficients in an arrangement that produces appropriate units? For instance, we expect to get a distance when we multiply velocity and time. Second, do the variables and coefficients, when multiplied, yield the correct specific units? For instance, do we get distance in miles when we multiply speed in miles per hour and travel time in hours? It helps to do a general-units check when developing the differential equation. Later, when we attempt to solve the differential equation, we perform the second-level check using specific units.

Table 2–1 shows the basic units, length, mass, and time, for the three major systems of measurement in use today (for electrical systems, we use voltage and current). For the general check, we use the symbols *L, M,* and *T* and express any other variables in their terms. For instance, if we are evaluating pressure, we would express

Unit	Symbol	English System	SI-Metric (MKS) System	Metric (CGS) System
Length	L	ft	m	cm
Mass	M	slug[1]	kg	g
Time	T	s	s	s
Voltage	V	V	V	V
Current	A	A	A	A
Temperature	D	degrees Rankine (R)	degrees Kelvin (K)	degrees Kelvin (K)

[1]slug = 32.2 lbm

Table 2–1 Basic units used in dimensional analysis (partial list)

Unit	Symbols	English System	SI-Metric (MKS) System	Metric (CGS) System
Velocity	L/T	ft/s	m/s	cm/s
Force	ML/T^2	lbf = slug · ft/s^2	newton (N) = kg · m/s^2	dyne
Torque	ML^2/T^2	ft · lbf	N · m	dyne · cm
Flow	L^3/T	ft^3/s	m^3/s	cm^3/s
Pressure	M/LT^2	lbf/ft^2	pascal (Pa) = N/m^2	$dyne/cm^2$
Energy	ML^2/T^2	ft · lbf	joule (J) = N · m	erg
Power	ML^2/T^3	ft · lbf/s	watt (W)	erg/s

Table 2–2 Derived units used in dimensional analysis (partial list) from International System of Units (SI)

it as M/LT^2, as seen in Table 2–2. This table is only a partial list of all the derived variables. Let us apply dimensional analysis to check the differential equation we derived. We will look at the units of each term in the differential equation and expect them to be the same.

Before we proceed, we should point out that dimensional homogeneity is a necessary condition, not a sufficient condition for the differential equation to be correct. In other words, it is possible for all the units of the terms to be correct yet the differential equation to be wrong. For instance, a dropped term in the derivation or a changed sign would certainly produce an incorrect equation.

Let us return to the task of checking the differential equation, which is rewritten next for continuity:

$$m\frac{d^2y}{dt^2} + b\frac{dy}{dt} = mg \qquad \textbf{Eq. 2.9}$$

The basic units of all terms are obvious, except for b, so let us return to the fundamentals. We said that:

$$\text{friction force} = b\frac{dy}{dt}$$

or

$$b = \frac{\text{friction force}}{\frac{dy}{dt}} \qquad \textbf{Eq. 2.10}$$

If we use square brackets [ ] to denote "the units of," then:

$$[b] = \frac{[\text{friction force}]}{\left[\frac{dy}{dt}\right]}$$

$$= \frac{\frac{ML}{T^2}}{\frac{L}{T}} = \frac{M}{T}$$

Now we can return to the differential equation, which we rewrite for clarity:

$$m\frac{d^2y}{dt^2} + b\frac{dy}{dt} = mg \qquad \textbf{Eq. 2.11}$$

We state that the three terms must have the same units:

$$\left[m\frac{d^2y}{dt^2}\right] + \left[b\frac{dy}{dt}\right] = [mg]$$

$$M\frac{L}{T^2} + \frac{M}{T}\frac{L}{T} = M\frac{L}{T^2} \qquad \textbf{Eq. 2.12}$$

Finally, we conclude that the differential equation is dimensionally homogeneous, or correct.

$$\frac{ML}{T^2} + \frac{ML}{T^2} = \frac{ML}{T^2} \qquad \textbf{Eq. 2.13}$$

We could make a mistake in each of the three terms and conclude that the equation is dimensionally correct, but such a mistake is unlikely to occur in a careful analysis.

One additional suggestion regarding units: Always use units that are consistent within a particular system of measurement (e.g., English, MKS Metric, or CGS Metric). Then, once you obtain a solution, convert it to the desired units. For instance, if the desired unit for a displacement is in inches, solve the differential equation using ft (the standard unit of length in the English System), then convert the result to inches.

The signs of the coefficients in differential-equation terms provide some insight into the type of solution that will result.[4] If there are any negative or missing terms, the solution would tend to oscillate or grow without bounds. Notice that in our sample differential equation, the y term is missing. Consequently, the displacement $y(t)$ would continue to grow indefinitely, which is true if the ball were to never hit the ground. Of course, in real life, at some point the ball will hit the ground. In other words, if the differential equation had a y term with $c > 0$, as in:

$$m\frac{d^2y}{dt^2} + b\frac{dy}{dt} + c\,y = mg \qquad \textbf{Eq. 2.14}$$

then $y(t)$ would approach 0, or a finite value. This topic, called stability analysis, is addressed in detail in Chapter 5.

2.6 The Laplace Transformation

One of the major uses of the Laplace transformation method is to solve linear differential equations. Even though there are other methods for solving them, Laplace's method is particularly attractive, because it converts the solution's process into a series of algebraic operations. A colleague explains the procedure in terms of walking next to a river, as shown in Figure 2–5. One side of the river represents the time domain; the other side the Laplace domain (in which functions are represented in terms of s, as will be seen later). Basically, our goal is the solution of a differential equation: to go from A to D in the figure. However, instead of walking on the south side of the river, we will cross and walk on the north side, then cross back again to the south side. Why? Going from A to D requires knowledge of classical techniques for solving differential equations. In addition, the procedure might be cumbersome. On the other hand, by using the Laplace method (going from A to B, B to C, and then C to D), the procedure primarily consists of a series of algebraic manipulations. As we discuss the procedure, we will refer back to this diagram and relate the steps to each leg of our trip.

For every function of time, there is a corresponding function of the new variable s, $F(s)$, on the other side of the river. Some functions are not permissible, but that will not be the case with the type of functions with which we will be working.

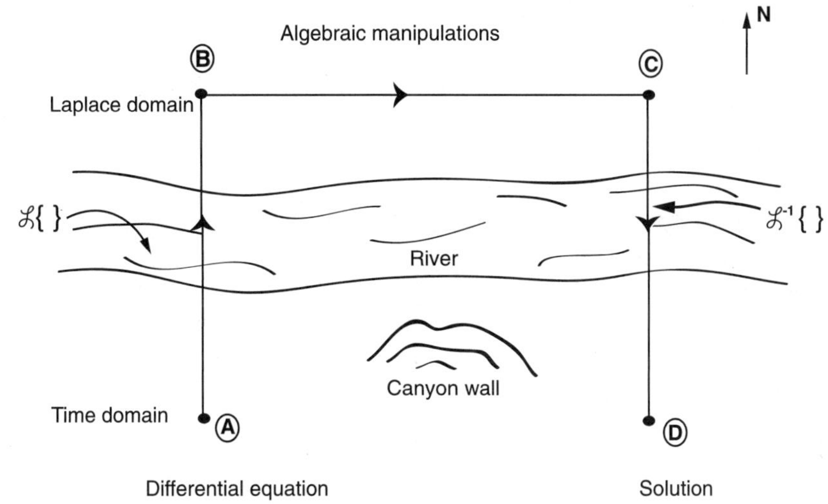

Figure 2–5 River analogy

Going from the south side of the river to the north side is equivalent to the following mathematical operation:

$$F = \mathcal{L}[f(t)] \triangleq \int_0^{\infty} f(t)e^{-st}\,dt \qquad \textbf{Eq. 2.15}$$

This is not an easy operation, except for very simple functions of $f(t)$. Fortunately, after the operation is performed once, the results can be recorded in a table. From then on, crossing the river becomes a simple table lookup operation.

To cross back over the river or, in other words, to obtain an $f(t)$ from an $F(s)$, we say that we are taking the inverse Laplace transform of $F(s)$. For this process, there is also a corresponding operation, namely:

$$f(t) = \mathcal{L}^{-1}[F(s)] = \frac{1}{2\pi j}\int_{\sigma-j\infty}^{\sigma+j\infty} F(s)e^{st}\,ds \qquad \textbf{Eq. 2.16}$$

This operation is even more complicated than the first, but, again, it can be done as another table lookup operation. Notice that $f(t)$ and $F(s)$ are not the same functions. We use lowercase letters for the time function, while uppercase for the Laplace-transformed function. Following the uppercase-lowercase notation becomes even more important if the (t) and the (s) notation are dropped for the sake of convenience.

If you looked at the definition of $F(s)$ closely, you probably noticed that the integration operation started at 0 and went to infinity. What about $t < 0$? Well, for this type of transformation, the value of the function $f(t)$ is 0 for $t < 0$. In other words, you always work with the portion of a function in which $t > 0$ and assume that $f(t) = 0$ for $t < 0$. Even though this is always understood, for emphasis reasons we often use the unity step function, $u(t)$ to multiply the function in question, defined as:

$$u(t) = \begin{Bmatrix} 0 \text{ for } t < 0 \\ 1 \text{ for } t \geq 0 \end{Bmatrix} \qquad \textbf{Eq. 2.17}$$

producing $f(t) = 0$ for $t < 0$.

Let us use the formal definition to calculate the Laplace transform of a simple function (e.g., cross the river from south to north).

Example 2.4

Calculate the Laplace transform $F(s)$ of the function $f(t) = 8u(t)$.

$$\begin{aligned} F(s) &= \mathcal{L}[f(t)] = \mathcal{L}[8u(t)] = \int_0^\infty 8e^{-st}\,dt \\ &= 8\left\{ -\frac{1}{s}e^{-st} \Big|_{t=0}^{t=\infty} \right\} \\ &= 8\left\{ -\frac{1}{s}\cdot 0 + \frac{1}{s}\cdot 1 \right\} \\ &= \frac{8}{s} \end{aligned} \qquad \textbf{Eq. 2.18}$$

Solution As previously mentioned, we will not be performing this operation often. Instead, we will use a table of transformations, as shown in Table 2–3. In addition, several relationships and operations are available to help us modify a function so that we can use Table 2–3. This table of relationships and operations is presented in Table 2–4.

Type	Time Function, $f(t)$	Laplace Transform, $F(s)$
Unit impulse	$\delta(t)$	1
Step	$A\,u(t)$	$\frac{A}{s}$
Delayed step	$A\,u(t-T)$	$\frac{A}{s}e^{-sT}$
Ramp	$A\,t\,u(t)$	$\frac{A}{s^2}$
Delayed ramp	$A(t-T)\,u(t-T)$	$\frac{A}{s^2}e^{-sT}$
Parabola	$At^2\,u(t)$	$\frac{2A}{s^3}$
General power function	$At^n\,u(t)$	$A\frac{n!}{s^{n+1}}$
Exponential decay	$e^{-bt}\,u(t)$	$\frac{1}{s+b}$
Sine wave	$\sin\omega t\,u(t)$	$\frac{\omega}{s^2+\omega^2}$
Cosine wave	$\cos\omega t\,u(t)$	$\frac{s}{s^2+\omega^2}$
Damped sine	$e^{-bt}\sin\omega t\,u(t)$	$\frac{\omega}{(s+b)^2+\omega^2}$
Damped cosine	$e^{-bt}\cos\omega t\,u(t)$	$\frac{s+b}{(s+b)^2+\omega^2}$
Smooth pulse	$te^{bt}\,u(t)$	$\frac{1}{(s-b)^2}$
Double exponential	$\frac{1}{b-c}(e^{bt}-e^{ct})u(t)$	$\frac{1}{(s-b)(s-c)}$
Growing sine	$\frac{t}{2\omega}\sin\omega t\,u(t)$	$\frac{s}{(s^2+\omega^2)^2}$
Growing cosine	$t\cos\omega t\,u(t)$	$\frac{s^2-\omega^2}{(s^2+\omega^2)^2}$

Table 2–3 Table of Laplace transform

Description	Time Function	Transformed Function
Multiplication by scalar	$a\,f(t)$	$a\,F(s)$
Superposition	$f_1(t) \mp f_2(t)$	$F_1(s) \mp F_2(s)$
Real differentiation (first derivative)	$\frac{df(t)}{dt} = \dot{f}$	$sF(s) - f(0+)$
Real differentiation (second derivative)	$\frac{d^2f(t)}{dt^2} = \ddot{f}$	$s^2F(s) - sf(0+) - \dot{f}(0+)$
Real differentiation (N^{th} order derivative)	$\frac{d^Nf(t)}{dt^N}$	$s^N F(s) - s^{N-1} f(0+) \ldots$ $-s^{N-2} f(0+) - \ldots f^{N-1}(0+)$
Real integration	$\int f(t)dt$	$\frac{1}{s}\left[F(s) + \int f(t)dt\Big\vert_{0+}\right]$
Initial value	$\lim_{t\to 0} f(t)$	$\lim_{s\to\infty} sF(s)$
Final value	$\lim_{t\to\infty} f(t)$	$\lim_{s\to 0} sF(s)$
Multiplication by decaying exponential	$e^{-at} f(t)$	$F(s + a)$
Real translation	$f(t - T)u(t - T)$	$e^{-sT} F(s)$

Table 2–4 Laplace transform relationships and operations

Multiplication by a Scalar

The multiplication by a scalar relationship is probably obvious, but it is used so often it deserves mention. It basically says that if we have a function multiplied by a scalar, all we need to do is to multiply the Laplace transform of the function by the scalar.

Delayed Functions and Translations

Often, we encounter physical phenomena that exhibit pure delays. For instance, if, when taking a shower, you change the valve setting, it takes a few seconds to notice the change in the water temperature, because the new-temperature water must travel from the valve, to the showerhead, and to your skin. Mathematical delays are

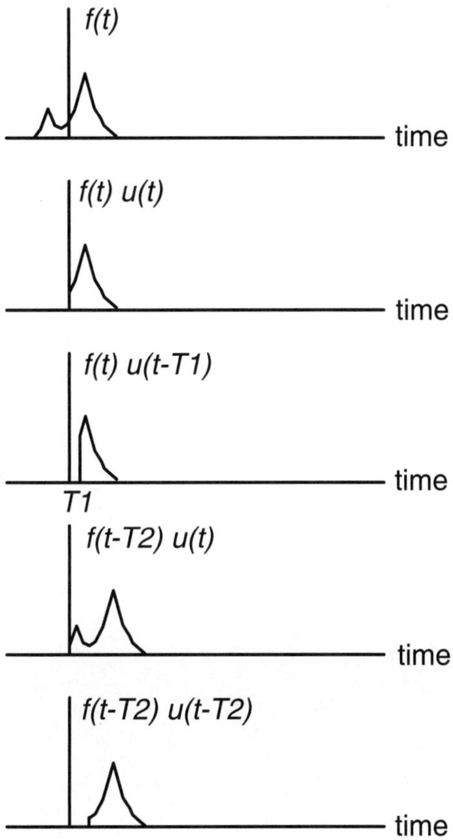

Figure 2–6 Delayed functions

useful in that you can synthesize a function by combining several functions that have been delayed. Before we apply delayed functions, it is important that we understand the mechanics of delaying and time translation. Consider a function $f(t)$, shown at the top of Figure 2–6, and all the variations obtained by delaying the function and/or multiplying $f(t)$ by delayed step functions.

Example 2.5

Obtain a time-domain expression for the function shown in Figure 2–7. Afterward, obtain the Laplace transform of $f(t)$.

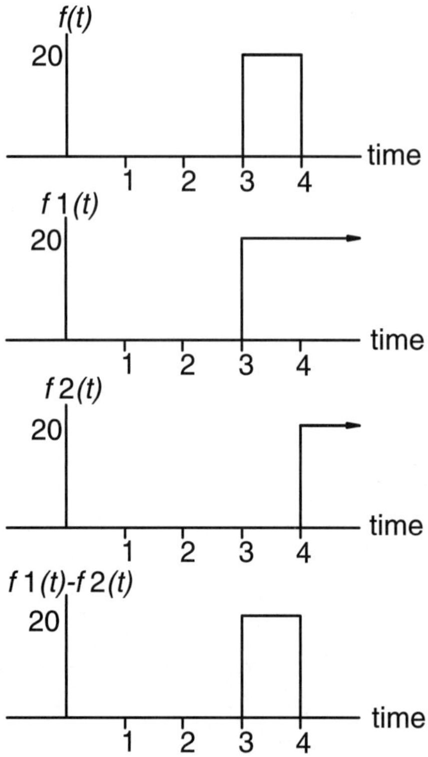

Figure 2–7 Example 2.5 Function

Solution Notice that this function can be synthesized by combining functions $f_1(t)$ and $f_2(t)$, also shown in Figure 2–7. Thus:

$$\begin{aligned} f(t) &= f_1(t) - f_2(t) \\ &= 20u(t-3) - 20u(t-4) \end{aligned} \quad \textbf{Eq. 2.19}$$

Now we can use the real translation relationship in Table 2–4:

$$\begin{aligned} F(s) &= \mathcal{L}[f_1(t) - f_2(t)] = F_1(s) - F_2(s) \\ &= \frac{20}{s}e^{-3s} - \frac{20}{s}e^{-4s} \end{aligned} \quad \textbf{Eq. 2.20}$$

$$F(s) = \frac{20}{s}(e^{-3s} - e^{-4s})$$

Multiplication by a Decaying Exponential

Example 2.6

Find the Laplace transform of:

$$f(t) = 2e^{-3t} \cos 5t\, u(t) \qquad \textbf{Eq. 2.21}$$

Solution Because we know that this function is 0 for $t < 0$, we will drop the step function $u(t)$ for the sake of simplicity:

$$f(t) = 2e^{-3t} \cos 5t$$

If $f(t)$ were simply cos $5t$, the Laplace transform would be $\frac{s}{s^2 + (5)^2}$. However, there are two additional components to the $f(t)$ at hand: a scalar (2) and multiplication by an exponential (e^{-3t}). Thus, using Table 2–3 and Table 2–4:

$$F(s) = 2 \cdot \frac{s+3}{(s+3)^2 + (5)^2}$$
$$F(s) = \frac{2s+6}{s^2 + 6s + 34} \qquad \textbf{Eq. 2.22}$$

Integral of a Function

Table 2–4 reveals that the Laplace transform of a derivative or an integral of a function simply results in an expression that is an algebraic operation of the Laplace transform of the original function. The example that follows illustrates the validity of the relationship. However, its primary use to us will be in the solution of differential equations, because it enables conversion of a complicated time relationship (an expression involving the function, its derivatives, and integrals) into a simple algebraic equation.

Example 2.7

Determine the Laplace transform of:

$$f(t) = \int t dt \qquad \textbf{Eq. 2.23}$$

Solution The function that is being integrated, call it $g(t)$, is t. Then, from Table 2–3:

$$G(s) = \frac{1}{s^2} \quad \textbf{Eq. 2.24}$$

Now, Table 2–4 tells us that:

$$\int g(t)dt = \frac{1}{s}G(s) \quad \textbf{Eq. 2.25}$$

because the finite integral term in the transformed expression is 0.

Thus:

$$F(s) = \frac{1}{s} \cdot G(s) = \frac{1}{s} \cdot \frac{1}{s^2} = \frac{1}{s^3} \quad \textbf{Eq. 2.26}$$

Notice that we could have obtained the same result by first performing the integration and then taking the Laplace transform:

$$f(t) = \int tdt = \frac{1}{2} \cdot t^2$$

then **Eq. 2.27**

$$F(s) = \frac{1}{2} \cdot \frac{2}{s^3} = \frac{1}{s^3}$$

which is the result previously obtained.

Laplace Transform of Derivatives

In the previous section, we showed how the Laplace transform allows the integration function to be converted into an algebraic operation. A similar transformation exists for the derivative, but it is not shown here.

Example 2.8

Obtain the Laplace transform of $x(t)$, which is a function of $f(t)$ and its derivative:

$$x(t) = 3f(t) + 4\frac{df(t)}{dt} \quad \textbf{Eq. 2.28}$$

Solution Taking the Laplace transform of both sides results in:

$$X(s) = 3F(s) + 4\mathcal{L}\left[\frac{df(t)}{dt}\right] \quad \textbf{Eq. 2.29}$$

Use the theorem from Table 2–4 to handle the derivative:

$$\mathcal{L}\left[\frac{df(t)}{dt}\right] = sF(s) - f(0) \quad \textbf{Eq. 2.30}$$

Thus:

$$\begin{aligned} X(s) &= 3F(s) + 4[sF(s) - f(0)] \\ &= 3F(s) + 4sF(s) - 4f(0) \\ X(s) &= F(s)\left[4s + 3\right] - 4f(0) \end{aligned} \quad \textbf{Eq. 2.31}$$

where $f(0)$ is the value of the function at $t = 0$, the initial value. Notice that the table also provides the expression for taking the Laplace transform of a higher derivative.

Final Value Theorem

The final value theorem will prove very useful later in the book. It allows us to predict the steady-state value (final value) of a time function by performing an operation in the s-domain: Returning to the river analogy, there are many cases in which one moves on one side of the river with more ease than on the other. The final value theorem can be stated as:

$$\lim_{t\to\infty} f(t) = \lim_{s\to 0} s \cdot F(s) \quad \textbf{Eq. 2.32}$$

Example 2.9

Find the final value of the following function, both in the time domain and in the s-domain:

$$f(t) = 5(1 - e^{-2t}) \quad \textbf{Eq. 2.33}$$

Solution—Time Domain

Because the second term is a decaying exponential, the function will approach 5 as $t \to \infty$.

Solution—*s*-Domain

The theorem tells us that:

Thus, we need to obtain $F(s)$.

$$f(t) = 5(1 - e^{-2t}) = 5 - 5e^{-2t}$$

$$F(s) = \frac{5}{s} - \frac{5}{s+2}$$

$$\lim_{t\to\infty} f(t) = \lim_{s\to 0} s \cdot F(s) = \lim_{s\to 0} s \cdot \left(\frac{5}{s} - \frac{5}{s+2}\right)$$

$$= 5 - \lim_{s\to 0} s \cdot \left(\frac{5}{s+2}\right) = 5 + 0 = 5$$

Which is the same result obtained by the time domain approach.

The Impulse Function

Often, one excites a system with an input in which the specific shape of the input is not as significant as the "weight" of the input. For instance, if we kick a ball, the shape of the force during the time that the shoe and ball are in contact is not as important as the area under the force versus time curve (called the impulse imparted to the ball). This area is equivalent to the change in momentum imparted as a result of the kick. For this reason, it is useful to use the impulse functions to analyze some dynamic systems. The impulse function is defined mathematically as a pulse whose width approaches 0 duration but nevertheless maintains an area equal to 1. The Laplace transform of such a function is simply 1.

Inverse Laplace Transformation by Table Lookup

Let us return to our river analogy regarding the use of the Laplace transform (i.e., Figure 2–5). We have seen that the Laplace transform allows us to convert a time-domain equation involving derivatives and integrals into an algebraic representation. This allows us to perform the necessary procedures using algebraic manipulations rather than calculus. However, at the end of these algebraic manipulations, we do need to return to the time domain: to go from C to D in Figure 2–5. This is where the inverse Laplace transform comes in. Basically, to perform this operation, we use

Table 2–3, but in the reverse direction as previously done. Unfortunately, the expression that results from the algebraic manipulation is often too complicated, and it does not appear in the table. It is therefore necessary to rearrange the result, typically a ratio of two polynomials, into simpler terms that can be found in the table. The procedure that is used is called a partial-fraction expansion. We will demonstrate it using abstract polynomials, then use it in the racquetball problem.

Poles and Zeros

Before we get involved in this process, it is necessary to define some terminology dealing with functions composed of the ratio of two polynomials. The roots of the numerator polynomial are called zeros, because the function becomes 0 when evaluated at these values. The roots of the denominator polynomial are called poles, because the function goes to infinity when evaluated at these values. A function approaching infinity at a certain value looks like a spike or a flagpole. From this, the name pole results. The following example illustrates them.

Example 2.10

Find the poles and zeros of the following function:

$$F(s) = \frac{6s + 18}{2s^2 + 12s + 16} \qquad \textbf{Eq. 2.34}$$

Solution First, let us factor out each polynomial (for a complicated polynomial, this might need the assistance of a computer program):

$$F(s) = \frac{6(s + 3)}{2(s^2 + 6s + 8)} = \frac{3(s + 3)}{s^2 + 6s + 8} = \frac{3(s + 3)}{(s + 4)(s + 2)} \qquad \textbf{Eq. 2.35}$$

Thus, this polynomial has a zero at $s = -3$ and poles at $s = -4$ and $s = -2$.

If we were to try to return $F(s)$ to the time domain by looking for an entry in Table 2–3, we would come to a dead end. However, if we could decompose it into:

$$F(s) = \frac{A_1}{s + 4} + \frac{A_2}{s + 2} \qquad \textbf{Eq. 2.36}$$

we certainly could use Table 2–3. The process of getting A_1 and A_2 such that we can write $F(s)$ in the preceding form is called partial-fraction expansion.

Partial-Fraction Expansion

There is a very simple procedure for decomposing $F(s)$, a ratio of two polynomials, into the summation of simpler polynomials. Generally speaking, $F(s)$ falls into one of the following categories, depending on the type of poles that make up the denominator polynomial of $F(s)$. The categories are:[5]

- Real poles of first order
- Real poles of multiple order
- Complex poles of first order
- Complex poles of multiple order

We will treat the first two categories here. For the third and fourth categories, see Reference 7 for this chapter, which treats the Laplace transform in more depth.

Category 1: Real Poles of First Order

A given $F(s)$ having distinct poles can be expressed as:

$$F(s) = \frac{A_1}{s+p_1} + \frac{A_2}{s+p_2} + \ldots + \frac{A_r}{s-p_r} + R(s) \qquad \textbf{Eq. 2.37}$$

where $R(s)$ is the remaining portion of the expansion that does not fall into this category.

Each A_i is given by:

$$A_i = (s-p_i)F(s)\Big|_{s=p_i} \qquad \textbf{Eq. 2.38}$$

Example 2.11

Use partial-fraction expansions and Table 2–3 to obtain $f(t)$ given $F(s)$.

Solution

$$F(s) = \frac{3(s+3)}{(s+4)(s+2)} \qquad \textbf{Eq. 2.39}$$

The goal here is to decompose $F(s)$ as:

$$F(s) = \frac{A_1}{(s+4)} + \frac{A_2}{(s+2)}$$

We can use the procedure given by the formula:

$$A_1 = (s+4)\frac{3(s+3)}{(s+4)(s+2)}\bigg|_{s=-4} = \frac{3(s+3)}{(s+2)}\bigg|_{s=-4} = 1.5$$

$$A_2 = (s+2)F(s)\bigg|_{s=-2} = (s-2)\frac{3(s+3)}{(s+4)(s+2)}\bigg|_{s=-2} = \frac{3(s+3)}{(s+4)}\bigg|_{s=-2} = 1.5$$

Thus, $F(s)$ can be expressed as:

$$F(s) = \frac{1.5}{s+4} + \frac{1.5}{s+2}$$

which can be written to fit the format of the table:

$$F(s) = \frac{1.5}{s+\frac{1}{0.25}} + \frac{1.5}{s+\frac{1}{0.5}}$$

Now we can use Table 2–3 to obtain $f(t)$:

$$f(t) = 1.5e^{-t/0.25} + 1.5e^{-t/0.5} = 1.5(e^{-4t} + e^{-2t}) \qquad \textbf{Eq. 2.40}$$

Again, this function has no meaning for $t < 0$.

Case 2: Real Poles of Multiple Order

If the denominator polynomial $F(s)$ has repeated roots at a given pole, the earlier procedure cannot be used. To deal with this case, we express $F(s)$ in the form:

$$F(s) = \frac{Q(s)}{(s-p)^r}$$

or

$$Q(s) = F(s)(s-p)^r \qquad \textbf{Eq. 2.41}$$

where $Q(s)$ includes everything in $F(s)$ besides the $(s-p)^r$ factor. The partial-fraction expansion of $F(s)$ would have the form:

$$F(s) = \frac{A_1}{(s-p)^r} + \frac{A_2}{(s-p)^r} + \ldots + \frac{Ar}{(s-p)} + R(s) \qquad \textbf{Eq. 2.42}$$

where $R(s)$ is the expansion due to all other poles. To obtain each A_i, we use the expression:

$$A_i = \frac{1}{(i-1)!} \frac{d^{i-1}}{ds^{i-1}} Q(s)\bigg|_{s=p} \quad \textbf{Eq. 2.43}$$

Use of the Laplace Transform to Solve a Differential Equation

Let us return to the differential equation derived earlier in the chapter, namely:

$$m\frac{d^2y}{dt^2} + b\frac{dy}{dt} = mg\,u(t) \quad \textbf{Eq. 2.44}$$

which can be expressed in shorthand notation as:

$$m\ddot{y} + b\dot{y} = mg\,u(t)$$

Now, let us take the Laplace transform to obtain:

$$m\{s^2Y(s) - sy(0) - \dot{y}(0)\} + b\{sY(s) - y(0)\} = \frac{mg}{s} \quad \textbf{Eq. 4.45}$$

Assume that at $t = 0$, the ball is released at the tell-tale reference ($y(0) = 0$) and with 0 velocity ($\dot{y}(0) = 0$). Thus, the previous equation becomes:

$$ms^2Y(s) + bsY(s) = \frac{mg}{s} \quad \textbf{Eq. 2.46}$$

or, solving for $Y(s)$:

$$Y(s) = \frac{mg}{s(ms^2 + bs)}$$

Factoring out an s, and dividing by m to make the highest-order coefficient of the denominator equal to 1:

$$Y(s) = \frac{g}{s^2(s + b/m)} \quad \textbf{Eq. 2.47}$$

We would like to decompose $Y(s)$ into the form:

$$Y(s) = \frac{A_1}{(s + b/m)} + \frac{B_1}{s^2} + \frac{B_2}{s} \qquad \text{Eq. 2.48}$$

Utilizing the first two categories, for partial-fraction expansion, we obtain A_1 B_1 and B_2.

First, let us obtain A_1:

$$A_1 = (s + b/m)\frac{g}{s^2(s + b/m)}\bigg|_{s=-b/m} = \frac{g}{(-b/m)^2} = \frac{gm^2}{b^2} \qquad \text{Eq. 2.49}$$

Now, for B_1 and B_2 we need to use the rules for repeated roots (Case 2):

$$B_2 = \frac{1}{(2-1)!}\frac{d}{ds}Q(s)\bigg|_{s=0} \qquad \text{Eq. 2.50}$$

where:

$$Q(s) = \frac{g}{s + b/m} = g(s + b/m)^{-1} \qquad \text{Eq. 2.51}$$

then:

$$\begin{aligned} B_2 &= \frac{1}{1!}\frac{d}{ds}\left(g(s + b/m)^{-1}\right) \\ &= g(-1)\frac{1}{(s + b/m)^2}(1)\bigg|_{s=0} = \frac{-gm^2}{b^2} \end{aligned} \qquad \text{Eq. 2.52}$$

Now calculating B_1:

$$B_1 = \frac{1}{(1-1)!}Q(s)\bigg|_{s=0} = (1)\frac{g}{0 + b/m} = \frac{gm}{b} \qquad \text{Eq. 2.53}$$

Thus, the coefficients are:

$$A_1 = \frac{gm^2}{b^2}$$

$$B_1 = \frac{gm}{b}$$

$$B_2 = -\frac{gm^2}{b^2}$$

Then:

$$Y(s) = \frac{gm^2/b^2}{(s+b/m)} + \frac{gm/b}{s^2} - \frac{gm^2/b^2}{s} \quad \text{Eq. 2.54}$$

Finally, we can use Table 2–3 to return to the time domain to obtain:

$$y(t) = \mathcal{L}^{-1}\{Y(s)\} = (gm^2/b^2)\ e^{-bt/m} + (gm/b)\ t - (gm^2/b^2) \quad \text{Eq. 2.55}$$

If we wanted to be more rigorous mathematically, we could place the step function after each term to emphasize the existence of the function for only $t > 0$.

Dimensional Homogeneity in the s-Domain

Quite a few algebraic manipulations can be involved in going from the differential equation to the s-domain representation. This is especially true if we are starting with several simple differential equations, a method that will be shown later. As a result, it is very easy to drop, gain, or transpose a term. For this reason, it is very useful to check the dimensional homogeneity of the transformed equations in the s-domain. As we mentioned before, dimensional homogeneity is necessary for the equation to be correct, but, again, it does not guarantee that the equation is without error. The approach presented earlier applies, but we have to determine how to handle the units of a Laplace-transformed variable, and s.

To determine the units of $Y(s)$, contemplate the Laplace transform definition. Notice that you are basically integrating with respect to time, because e^{-st} is unitless. Thus, if $y(t)$ has units of L (length), then $Y(s)$ has units of LT. In summary, the units of the transformed variable $F(s)$ are:

$$\{F(s)\} = \{f(t)\}T \quad \text{Eq. 2.56}$$

where the curly brackets { } denote "the units of." To determine the units of *s*, focus on the Laplace transform definition, and notice that the exponent of the natural logarithm, *e*, should also be unitless. This means that:

$$\{-st\} = 1$$

or

$$\{s\}T = 1 \qquad \text{Eq. 2.57}$$

$$\{s\} = \frac{1}{T}$$

Example 2.12

What are the units of the term $s^2 I(s)$, given that $I(s)$ is the Laplace transform of a variable $i(t)$, representing a current?

Solution The units of $I(s)$ are $A \cdot S$ (due to the Laplace transformation). Then:

$$\{s^2 I(s)\} = \frac{1}{T^2} \cdot A \cdot T = \frac{A}{T} \qquad \text{Eq. 2.58}$$

Thus, the units of the term in question are A/T, or amps/s.

2.7 Transfer Functions

Transfer functions are excellent tools for describing the characteristics of dynamic components and systems. Control engineers come from many backgrounds: Some have electrical or electronics background; others might have mechanical or chemical backgrounds. In the aerodynamics field, control engineers investigate the stability of wings and control surfaces in an airplane. With block diagrams, transfer functions make up the language control engineers commonly use. These two basic communication tools offer a common language that allows engineers to communicate the behavioral aspects of a component or system, regardless of what that component or system looks like or how it is built.

To illustrate the transfer-function concept, let us return to the differential equation of the falling racquetball, which we derived. First, we need to determine the output and input of the system. Most systems have more than one output, so it

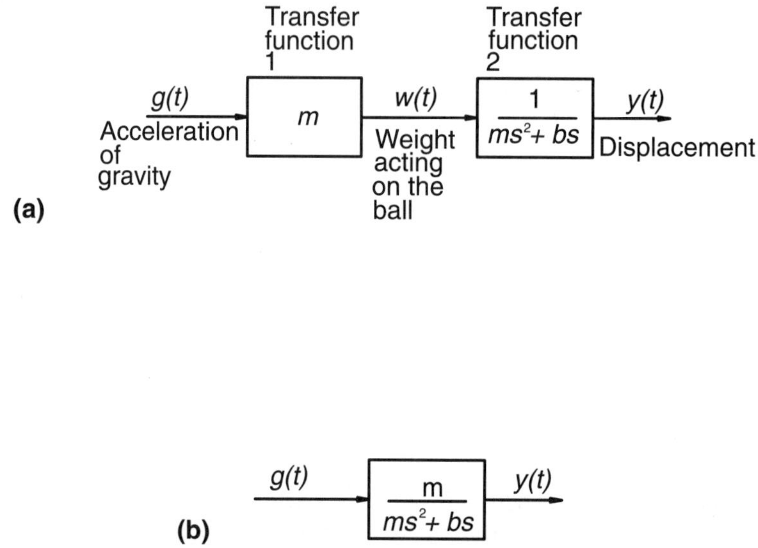

Figure 2–8 Transfer function for ball problem (a) separated and (b) combined

is a matter of choice: Which is most important to the control engineer? We will consider the output to be the displacement of the racquetball, $y(t)$, but we could have also considered the ball's velocity as the output.

Now, let us focus on the input. For our system, the input is the "driver." It is the pull or weight exerted by gravity on the ball. To embellish the example, assume that somehow the acceleration of gravity is a quantity that changes. As a result, it is not simply g = constant, but $g(t)$. Looking at Figure 2–8a, we see the first block with an m inside. This block conveys that if a mass, m, is subjected to a gravitational field, $g(t)$, then a weight (or force) of magnitude $w(t) = mg(t)$ results. The element inside the block represents a constant that, when multiplied by the input, produces the output. In another way of looking at it, the block represents the ratio of the output to the input. For this case, the transfer function is simply a constant, but it could be a complex differential equation. In most transfer functions, we choose to represent the dynamic characteristics in the s-domain instead of in the time domain.

Let us focus now on the second block. First, let us rewrite the differential equation, which we had derived:

Eq. 2.59

$$m\frac{d^2y}{dt^2} + b\frac{dy}{dt} = mg\,u(t)$$

Because we are assuming that the pull of gravity is not constant, we rewrite it as:

$$m\frac{d^2y}{dt^2} + b\frac{dy}{dt} = w(t) \qquad \text{Eq. 2.60}$$

Taking the Laplace transform, assuming that the initial velocity and displacement are 0 results in:

$$ms^2Y(s) + bsY(s) = W(s) \qquad \text{Eq. 2.61}$$

where $W(s)$ is the Laplace transform of $w(t)$. Now we can factor out the $Y(s)$ and express the result as the ratio of the Laplace transform of the output to the Laplace transform of the input or:

$$\frac{Y(s)}{W(s)} = \frac{1}{ms^2 + bs} \qquad \text{Eq. 2.62}$$

Thus, the first block describes the acceleration of gravity-weight phenomenon, while the second describes the displacement of the ball given the weight acting on it. Notice that both blocks have to do with the ball, so we might choose to look at the input as the acceleration of gravity, $g(t)$, and the output the displacement of the ball, $y(t)$. This perspective is shown in Figure 2–8b, where we have combined both blocks. Let us build our vocabulary as control engineers and learn the characteristics of block diagrams, as well as the rules for combining and simplifying them.

2.8 Block Diagrams

Transfer functions and block diagrams have certain characteristics and properties we must keep in mind when working with them. The first is their unidirectionality, which means that information flows in one direction only. An input produces an output, not vice versa. Consider a lightbulb. When we apply a voltage, light is produced. Notice that if we shine light on the lightbulb, we do not obtain an electrical voltage. Nevertheless, we can use the transfer function another way. If we know the output, we can determine what type of input causes the observed output.

A second property is that the block diagram does not convey the power associated in maintaining the operation, or the power associated with the variable. For instance, if a block diagram has a current as an input, it does not convey the voltage necessary to produce that current. Third, the block diagram does not communicate the type of system involved: It does not indicate whether the component being described is a mechanical, an electrical, a fluid, or a software module. For that reason, one should choose the variable names appropriately and put labels outside the block. A fourth property associated with blocks is that they do not load each other: The output level of one block is not brought down by connecting it to the other block. This does not mean that loading conditions cannot be described by means of block diagrams. They are handled by additional blocks and paths, as will be seen later.

The Terms of Block Diagrams

Following are some of the definitions associated with block diagrams:

Summing junction—a device that adds or subtracts a series of signals coming together at that point.

Pick-off point—a point having multiple connections that take the signal at the point to additional places in the diagram.

To be mathematically proper in a block diagram, all blocks (representing dynamic components) should be represented either in the time domain or in the *s*-domain, rather than using both methods simultaneously. However, in practice, both forms of representation coexist in the same block diagram.

Example 2.13

Represent the following mathematical expression using block diagrams in both the (1) time domain and the (2) *s*-domain:

$$y = 20x_1 + 4\frac{dx_2}{dt} - 9\int x_3 dt \qquad \textbf{Eq. 2.63}$$

Solution The solution to this example appears in Figure 2–9. Notice that, in the time domain, variables are represented using lowercase, while in the *s*-domain, they are in uppercase. Often, the (*t*) and (*s*) components of the variables are dropped for the sake of expediency.

(a)

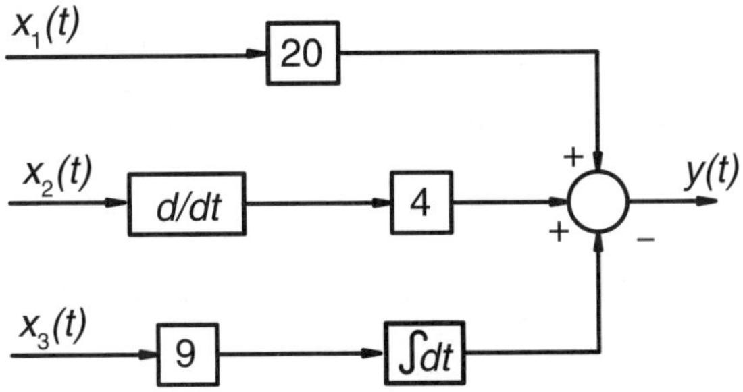

(b)

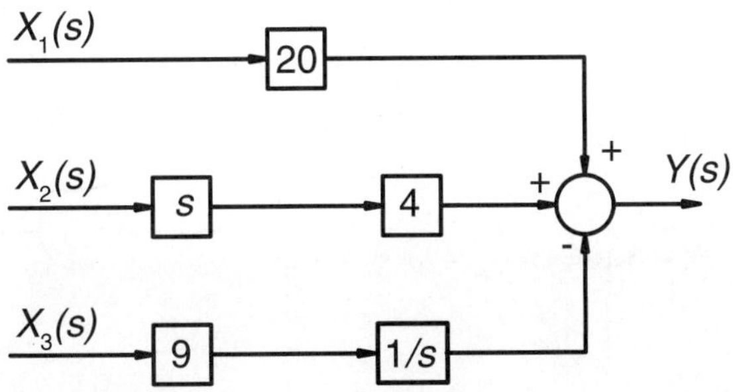

Figure 2–9 Block diagrams for Example 2.13 (a) time domain representation and (b) s-domain representation

Reduction of Block Diagrams

It often helps to condense a series of blocks and transfer functions in a block diagram into a single transfer function. Even though this process eliminates accessibility to the intermediate variables, and excludes the possibility of introducing initial conditions, it compacts the representation. Note that a linear system's stability characteristics depend on system structure, which can be assessed from the transfer function coefficients and not the initial conditions or type of inputs. Figure 2–10 shows graphically some of the rules for reducing a block diagram. An implied assumption with these rules is that the blocks represent linear relationships. When block diagrams include nonlinear blocks, one must be careful in treating them. Notice the block diagram in Figure 2–11. The first block is linear, while the second is nonlinear. Because of the nonlinear block, the second representation does not equal the first.

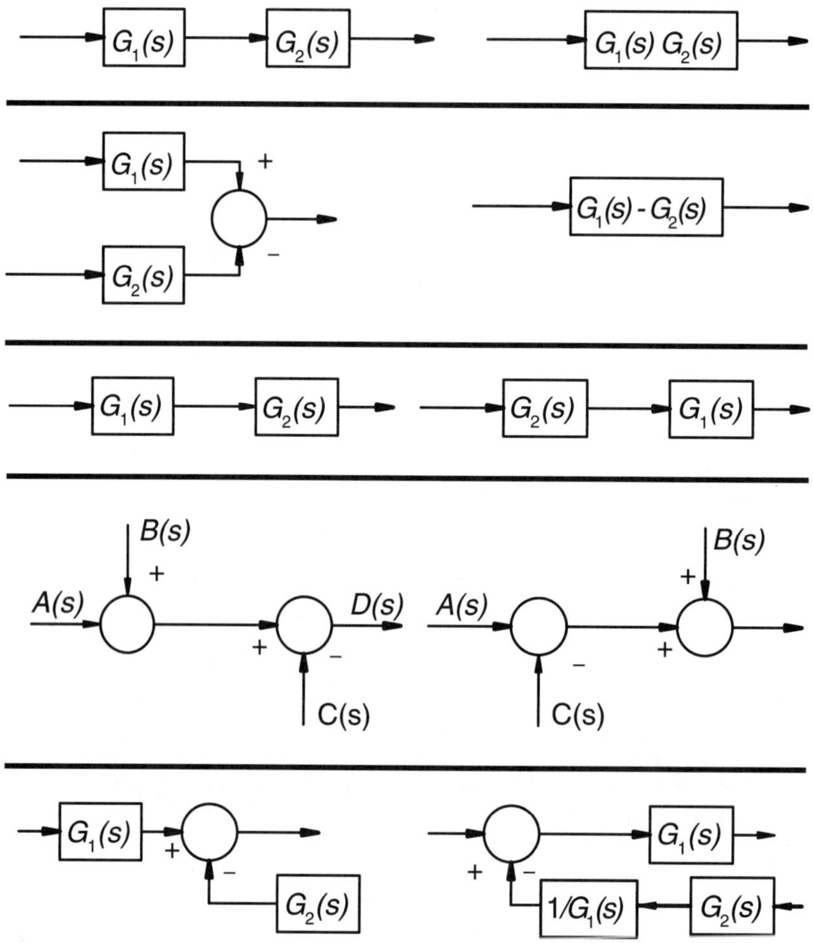

Figure 2–10 Block diagram reduction rules

One general type of block diagram occurs every time there is a closed-loop system. Because it is so common, we want to derive and memorize the result. Consider the block diagram in Figure 2–12a. It can be reduced to the transfer function in Figure 2–12b. Notice that, for compactness, the (s) subscripts have been dropped from all the transformed variables. Visually analyze the following relationships, and confirm their validity.

$$F = H(s)\,C$$

$$E = R - F = R - H(s)\,C \quad \text{Eq. 2.64}$$

$$C = E\,G(s)$$

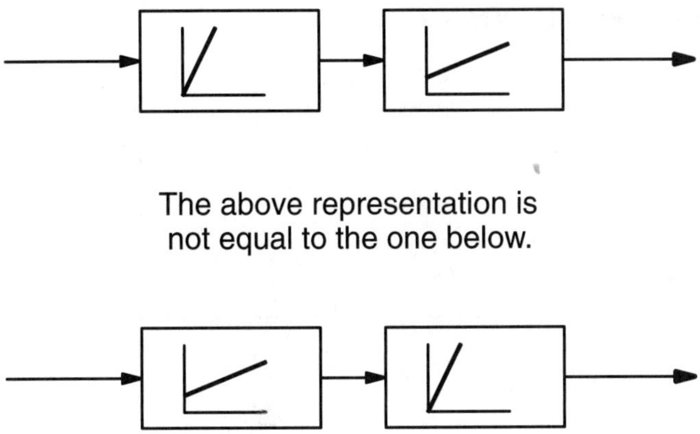

Figure 2–11 Linear and nonlinear blocks

Now substitute the second equation into the third equation (eliminating E):

$$\begin{aligned} C &= (R - H(s)\,C)\,G(s) \\ &= G(s)\,R - H(s)G(s)\,C \end{aligned} \quad \text{Eq. 2.65}$$

and, finally, expressed as a transfer function:

$$\frac{C}{R} = \frac{G(s)}{1 + G(s)H(s)} \quad \text{Eq. 2.66}$$

This expression will be used many times throughout this book.

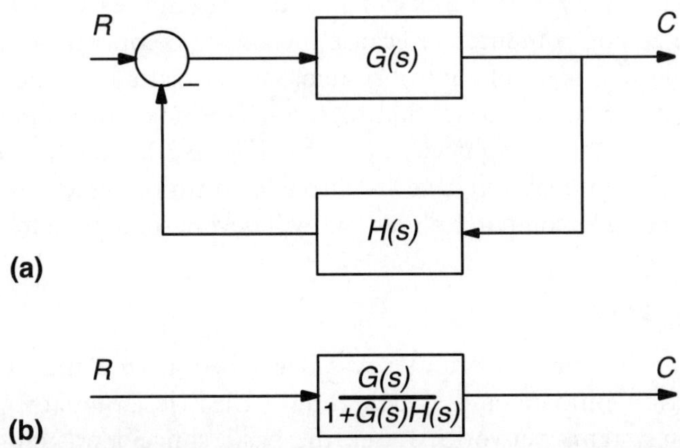

Figure 2–12 Block diagram simplification for typical negative feedback representation

Simplifications of Block Diagrams

As mentioned previously, the procedure for simplifying a block diagram consists of reducing a series of blocks into one transfer function. The basic approach attempts to rearrange the components into a form in which the reduction rules can be applied.

Example 2.14

Reduce the block diagram in Figure 2–13.

Solution The blocks are gradually combined to produce the transfer function shown at the botttom of Figure 2–13.

2.9 MATLAB/SIMULINK

MATLAB is a computer program that facilitates the analysis and design of dynamic systems. Used heavily in industry, it is credited with reducing the development time of most control system projects. The company that produces MATLAB has also developed "toolboxes," or application packages, that use and enhance the power of MATLAB. SIMULINK is designed to analyze and simulate dynamic systems that can be represented in block diagrams and transfer functions. It is like a virtual analog computer.

Even though most teaching institutions no longer use analog computers, analog computers were the main workhorses of early control engineers. Many of the complex dynamic systems used today (e.g., autopilots and reactors) were originally simulated in analog computers. Because the analog computer played such a major role in the development of control systems, and because most control engineers were familiar with it, it appears that software developers have emulated the analog computer's operation graphically (as in SIMULINK). An engineer who is familiar with analog computers will have little trouble transitioning to SIMULINK. Due to the role the analog computer played, we will devote some time to it.

The Analog Computer

The predecessor of the analog computer was developed around the early 1900s by General Electric Company and Westinghouse Electric Company, primarily to simulate power systems networks. Later, the basic concept was enhanced to

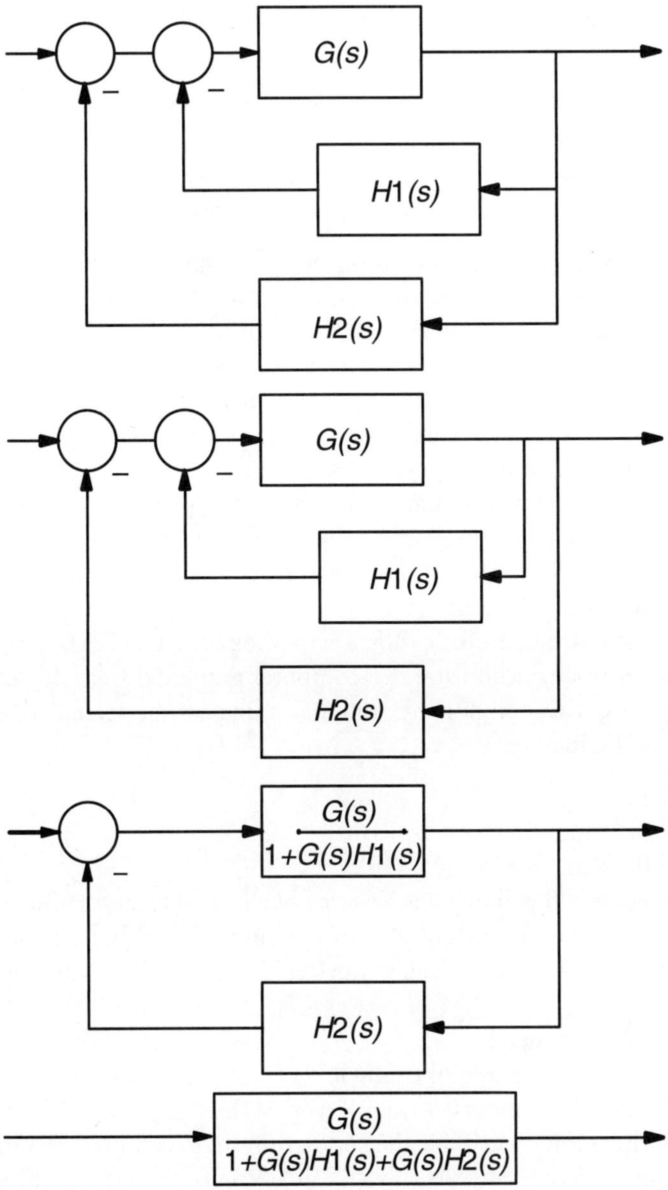

Figure 2–13 Block diagram simplification example

perform the integration task required to solve differential equations. The basic components in an analog computer are:

- the inverting op-amp, which sums several voltages and negates their output
- the negative integrator, an op-amp with a resistor and capacitor that integrates the sum of various inputs, integrates it, and reverses the sign
- potentiometers, which provide a selectable attenuation to a given signal

The analog computer representation of a differential equation consists of a board with a multitude of wires connecting op-amps to potentiometers and integrators. As the simulation process begins, the solution variables appear as analog voltages, which can be recorded on a plotter or an oscilloscope.

Magnitude scaling techniques allow the simulation of physical variables whose real magnitudes vary by several orders. Here the variables are represented by equivalent variables with values between +/− 10V. Similarly, time-scaling techniques allow solutions that have dynamics lasting a few microseconds to be realized in minutes so that they can be observed by the naked eye and/or recorded on a plotter. Figure 2–14 shows an analog computer and common recording instruments, the oscilloscope and the X-Y recorder.

As mentioned earlier, SIMULINK is a toolbox for MATLAB that re-creates an analog computer wiring schematic in a computer graphic display. Instead of having a wire connecting two components, one has a line on the screen. The steps for running SIMULINK include:

1. Load MATLAB.
2. Establish the working directory.
3. Invoke SIMULINK (type >>simulink).
4. Open a new window for drawing a series of blocks and transfer functions corresponding to the dynamic system you are trying to solve. Include devices that excite your system (sources), as well as devices that allow you to display the solution (sinks).
5. Connect all the components.
6. Establish numerical values for all coefficients.
7. Select “Simulation Parameters” to set details of the simulation. The main ones are stop time (how long you want the simulation to go) and maximum step size (the integration step that SIMULINK/MATLAB use to perform the numerical integration). For most problems, the “Auto” setting will be adequate.
8. Start the simulation by pressing the small triangle pointing to the right.
9. Look at the solution. Make changes in one of the parameters as desired, and restart the simulation to see how the change affects the parameter.

MATLAB/SIMULINK performs its task by performing numerical integration. This process was performed in the analog computer with op-amps.

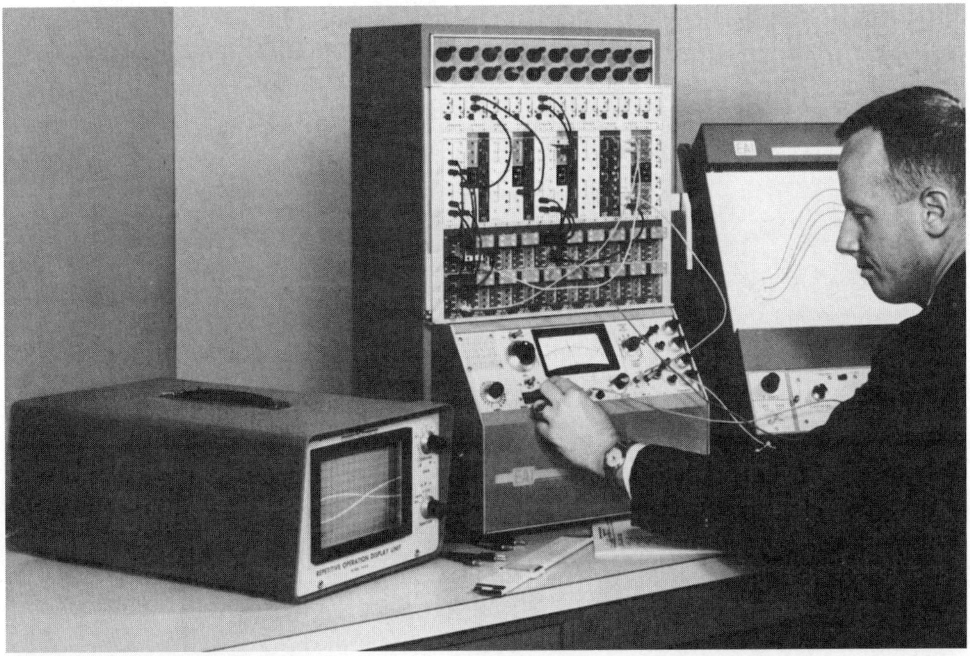

Figure 2–14 Analog computer and associated display devices (oscilloscope and X-Y plotter) (Courtesy of Electronic Associates, Inc.)

The user might want to change other settings in the simulation, like those on the "Parameters" menu. They are explained next with those mentioned previously:

Start time—Time when the simulation begins; usually set to 0.

Stop time—Time when the simulation is to stop.

Solver options—Variable-step or fixed-step; select variable-step.

Max step size—For the variable-step selection, the numerical integration procedure will try to use the largest step size possible so that simulation time is minimized. However, instead of leaving it on AUTO, the user might want to use a smaller step size to obtain a smoother curve.

Initial step size—Leave as indicated, AUTO.

Relative tolerance—Leave as indicated, 1*e*-3.

Absolute tolerance—Leave as indicated, AUTO.

You will also see a pull-down menu for "Solver Options." These options correspond to specific numerical integration routines. Some integration routines work better than others in specific types of problems. For the type of problems we will be doing in this book, most will work. For now, use the default, "ode45."

Example 2.15

Using MATLAB/SIMULINK, solve the distance-traveled problem, whose velocity profile is shown in Figure 2–1. Obtain plots of both velocity and displacement versus time.

Solution To solve this problem, we need to implement the differential equation in SIMULINK. This presentation assumes that you are in the lab and storing your files on the A: drive. Go back to the example, and review the solution process you carried out previously. Now, you will be implementing the problem using MATLAB/SIMULINK.

1. Assuming that you are in the "Desktop Windows" screen, click on the MATLAB icon. The main MATLAB screen appears.
2. Type

   ```
   cd  a:\ myfiles
   ```

   to establish the directory MYFILES in the floppy as your default directory. Any files you generate will be stored here. This assumes that there is already a subdirectory in your floppy (A:) called MYFILES. If there is not one, create it.
3. Invoke the SIMULINK toolbox by typing SIMULINK. The SIMULINK window appears.
4. Move the mouse to point to the window and click anywhere. The top bar on the window darkens (typically dark blue), indicating that the window is active (not necessary in the newer versions).
5. Click on FILE, then NEW, the MODEL. This opens a blank window (titled "Untitled"), ready for you to start adding components. You might want to go to FILES in the new window, then click on SAVE and call the schematic something like "example 9." (Note: Some early versions of MATLAB do not allow you to start a file name with a numeric character.) Now your SIMULINK schematic has a name. Whenever you work with computers, save your work regularly.
6. Now let's start bringing the various components into the window. Returning to the SIMULINK window, double-click on SOURCES, you will see a window that has a series of blocks that originate signals. Place the pointer on top of STEP, click it and, while holding it down, drag it into your recently created SIMULINK window.

7. Repeat Step 6 three more times, because we will need four step functions to produce the speed curve of Figure 2–1. Eventually, we will end up with the schematic shown in Figure 2–15.
8. Click on the "Step" label under the step function block. You will see a blinking cursor. Type a label descriptive of the step function (i.e., $60u(t)$).
9. Now click on the MATH OPERATIONS icon of the SIMULINK window.
10. Bring the SUM icon into your window and double-click on it so that you can change the number of inputs to 4 " + ."
11. Change the shape to rectangular.
12. Return to the SIMULINK list and click on the CONTINUOUS icon.
13. Click the INTEGRATION icon and bring it into your window (see Figure 2–15 to confirm what the block looks like).
14. Double-click on the INTEGRATOR icon to confirm that the initial condition is zero.
15. Return to the SIMULINK list and click on the SINKS icon.
16. Bring two SCOPEs into your window.
17. Bring a DISPLAY into your window.
18. Move the blocks so that they are in the approximate locations shown in Figure 2–15. One of them will be used to show a graph of speed versus time, the other one of distance versus time. You can also use a display that indicates the numerical value of the chosen variable. Thus, at the end of the simulation it will display the final value.

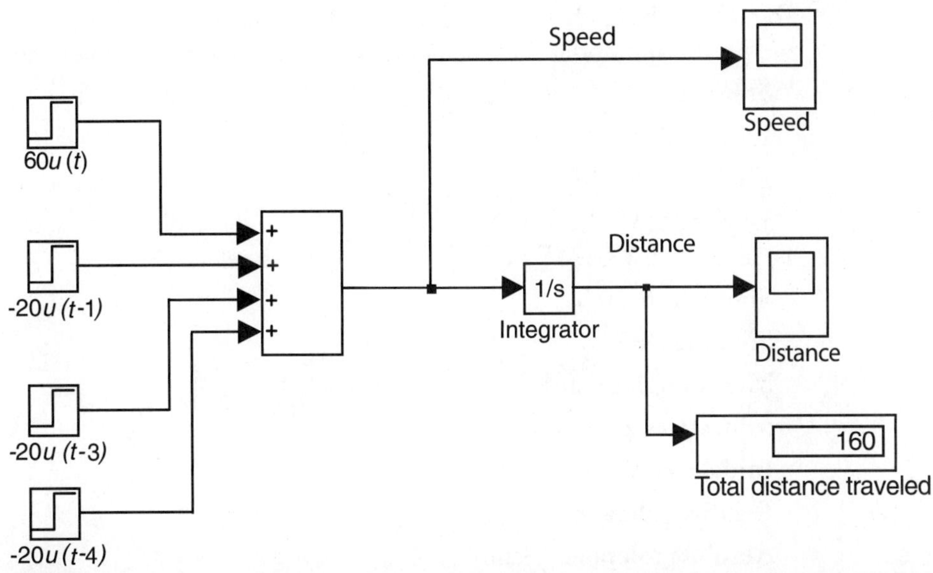

Figure 2–15 SIMULINK solution of Example 2.1—total distance traveled by a car given its velocity profile

19. The next task consists of interconnecting all the blocks you have brought in. Do this by going to the connection points on the blocks, left-clicking, and moving the connection path (wire) to the other block (with the left mouse button depressed). If you need a connection that is not straight, just release the left mouse button at the end of the straight section, and repeat the procedure for the next leg of the connection. Using these procedures connect all the blocks. You can also use an automated approach where you highlight the source block then you control-click the destination block. SIMULINK automatically places a connection between the two blocks.
20. Go to the label associated with the block and left-click on it. You will see that it becomes highlighted. If you begin to type, you will see the new text replace the existing text. It is very helpful to use names that will help you remember the function occurring in the block. You can also place text around your SIMULINK schematic by double-clicking where you want the text to appear.
21. Now let us select the specifics of the step functions to obtain the speed step functions to obtain the speed graph in Figure 2–1. We will obtain this by adding the following functions:

    Step 1—starts at $t = 0$, has a weight of 60

    Step 2—starts at $t = 1$, has a weight of -20

    Step 3—starts at $t = 3$, has a weight of -20

    Step 4—starts at $t = 4$, has a weight of -20
22. To make each step have these characteristics, double-click on each one of them and set the appropriate values. For example, for Step 1 the values are:

    Step time—0

    Initial value—0

    Final value—60
23. Now go to SIMULATION in the top menu, and double-click on it.
24. Then, go to PARAMETERS. For this problem select:

    Simulation—0.0

    Stop time—5

    Max. step size—auto

    Min. step size—auto

    Initial step size—auto

    Relative tolerance—1*e*–3

    Absolute tolerance—auto
25. After entering these values, click OK.

26. Now you are ready to begin the simulation. Click on SIMULATION in your window, then click on START (or simply click on the triangle pointing to the right). You will hear a "bip," which means the simulation has terminated.
27. Double-click on one of the SCOPES. A graph (of distance versus time or speed versus time) appears. The display block will be showing the total distance traveled. If you click on the binoculars icon, the graph will be repositioned so that you can see the entire plot.

Using SIMULINK to Solve a Differential Equation

Even though so far we have not developed many differential equations ourselves, we will use SIMULINK to solve one of them. Another way of saying "solving a differential equation" is "finding the time response" or "propagating the solution." We will demonstrate the procedure with an example.

Example 2.16

Solve the following differential equation in terms of $x(t)$, the position of a component in a system.

$$2\ddot{x} + 3\dot{x} + 45x = f(t) \quad \textbf{Eq. 2.67}$$

where $f(t)$ is the forcing function with a step of weight 90, or:

$$f(t) = 90u(t)$$

and having the initial conditions corresponding to 3 ft/s and 0.5 ft.:

$$\dot{x}(0) = -3$$
$$x(0) = 0.5$$

Solution First, solve for the highest derivative term:

$$2\ddot{x} = 90u(t) - 3\dot{x} - 45x$$

The scheme is to assume that one has the quantity $2\ddot{x}$ available. We move forward in obtaining all other lower derivative variables and, once we have them, we see that we can create the $2\ddot{x}$ term, which we had assumed we had available. Figure 2–16 shows the development in obtaining the complete SIMULINK diagram. The simulation parameters from the previous simulations happen to work

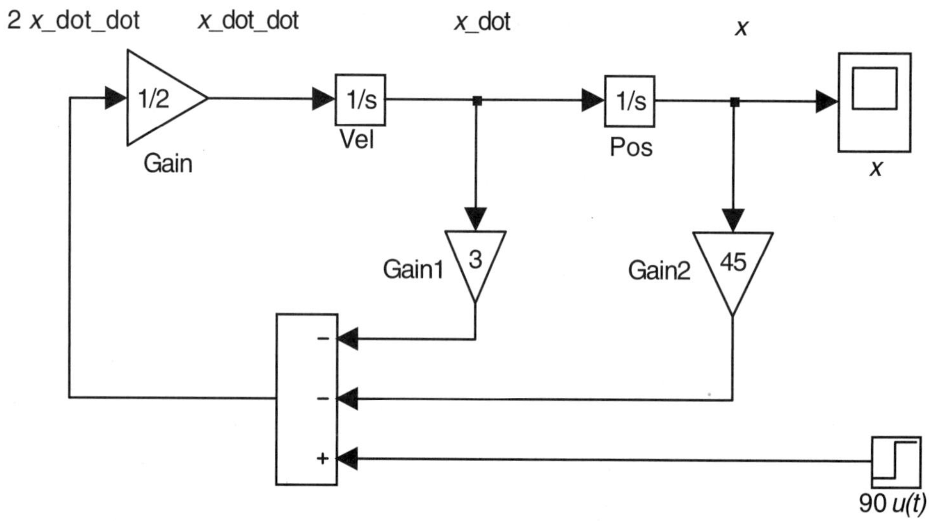

Figure 2–16 SIMULINK diagram for solving a differential equation

pretty well for this example, with the exception of the final time, which should be changed to 75.

Before we initiate the solution, we need to set the initial conditions. This is done by double-clicking on the integrators: integrator "vel" will hold the initial velocity and integrator "pos" the initial position. Notice that this SIMULINK diagram makes all variables easily accessible for plotting or recording. For instance, if one wants to plot the acceleration, $\ddot{x}$, one simply connects SCOPE to that point in the diagram. At this point, we are ready to begin the simulation by going to SIMULATION, then to START. Figure 2–17 shows the time response.

Using SIMULINK to Determine the Output of a Transfer Function

SIMULINK not only handles simple blocks, such as integrators and constants, it can work directly with large transfer functions. The SIMULINK diagram of an earlier example can be reduced to one transfer function using the rules previously described (this is left as an exercise). The resulting transfer function is:

$$\frac{X(s)}{F(s)} = \frac{1}{2s^2 + 3s + 45} \qquad \text{Eq. 2.68}$$

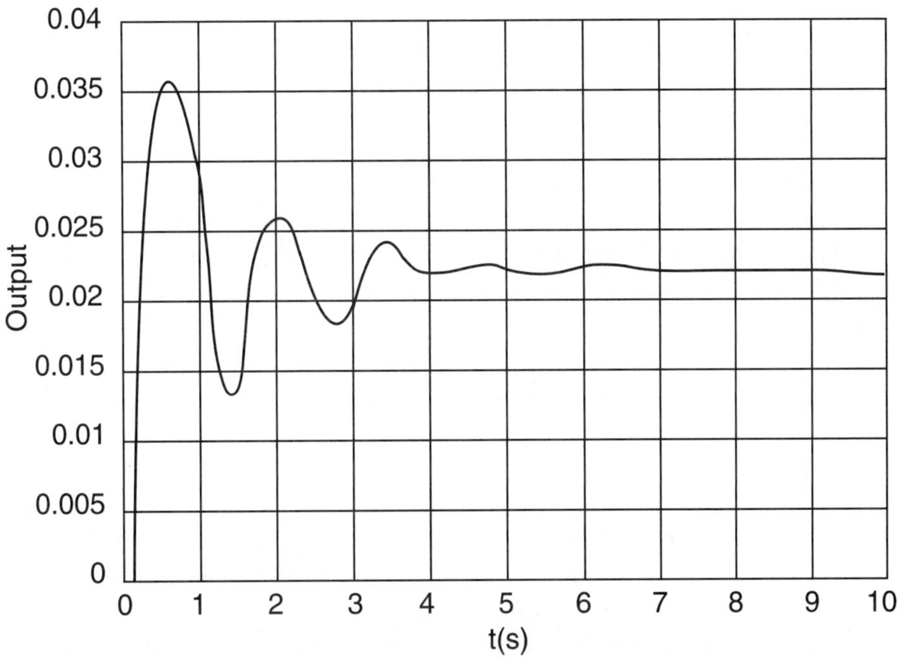

Figure 2–17 System response obtained from SIMULINK (Example 2.17)

Example 2.17

Determine the response of the previous transfer function to a step with weight 90.

Solution This problem has the same type of input as an earlier example. To model the transfer function, double-click SIMULINK. When the drop-down menu appears, click CONTINUOUS, then TRANSFER FUNCTION. Drag this to your new window and double-click on it. You will see two locations in which to enter information. These locations correspond to the coefficients of the numerator and denominator polynomial coefficients of the transfer function. MATLAB/SIMULINK handles polynomials by means of an array containing the coefficients of the polynomials. Coefficients are entered in descending powers of s. Thus, for the numerator, enter 1. For the denominator, enter 2, 3, and 45. Make sure to leave the square brackets around the set of coefficients. As done in the previous example, connect the STEP (forcing function) and the SCOPE to the input and output of the transfer function, respectively. Now you are ready to simulate the system. Notice

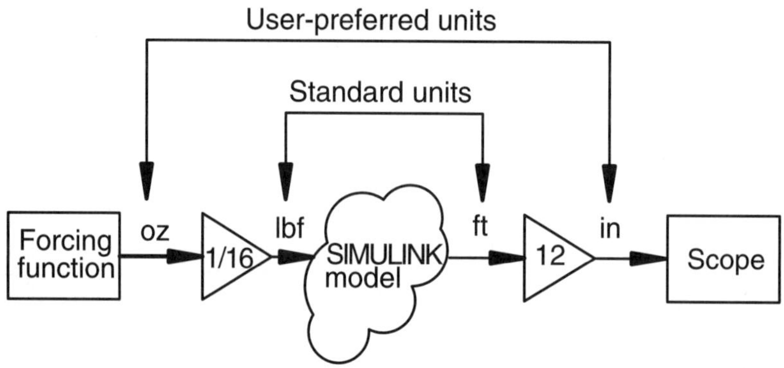

Figure 2–18 Recommended procedure for dealing with nonstandard units

that in this transfer-function representation, you have lost the ability to dictate initial conditions, as you were able to do previously. Now, to simulate the system, go to SIMULATION and then to START.

Different Units

It is best to derive the transfer function in terms of standard units. However, one may want the results in other units, such as the units with which the user is familiar. A procedure that allows the use of standard units in the transfer function while simultaneously having inputs and outputs in any desired units is shown in Figure 2–18. Here we start with a forcing function in ounces, then convert it to pounds (lbf) and apply it to the transfer function. Afterward, the transfer function produces its output in feet. We convert it into the desired units, inches.

Summary

1. Differential equations are mathematical devices used to capture the dynamic characteristics of physical systems.
2. The Laplace transformation provides an algebra-based method for dealing with problems involving differentiation and integration.
3. SIMULINK provides a graphical environment for capturing the dynamic system and also a computational interface to MATLAB, which solves the equivalent differential equation.

4. Dimensional homogeneity is a useful way to determine the necessary correctness of a differential equation.
5. Transfer functions are another vehicle for capturing the dynamic characteristics of physical systems.
6. Block diagram reduction techniques simplify a multitransfer function block diagram into one transfer function.

Idea Portfolio

1. What are the variables involved in the dynamic system under question? Which variables affect the system operation? Which variables need to be controlled? Which input variables (or independent variables) can be controlled to cause the output (or dependent variables) to change? What are the relationships between inputs and outputs? Are the variables related in an algebraic relationship or in a dynamic one? How would you measure the relevant variables? What instrumentation would be necessary? What are the appropriate units? Do they form a consistent set of units?
2. How do various parameters tend to affect output? Create a table that shows the relationship between input and output, using the following notation:

   ↑: increase; ↓ decrease; NE no effect
3. Notice that, due to the inherent characteristic of most closed-loop control systems, variations in some variables tend to be nulled.

Increasing Value of the Following Parameter	Effect on Output 1	Effect on Output 2
Causes the outputs to:		
Parameter 1		
Parameter 2		
. . .		

Problems

1. Consider an independent voltage source, shown in Figure 2–19, driving a 5-ohm resistor.

   Obtain mathematical expressions for each of the following quantities:
   - Voltage across resistor
   - Current through resistor
   - Power (heat) produced by resistor
   - Total energy dissipated by resistor

   in terms of the input voltage $v(t)$. Use a combination of delayed step functions to produce the input voltage waveform. Implement them in SIMULINK, and simulate. Provide plots for each of the preceding quantities. Include your SIMULINK diagram in your results.

2. Determine the Laplace transform of the following functions:

   $f(t) = (120 \cos 377t)\, u(t)$

   $f(t) = (89e^{-2t} \sin 1000t)\, u(t)$

   $f(t) = t\, u(t-3)$

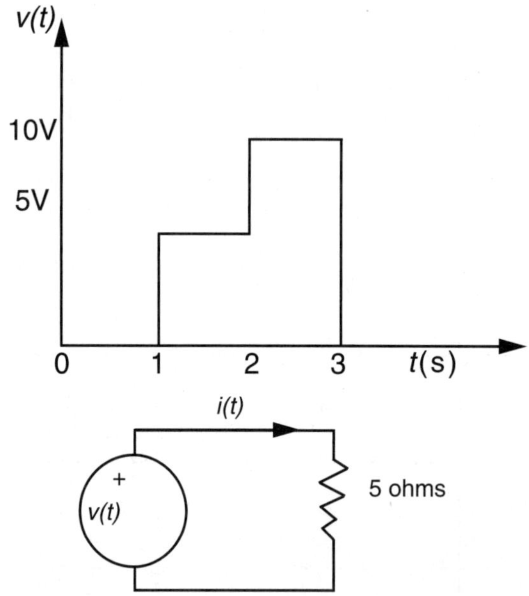

Figure 2–19

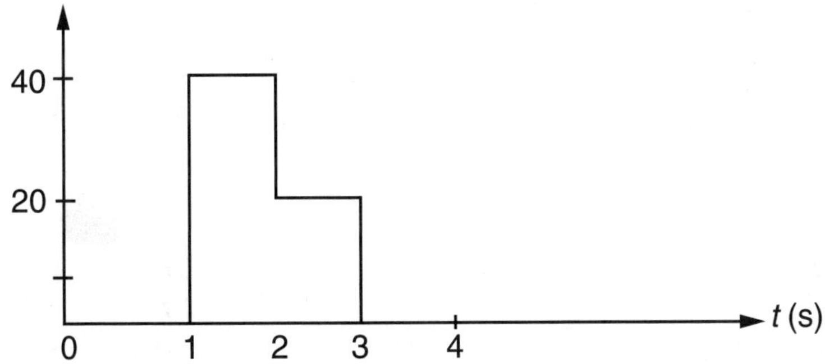

Figure 2–20

3. Represent the function shown in Figure 2–20 as combination of delayed step functions.
4. Obtain a mathematical representation for the function shown in Figure 2–21.
5. Sketch the output, $c(t)$ for the block diagram shown in Figure 2–22a, given the input, $r(t)$ shown in Figure 2–22b. Use the graph of $r(t)$ for showing $c(t)$.

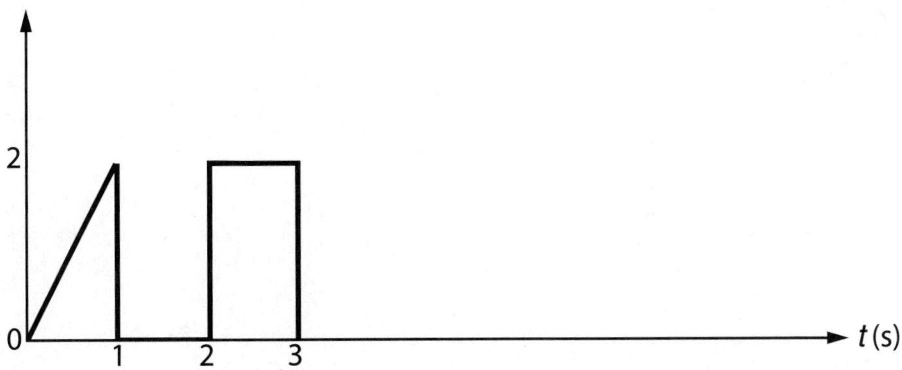

Figure 2–21

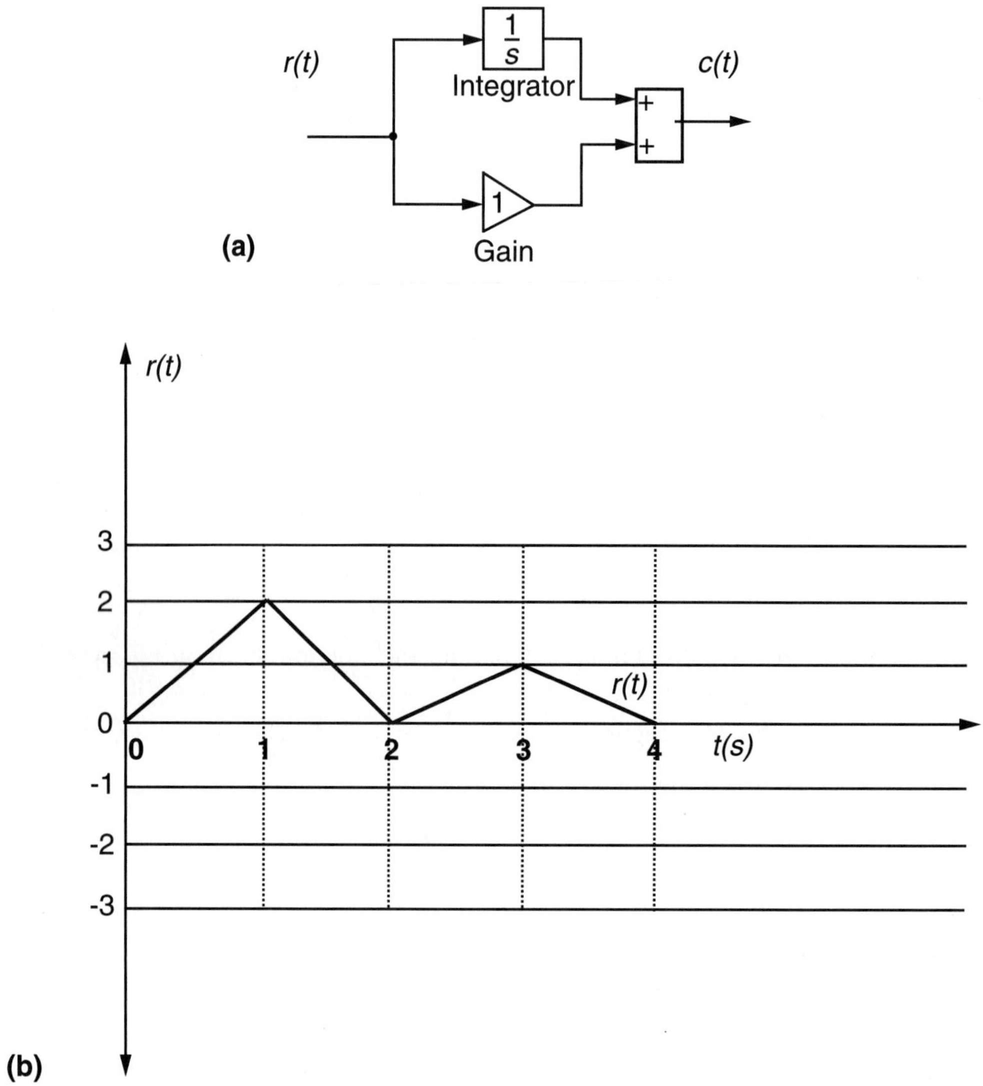

Figure 2–22

6. Consider the curves $r(t)$ and $c(t)$ shown in Figure 2–23, representing the input and output of several "black boxes." Provide an appropiate block diagram of what is happening inside the black box. Note: The solution for the first problem is given as an example.

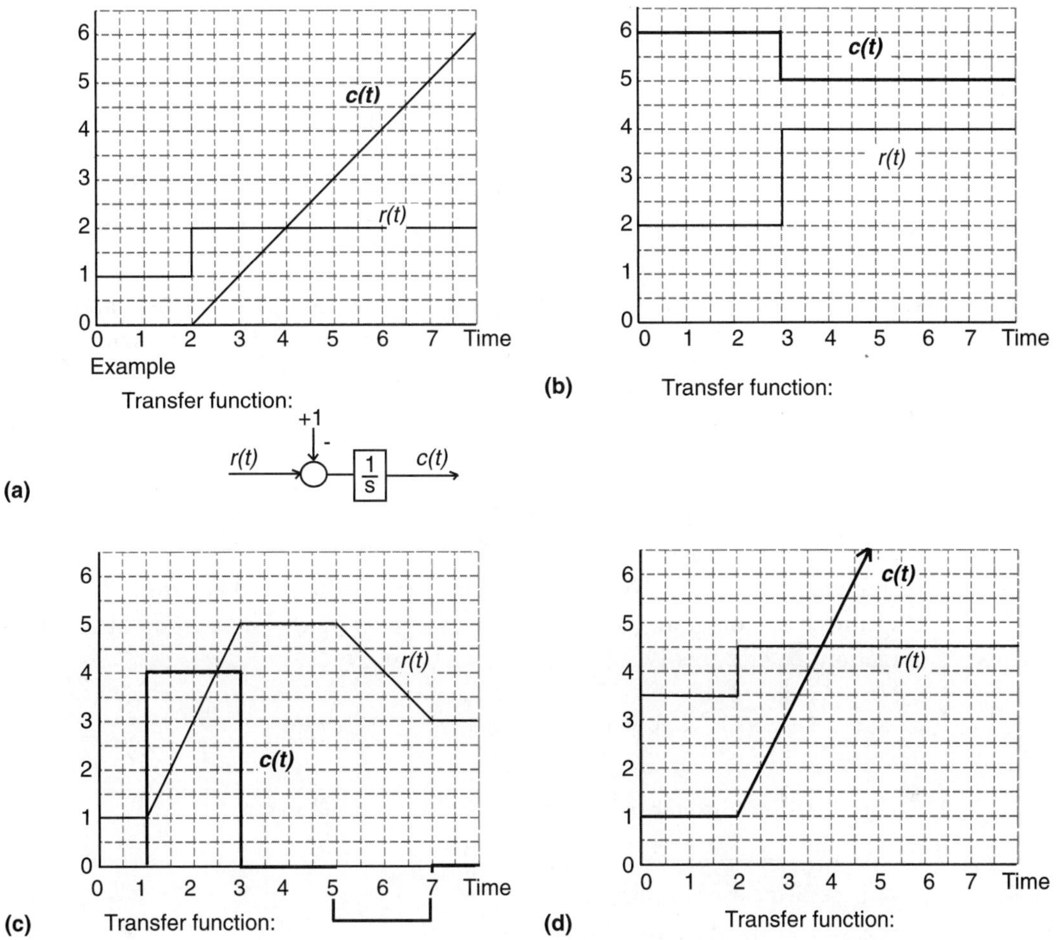

Figure 2–23

7. Perform dimensional analysis (i.e., determine if there is unit agreement among the different terms) on the following differential equation:

$$m\ddot{x} + k^2 x = f(t)$$

where:

m = mass
$\ddot{x}$ = acceleration
x = displacement
k = spring constant (elemental equation: force = k · displacement)
$f(t)$ = force

8. Find $f(t)$ given $F(s)$:

$$F(s) = \frac{9s + 23}{(s + 2)(s + 3)}$$

$$F(s) = \frac{4s^2 + 6s + 3}{s(s^2 + 4s + 3)}$$

9. In class, we said that there is an analogy between multiplication and logarithms, similar to the river analogy that was used for the time domain and the Laplace domain. Indicate which of the following procedures is analogous to multiplying A and B:
   - Adding log(A) to log(B)
   - Multiplying log(A) and log(B)
   - Subtracting log(B) from log(A)
   - Taking the log of A and taking the log of B
10. Implement a SIMULINK diagram for solving differential equation. Note: All you need is a diagram (computer or hand-generated). There is too little information to solve this DE.

$$8\dddot{y} + 4\ddot{y} + 2\dot{y} + 7y = 5\,y\,f(t)$$

11. For preceding differential equation, indicate the following:

    Order __________

    Type: Linear or nonlinear

    Name of the dependent variable __________

    Name of the independent variable __________
12. Implement a SIMULINK diagram for solving the following differential equation:

$$4\ddot{y} + 233\dot{y} + 27y = 5t^2u(t)$$

13. For the differential equation in Problem 12, indicate the following:

    Order __________

    Type: Linear or nonlinear

14. Find the final value of:

$$X(s) = \frac{(5s + 12)}{(8s + 2)(2s + 3)}$$

- Using the Final Value theorem
- By obtaining $x(t)$ (using partial-fraction expansions and then taking the inverse Laplace transform)

15. Determine the appropriate units for the constant C in the following differential equation, which describes a thermal system. The units of T_y and T_∞ are in °F and $\dot{T}_y$ is in °F/min.

$$C\dot{T}_y + T_y = T_\infty$$

16. Determine the appropriate units for the constant b in the following differential equation, which describes a falling racquetball. Use the M,L,T nomenclature to express your results. Use the unit information in the differential equation to arrive at your answer. Assume that the differential equation is correct.

$$m\ddot{y} + b\dot{y} = mg$$

17. A skydiver jumps from an airplane flying at 5,000 ft. The skydiver freefalls for 10 seconds, achieving a speed of 120mi/hr. Then her parachute opens, slowing the diver to 12 mi/hr. Eventually, she lands safely, 30 seconds after her parachute opens. Sketch the $\ddot{y}$, $\dot{y}$, and y (acceleration, velocity, and displacement of the skydiver) as a function of time. The tell-tale (reference direction indicator) is attached to the airplane and pointing downward.
18. Consider two marbles of different sizes and initially different temperatures, T_{10} and T_{20}. Marble 1 has a diamater that is twice that of Marble 2. Both marbles are dropped inside a perfectly insulated container full of water, at temperature T_{water0}. Immediately after being dropped into the container, a top (also a perfect insulator) is placed on the container. Produce a graph showing what you expect the temperatures of the marbles (T1 and T2) and the water to do. Pay special attention to the speed of response and the final value of each of the bodies.

19. Simplify the block diagrams in Figure 2–24 into one transfer function. Express the resulting transfer function as a ratio of two polynomials.

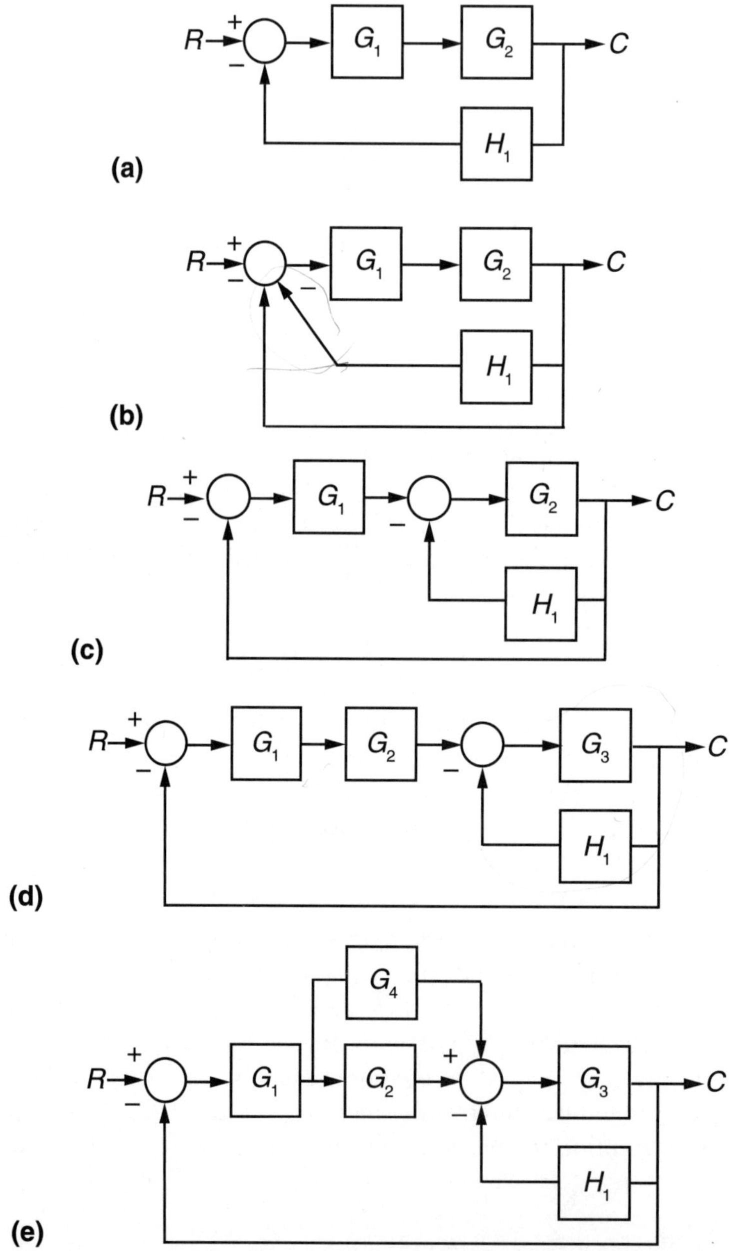

Figure 2–24

20. Consider two block diagrams as shown in Figure 2–11. The first block (at the top figure) is characterized by:

$$f_1(x) = 5x$$

and the second block (on the top figure) by:

$$f_2(x) = 3x + 2$$

Apply a signal y spanning from 0 to 5 to both the top and bottom block diagrams and compare their outputs. Are the outputs the same?

21. Using block diagram reduction techniques, find the transfer function from each input to the output $C(s)$ for the block diagram in Figure 2–25.
22. Obtain an s-domain block diagram representation for the block diagram in the time domain in Figure 2–26.
23. For Question 22, obtain a mathematical expression for:
    - $y(t)$
    - $Y(s)$
24. The differential equation for a racquetball dropped from the top of a building is:

$$m\frac{dv}{dt} + bv = mg\,u(t)$$

where m is the mass, b is an equivalent of friction, and g is the acceleration of gravity.

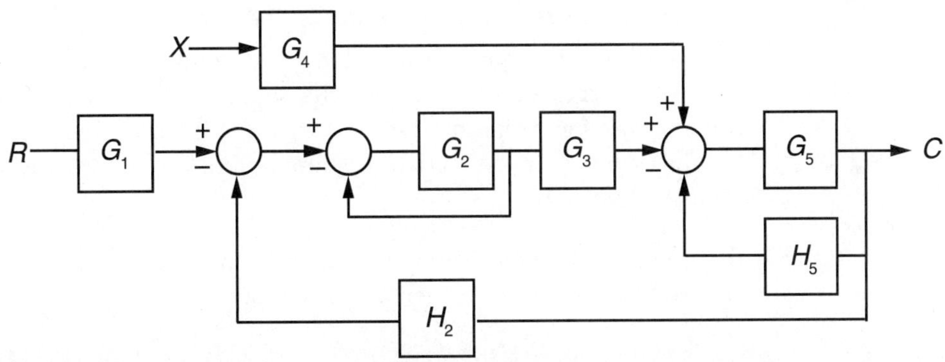

Figure 2–25

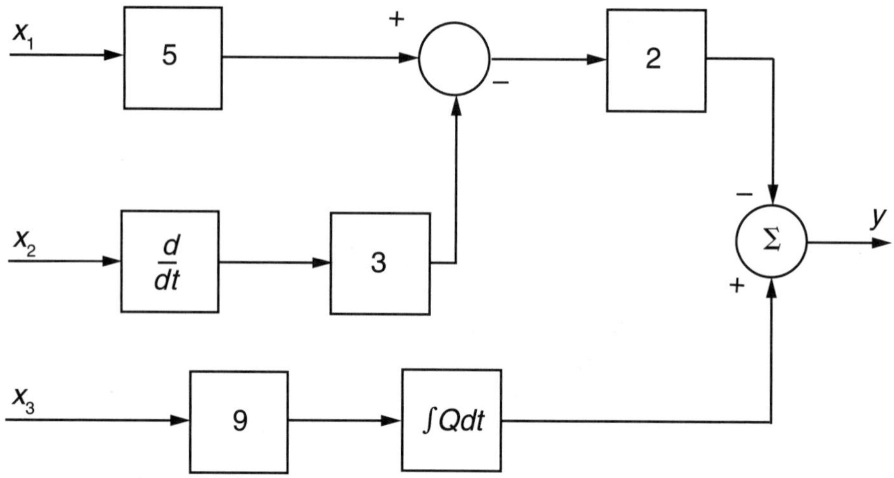

Figure 2–26

- Take the Laplace transform of the differential equation, and solve for $V(s)$. Using the approach presented in class, confirm the dimensional homogeneity of $V(s)$.
- Using the Laplace method, obtain an expression for $v(t)$ in terms of g, m, and b. Confirm the dimensional homogeneity of $v(t)$.
- Using the following parameter values, solve the differential equation using SIMULINK. Simulate until $t = 7$s.

  b = 0.00097 lbf·s/ft (pounds-force seconds per foot)

  m = 0.0020556 slugs (which is 30 g, the mass of the racquetball)

  g = 32.2 ft/s^2 (the acceleration of gravity)
- Substituting the numerical values just found, with the solution obtained in the second part, plot the solution, $v(t)$, using Excel or any other graphing program, for $0 < t < 7s$. Compare with the result obtained in your SIMULINK simulation. What is the terminal ($\lim_{t\to\infty} v(t)$) velocity?

25. Say that you are modeling the population of fish in a bay. Assume that the fish are so numerous that we can regard their number to be changing continuously

rather than with time. If the variable $x(t)$ represents the number of fish in the bay:

- Represent the number of fish as a differential equation.
- How would you determine the relevant parameters in the preceding differential equation?
- Indicate some limitations and/or shortcomings of the resulting differential equation.

3

Modeling of Dynamic Systems

Learning Objectives

In this chapter, you will learn how to:

- Model the following dynamic systems: electrical, fluid, mechanical, and thermal.
- Recognize the many similarities of various systems.
- Solve a differential equation of a physical system using MATLAB/SIMULINK.
- Combine elemental and structural equations into one transfer function or differential equation.

3.1 Historical Perspective

Modeling is the process whose goal is to develop a mathematical description (called a mathematical dynamic model) of the process to be analyzed or controlled. The term *model,* as it is used and understood by control engineers, means a set of differential equations that describe the dynamic behavior of the system.[1]

It is interesting to observe that, in the earliest occurrences of closed-loop controls, no analysis was performed. It was the creativity and intuition (and probably persistence) of groups of innovative individuals that allowed those individuals to transform their ideas into useful devices. A large group of inventions and working devices followed, also using little or no analysis, as described in a book that focuses on the historical development of feedback control by O. Mayr (1970)[2] or briefly presented by Franklin et al. Not even the well-known speed governor used in steam engines, or the fly-ball governor used by several early pioneers, including James Watt, originated with significant mathematical support.

As mechanisms became more complicated, their desired behavior was difficult to achieve through simple trial-and-error. The historical record shows how mathematicians and engineers came up with novel analysis and design tools that allowed their ideas to become realities.

Apparently, in the time of Orville and Wilbur Wright, there were more capable aerodynamicists than the two brothers as well as experimenters with better resources. Many considered the Wright brothers' greatest assets their ability to find, integrate, and use the best theory and to perform critical experiments. This combination of skills allowed the brothers to test and arrive at a craft that could sustain flight.

Some dynamic systems tend to be stable, even though at first they may appear to be unstable, and vice versa. Interestingly, a ping-pong ball suspended in air with a blower does not fall to the side, as illustrated in Figure 3–1. The underlying dynamic relations predict this behavior (Bernoulli's Law). As another example, have you wondered why almost every airplane has swept-back wings, instead of forward-swept wings? The reason is one of stability rather than cosmetics. In forward-swept wings, the natural bending of the wings increases the angle of attack, which in turn further increases bending. This relationship could easily snap a wing. On the other hand, in a swept-back wing, the bending decreases the angle of attack, which in turn decreases lift and consequently decreases bending. Thus, the airplane settles into an equilibrium position.[3]

A similar relationship is found in a plowshare pulled by a tractor. The disks that break the soil are arranged in a swing-back structure. Assume that one side of the plowshare encounters some very hard soil, which makes the plowshare rotate. The resulting rotation decreases the effective arm of the plowshare on the hard-soil side while increasing the effective arm on the other side. Consequently, two mechanisms become active and straighten the plowshare.

Modeling is an essential step when analyzing and designing a control system. Modeling provides the connection between the real or physical system and the control engineer's mathematical model. Figure 3–2 shows a simplified flowchart of the modeling process. Modeling is a skill, a way of thinking that proves useful regardless of the field of study or work.

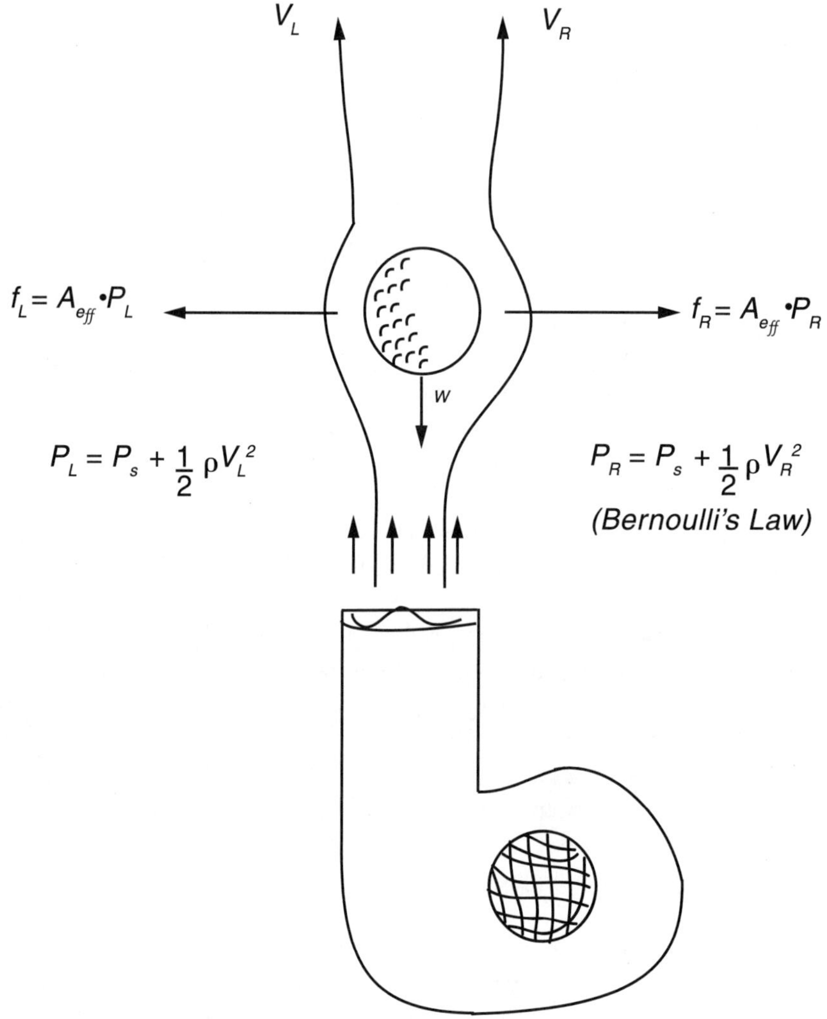

Figure 3–1 Inherently stable system: ping-pong ball suspended by blower

3.2 Modeling Principles

The modeling process might appear at first as art, because it is hard to always follow a specific procedure. However, as you begin to gain experience modeling systems, you will detect a general methodology that makes sense. We will present modeling in two phases: (1) the first phase, in which you develop the skills for modeling, and (2) the second phase, in which you begin to model real systems.

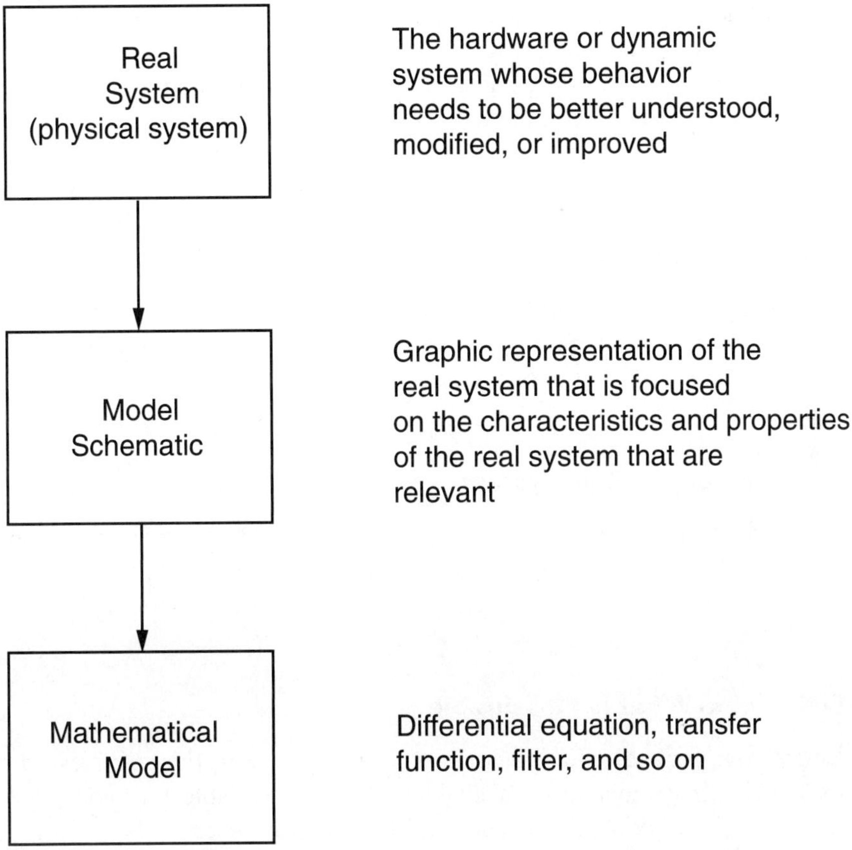

Figure 3–2 Flowchart of the simplified modeling process

The first phase consists of:

1. Learning the characteristics of simple components (by themselves) and the mathematical relationships describing them.
2. Being able to predict the behavior of systems made of these simple components when grouped. After making a prediction, you will simulate the system to confirm or enhance the accuracy of your prediction.
3. Arranging these simple components to obtain a specific behavior. Again, you will use simulation to confirm and/or tune your design.
4. Learning to recognize analogies between different systems: electrical, mechanical, fluid, and thermal.

Once you have established a certain level of comfort in these four areas, you are ready to start modeling real systems.

The process for modeling a real system has more steps than shown previously in Figure 3–2. A more complete process, which is shown in Figure 3–3, is discussed next.

Define the Goals of the Effort

Defining goals is very important, because it is easy to develop model items that do not relate to the problem. As you learn more about the system, you might choose to model it using a different approach.

Learn about the Real System

Watch the system operate. What is peculiar about it? Try to relate your experiences with the ideal components (which you will learn about shortly) to the real system. Subject the system to various commands, and observe the system's response.

Determine What Is Measurable

Relate your goals to what you can easily measure. Are the variables accessible? Is the needed instrumentation available? Does the variable that you plan to use as the control variable exhibit enough change to be usable? Is there noise in the signal?

Create a Model Schematic

A real system exhibits many characteristics that do not relate to the control problem. For this reason, it is important to draw a model schematic of ideal components (which we will learn about in detail shortly). Does this model schematic, in the eyes of the control engineer, behave in the same way as the real system (within acceptable accuracy), provided that operation is confined to a specific operating range? This step in the process requires practice and experience. It is also an iterative process: A tentative model schematic might change as the control engineer realizes that there is a better model. Consequently, the engineer will enhance the model schematic for a better description of the real system.

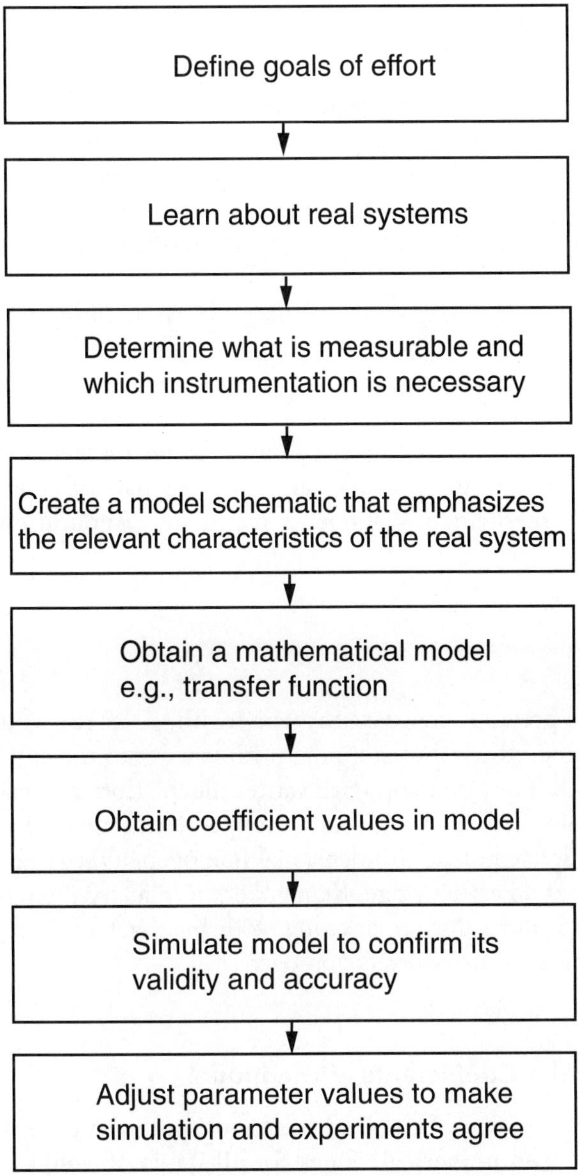

Figure 3–3 Expanded schematic of the modeling process

Obtain the Mathematical Model

Once the model schematic is completed, the step of obtaining the mathematical model is more straightforward than the previous one, even though it might involve a great deal of work. There are several approaches to obtaining the mathematical model. Some have little structure, while others are very formal. Even though these approaches produce different mathematical representations, they are equivalent: That is, their predicted behavior (simulation or solution of the differential equation) is the same.

The two most common forms of mathematically representing real systems are the differential equation and the block diagram. The block diagram, the common language of the control engineer, represents components (e.g., hardware, sensors, controllers) with their appropriate transfer functions. The block diagram also shows how these components connect. Other mathematical forms of representing the real system are available, as we will see later. Sometimes, the process of obtaining the mathematical model is called the "Application of First Principles."

Check the Model

As was suggested in the previous chapter, it is worthwhile to check your work as you go, rather than to proceed without stopping until you get to the end and then begin checking your work. The latter approach wastes much effort, especially if one finds a problem at the end. Erroneous units and dropped terms are common problems, so the techniques derived earlier (dimensional homogeneity) can be very useful in spotting problems at an earlier stage. Remember, in industry there is no such thing as partial credit. If your solution is wrong, your boss will not care if it was caused by a missing term or by an incorrect approach.

Obtain the Values of the Coefficients in the Model

Before a differential equation can be solved (i.e., propagated in time to predict dynamic behavior), we need numerical values for all its coefficients. This step requires experimentation. For example, assume that a component in the system is modeled as a spring. We must determine the component's spring coefficient, a number that indicates the component's stiffness. To determine the spring coefficient, we must apply a known force and measure the component's deflection. Dividing the magnitude of the force applied to the spring by the displacement yields the spring coefficient. When a spring tester is available, simply place the spring in the machine. This machine performs a test procedure on the spring automatically and displays the spring coefficient.

In some cases, the determination of the coefficient's value in a particular kind of spring might be beyond our expertise, even though the spring is used in a system for which you are in charge. In such cases, consider contacting another engineer or a firm with expertise in the study area. In some cases, your facility may lack the equipment needed to perform the test. For instance, assume that we have a hydraulic device with the characteristics of an orifice (a fluid restriction), and we lack the equipment to experiment on this device (e.g., pumps, flow meters, and other type of instrumentation). In a case like this, go to a firm or a specialist to perform the required test, or get the information from the manufacturer.

In some cases, it might be possible to compute the value of the parameter analytically. For instance, if we know the shape and kind of material (and consequently the density), we can calculate the mass of a component. In the spring example, we can use a formula in terms of wire diameter, spring diameter, number of turns, and elastic modulus of the wire to calculate the numerical value of the spring constant.

Simulate to Confirm the Validity of the Model

During simulation, we evaluate how well the mathematical model predicts the behavior of the real system. We do this by subjecting the model to specific inputs, such as step, pseudo-impulse, and sinusoidal functions. Then, we compare responses. We can also perform equivalent tests on the real system and record the system behavior. Then, we compare both sets of responses to determine if coefficients need adjustment. Agreement at this step is very important, because it provides a foundation for the work that follows. Assume that the coefficients and model are wrong until proven correct.

Modify the Model and Run Additional Tests (If Needed)

We might observe discrepancies between the model and the real system and conclude that the model must be reformulated. Thus, we begin an iterative process whereby we modify the model to make it agree with the real system. We must identify which of the excitation tests used in the simulation exhibited agreement and which exhibited disagreement, and we must identify which ideal components must be added or eliminated to achieve agreement.

Once there is agreement, you can proceed to the design phase. Some systems do not allow testing in open-loop manner because of danger to the operator or damage to the system. In these cases, it is necessary to connect the system in a simplified closed-loop configuration (i.e., negative feedback system) so that the tests can be carried out. In other cases, it might be necessary to install sensors, which would disable the driving mechanism (which provides the power for the system to move), if the system were to operate outside certain safe limits. A high-force hydraulic actuator, in which position feedback ensures proper operation, is such a system.

3.3 Ways to Represent the Mathematical Model

There are various ways to represent a dynamic system mathematically. Often, personal preference or the software package determines how a physical system is represented. The following section presents some of the most common ways.

Differential Equations

Differential equations, the most common way to represent a dynamic system, are understood universally by the mathematical and engineering community. Ordinary differential equations generally describe systems whose behavior can be compartmentalized into "lumps," each with unique properties. Often, it is said that ordinary differential equations describe systems made of lumped parameters.

Partial differential equations describe systems with distributed properties. Examples of distributed systems include a wave in a pool and the movement of a rope tied to a wall: The untied end moves upward and downward. A partial differential equation contains the partial derivative $\partial y/\partial t$, instead of regular derivatives, dy/dt.

Differential equations can be linear or nonlinear. When they are linear, all terms have constant coefficients rather than products of terms or time functions. Notice that a linear differential equation can have a forcing function (input) that is a nonlinear function of time.

The order of the differential equation is typically the highest-order derivative, and it indicates the complexity of the system. Perhaps a more precise way of referring to the order of the differential equation is the difference between the order of the highest and lowest derivative terms. In systems with elements with the ability to store energy, the differential equation indicates the number of energy-storage components. Another related aspect, the number of integrators required to solve the differential equation using a program such as MATLAB/SIMULINK, equals the differential equation's order.

Following are some examples of differential equations.

When the differential equation includes integrals, it is called an integro-differential equation.

Following is a second-order, linear differential equation:

$$8\frac{d^2y}{dt^2} + 3\frac{dy}{dt} + 5y = 3t^2u(t) \qquad \text{Eq. 3.1}$$

Following is a first-order, nonlinear, ordinary differential equation:

$$5\dot{x} + 4\dot{x}x = 5u(t) \qquad \text{Eq. 3.2}$$

Following is a first-order, nonlinear, partial differential equation:

$$u\left(\frac{\partial u}{\partial x} + \frac{\partial u}{\partial y}\right) = 0 \qquad \text{Eq. 3.3}$$

Following is a second-order, integro-differential equation:

$$3\frac{dv}{dt} + 8\int v\,dt = 5tu(t) \qquad \text{Eq. 3.4}$$

Transfer Functions

Transfer functions are probably the most common form of system representation, because they are understood readily by technologists from different disciplines. Transfer functions are used throughout this book.

Procedural Approaches

Several approaches start with the model schematic, then offer detailed procedures for producing the differential equation. Among these approaches are the linear graph, the flow graph, and the bond graph. Figure 3–4 shows the symbols and diagrams used to represent the model schematic. While these approaches are not difficult, they require a significant amount of time to obtain the mathematical representations. They are particularly useful when the system has many components. Because procedural approaches are beyond the scope of this book, references are given at the back of this book for the interested reader.

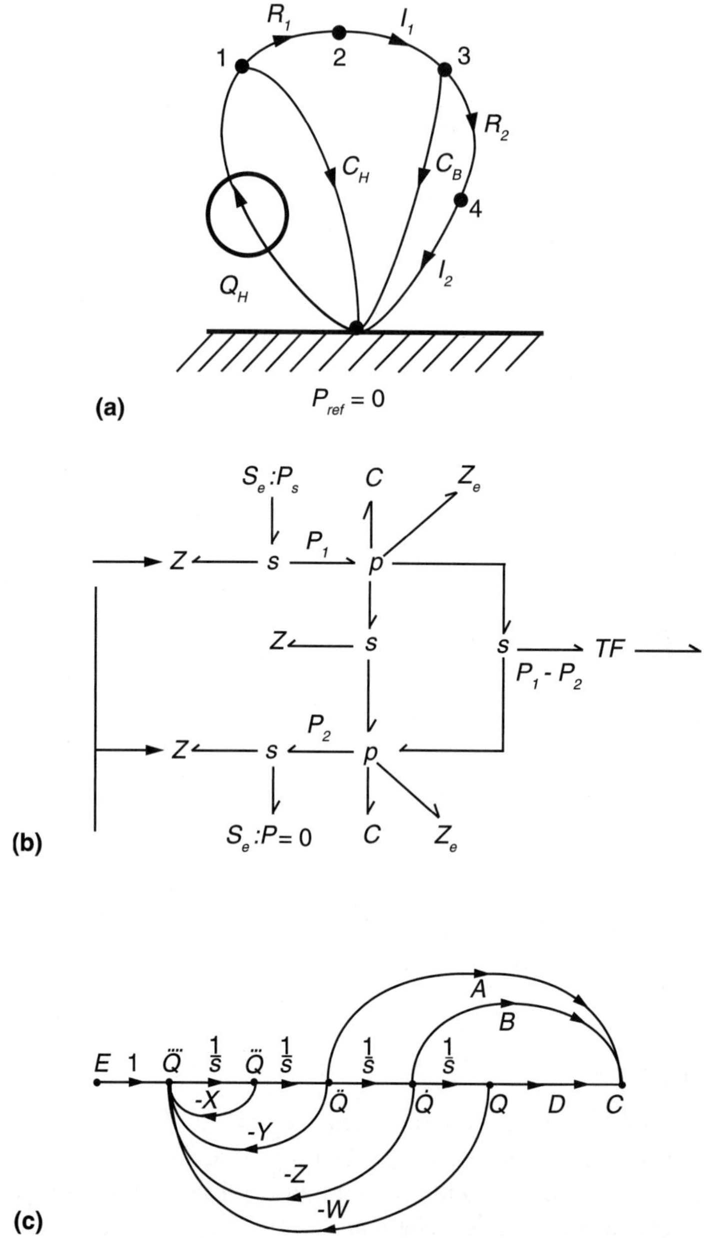

Figure 3–4 Examples of re-creating model schematics using a (a) linear graph, (b) bond graph, and (c) flow graph

Other Representations

Other graphic representations communicate the dynamics of a system or component. We will learn them and use them in subsequent chapters. Some of these include frequency, Bode, Nyquist, and polar plots.

3.4 Pedagogical Suggestion

What is the best order in which to present this material? One opinion is that the best approach is to present the various types of physical-system components (e.g., electrical, hydraulic) as quickly as possible so as to create a framework that illustrates the relationships and analogies of different types of physical systems. Once this has been done, go back to each category in more detail and proceed with the procedure for obtaining the system equation. Once a category has been completed, go to the next category and repeat the procedure. Finally, consider a system that is made of several types of physical systems (i.e., a hybrid system).

3.5 Modeling Physical Components

It is amazing to note the many similarities (mathematical relationships) among different types of systems. Even more interesting are the common behavioral modes and variables in these systems.

Analogies

Often, a system can be represented by an analogous one (one of a different type but with the same behavior as the original system). An analogous system reveals characteristics not obvious in the original system.

Across- and Through-Variables

It is possible to identify analogous variables that have much in common in all types of systems. Because of the way in which analogous variables manifest themselves in systems, they are called across-variables (A-variables) and through-variables (T-variables). To distinguish them, it helps to remember the ways in which they are

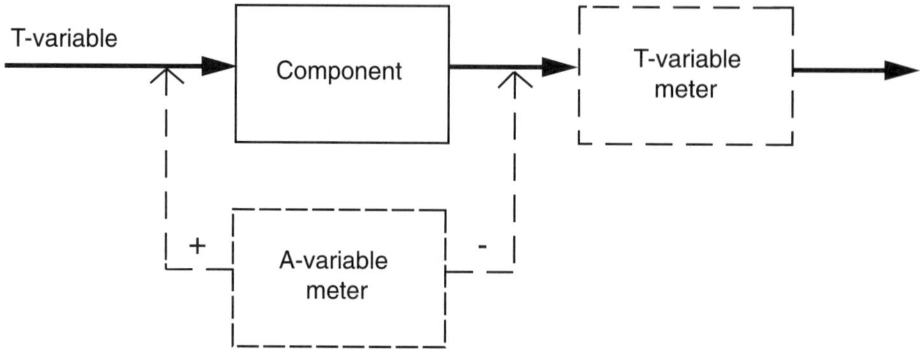

Figure 3–5 General schematic for T- and A-variable measurement

measured in a component. Figure 3–5 shows a general schematic of a component and its associated T- and A-variables. Notice that the T-variable flows through the component and, to measure it, the flow path must be interrupted so that a T-meter can be inserted. Similarly, to measure the associated A-variable, we measure across the component using an A-meter. Table 3–1 shows the T- and A-variables for electrical, hydraulic, pneumatic, and mechanical systems. There are other ways of matching variables from different systems. The choices presented here provide similar mathematical relationships. Consider the two analogous components in Figure 3–6 (each of the equations associated with these components is presented in more detail later). Notice that in both equations describing the component, the A-variable appears differentiated. Another similarity is that the product of the T- and A-variables yields power, except in thermal systems where heat flow, Q, is power.

Another interesting fact about components is that their physical size tends to be relative to the magnitude of the maximum T-variable the component can

Type of System	T-variable	A-variable
Electrical	*i*, current	*V*, voltage
Fluid-hydraulic	*q*, volumetric flow	*p*, pressure
Fluid-pneumatic	*q*, volumetric flow	*p*, pressure
Mechanical-translational	*f*, force	*v*, velocity
Mechanical-rotational	*T*, torque	ω, angular velocity
Thermal	*Q*, heat flow	*T*, temperature

Table 3–1 T- and A-variables in electrical, fluid, mechanical, and thermal systems

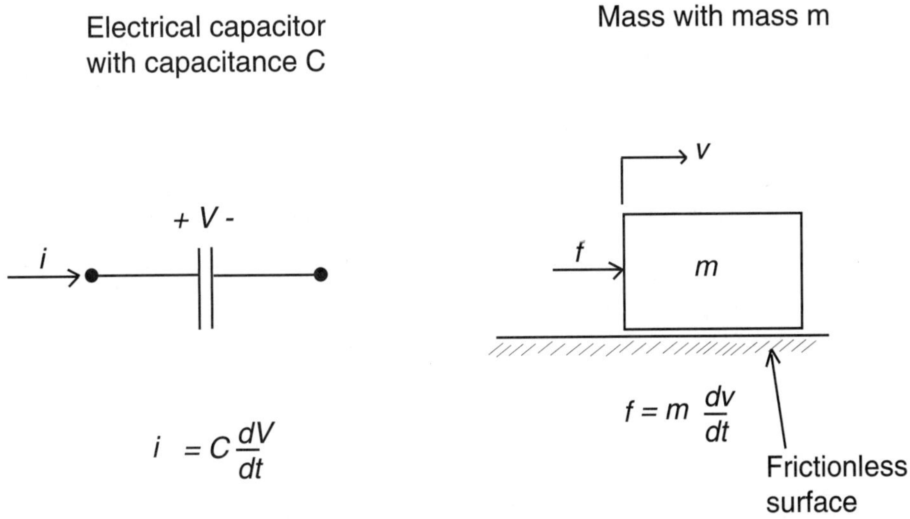

Figure 3–6 Example of two analogous components

handle. For instance, a physically large electrical resistor is one that can conduct a large current (T-variable). Similarly, a large mechanical component is one that can transmit a large force. On the other hand, notice that, as long as the resistance is high, a very small resistor can be subjected to a large voltage (A-variable). Thus, there is no correlation between the size of the component and the maximum allowable A-variable. In the same vein, a small mass can be moved at high velocities (A-variable). This information often helps relate the model to the characteristics of the device.

Lumped Components and Their Elemental Equations

We will be using ideal "lumped parameter" components as we try to model real physical systems. These building blocks have only one property, which is why they are often called lumped components. Each type of physical system class has its own set of components. Each component is analogous to other components in other classes of physical systems.

Elemental equations are the equations (algebraic and differential equations) that describe the ideal building blocks or components in modeling physical systems. These components fall into two major classifications: (1) energy dissipating and (2) energy storing. The energy-storing components can be further subdivided into

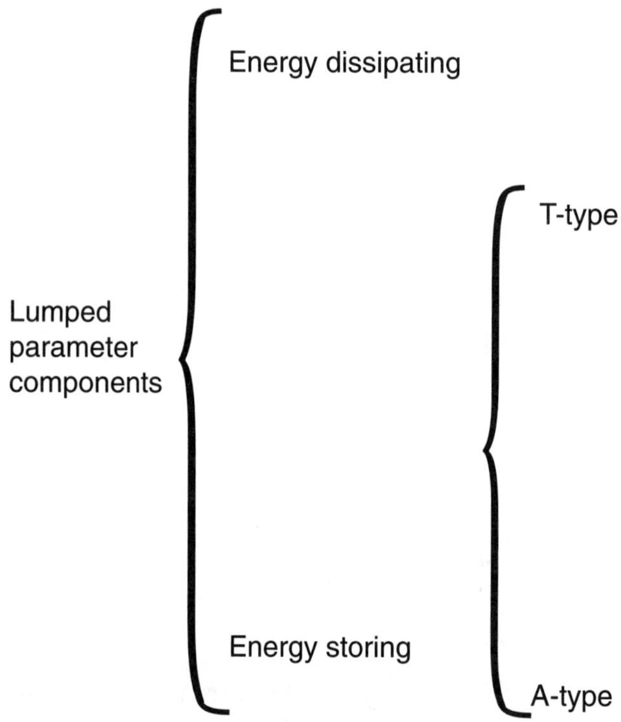

Figure 3–7 Classification of lumped parameters

two classes: (1) T-type and (2) A-type. Figure 3–7 illustrates this classification. As you will see shortly, two tasks determine whether an energy-storage element is a T- or an A-type: (1) determining which variable appears differentiated in the elemental equation and (2) by determining the variable used in the instantaneous energy-stored equation.

The following sections discuss the components or elements associated with each type of physical system. We will describe each of them in detail.

3.6 Electrical Systems

The components of electrical systems are the best introduction to the process of modeling physical components.

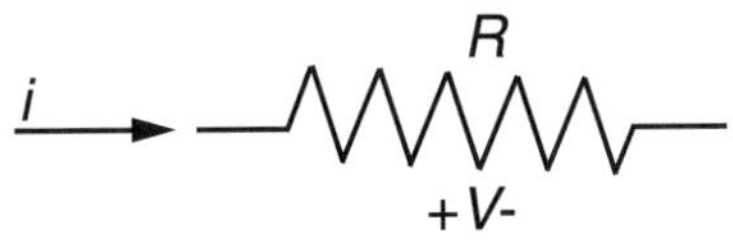

Figure 3–8 Symbol for electrical resistor

Energy-Dissipating Component—The Resistor

The resistor is a very common dissipative component, described mathematically by the elemental equation known as Ohm's Law:

$$V = iR \qquad \text{Eq. 3.5}$$

where V is the voltage (A-variable), i is the current (T-variable), and R is the resistance, in ohms (Ω). Its schematic symbol, with the associated variables, is shown in Figure 3–8. Power dissipated in a resistor is given by:

$$P = iV = i^2R = \frac{V^2}{R} \qquad \text{Eq. 3.6}$$

A-Type Energy-Storing Device—The Capacitor

The capacitor stores energy by storing electrical charge. The current is given by the elemental equation:

$$i = C\frac{dV}{dt} \qquad \text{Eq. 3.7}$$

where C is the capacitance, in farads (F). It is represented in Figure 3–9. The energy stored in the capacitor, E, depends on voltage only, not current, and is given by:

$$E = \frac{1}{2}CV^2 \qquad \text{Eq. 3.8}$$

It is beyond the scope of this book to derive this equation, so confirm it with other sources. However, what is proposed here is the general form of the energy equation:

$$E = {}^1\!/_2\,(\text{coefficient from elemental equation})(\text{T- or A-variable})^2 \qquad \text{Eq. 3.9}$$

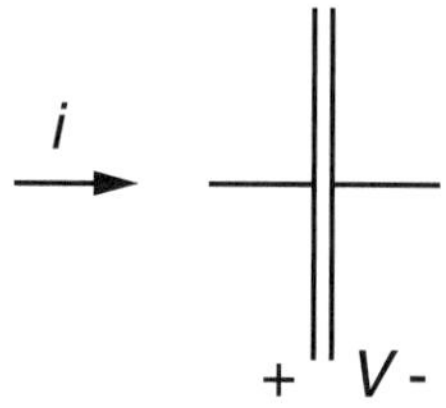

Figure 3–9 Symbol for electrical capacitor

The following observation supports the concept that the capacitor is an A-type of energy-storing element. Notice that voltage, the A-variable, appears differentiated in the elemental equation. It also appears explicitly in the energy equation. For this reason, the capacitor is designated as an A-type of energy-storage element.

T-Type Energy-Storing Component—The Inductor

The inductor stores energy in a magnetic field. The voltage across the inductor is given by the elemental equation:

$$V = L\frac{di}{dt} \qquad \textbf{Eq. 3.10}$$

where L is the inductance in Henrys (H). It is represented in Figure 3–10. The energy stored in an inductor depends on the current only, not the voltage, and is given by:

$$E = \frac{1}{2}Li^2 \qquad \textbf{Eq. 3.11}$$

By observation of the elemental and energy equations, let us confirm that the inductor is properly designated as the T-type energy-storing element. Notice that the T-variable, the current, i, appears differentiated in the elemental equation and that the energy equation depends explicitly on the T-variable, i.

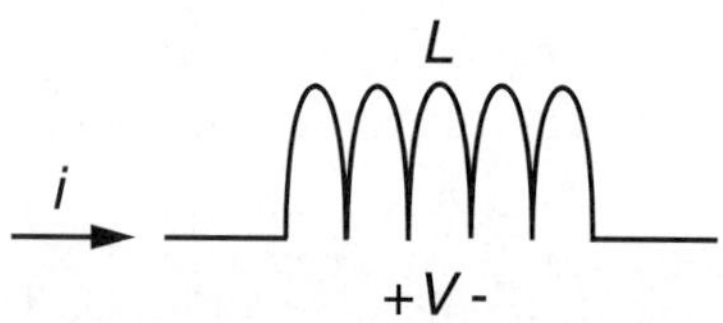

Figure 3–10 Symbol for electrical inductor

3.7 Fluid Systems—Hydraulic

Even though you might be unfamiliar with the components of fluid systems, you will see that they are similar to the components of electrical systems.

Energy-Dissipating Device—The Fluid Resistor

The fluid resistor is a component that offers resistance to flow. It can be constructed by inserting a series of small tubes (capillaries or straws) inside the fluid conductor. It is shown in Figure 3–11. Its elemental equation is given by:

$$p = R_f q \qquad \textbf{Eq. 3.12}$$

where p is the pressure difference, q is the flow, and R_f is the fluid resistance. The energy dissipated in a fluid resistor is given by:

$$E = pq = q^2 R_f = \frac{p^2}{R_f} \qquad \textbf{Eq. 3.13}$$

Often, you will find orifices that are constructed differently, such as shown in Figure 3–12. Here, the pressure and flow are not related linearly. In cases like this, to perform a linear analysis on the system one must pass a straight line through the p-q curve around the operating point (often called the Q-point) in an effort to approximate the characteristics of a linear resistor $R_{f\text{-}equi}$.

A-Type Energy-Storing Component—The Storage Tank

A water-storage tank on a hill is an A-type energy-storing component. Figure 3–13 will be used to derive the elemental equation.

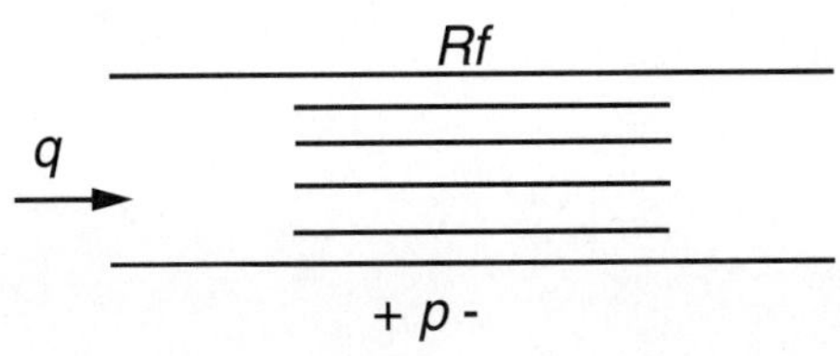

Figure 3–11 Symbol for hydraulic resistor

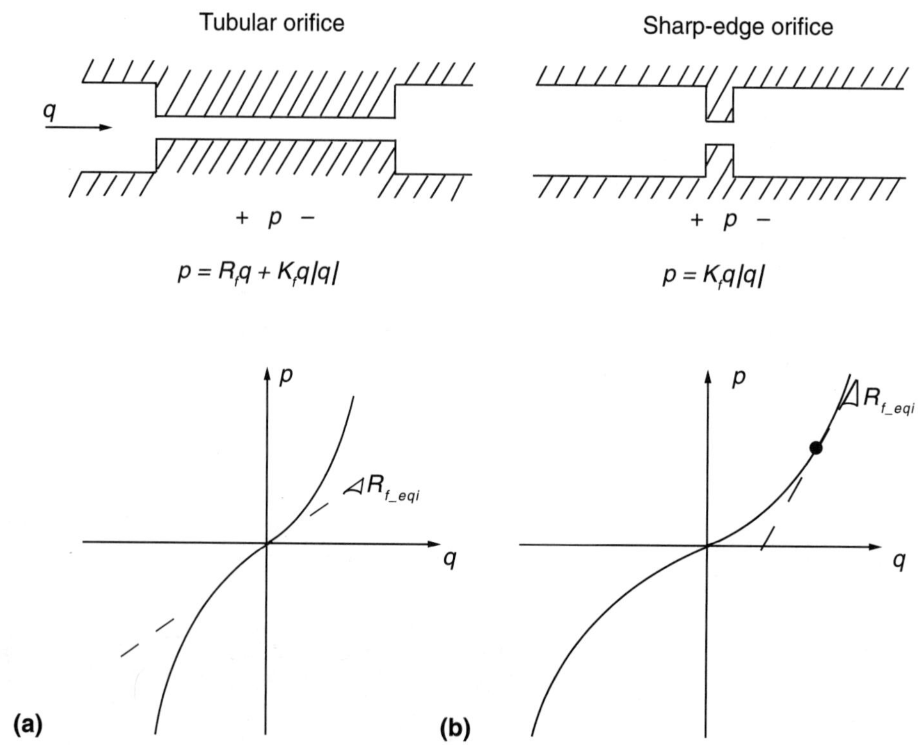

Figure 3–12 Fluid resistance: (a) pseudolinear and (b) nonlinear

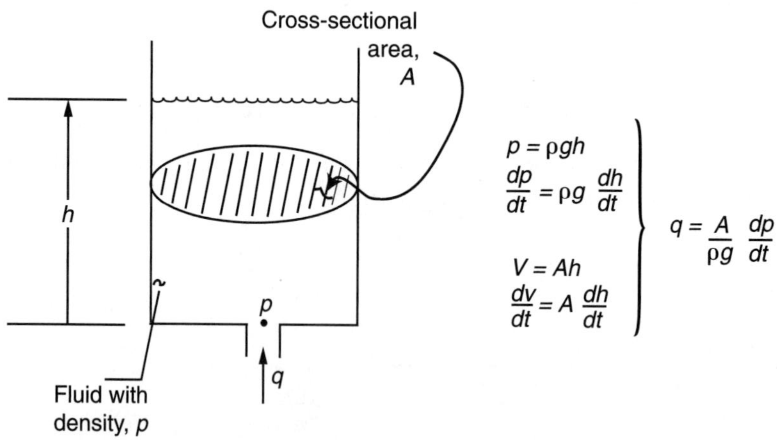

Figure 3–13 Hydraulic accumulator

Consider the flow entering through an opening at the bottom of the tank (without any pressure drop). The pressure at the bottom of the tank is given by:

$$p = \rho g h \qquad \textbf{Eq. 3.14}$$

where ρ is the fluid density, g is the acceleration of gravity, and h is the height of fluid in the tank. Taking the derivative of this equation produces:

$$\dot{p} = \rho g \dot{h} \qquad \textbf{Eq. 3.15}$$

This result will be used for substituting $\dot{h}$. The volume of fluid is given by:

$$V = Ah \qquad \textbf{Eq. 3.16}$$

where A is the cross-sectional area of the tank. Taking the derivative of this equation yields:

$$\dot{V} = A\dot{h} \qquad \textbf{Eq. 3.17}$$

and because:

$$q = +\dot{V} \qquad \textbf{Eq. 3.18}$$

then

$$q = A\dot{h} \qquad \textbf{Eq. 3.19}$$

Solving the differential pressure equation for $\dot{h}$, and substituting it into the previous equation, results in:

$$q = \frac{A}{\rho g}\dot{p} \qquad \textbf{Eq. 3.20}$$

Defining C_f

$$C_f = \frac{A}{pg}$$

or

$$q = C_f \dot{p}$$

where C_f is the fluid capacitance. A storage tank possesses energy given by:

$$E = \frac{1}{2} C_f p^2 \qquad \textbf{Eq. 3.21}$$

Again, notice the presence of p in the elemental equation and in the energy equation.

T-Type Energy-Storing Component—Hydraulic Line with Inertia

A fluid moving inside a pipe has a certain amount of inertia and, consequently, energy. Have you ever tried to close a valve quickly? Have you noticed how the rest of the plumbing system rattles, which is due to an effect called water hammer? The inertial characteristic is described by the elemental equation:

$$p = \frac{\rho L}{A} \dot{q} \qquad \textbf{Eq. 3.22}$$

where L is the length of the conduit, A is its cross-sectional area, and ρ is the fluid's density. Thus, we can define fluid inertia, I_f:

$$I_f = \frac{\rho L}{A} \qquad \textbf{Eq. 3.23}$$

and rewrite the equation as:

$$p = I_f \dot{q} \qquad \textbf{Eq. 3.24}$$

where I_f is the fluid inertia.

The energy contained in this moving fluid is given by:

$$E = \frac{1}{2} I_f q^2 \qquad \textbf{Eq. 3.25}$$

Additional A-Type Energy-Storing Component—The Accumulator

Some hydraulic systems use what are called accumulators. Accumulators are constructed by connecting a piston with a spring, as shown in Figure 3–14. The

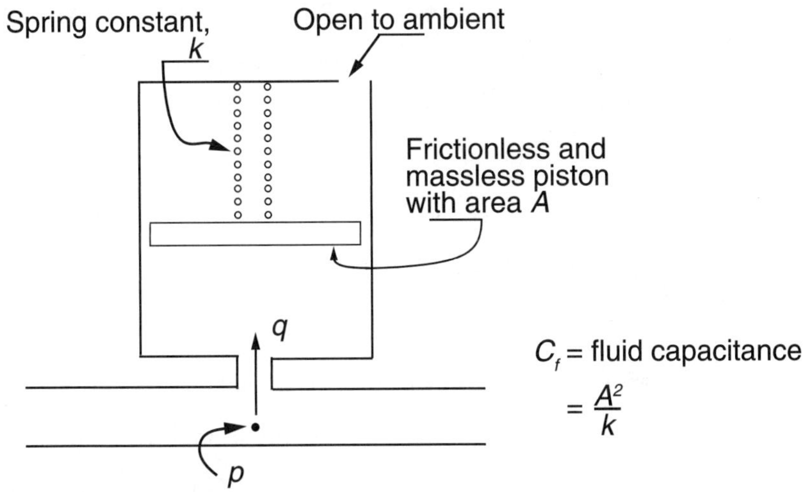

Figure 3–14 Bladder/piston accumulator

same effect is achieved with a pneumatic bladder. The fluid capacitance of this device is given by:

$$C_f = \frac{A^2}{k} \quad \text{Eq. 3.26}$$

where A is the area of the piston and k is the spring constant (stiffness) of the supporting spring. Often, accumulators are built by using an elastic container (i.e., a heavy-duty balloon) inside a metal container.

3.8 Fluid Systems—Pneumatic

High-pressure pneumatic systems are very nonlinear and do not lend themselves well to linearized analysis, except in situations in which the range of operation revolves around a narrow window. For this reason, these systems will not be discussed in this text. However, there is a special class of pneumatic systems that operate at relatively low gauge pressures, less than +/−2 psi gauge. In this type of system, air can be considered an incompressible fluid in most equations. One example is the plastic tube banks use to transport deposits/withdrawals between the cashier inside the bank and the customer at the drive-through window. We will next discuss the elements associated with this special class of pneumatic systems.

Energy-Dissipating Component—The Pneumatic Fluid Resistor

Just as we found in hydraulic systems, we can build pneumatic fluid resistors by inserting small-diameter capillaries inside the conducting vessel. Again, the elemental equation becomes:

$$p = R_f q \quad \textbf{Eq. 3.27}$$

where R_f is the pneumatic fluid resistance.

A-Type Energy-Storing Component—Pneumatic Tank

In pneumatic systems, in addition to the accumulator there is another form of energy-storing component, made up of a large tank. Because air is a compressible fluid, a large volume of air embodies this compressibility effect, making a rigid tank act as an accumulator. The capacitance of this tank is given by:

$$C_f = \frac{V_{tank}}{n\bar{p}} \quad \textbf{Eq. 3.28}$$

where V_{tank} is the volume of the tank, $\bar{p}$ is the average pressure (in absolute units), and n is the polytropic exponent. When pressure changes rapidly, $n = 1.4$. This behavior describes an adiabatic process, which means there is no heat transfer in and out of the tank as the air is compressed (and, as a result, small changes in temperature occur). When pressure inside the tank changes very slowly, the process is said to be isothermal, indicating that the temperature inside the tank is constant, as well as the same as that outside the tank. In this case, $n = 1.0$. In this situation, changes are happening so slowly that heat transfer occurs to maintain constant temperature.

Some pneumatic systems require a fluid capacitance, C_f, that can only be obtained with very large volume. Often, there is no room to accommodate a vessel this large. In cases like this, the overall capacitance can be increased using a pneumatic piston-spring accumulator. This is described in Figure 3–14 and by Equation 3.26.

T-Type Energy-Storing Component

Because air is much less dense than water, this type of effect is negligible and rarely considered. In other words, inertial effects in pneumatic systems are almost always insignificant.

3.9 Mechanical Rectilinear Systems

Mechanical systems are perhaps the most visual systems that we will study, but their analog counterpart is not always clear. We will compare the mechanical components' elemental equations with those that we have derived to determine their proper analog.

Energy-Dissipating Component—The Dashpot or Damper

A dashpot is a component that offers resistance to movement, very much like a shock absorber in a car, except that it lacks a spring. A way of realizing a hydraulic dashpot is shown in Figure 3–15. Notice that an effort to move the piston to the

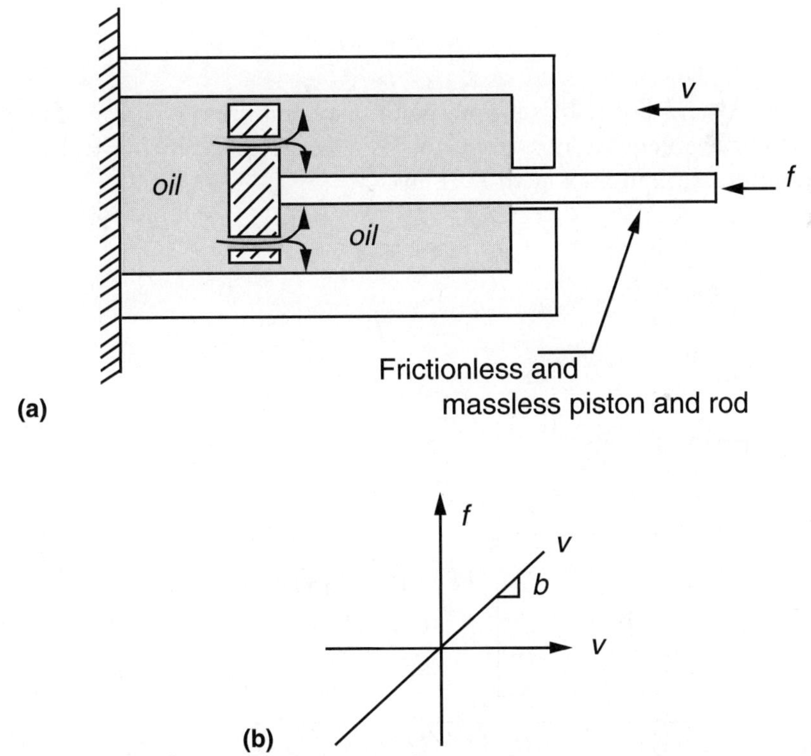

Figure 3–15 Hydraulic dashpot/damper: (a) schematic and (b) characteristics

left causes the oil on the left side of the piston to move to the right. As the oil passes through the orifices in the piston, a pressure differential develops across the two sides of the piston. This pressure differential is manifested as an opposing force. As long as the piston does not contact the ends of the cylinder, it can be said that the force required to move the piston is proportional to the speed of movement, as shown in Figure 3–15b.

The coefficient b is called the damping coefficient. As you can imagine, the value of the coefficient is proportional to the area of the piston and the viscosity of the fluid, inversely proportional to the size of the orifice and the number of orifices in the piston.

A-Type Energy-Storing Component—The Mass

Any object having mass is an A-type energy-storing component. Consider the mass in Figure 3–16. If a force is applied to a mass on a frictionless surface, its acceleration is proportional to the force, as given by Newton's Second Law:

$$f = ma \qquad \textbf{Eq. 3.29}$$

Now the A- and T-variables for mechanical systems are velocity and force, so if we want to have an equation in terms of these variables (or its derivatives) we must replace the acceleration, a, with $\dot{v}$. Thus:

$$f = m\dot{v} \qquad \textbf{Eq. 3.30}$$

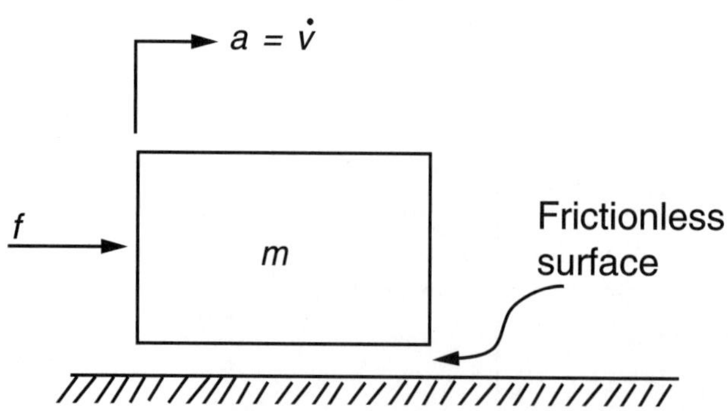

Figure 3–16 A-type, mechanical, energy-storing component, the mass

The energy of a moving object, called kinetic energy, is given by:

$$E = \frac{1}{2}mv^2 \qquad \text{Eq. 3.31}$$

Notice that both the elemental equation and the energy equation are in terms of v, the A-type variable. This certainly defines the mass as an A-type energy-storing device.

T-Type Energy-Storing Device—The Spring

Consider Figure 3–17, which shows a mechanical spring and its characteristics. These characteristics assume that the forces acting on the spring are small enough so that the spring neither bottoms out nor is stretched beyond its elastic range. The slope of the force versus the displacement curve is called the spring constant, k, and its corresponding equation is:

$$f = kx \qquad \text{Eq. 3.32}$$

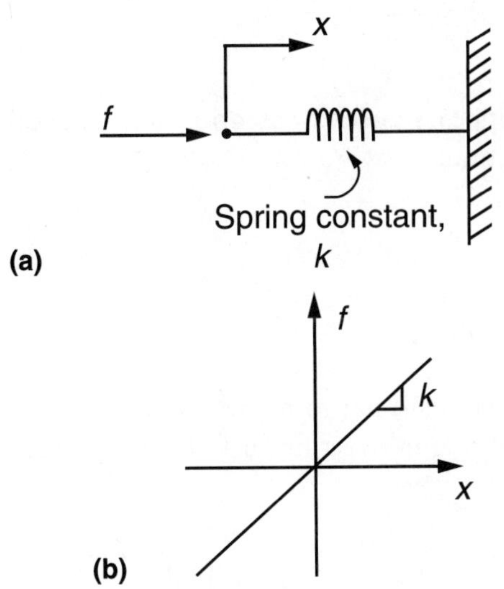

Figure 3–17 T-type, mechanical, energy-storing component, the spring: (a) schematic and (b) characteristics

Again, the equation is not written in terms of the desired mechanical variables (f, v, or its derivatives). This problem can be remedied by differentiating the previous equation, or:

$$\dot{f} = k\dot{x}$$
$$\dot{f} = kv \qquad \text{Eq. 3.33}$$

which can be written as:

$$v = \frac{1}{k}\dot{f} \qquad \text{Eq. 3.34}$$

This elemental equation makes it clear that the spring is a T-type energy-storing component. The energy stored in a spring is given by:

$$E = \frac{1}{2}kx^2 \qquad \text{Eq. 3.35}$$

but, again, x is not one of the desired variables. However, we can use the basic definition to rewrite that equation as:

$$E = \frac{1}{2}\frac{1}{k}f^2 \qquad \text{Eq. 3.36}$$

Again, notice that the T-variable appears differentiated in the elemental equation and also appears in the energy equation.

3.10 Mechanical Rotational Systems

Mechanical rotational systems are very similar to rectilinear mechanical systems, except that they rotate instead of moving in a rectilinear fashion. The variables used to denote rotational motion are:

C—for angular displacement

$\dot{C}$ or ω—for angular velocity

$\ddot{C}$ or α—for angular acceleration

Mechanical rotational systems exhibit the same type of components found in regular mechanical systems: dashpots, rotational springs, and masses that are rotating (which process inertia).

3.11 Thermal Systems

Thermal systems occur in many forms, from the heating and cooling of your home to the reentry of the space shuttle into the earth's atmosphere. Most thermodynamics courses focus on the static aspects of thermodynamics. The tools you will pick up here will help you deal with the dynamic aspects of thermodynamics and heat transfer.

Thermal Processes/Modeling Elements

As in the previous examples, two variables are analogous to the previously studied variables:

1. A-variable T (temperature)
2. T-variable q (heat flow)

Unfortunately, the analogy stops here, because q is in units of power. Thus, the previously observed feature that the product of the A-variable and the T-variable produces power is not true in thermal systems. Nevertheless, most of the insights that can be gathered from the analogous nature of the systems' relationships are still valid.

Let us first deal with the thermal process that resembles resistive-type elements. In thermal systems, we have three such processes. Notice that it would be improper to say that these are energy-dissipating elements.

Conduction

Conduction occurs when heat travels through solid media. For instance, heat that enters a house through the walls is a type of conduction, as shown in Figure 3–18a. It is very beneficial to represent the process with an electrical equivalent circuit (a resistor), as shown in Figure 3–18b. The fundamental equation is:

$$T_1 - T_2 = R_c q \qquad \text{Eq. 3.37}$$

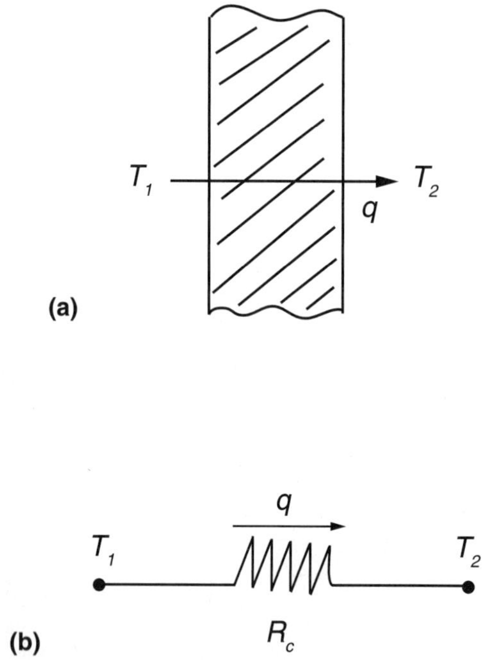

Figure 3–18 Thermal conduction: (a) thermal process and (b) electrical equivalent

where:

$$R_c = \frac{d}{KA} \qquad \textbf{Eq. 3.38}$$

and:

d = thickness of the wall
A = area of the wall
K = thermal conductivity of the wall's material

Convection

Convection is similar to conduction, except that the heat is transmitted through a fluid (liquid or gas) medium. The fact that the medium is a fluid allows mobility within the fluid, as shown in Figure 3–19. The fundamental equation is:

$$T_1 - T_2 = R_h q \qquad \textbf{Eq. 3.39}$$

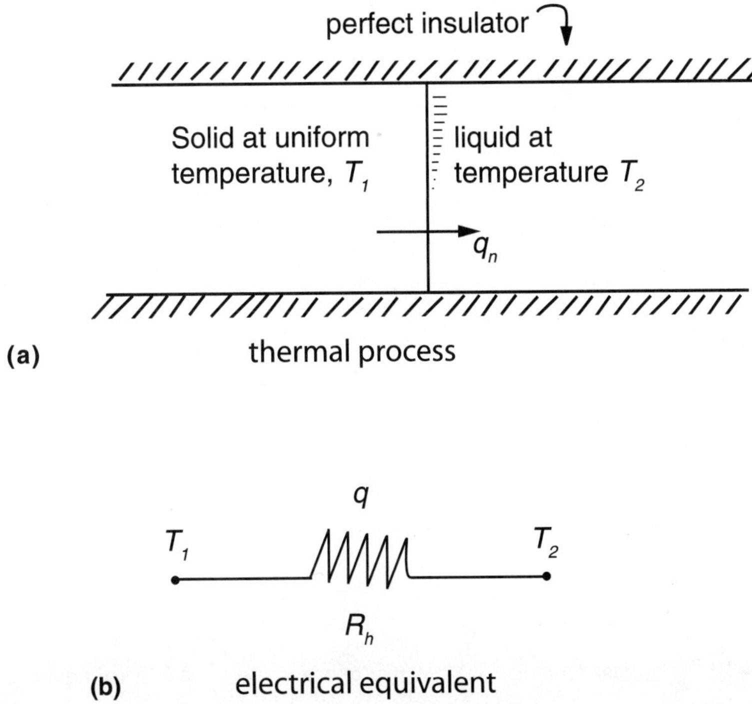

Figure 3–19 Thermal convection: (a) thermal process and (b) electrical equivalent

where the convection resistance, R_h, is grown by:

$$R_h = \text{convection resistance} = \frac{d}{hA}$$

and:

$$d = \text{thickness}$$

$$h = \text{convection coefficient}$$

$$A = \text{area}$$

Notice that this process is also represented with an electrical resistor.

Radiation

Radiation is the method of heat transfer we experience when we feel the sun's warming rays or when we open a hot oven. All objects radiate heat, from a hot one

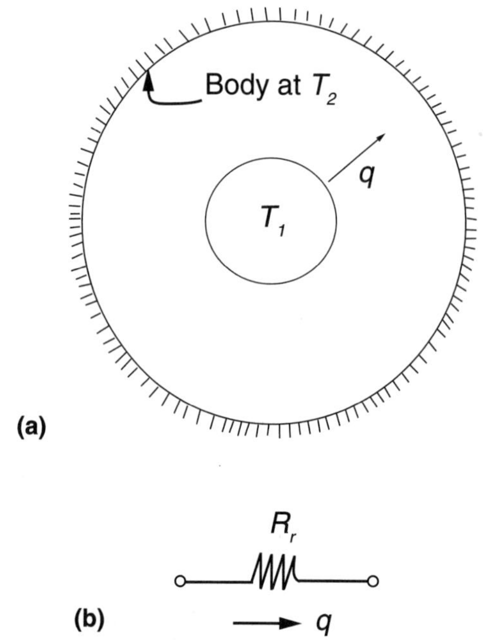

Figure 3–20 Thermal radiation: (a) thermal process and (b) electrical equivalent

to a cold one and vice versa. However, the net effect is that more heat radiates from the hotter object to the colder one, as shown in Figure 3–20. It is defined by:

$$q_r = A\varepsilon F\sigma(T_1^4 - T_2^4)$$

A = area of the surface of the object of interest

ε = parameter descriptive of the emission properties of surfaces 1 and 2

σ = Stefan-Boltzman Constant

F = geometrical shape factor

T_1, T_2 = absolute temperatures of bodies 1 and 2

Even though this process is a nonlinear, the equation can be approximated as a linear relationship, especially if the temperatures change little.[4] Notice that:

$$\begin{aligned}(T_1^4 - T_2^4) &= (T_1^2 + T_2^2)(T_1^2 - T_2^2) \\ &= (T_1^2 + T_2^2)(T_1 + T_2)(T_1 - T_2)\end{aligned} \qquad \textbf{Eq. 3.40}$$

In many cases, significant variation in $(T_1 - T_2)$ has an insignificant effect on $(T_1^2 + T_2^2)(T_1 + T_2)$. Consequently, we can approximate the following term:

$$(T_1^2 + T_2^2)(T_1 + T_2) \quad \text{Eq. 3.41}$$

as a constant, and define the radiation resistance, R_r:

$$\frac{1}{R_r} \cong A\varepsilon F\sigma(T_1^2 + T_2^2)(T_1 + T_2) \quad \text{Eq. 3.42}$$

therefore:

$$(T_1 - T_2) = q\,R_r \quad \text{Eq. 3.43}$$

Notice that this process is also represented with an electrical resistor.

Thermal Capacitance

We will now focus on the characteristics of thermal systems to store heat. One can relate this phenomenon to an electric capacitor. If the voltage is increasing, then the current through the capacitor must be changing. Similarly, if the temperature of an object is increasing, heat must be entering the object. This relationship can be expressed mathematically as:

$$q = C\dot{T}$$
$$C = mc_x \quad \text{Eq. 3.44}$$

where:

q = heat flow into the object
T = temperature of the object
C = thermal capacitance of the object
m = mass of the object
c_x = specific heat of material

Notice that the variable that appears differentiated is temperature, the A-variable. Thus, we consider this element an A-type energy-storing component.

Thermal Inertness

In other systems we saw the occurrence of both A-type and T-type energy storing elements. However, no thermal element with the characteristics of a T-type energy-storing element has been identified.

Summary of Lumped Parameters

You probably find it amazing to realize the high level of similarity among the different types of systems. Table 3–2 shows electrical, hydraulic, pneumatic, and mechanical systems. These systems faithfully obey the characteristics of A- and T-variables. Table 3–3 summarizes the characteristics of thermal systems.

3.12 Modeling Physical Systems

So far, we have looked at individual components in different types of physical systems: electrical, fluid, mechanical, and thermal. We have looked at each of these components in isolation, that is, one at a time. In real life, however, a system is rarely made of one component. Instead, most real-life systems have many components arranged in many ways. Because modeling is a mature field, many modeling methods have been developed (for more information, see the references at the back of the book). The next step in the process is to learn how to transform a system composed of several interconnected components into a desired mathematical representation. We are in the last step of Figure 3–2.

Before we proceed, we must establish some terminology. There are two types of equations: (1) elemental and (2) structural. Elemental equations describe the behavior of a specific component, and they were used in previous sections. For instance, Ohm's Law, $V = i^*R$, is an elemental equation. It relates the two variables, i and V, by means of the proportionality constant, R. This relationship is always true, regardless of how the component behaves or where the component is located in the circuit, or what type of excitation it is experiencing. Notice that this equation says nothing in regard to how the component is connected to other components in the circuit.

Structural equations, on the other hand, contain information on how the components are connected to other components in the circuit. Basically, two types of equations are obtained while obtaining the structural equations of a system. Structural equations have specific names, depending on the kind of system, but they have two general names: (1) continuity equations and (2) compatibility equations. Some type-specific names are:

Electrical systems—Kirchhoff's voltage and current laws

Hydraulic and pneumatic systems—conservation of mass

Variable	System Type	Electrical	Hydraulic	Pneumatic	Mechanical Translation	Mechanical Rotation
VARIABLE NAME	A-Variable	V (voltage)	p (pressure)	p (pressure)	v (velocity)	ω (angular velocity)
	T-Variable	i (current)	q (flow)	q (flow)	f (force)	T (torque)
A-TYPE ENERGY STORAGE ELEMENT	Name	Capacitor	Storage Tank	Pneumatic Tank	Inertia	Angular Inertia
	Model		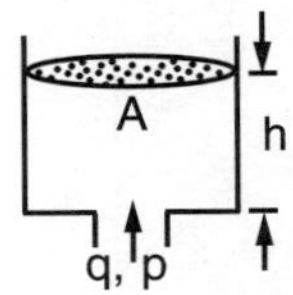	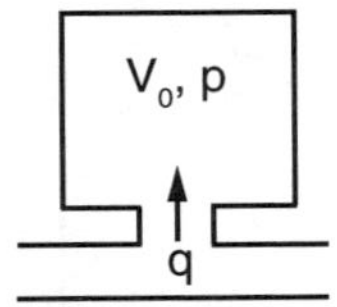	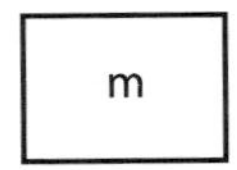	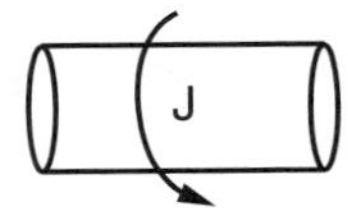
	Elemental Equation	$i = C\,(dV/dt)$	$q = (A/\rho g)(dp/dt)$	$q = (V_0/n\,\bar{p}\,)(dp/dt)$	$f = m\,(dv/dt)$	$T = J\,(d\omega/dt)$
	Energy Stored	$E = (1/2)CV^2$	$E = (1/2)(A/\rho g)p^2$	$E = (1/2)(V_0/n\,\bar{p}\,)p^2$	$E = (1/2)\,mv^2$	$E = (1/2)\,J\omega^2$
T-TYPE ENERGY STORAGE ELEMENT	Name	Inductor	Long Pipe	Long Pipe	Spring	Torsional Spring
	Model		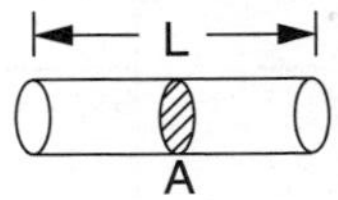	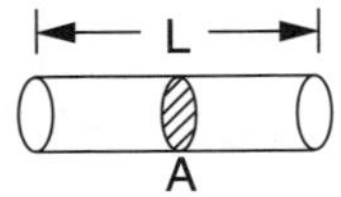	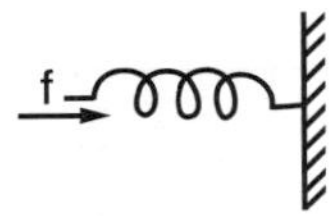	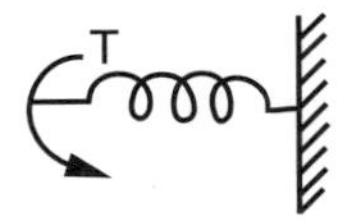
	Elemental Equation	$V = L(di/dt)$	$p = (\rho L/A)(dq/dt)$	$p = (\rho L/A)(dq/dt)$	$v = (1/k)(df/dt)$	$\omega = (1/k_\theta)(dT/dt)$
	Energy Stored	$E = (1/2)Li^2$	$E = (1/2)(\rho L/A)q^2$	$E = (1/2)(\rho L/A)q^2$	$E = (1/2)(1/k)\,f^2$	$E = (1/2)(1/k_\theta)T^2$
D-TYPE ENERGY DISSIPATOR	Name	Resistor	Fluid Restrictor	Fluid Resistor	Dashpot	Dashpot
	Model		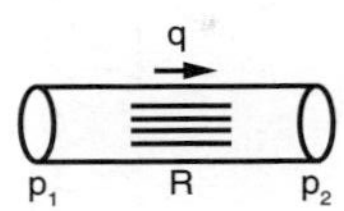	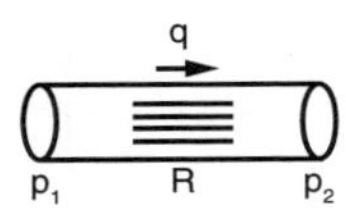	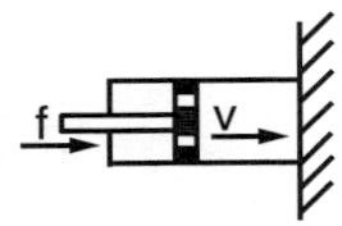	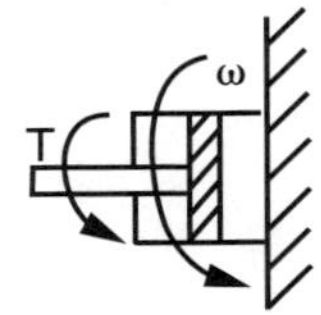
	Elemental Equation	$V = Ri$	$p = p_1 - p_2 = Rq$	$p = p_1 - p_2 = Rq$	$v = (1/b)f$	$\omega = (1/b_\theta)T$
	Power Stored	$P = Vi = V^2/R = i^2R$	$P_q = pq = p^2/R = q^2R$	$P_q = pq = p^2/R = q^2R$	$P = fv = bv^2 = f^2/b$	$P = T\omega = b_\theta\omega^2 = T^2/b_\theta$

Table 3–2 Electrical, fluid, and mechanical systems

Process	Name		Symbol	Elemental Equation
D-TYPE PROCESS	HEAT FLOW	CONDUCTION	R_c	$\Delta T = R_c q$ $R_C = \frac{d}{kA}$ d = thickness k = thermal cond.
		CONVECTION	R_h	$\Delta T = R_h q$ $R_h = \frac{1}{hA}$ h = convect. coef. A = area
		RADIATION	R_r	$\Delta T = R_r q$ R_r = see Eq. 3.24
A-TYPE PROCESS	THERMAL CAPACITANCE		C	$q = C\dot{T}$ $C = mc_x$ T = temp of mass m = mass, c_x = specific heat of material
T-TYPE PROCESS	HAS NOT BEEN IDENTIFIED			
	A-VARIABLE: T (temperature)			
	T-VARIABLE: q (heat flow)			

Table 3–3 Characteristics of thermal systems

Mechanical systems—Newton's Second Law (for the approach used in this book)

Thermal systems—conservation of energy

We will present several informal approaches for obtaining mathematical representations for dynamic systems. These methods are particularly suited to electrical, hydraulic, and thermal systems. Once the model schematic is agreed upon, it can be represented with a pseudo-electrical circuit analog representation. For mechanical systems, however, an electric circuit is usually not used, unless the circuit is to be simulated with an electrical current simulator (e.g., PSPICE, Electronic

Workbench, MicroCap). In any case, each of the system categories will be addressed individually in following sections.

Before we get involved in the specific process for obtaining the dynamic equations, it is helpful to see and compare the general approaches we will be taking as we model the different system types (electrical, fluid, mechanical, and thermal). The approaches for the different types of systems are shown in Figure 3–21. Approaches vary slightly from one type of system to another.

Remember that modeling is the process of taking a real-life physical system and representing it in a simplified, identifiable, and concrete model schematic. One of the benefits of a model schematic is that it provides a vehicle for objective communication between engineers. You might find that, initially you disagree with a colleague on the "proper" model schematic. The model schematic provides a specific, concrete representation of the real-life physical system and facilitates discussion between you and your colleague. Once the "correct" model schematic is agreed upon, you can proceed to a mathematical representation, which can be used for further analysis and design. Notice that all the recommended approaches shown in Figure 3–21, except mechanical systems, produce an equivalent electrical circuit. This does not imply that an equivalent electrical circuit is not available for mechanical systems. Such could be used later, however, from a perspective of simplicity in the process the suggested approach, namely, the application of Newton's Second Law, provides a much more streamlined process.

There are two major methods for modeling system types. Method 1 is aimed at obtaining one transfer function, which can be used in SIMULINK, MATLAB, or noncomputer-based analysis. Method 2 produces an unsimplified block diagram (i.e., the blocks are not combined into one transfer function that can be simulated using SIMULINK). SIMULINK provides the framework for transforming this multiblock representation into one transfer function using modern control techniques (to be presented in Chapter 9). For facilitating the process of transforming a model schematic into a mathematical representation, Method 1 and Method 2 are presented in outline form next.

METHOD 1: Obtain One Transfer Function for Simulation Using MATLAB or SIMULINK

1. Represent the system using a model schematic.
2. Define variables and coefficients.
3. Write elemental equations.
4. Write structural equations. Include auxiliary equations, if necessary.
5. Take the Laplace transform of all the equations.

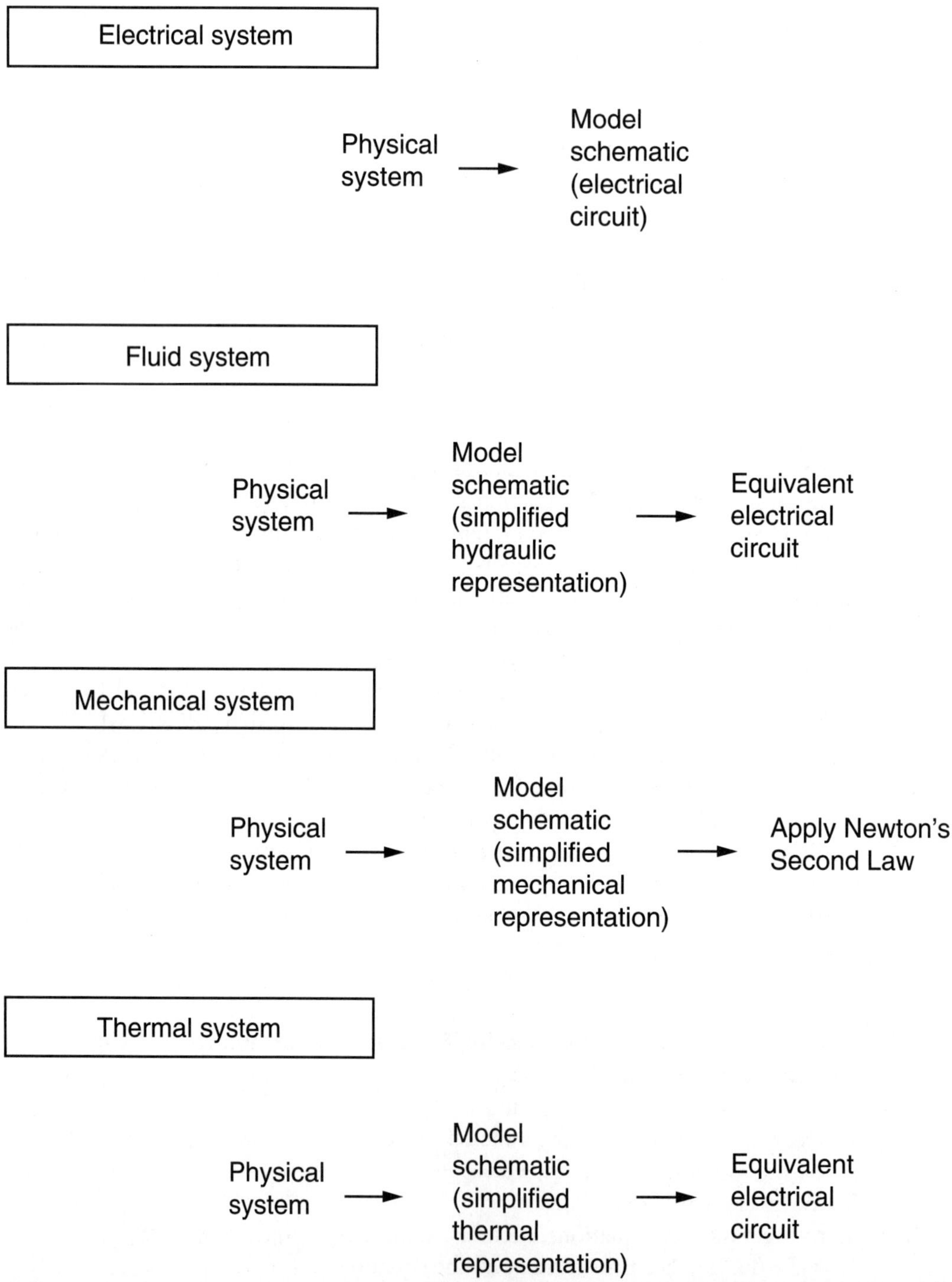

Figure 3–21 Comparison of modeling methods for different types of systems

6. Determine the output variable of interest.
7. Determine the input variable of interest.
8. Assess/check work.
   - Count the dependent variables (unknown) = u.
   - Count the equations (Kirchhoff's Current Law, Kirchhoff's Voltage Law, or auxiliary) = e.
9. Confirm that $u = e$.
10. Eliminate the variables of no interest, one at a time.
11. Render equations in transfer function form:

$$\frac{C(s)}{R(s)} = \frac{\text{Laplace transform of output variable}}{\text{Laplace transform of input variable}}$$

12. Enter into SIMULINK using the transfer function block.
13. Select numeric values for the coefficients (either in SIMULINK blocks or the MATLAB workspace).
14. Select the input type (e.g., step, sine, wave) and simulation parameters (e.g., integration routine, end time).
15. Simulate.

Notice that you can enter no initial conditions using this approach. That is, all the variables take 0 for their initial condition.

METHOD 2: Obtain a Group of Differential Equations for Simulation Using SIMULINK

1. Represent the system using a model schematic.
2. Define variables and coefficients.
3. Write elemental equations.
4. Write structural equations.
5. Solve each equation for the highest derivative term.
6. Implement in SIMULINK each equation that was solved for its highest derivative term. (This is same procedure used in Chapter 2.)
7. Connect all the blocks, ensuring that each block is connected.
8. Connect the display devices to the variables of interest.
9. Select numeric values for the coefficients.
10. Select inputs, initial conditions, and simulation parameters (e.g., integration routine, end time).
11. Simulate.

Notice that, unlike in the previous approach, you can enter any initial conditions. Enter those conditions by double-clicking on the integrators whose output is the variable. For instance, the integrator that has velocity as its input and position as its output is the one that captures the initial position. Notice that having a model schematic that represnts the physical system is the first step in both Methods 1 and 2.

In addition to Method 1 and Method 2, other approaches can be used to obtain a mathematical representation of the dynamic system and, subsequently, to simulate the system (i.e., obtain the dynamic response). There are some methods that are recommended for one type and not for others. The recommended methods for each type of system are shown in Figure 3–22, Figure 3–23, Figure 3–24, and Figure 3–25 and will be demonstrated in the following sections.

3.13 Modeling Electrical Systems

When the circuit/dynamic system under question contains only a few variables of interest, it is possible to use less formal methods for obtaining the differential equation. The presentation that follows addresses such a situation. This approach does not limit the power of the method, because once a differential equation is available, it is possible to obtain all other voltages and currents in the circuit with some additional work. This is possible because linear relationships exist that relate the unknown variable to those variables that have already been captured in the differential equation.

Remember that a physical system is completely described by its corresponding model schematic. Once the model schematic is agreed on, there is no freedom regarding the choices of elemental equations, because the modeling task is complete. However, we have seen that there is leeway with respect to the structural equations and how we obtain them. Electrical circuits must obey two sets of laws, which lead to the structural equations. The two sets are called Kirchhoff's Current Law (KCL) and Kirchhoff's Voltage Law (KVL). Before we proceed, we need to define some of the following nomenclature used in electrical circuits:

Node—The connection points between two or more elements are joined.

Junction—The connection points between three or more elements are joined.

Branch—The path between two different junctions with no additional junctions between. The path may consist of one or more elements in series.

Loop—One closed path.

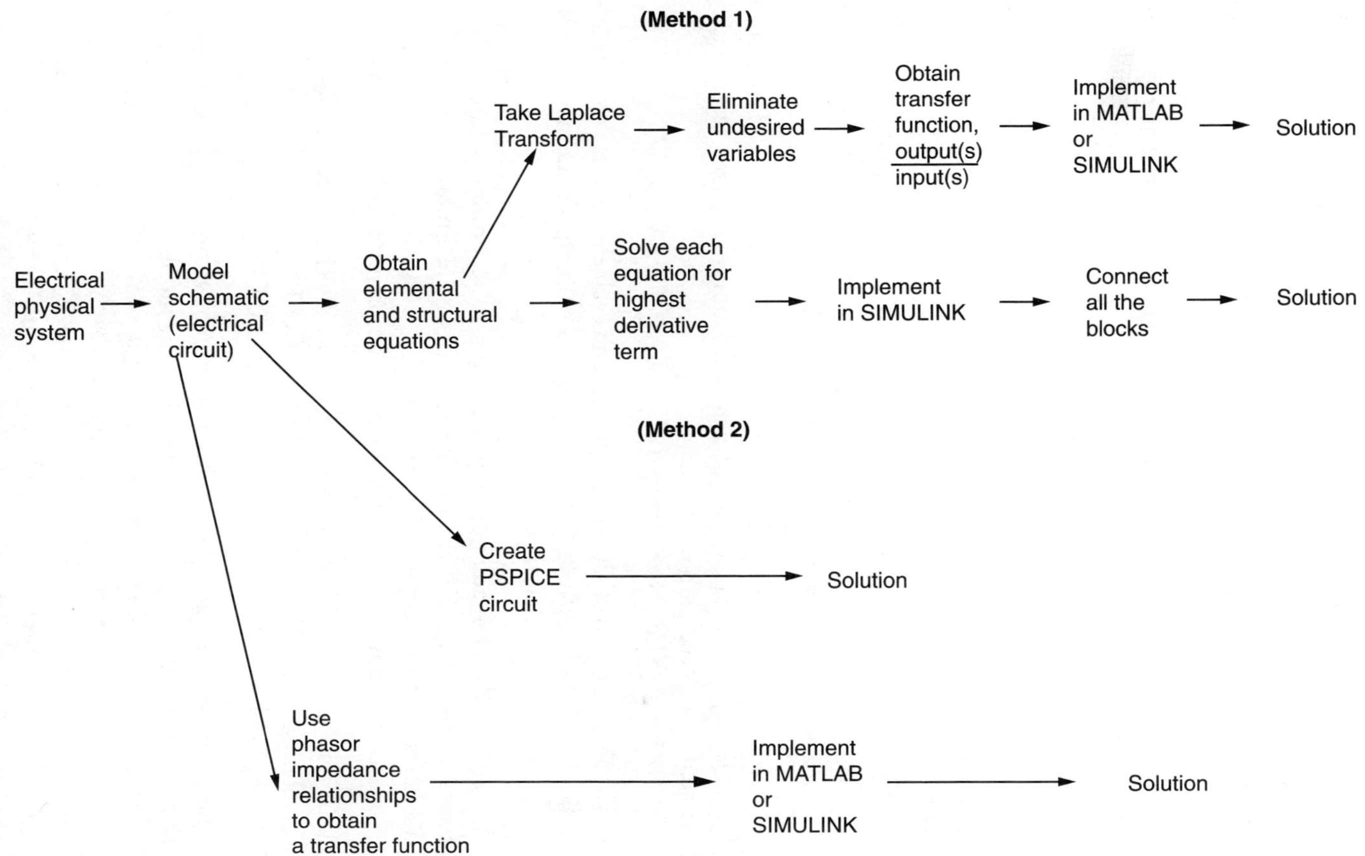

Figure 3–22 Procedure for obtaining a mathematical model and a time response of an electrical system

Hydraulic physical system → Model schematic (simplified hydraulic representation) → Electrical circuit equivalent → Use approaches/methods for electrical circuits

Figure 3–23 Procedure for obtaining a mathematical model and time response of a hydraulic system

Figure 3–26 shows a circuit and the associated nomenclature.

Now let us define KCL and KVL. KCL states that the algebraic sum of currents entering a node equals 0. However, there are other valid ways of conveying the same concept (and writing the equation):

- The sum of currents entering the node equals the sum of currents leaving the node.
- The algebraic sum of current leaving the node is 0.

It is a matter of personal preference which realization of the KCL is used. The other law is KVL, which states that the algebraic sum of voltages across each electrical element in a closed path equals 0.

As with KCL, there are two other ways of realizing KVL:

- The algebraic sum of all voltage rises around each element in a closed path equals 0.
- The sum of all the voltage rises in a closed path equals the sum of voltage drops around the same closed path.

A point should be made before proceeding. KCL produces equations that are in terms of the voltages of the different nodes in the circuit. In contrast, KVL produces equations in terms of the currents.

At this point, we have three choices before proceeding. We can find:

1. Approach A—All currents and voltages in the circuit (KVL + KCL)
2. Approach B—All currents in the circuit (KVL)
3. Approach C—All voltages in the circuit (KCL)

The initial task, regardless of which approach we choose to follow, is to perform some accounting in which we try to relate the number of unknowns to the number of equations. This allows us to determine if we have all the information we need.

First, assume that we are using Approach A. Assume that we have a circuit with k elements. This means that we will have $2k$ unknowns (k currents and k voltages). We have three sources for the equations. They are shown in Table 3–4 and can be used when all voltages and currents are included.

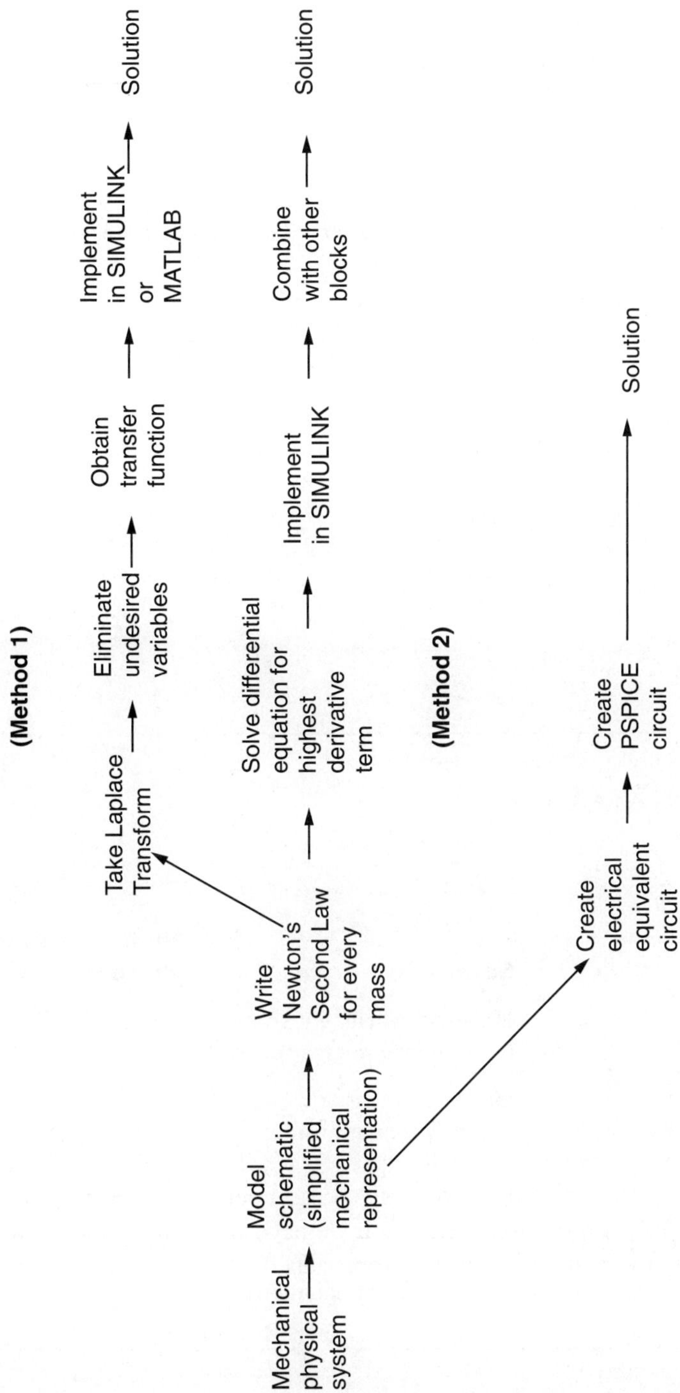

Figure 3–24 Procedure for obtaining a mathematical model and a time response of a mechanical system

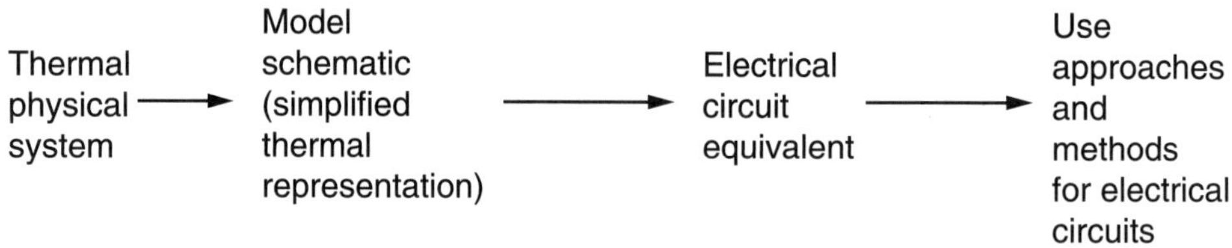

Figure 3–25 Procedure for obtaining a mathematical model of a thermal system

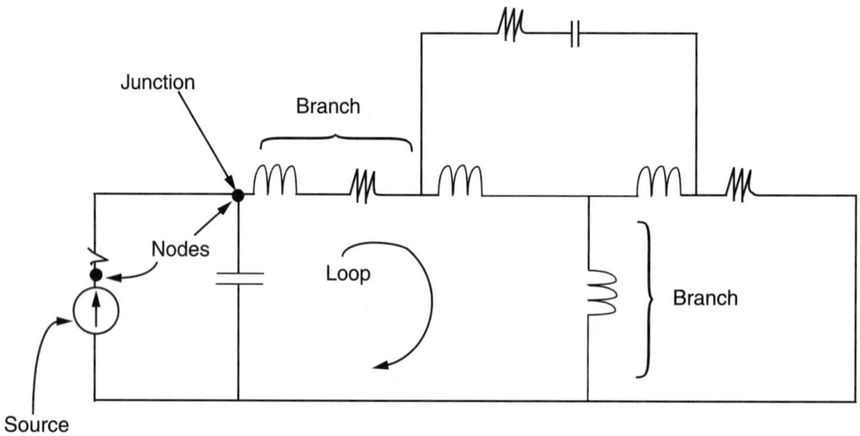

Figure 3–26 Parts of a network

Type of Relationship	Source	Number Present	Equations Contributed
Direct specification	Independent current sources, independent voltage sources	k_s	k_s
Elemental voltage-current relationship or dependent voltage (current)	Elemental equations, dependent relationships	$k - k_s$	$k - k_s$
Structural current-current relationships (obtained from KCL)	Node equations	$x_n + 1$	x_n
Structural voltage-voltage relationships (obtained from KVL)	Mesh (or loop) equations	x_m	x_m

Table 3–4 Sources of equations

To be able to solve the system of equations, the number of equations must equal the number of unknowns. This means that:

$$(k_s) + (k - k_s) + (x_n + x_m) = 2k \qquad \text{Eq. 3.45}$$

or:

$$x_n + x_m = k$$

which means that the total number of structural equations must equal the number of elements (k) in the circuit. Say that the circuit has n nodes and m mesh (or loops). We will need:

$x_n = n - 1$ node equations (obtained by applying KCL)

$x_m = m$ mesh equations (obtained by applying KVL)

Notice that the number of equations $((n - 1) + (m))$ equals k, the number of equations we need to solve our system of equations. Why not use n node equations instead of $n - 1$ equations? If an additional node equation were to be used, it would provide no additional new information.

Approach A: All the Current and Voltage in the Circuit (KVL + KCL)

The main benefit of Approach A is that it provides a structured framework that allows the solution regardless of the size/complexity of the dynamic system. It also captures every current and voltage in the circuit. Therefore, once a solution of the unknowns is obtained, all other variables can be easily obtained. On the other hand, this procedure can be quite time-consuming. For examples of this approach, see the references at the back of the book.

More often than not, it is not necessary to obtain every current and voltage in the circuit, because often we need only specific currents or voltages. This is not a major limitation. Assume that once we have obtained the needed current, we realize we need a voltage. We can go back to the circuit and obtain an expression that allows the calculation of the needed voltage from the obtained current.

Approach B: All Currents in the Circuit (KVL)

Assume that we only need to determine currents in the circuit. We can obtain these equations by applying KVL around meshes (m is the number of meshes in the circuit). Remember that each of the voltage drops is expressed in terms of current, by means of the elemental equation.

Consider the circuit in Figure 3–27. Notice that the circuit has one mesh. As a result, we must obtain one equation. Applying KVL, we obtain:

$$-v_s + iR + \frac{1}{C}\int idt = 0 \qquad \text{Eq. 3.46}$$

or:

$$iR + \frac{1}{C}\int idt = v_s$$

Notice that we have obtained an integral equation instead of a differential equation. If we wanted to solve this integral equation using SIMULINK, we would follow the procedure in Chapter 2, namely isolate to the right the highest derivative term (in this case, iR):

$$iR = v_s - \frac{1}{c}\int idt \qquad \text{Eq. 3.47}$$

and progressively integrate the terms. The complete SIMULINK implementation is shown in Figure 3–28a.

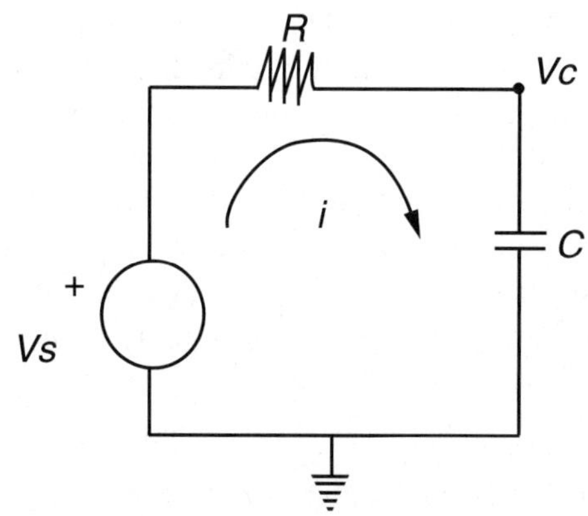

Figure 3–27 Example of Approach B: obtaining a mathematical representation in terms of currents in the circuit (KCL)

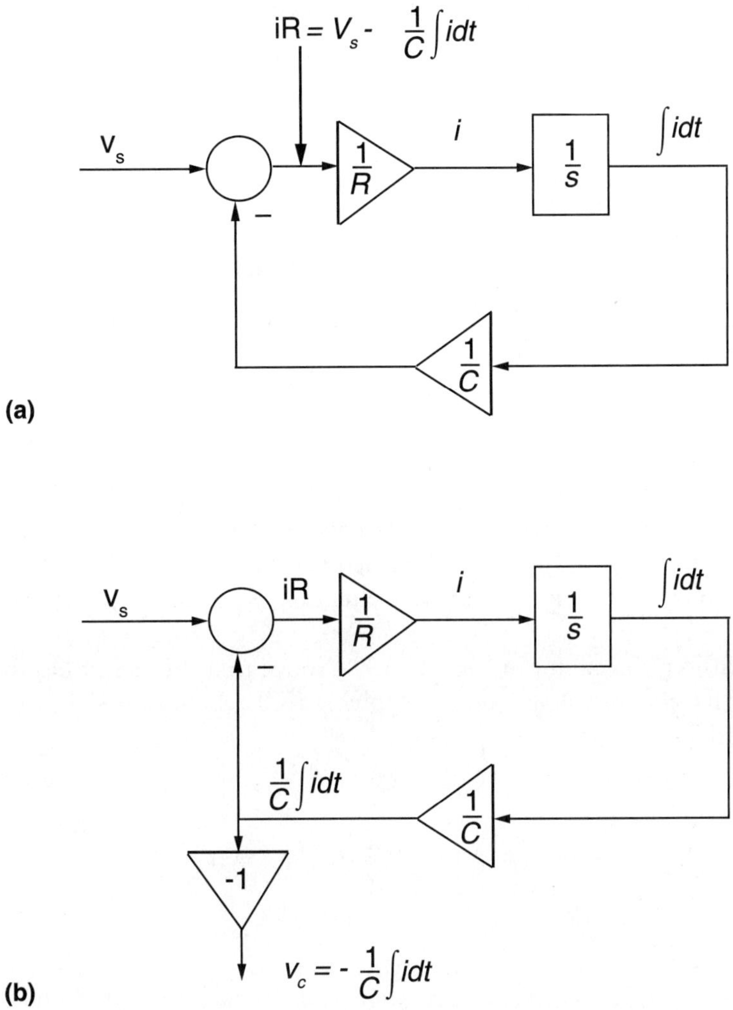

Figure 3–28 Solution implementation using SIMULINK: (a) v_c not included and (b) v_c included

Now, assume that we realize afterward that we also need the voltage at the end of the capacitor (instead of the current). The elemental equation for the capacitor is:

$$i = C\frac{d}{dt}(0 - v_c) \qquad \textbf{Eq. 3.48}$$

which can be integrated, and rearranged to look like:

$$v_c = \frac{1}{C}\int idt \qquad \textbf{Eq. 3.49}$$

We can modify the SIMULINK simulation as shown in Figure 3–28b to obtain v_c.

Approach C: All Voltages in the Circuit (KCL)

Before we proceed, we need to clarify the nomenclature associated with voltages. There are two types of voltages in a circuit:

1. Voltage across an element
2. Voltage at a specific node in the circuit

The voltages at the various nodes of the circuit are often called "node-to-datum" (NTD) voltages. In a circuit with *n* nodes, one node is selected as the datum, or reference node, typically ground.[5] Remember that, as we apply KCL, we will use the NTD voltage for each node, thereby obtaining equations in terms of NTD voltages.

Consider the circuit in Figure 3–29. Notice that the circuit has three nodes and one independent voltage source. Applying KCL, the sum of the currents entering the node is 0.

At mode v_c:

$$\frac{v_s - v_c}{R} + C\frac{d}{dt}(0 - v_c) = 0 \qquad \textbf{Eq. 3.50}$$

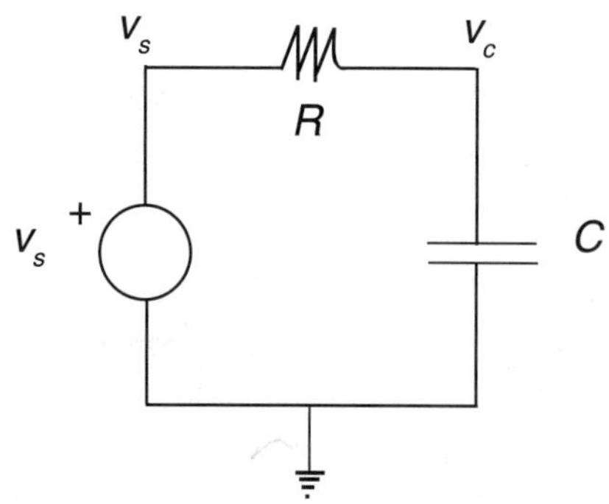

Figure 3–29 Example of Approach C: obtaining a mathematical representation in terms of voltages in the circuit (KVL)

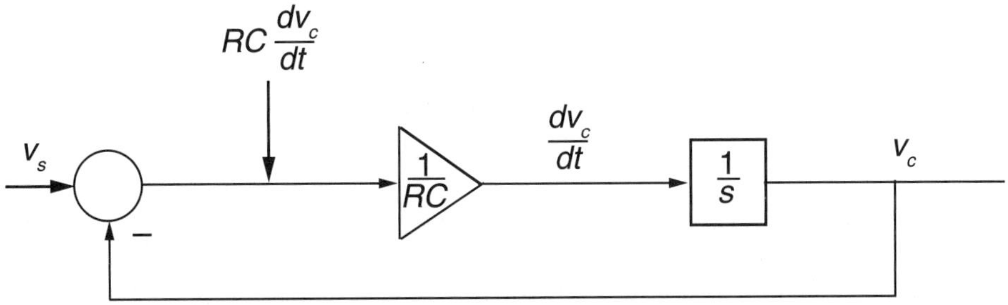

Figure 3–30 Solution implementation using SIMULINK

which can be arranged as:

$$v_s = RC\frac{dv_c}{dt} + v_c$$

If we want to solve this using SIMULINK, we isolate to the right of the equation the highest derivative term $\left(RC\frac{dv_c}{dt}\right)$ and deal with the rest of the differential equation as shown in Chapter 2. The SIMULINK implementation is shown in Figure 3–30.

You will encounter situations in which independent sources complicate the process of obtaining the structural equations. For instance, say that you are dealing with a circuit that has several voltage sources, and you want to use KCL to obtain equations in terms of node voltages. It is important to realize that these independent sources introduce additional unknowns. An independent current source introduces an unknown voltage, and an independent voltage source introduces an unknown current. However, they also contribute auxiliary relationships that can be captured with an additional equation and thereby contribute to the solution. In cases like these, you can use methods that distinguish between regular nodes and essential nodes.[6]

Consider the circuit in Figure 3–31. This circuit has five nodes; consequently, we should apply KCL at four nodes. However, notice that we cannot state the current through voltage sources. Consequently, we need to introduce two new variables, i_s and i_{s2}. The assumed directions are unimportant. If we are wrong, the solution will yield a negative current. The KCL applied at the four nodes yields:

at node v_1:

$$i_{s1} + \frac{v_4 - v_1}{R_2} + \frac{0 - v_1}{R_1} = 0 \qquad \text{Eq. 3.51}$$

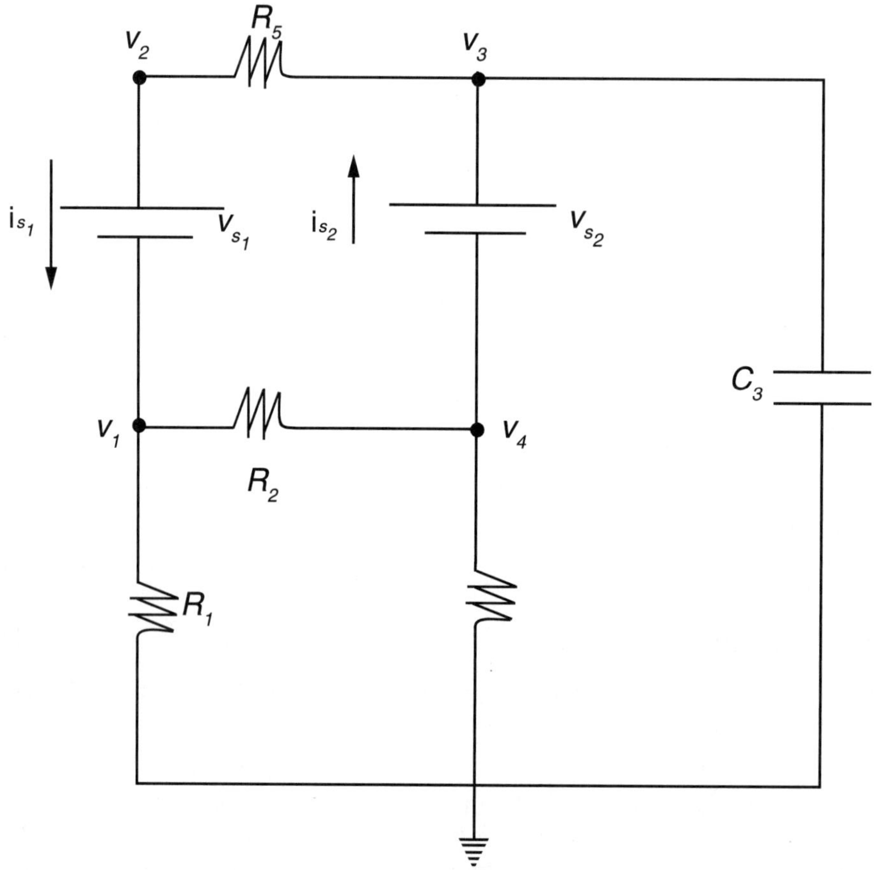

Figure 3–31 Example of circuit with several independent sources

at node v_2:

$$-i_{s1} + \frac{v_3 - v_2}{R_5} = 0 \qquad \textbf{Eq. 3.52}$$

at node v_3:

$$\frac{v_2 - v_3}{R_5} + i_{s2} + C\frac{d}{dt}(0 - v_3) \qquad \textbf{Eq. 3.53}$$

at node v_4:

$$\frac{v_1 - v_4}{R_2} + \frac{0 - v_4}{R_4} - i_{s2} = 0 \qquad \textbf{Eq. 3.54}$$

Fortunately, the independent sources add two additional equations, namely auxiliary equations:

$$v_3 - v_4 = v_{s2}$$ **Eq. 3.55**

$$v_2 - v_1 = v_{s1}$$ **Eq. 3.56**

Check:

Unknowns: v_1, v_2, v_3, v_4, i_{s2}, i_{s1}

Equations: Six equations

Conclusion: Solution is possible

Example 3.1

Consider the circuit in Figure 3–32.

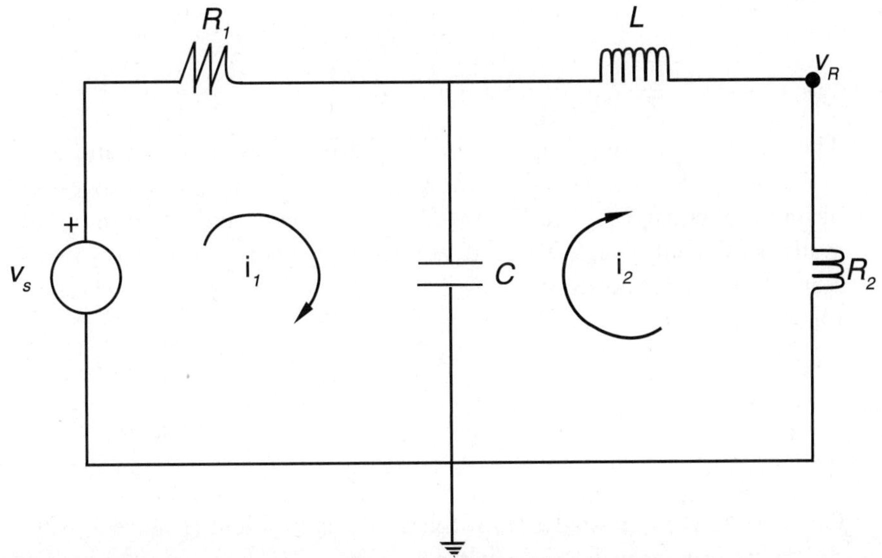

Figure 3–32 Example circuit whose mathematical representation is to be obtained by mesh currents (KVL)

Solution Since the circuit has two meshes, it will produce two equations. For the first mesh:

$$-v_s + i_1 R_1 + \frac{1}{C}\int (i_1 - i_2)dt = 0 \qquad \textbf{Eq. 3.57}$$

For the second mesh:

$$\frac{1}{C}\int (i_2 - i_1) + L\frac{di_2}{dt} + i_2 R_2 = 0 \qquad \textbf{Eq. 3.58}$$

Solve the first equation for the highest-order derivative:

$$iR_1 = v_s - \frac{1}{C}\int (i_1 - i_2)dt \qquad \textbf{Eq. 3.59}$$

And the second equation:

$$L\frac{di}{dt} = \frac{1}{C}\int (i_2 - i_1) - i_2 R_2 \qquad \textbf{Eq. 3.60}$$

These equations can be implemented in SIMULINK, as shown in Figure 3–33.

In short, we advise that whenever you embark on the process of obtaining differential equations, you keep track of unknowns and the number of equations available. When you are short (i.e., when you have more unknowns than equations), seek additional relationships, which will allow you to obtain a corresponding equation.

Example 3.2

Consider the circuit analyzed in the previous problem (Figure 3–32). This time, analyze the circuit using node voltages.

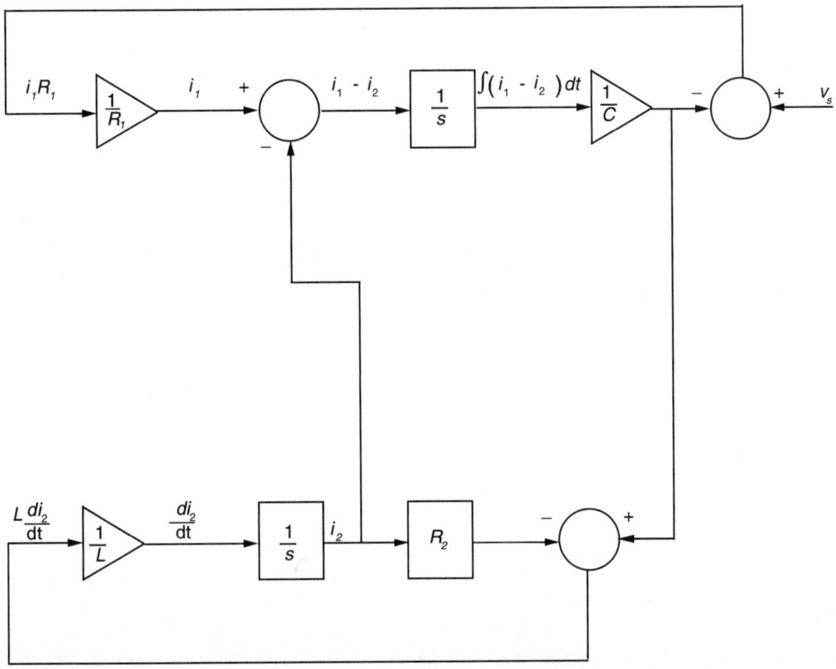

Figure 3–33 Solution implementation using SIMULINK

Solution The resulting (integral) equations are:

$$\frac{v_s - v_c}{R_1} + C\frac{d}{dt}(0 - v_c) + \frac{1}{L}\int (v_r - v_c)dt = 0 \qquad \textbf{Eq. 3.61}$$

$$\frac{1}{L}\int (v_c - v_r) + \frac{0 - v_r}{R_2} = 0$$

Solving the equations for the highest-order derivative:

$$\frac{v_c}{R_1} = -C\frac{dv_c}{dt} + \frac{1}{L}\int (v_r - v_c)dt + \frac{v_s}{R_1} \qquad \textbf{Eq. 3.62}$$

Solving the second equation:

$$\frac{v_R}{R_2} = -\frac{1}{L}\int (v_r - v_c)dt \qquad \textbf{Eq. 3.63}$$

The SIMULINK implementation is shown in Figure 3–34.

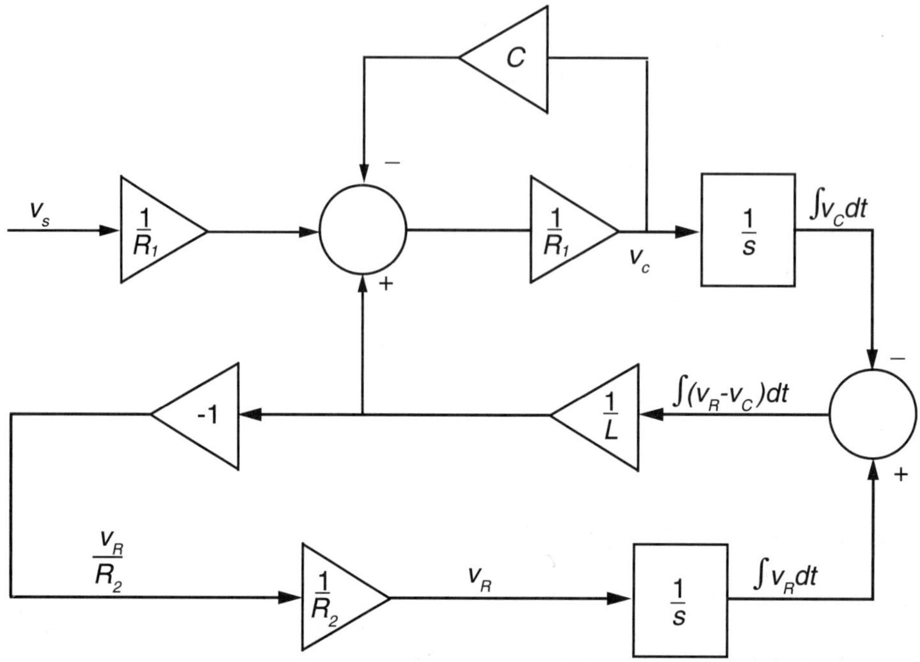

Figure 3–34 Solution implementation using SIMULINK for example problem

3.14 Obtaining a Transfer Function for Analysis or Simulation with SIMULINK

Often, all that you want is a transfer function representing the system. As an example, perform the following procedure on the circuit we have already analyzed. This is similar to the approach presented in Method 1.

Example 3.3

Obtain a transfer function relating output, i_2, to the input, v_s, for the circuit in Figure 3–32.

Solution For convenience, write the two previously obtained mesh (or loop) equations.

First mesh:

$$-v_s + i_1 R_1 + \frac{1}{C}\int (i_1 - i_2)dt = 0 \quad \text{Eq. 3.64}$$

Second mesh:

$$\frac{1}{C}\int (i_2 - i_1) + L\frac{di_2}{dt} + i_2 R_2 = 0 \quad \text{Eq. 3.65}$$

Taking the Laplace transform, we obtain:

$$-V_s + I_1 R_1 + \frac{1}{C}(I_1 - I_2) = 0 \quad \text{Eq. 3.66}$$

and:

$$\frac{1}{C}(I_2 - I_1) + LsI_2 + I_2 R_2 = 0 \quad \text{Eq. 3.67}$$

The first equation can be rearranged as:

$$-V_s + I_1(R_1 + \frac{1}{Cs}) - I_2\frac{1}{Cs} = 0 \quad \text{Eq. 3.68}$$

and the second one as:

$$I_1 = I_2(LCs^2 + R_2Cs + 1) \quad \text{Eq. 3.69}$$

After substituting the last equation into the previous one, (i.e., eliminating I_1) and collecting like terms:

$$\frac{I_2}{V_s} = \frac{1}{\left(LCR_1s^2 + (L + R_1R_2C)s + (R_1 + R_2)\right)} \quad \text{Eq. 3.70}$$

3.15 Modeling Fluid Systems

Fluid systems appear in many facets of life, from the water-distribution systems in cities to the respiratory systems in our bodies. With their clear relationship between current and flow, such systems are the best vehicles for seeing the analogy to electrical systems. For this reason, when obtaining a mathematical representation of a hydraulic or pneumatic system, create an equivalent electrical circuit. Instead of using voltages throughout the circuit, use pressure for the A-variable. Similarly, use flows instead of currents. Hydrostatic heads are represented as independent voltage sources.

Example 3.4

Consider the hydraulic system in Figure 3–35a. An independent flow fills the reservoir, which then empties through two paths: (1) one close to the bottom of the reservoir and (2) one farther away. Model the hydraulic system using an electrical analog, and obtain the dynamic equations.

Solution The corresponding analog electrical circuit is shown in Figure 3–35b. It helps to envision the independent flow source as entering at the bottom of the tank. Notice that, because the flow is an independent source, it will inject Q_i whether it is easy (by having to overcome no head, as in the top) or difficult (by having to overcome the pressure at the bottom of the tank). The resulting equations are:

$$Q_i + C\frac{d}{dt}(0 - p_T) + \frac{0 - p_T}{R_1} + \frac{1}{L}\int (p_0 - p_t)dt \qquad \textbf{Eq. 3.71}$$

$$\frac{1}{L}\int (p_0 - p_t) + \frac{0 - p_0}{R_2} = 0 \qquad \textbf{Eq. 3.72}$$

Multiply the first equation by R_1 to get:

$$R_1C\frac{dp_T}{dt} = R_1Q_1 - p_T + \frac{R_1}{L}\int (p_0 - p_T)dt \qquad \textbf{Eq. 3.73}$$

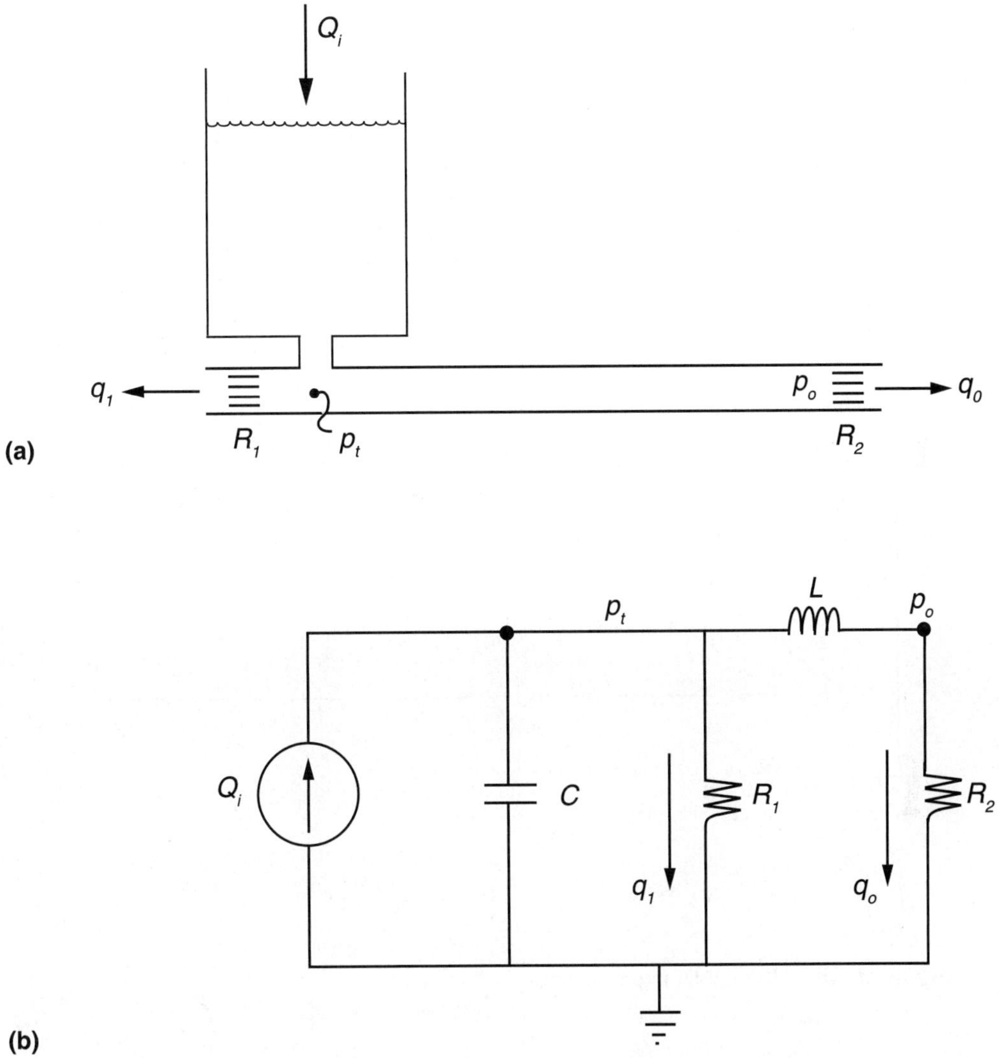

Figure 3–35 (a) Hydraulic model schematic and (b) electrical circuit equivalent

Multiply the second equation by R_2 to get:

$$p_0 = \frac{R_2}{L} \int (p_T - p_0)dt \qquad \textbf{Eq. 3.74}$$

The SIMULINK implementation of these equations is shown in Figure 3–36.

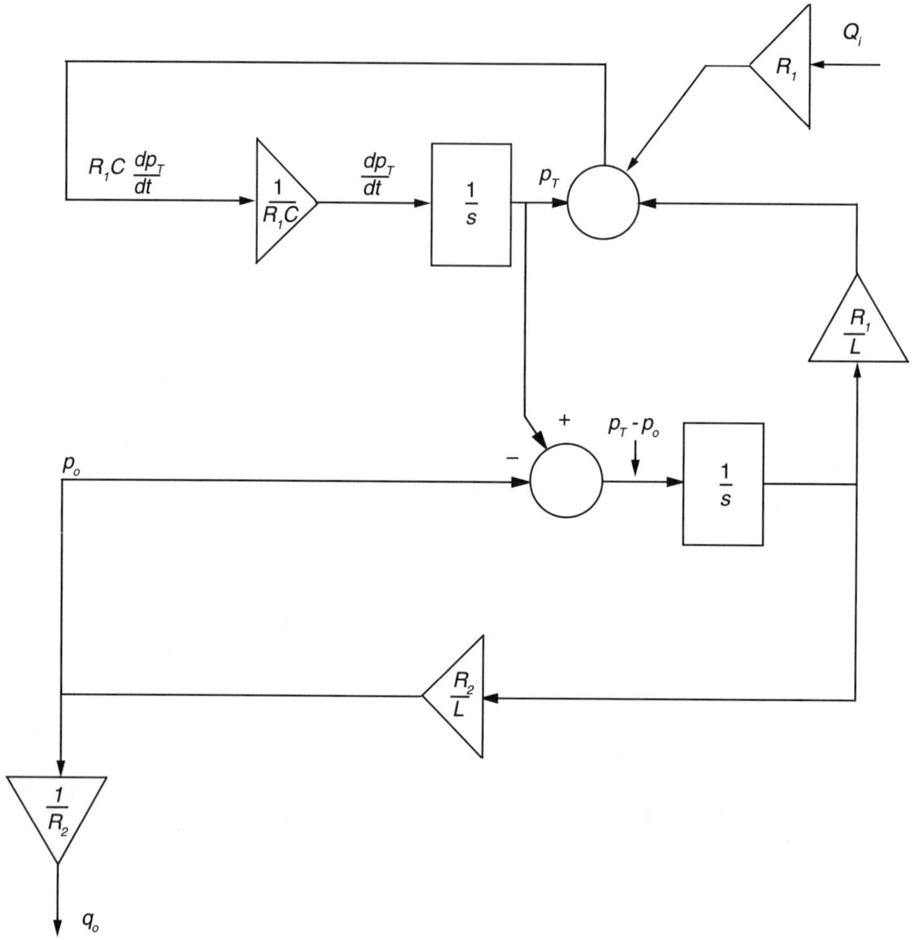

Figure 3–36 SIMULINK implementation of solution

3.16 Modeling Pneumatic Systems

Pneumatic systems are modeled in the same manner as hydraulic systems, except that there are no hydrostatic drops or rises, because gases are so much lighter than fluids. The other basic difference, as presented in the section dealing with pneumatic components, is that a larger volume represents an equivalent capacitor due to the compressibility of gas.

Notice that lumped parameters can be used to represent pneumatic components, provided that the distances between components are not too great and the frequencies are relatively low. The situation can be improved by subdividing each component into a series of lumped components.

Example 3.5

Consider the pneumatic system in Figure 3–37a. Represent the system with an electrical equivalent, and obtain a mathematical representation. Implement the resulting equations into a SIMULINK block diagram.

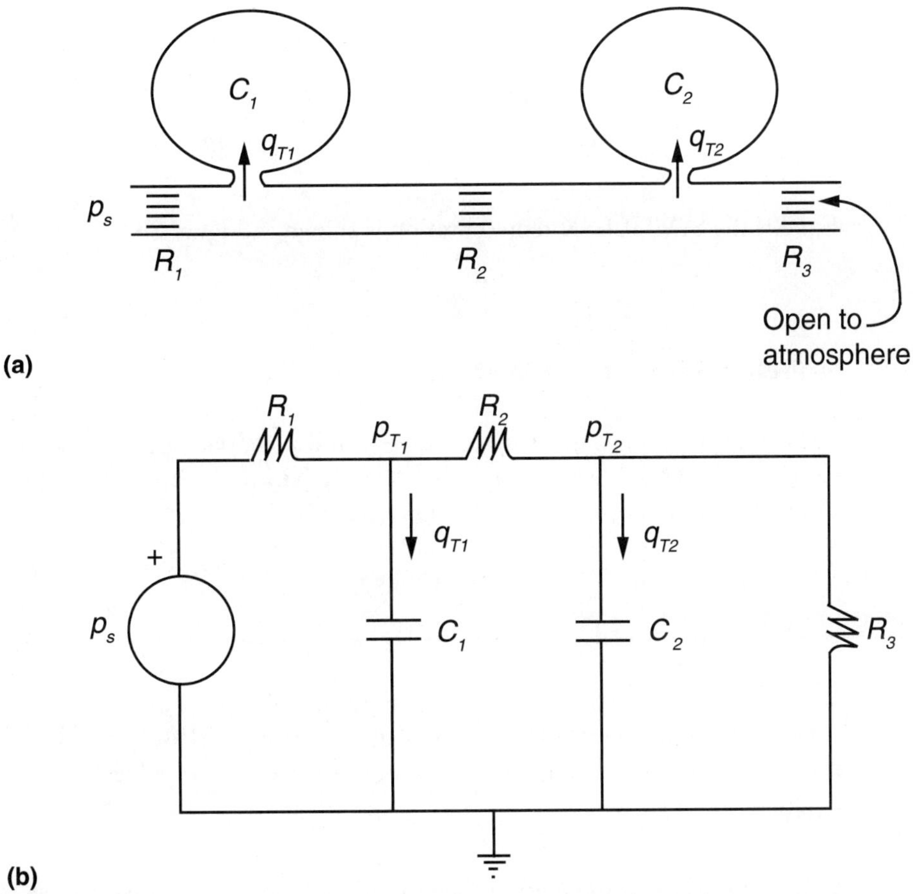

Figure 3–37 (a) Pneumatic circuit and (b) electrical circuit equivalent

Solution The equivalent circuit is shown in Figure 3–37b. Using the pneumatic-equivalent KCL, we obtain the following equations:

$$\frac{p_S - p_{T1}}{R_1} + C_1\frac{d}{dt}(0 - P_{T1}) + \frac{p_{T2} - p_{T1}}{R_2} = 0 \qquad \text{Eq. 3.75}$$

$$\frac{p_{T1} - p_{T2}}{R_2} + C_2\frac{d}{dt}(0 - p_{T2}) + \frac{0 - p_{T2}}{R_3} = 0 \qquad \text{Eq. 3.76}$$

These equations can be rearranged as:

$$R_1C_1\frac{dp_{T1}}{dt} = \frac{R_1}{R_2}p_{T2} - \left(\frac{R_1}{R_2} + 1\right)p_{T1} \qquad \text{Eq. 3.77}$$

$$R_2C_2\frac{dp_{T2}}{dt} = p_{T1} - \left(\frac{R_2}{R_3} + 1\right)p_{T2} \qquad \text{Eq. 3.78}$$

Finally, the SIMULINK representation is shown in Figure 3–38.

3.17 Modeling Mechanical Systems

In dealing with mechanical systems, we can pursue two approaches: (1) create an electrical circuit or a linear graph or (2) apply Newton's Second Law to each mass. We will do an example with each approach.

Example 3.6

Represent the mechanical systems in Figure 3–39a with an electrical circuit, and simulate the system for a 20-lbf step force input using PSPICE. PSPICE is an electrical circuit simulator that requires as inputs the circuit topology and forcing voltages/currents and produces corresponding voltages and current in the circuit.

Solution It helps to envision the flow of force in a mechanical system. A portion of the force applied to a mass accelerates the mass. For that reason, every time you

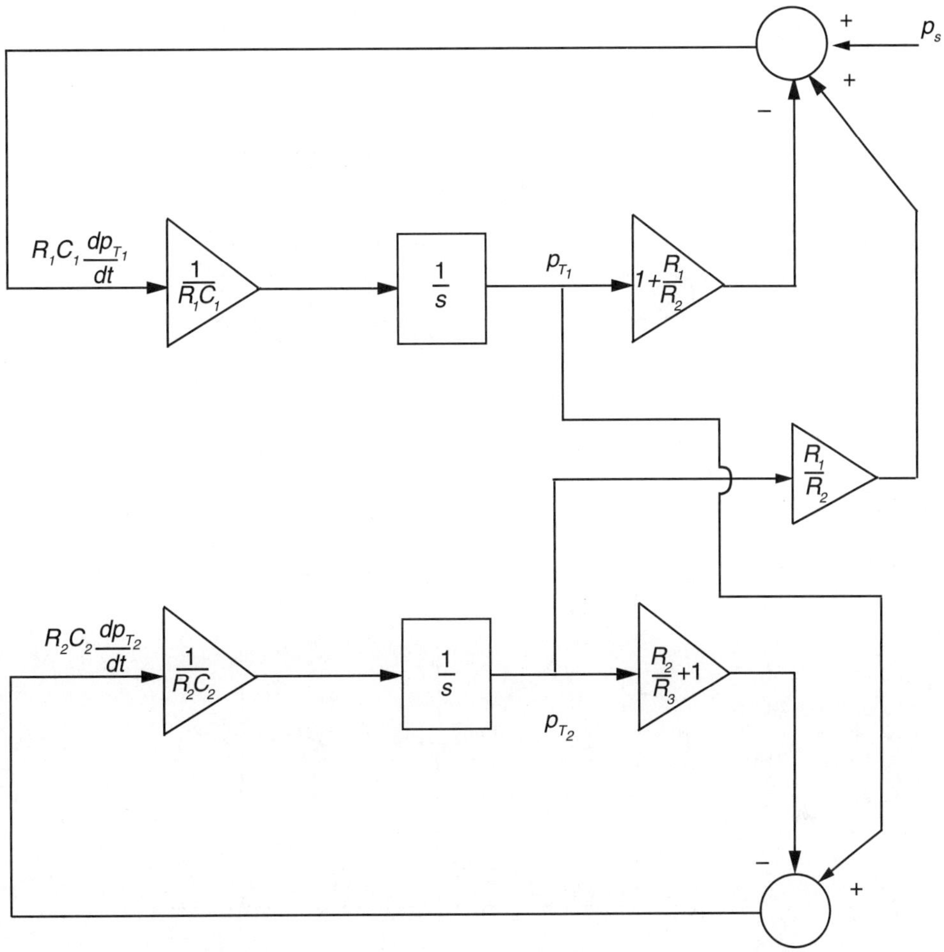

Figure 3–38 SIMULINK implementation of pneumatic problem

have a mass, you have an equivalent capacitor connected to the ground. The remainder of the force flows through the dashpot and through the spring to the reference frame. Figure 3–39b shows the equivalent electrical circuit. Notice that the equivalent resistance, R, is $1/b$. The reason for this can be ascertained by comparing the D-type equations for mechanical and electrical systems in Table 3–2. Similarly, the spring is represented by its analogous counterpart, the inductor (as a memory aid; think of the similarity of a spring and the graphical representation of an inductor).

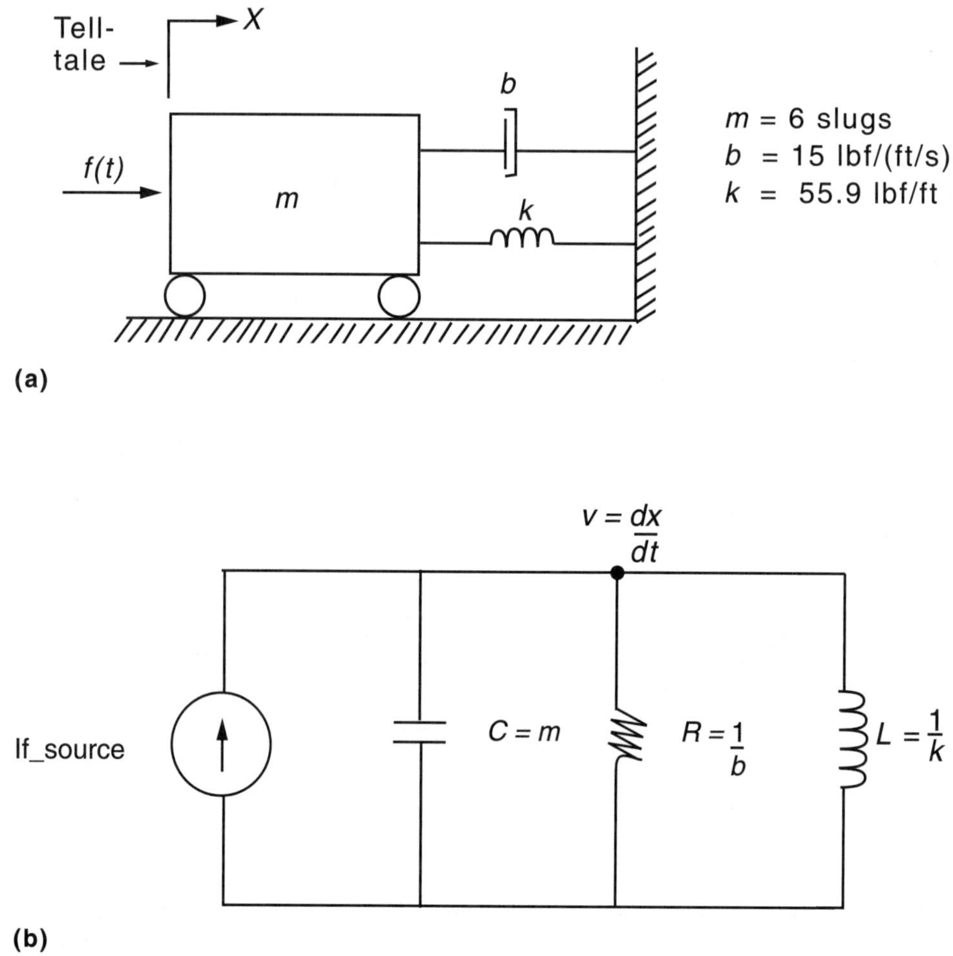

Figure 3–39 (a) Mechanical system and (b) PSPICE electrical circuit equivalent

This circuit is implemented with PSPICE in Figure 3–40. Notice that the voltage at the top of the circuit represents velocity. However, our primary interest is displacement. Recall the relationship between displacement and velocity:

$$x = \int \surd \, dt \quad \textbf{Eq. 3.79}$$

This can be easily implemented in PSPICE's graphical display utility PROBE. This is done by applying the integral operator (S ( ) in PROBE) to the variable that is to

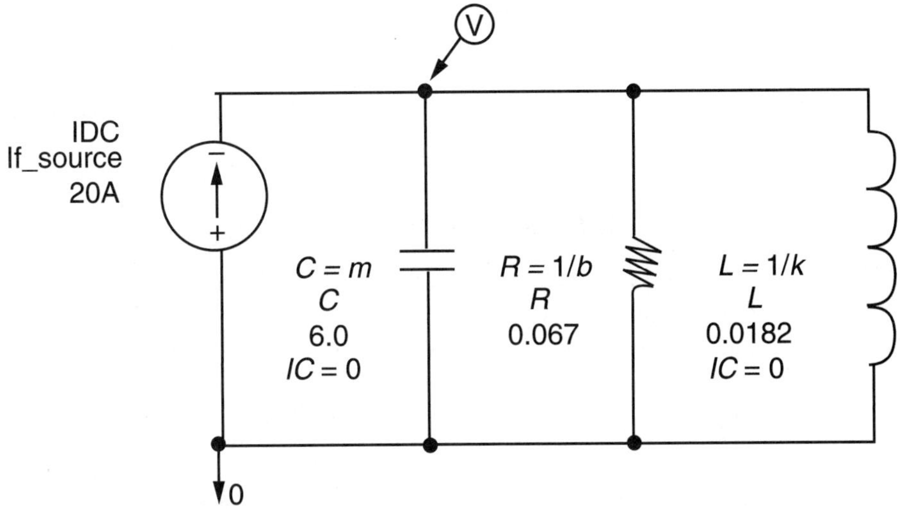

Figure 3–40 Electrical equivalent circuit for mechanical system

be integrated. The resulting velocity and displacement are shown in Figure 3–41. In the following example, we will use the other approach.

The next approach, which is recommended, departs significantly from what has been done with nonmechanical systems. The following outline facilitates the procedure for deriving the equations. Once the model schematic is agreed on do the following:

1. Define the direction of the tell-tales.
   - Place a tell-tale for each mass in the favored direction of movement (not necessary, but helpful).
   - Make all the tell-tales have the same direction (recommended to minimize errors).
   - Place independent force sources in the direction that causes positive movement (as previously defined by the tell-tale).
2. Assume that the mass under investigation is moving in the positive direction and with a positive velocity and that all the other masses are not moving, even if there are other independent sources acting on them (determining how this condition actually happens is not necessary).
   - It helps to redraw each mass and the forces acting on it (like a free-body diagram). This step is not necessary once you gain some experience.

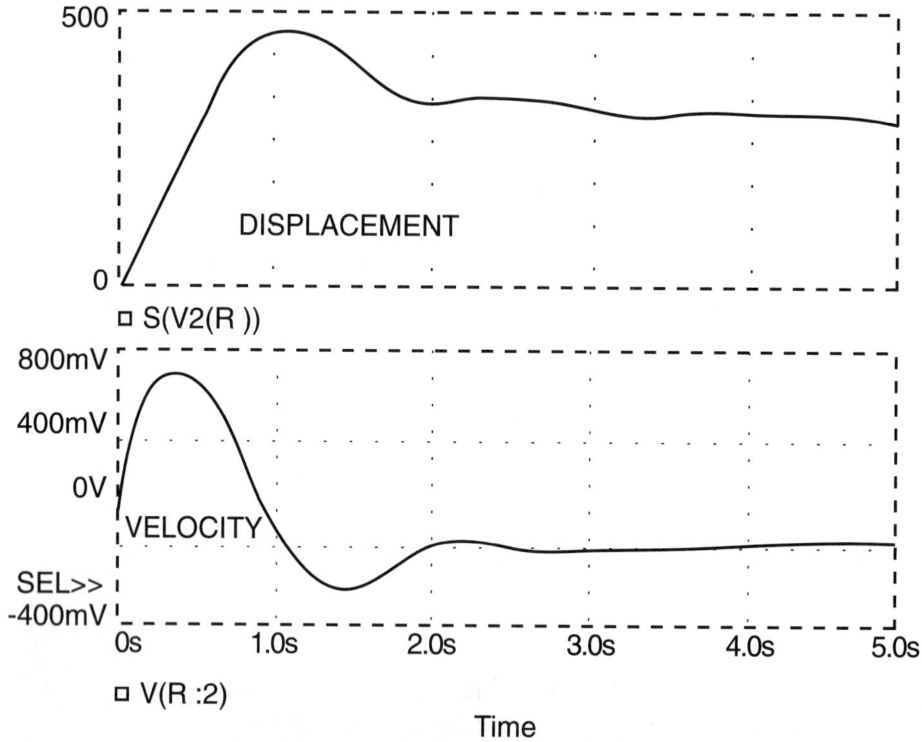

Figure 3–41 PSPICE's probe display of simulation

3. Incorporate all the forces into Newton's Second Law, or:

$$\sum_{\substack{\text{acting on} \\ \text{mass } m}} F = m\ddot{x} \qquad \text{Eq. 3.80}$$

4. Move all the terms associated with the mass under evaluation (velocities and displacements) to the side of the inertia term (i.e., $m\ddot{x}$). If any of the terms are negative, an error has been made.
5. Implement in SIMULINK.

Example 3.7

Obtain a mathematical representation for the mechanical system in Figure 3–42a. Implement the situation in SIMULINK.

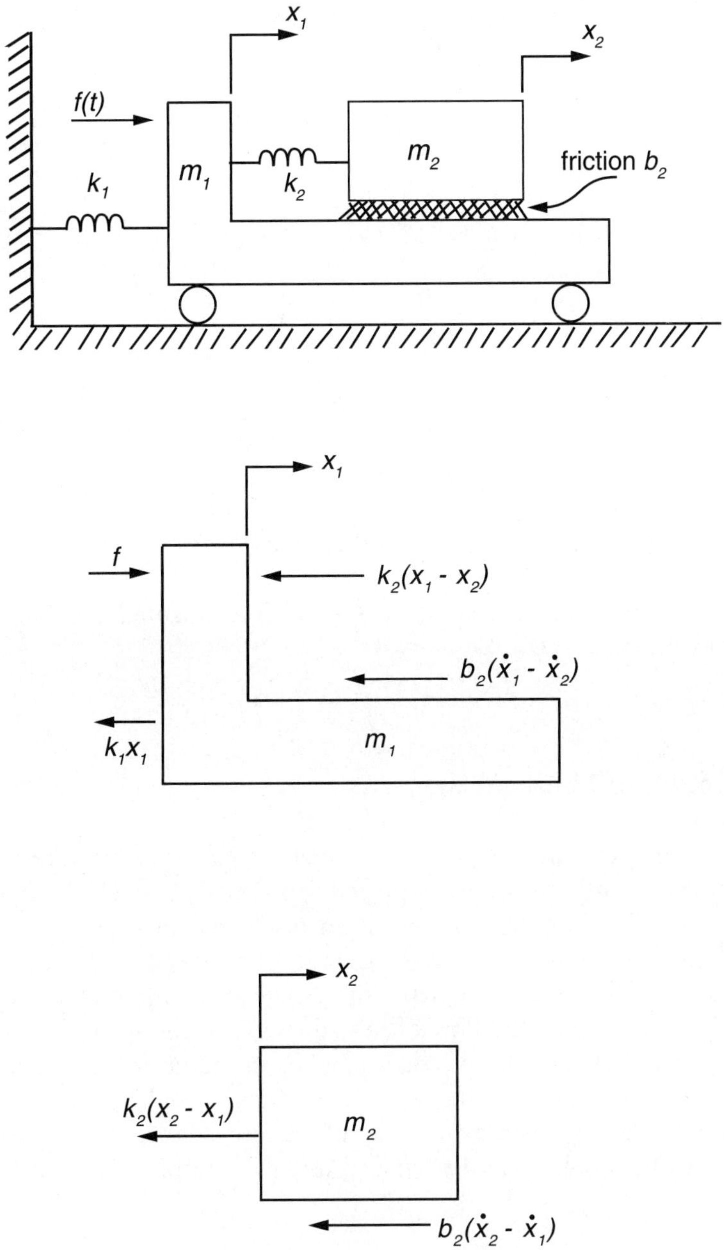

(a)

(b)

Figure 3–42 (a) Mechanical system and (b) free-body diagram of forces acting on masses

Solution The resulting free-body diagram is shown in Figure 3–42b. Applying Newton's Second Law to the two masses results in:

$$\sum_{\substack{on\\m_1}} F = m_1\ddot{x}_1 \qquad \textbf{Eq. 3.81}$$

$$f - k_1x_1 - b_2(\dot{x}_1 - \dot{x}_2) - k_2(x_1 - x_2) = m_1\ddot{x}_1$$

$$\sum_{\substack{on\\m_2}} F = m_2\ddot{X}_2 \qquad \textbf{Eq. 3.82}$$

$$-k_2(x_2 - x_1) - b_2(\dot{x}_2 - \dot{x}_1) = m_2\ddot{x}_2$$

The two differential equations have the highest-order derivative terms isolated:

$$m,\ddot{X}, \text{ and } m_2\ddot{X}_2$$

Similar terms can be combined and then implemented in SIMULINK, as shown in Figure 3–43.

3.18 Modeling Thermal Systems

The first step in modeling a thermal system is to produce a model schematic, which in this case is a simplified representation of the physical system. This model schematic depicts what is an insulator and what is not. It also shows the boundary conditions: regions of constant temperatures. Then the next step is to translate the model schematic into an electrical equivalent circuit. It is also beneficial to imagine the flow of heat throughout the system and translate that flow into an electrical current equivalent. In the resulting equivalent electrical circuit, instead of showing voltages, temperatures should be used. Similarly, heat flow should be represented in terms of currents. Once this representation is complete, the problem can be pursued using the techniques used to analyze electrical circuits.

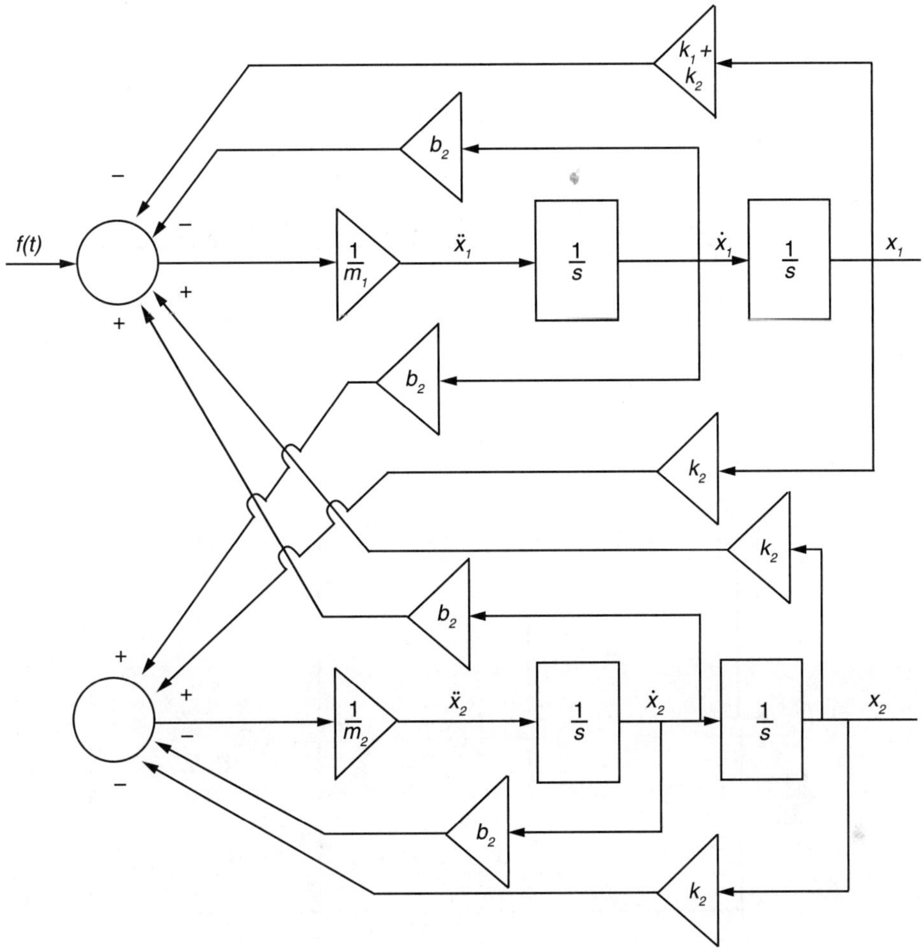

Figure 3–43 SIMULINK implementation of mechanical system

Example 3.8

Consider the thermal system in Figure 3–44a. In this problem, a rod is partially inside the pipe (where water is flowing) and partially outside. The temperature of the rod is the variable of interest. Produce an electrical circuit equivalent of the system.

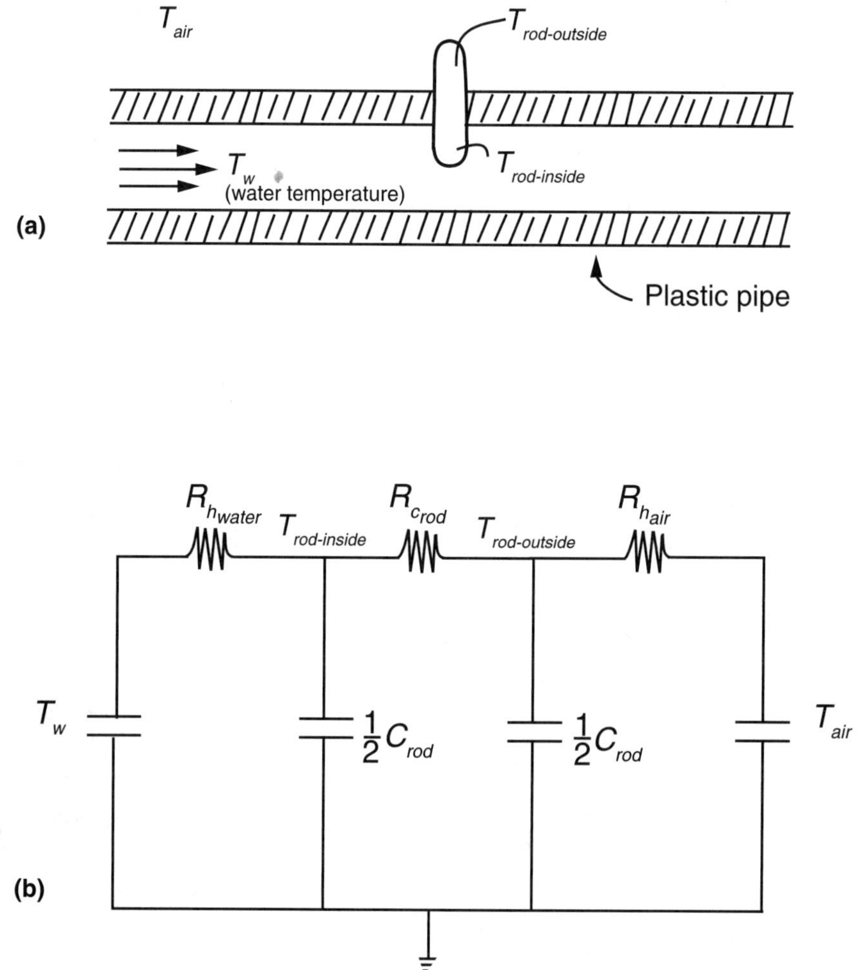

Figure 3–44 (a) Thermal system and (b) electrical equivalent

Solution Notice that the water flowing past the rod is going to act as a constant temperature source. The rod will probably experience two major temperatures: (1) one corresponding to the portion inside the pipe and (2) one corresponding to the portion outside. The equivalent circuit is shown in Figure 3–44b.

As mentioned earlier in this chapter, the modeling process is somewhat of an art. Because many assumptions are made, not all engineers will agree on the appropriateness of a model. For that reason, control engineers dedicate a significant amount of their time to developing models and performing experiments to verify those models' validity.

3.19 Modeling Hybrid/Mixed Systems

Hybrid/mixed systems are composed of more than one type of physical system. For instance, a DC motor is, in reality, a hybrid system, because it is made of electrical and mechanical components. Basically, we need to obtain as many equations as we have unknowns, and then combine them.

Example 3.9

Consider a centrifugal pump driving an actuator connected to a mass with friction, as shown in Figure 3–45a. Pump output characteristics can be approximated with a straight line, as shown in Figure 3–45b. Obtain a block diagram representation that can be implemented in SIMULINK.

Solution The pump can be characterized as:

$$p_p = P_{p\infty} - q_p R_p \quad \textbf{Eq. 3.83}$$

Flow and piston movement:

$$\frac{q_p}{A} = \dot{x} \quad \textbf{Eq. 3.84}$$

Pressure—force:

$$f_{act} = P_p A \quad \textbf{Eq. 3.85}$$

Newton's Second Law:

$$\sum F = f_{act} - b\dot{x} = m\ddot{x} \quad \textbf{Eq. 3.86}$$

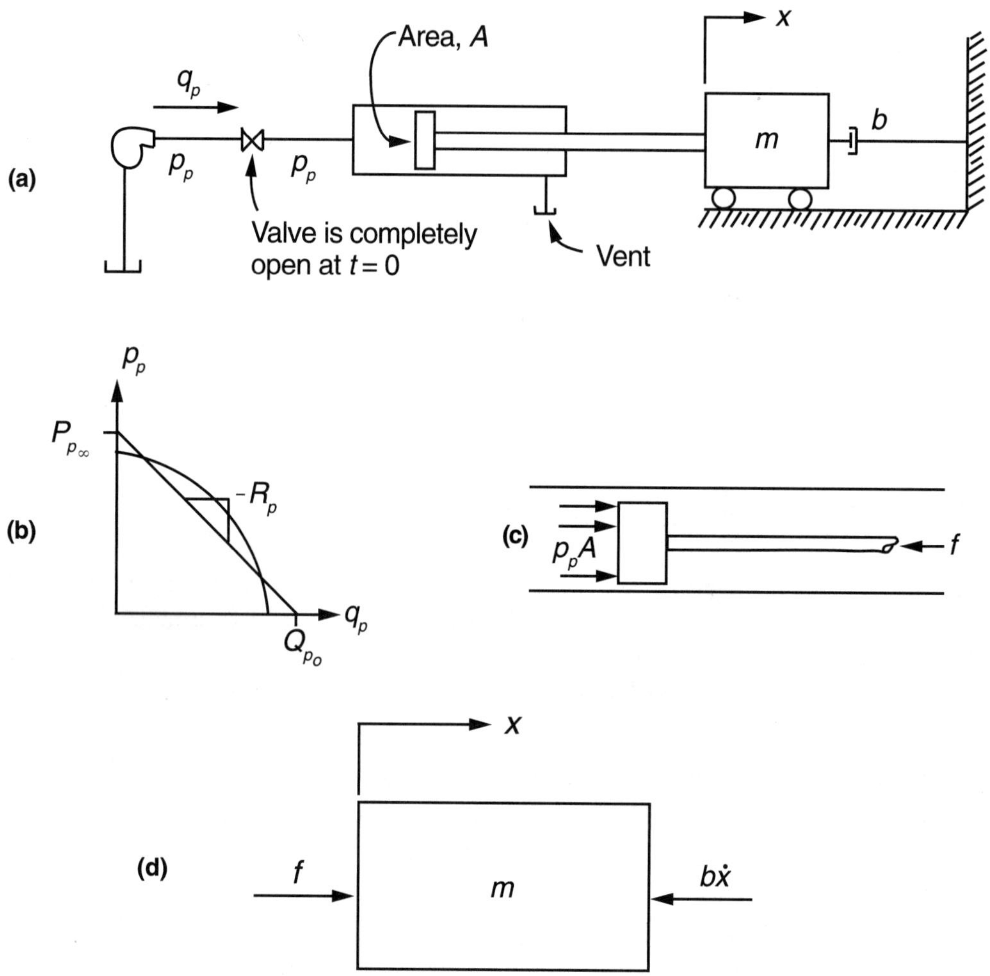

Figure 3–45 (a) Hydraulic-mechanical system, (b) pump characteristics, (c) actuator force balance, and (d) forces acting on mass

Number of unknowns:

Unknown:

$\dot{x}, \ddot{x}, q_P, P_P$

Equations:

Four

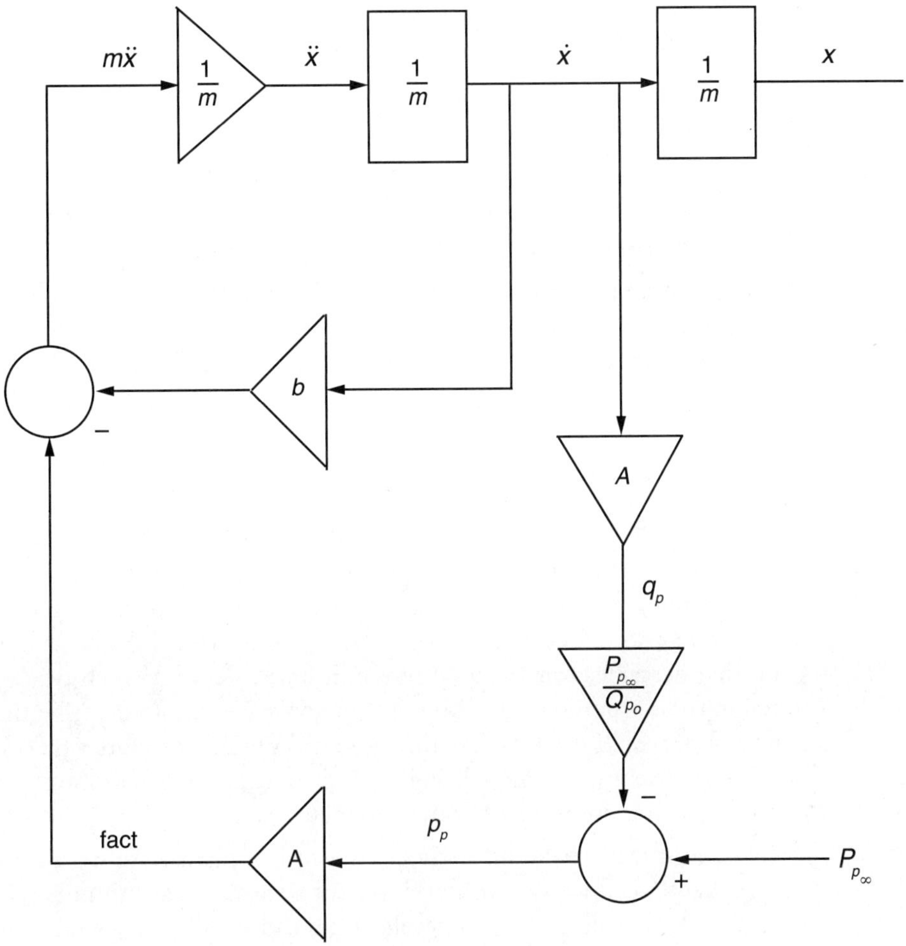

Figure 3–46 SIMULINK implementation

Consequently, the system has a solution.

These equations are then implemented in SIMULINK, as shown in Figure 3–46.

Summary

1. There are several procedures for obtaining a mathematical relationship (differential equation or transfer function) that describes a set of interconnected components.

2. Most physical variables can be defined as through-variables (T-variables) or across-variables (A-variables).
3. Elemental equations describe a component's characteristics. Structural equations describe how components are connected.
4. Analogies exist among different types of physical systems. These analogies are established by components having similar elemental equations.
5. Most components can be classified as energy storing or energy dissipating.
6. Ordinary differential equations describe physical systems whose characteristics can be lumped into subcomponents. Partial differential equations describe systems whose characteristics are distributed over an infinite number of subcomponents.

Idea Portfolio

1. Which type of dynamic systems are involved in your project (electrical, mechanical, hydraulic)? Which systems are really relevant? Propose a high-level block diagram that describes what is happening in this system. Propose a block diagram that describes the dynamic system in more detail. Which type of transfer function (first order low pass, integrator) would you use to describe the components? How would you simulate this system? Which procedure would you use to obtain a more complete model? Which type of experiments or measuring procedures would you use to identify the parameter values?
2. Create two spreadsheets that estimate the cost of manufacturing such a system. Spreadsheet 1 is for a system/development program that minimizes cost (even though it might take longer to develop). Spreadsheet 2 is for a system/development approach that minimizes development time (even though it would cost more). Make sure to include component and labor costs (e.g., for design, construction, testing).

Problems

1. Obtain a differential equation and a transfer function for the electrical circuit in Figure 3–47.
2. Obtain a transfer function for the circuits in Figure 3–48.
3. Determine the order of the electrical system in Figure 3–49.
4. Apply KVL to the electrical circuit in Figure 3–50 to obtain an integro-differential equation.

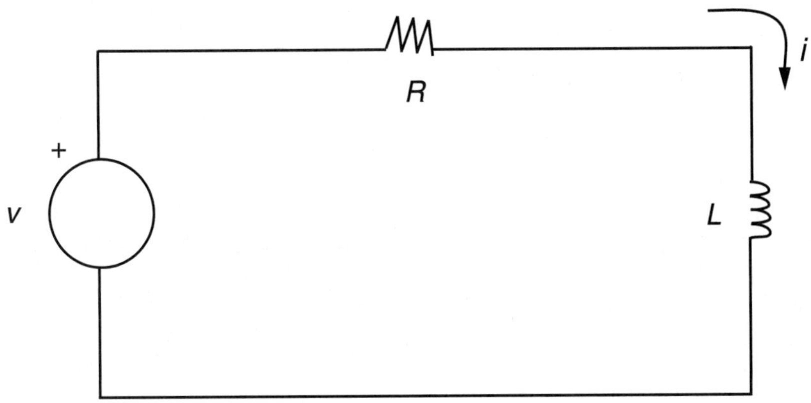

Figure 3–47

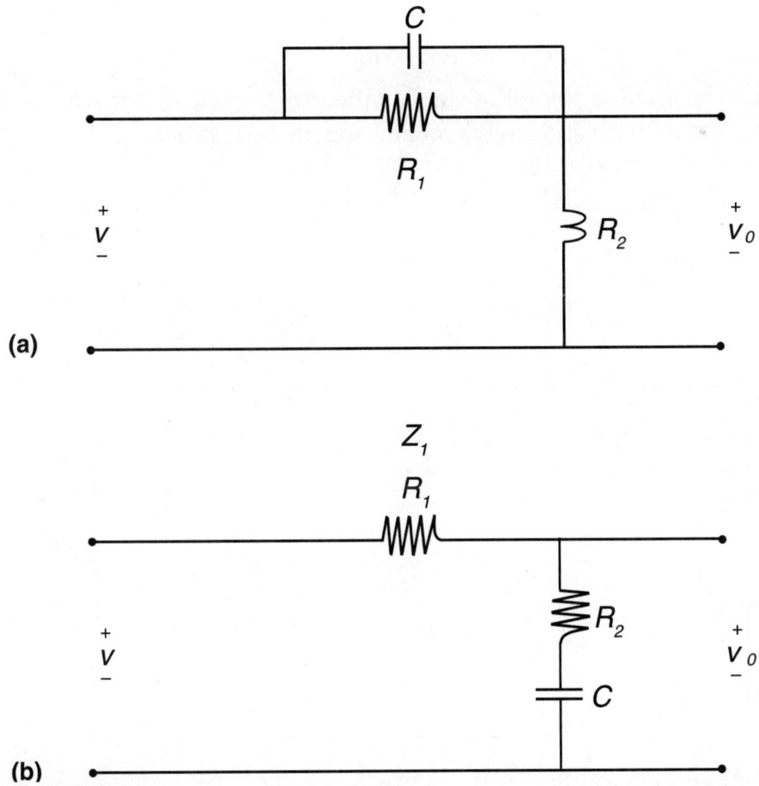

Figure 3–48

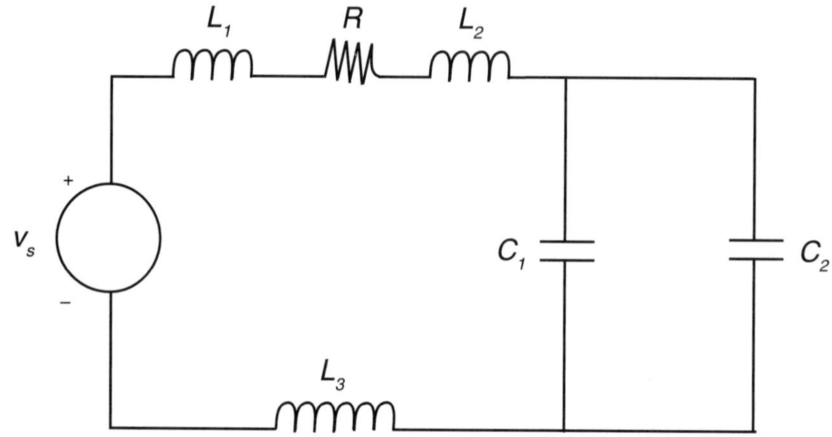

Figure 3–49

5. Write loop equations for the electrical circuit in Figure 3–51.
6. Voltage is a T-variable or an A-variable ?
7. Indicate whether the following equations are elemental or structural. Convey your answer with an E for elemental and an *S* for structural.
   - $v_1 + v_2 + v_3 = 0$
   - $v_1 = Ri_1$

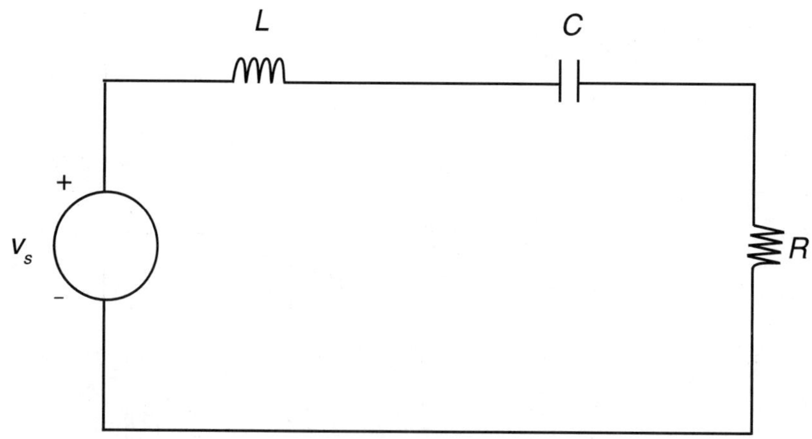

Figure 3–50

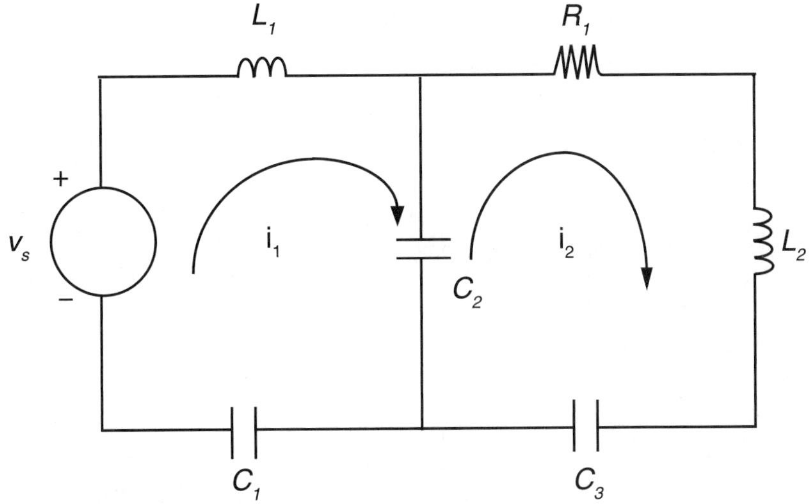

Figure 3–51

8. A low-pass filter is made of a 10K ohm resistor and a 0.1uF capacitor. Obtain a transfer function for the system, and confirm that it is really a low-pass system. What is the time constant associated with the system?
9. A 12V DC solenoid has an internal resistance of 5 ohms and effective indictance of 0.45 mH. What is the time constant of the system? When a 12V signal is applied, what will be the final current? How long will it take for the current to become half its final value?
10. Assume that you are to represent a mechanical damper (dashpot) with a damping coefficient, *b*, of 2 lbf/(ft/s) with an electrical analog. The corresponding electrical component would be a(n) ___________ with a value of ___________.
11. Obtain differential equations for the mechanical systems in Figure 3–52.
12. Obtain the differential equation and transfer function for the mechanical system in Figure 3–53, where x_i is the input and x_o is the output.
13. Create a model schematic for a shock absorber in a car. What happens when the shock absorber gets old?
14. Indicate the mechanical component that is equivalent to the electric capacitor.
15. An F18 is accelerated on the top of an aircraft carrier using a steam catapult. The aircraft, which at full capacity weighs 35,800 lbm, is accelerated from 0 to 160 mph in $2^1/_2$ s. Assuming a constant acceleration, determine the maximum

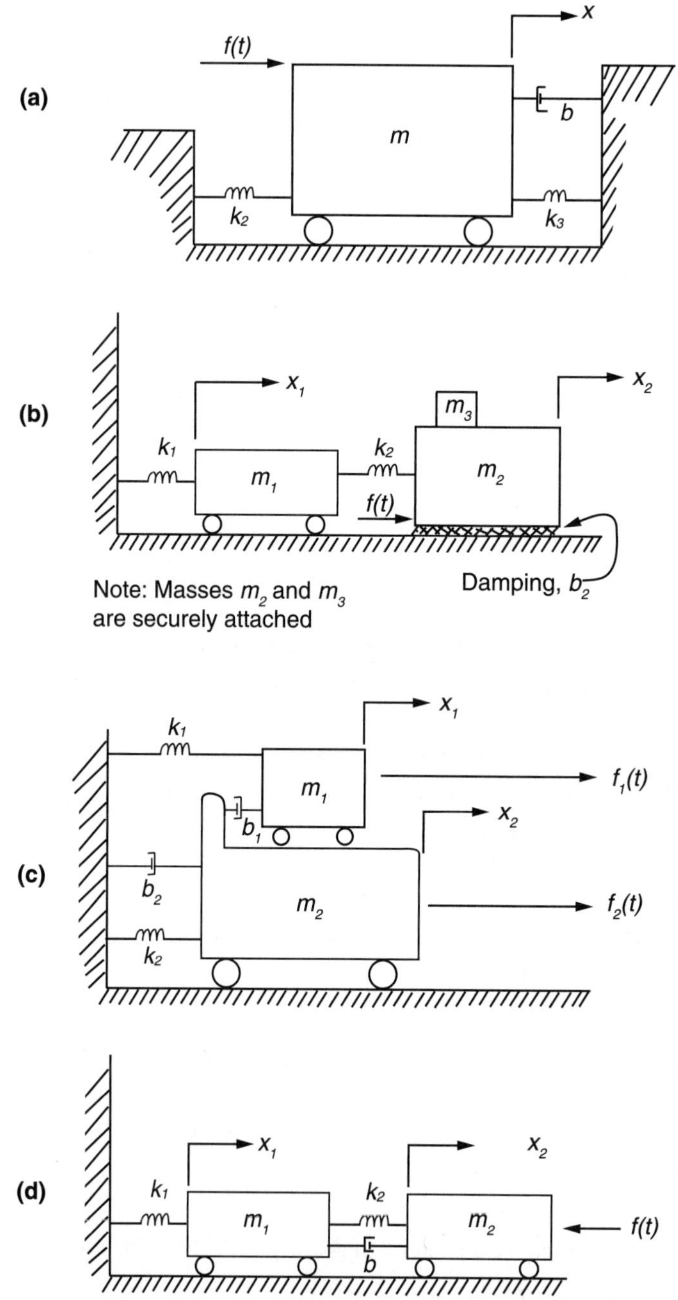
(a)
f(t)
x
m
b
k2
k3
(b)
x1
x2
m3
k1
k2
m1
m2
f(t)
Note: Masses m2 and m3 are securely attached
Damping, b2
(c)
k1
x1
m1
f1(t)
b1
x2
b2
m2
f2(t)
k2
(d)
x1
x2
k1
k2
m1
m2
f(t)
b

Figure 3–52

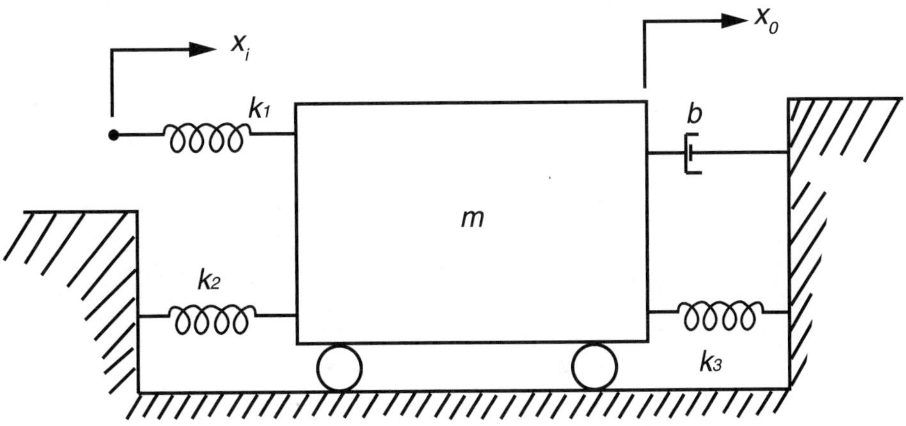

Figure 3–53

force the catapult exerts on the aircraft. Assume no air friction. Perform your calculation for the following conditions:

- Engine contributes no thrust
- Engine contributes full thrust (2 × 16,000 lbf = 32,000 lbf)

16. Rework Problem 15 using metric units.
17. Consider a ship moving on a calm water surface at a constant velocity V_1, as shown in Figure 3–54; δ is the rudder angle (referenced to the ship midline), and ψ is the ship's heading, measured with respect to a stationary, geographic coordinate system.
    - Obtain a general transfer function that relates the output ψ to the input δ. Assume that the ship continues to move at a steady V_1.
    - Modify/augment the transfer function (obtained in the preceding part) if the new velocity of the ship is 0.
18. Determine the order of the hydraulic system in Figure 3–55.
19. Provide an electrical equivalent circuit for the hydraulic system in Figure 3–56.
20. Obtain a SIMULINK diagram that would enable the determination of q_1 and q_2 for the hydraulic system of Figure 3–56.
21. Obtain a transfer function relating $Qo(s)$ and $Pi(s)$ for the low-pressure pneumatic system in Figure 3–57.
22. Water flows from two canals at a 100-ft elevation difference through an 8″ PVC pipe (inside diameter = 9.94″). The resulting flow under this condition is

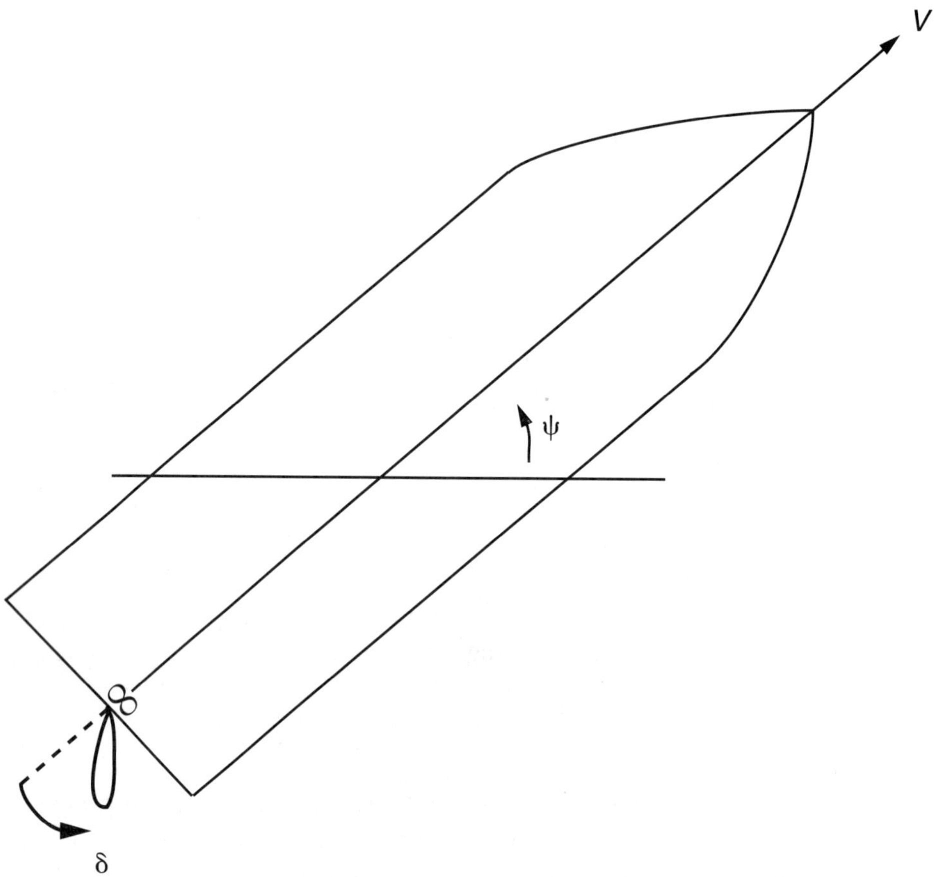

Figure 3–54

1,000 gpm. If the valve is closed in 2s (full open to full closed), what is the maximum pressure seen by the valve? Assume that the flow decreases linearly from 1,000 gpm to 0 in 2s.

23. Model the thermal systems in Figure 3–58 by suggesting an equivalent electrical circuit.

24. Consider two marbles of different sizes and initially at different temperatures, $T1_0$ and $T2_0$. Marble 1 has a diameter that is twice that of Marble 2. Both are dropped inside a perfectly insulated container full of water at temperature T_{water0}. Immediately after being dropped into the container, a top (also a perfect insulator) is placed on the container.

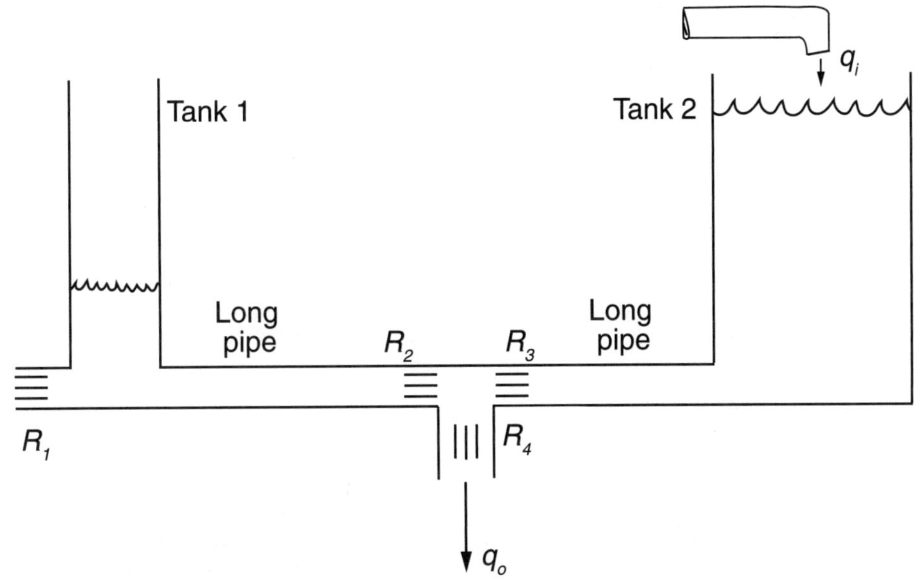

Figure 3–55

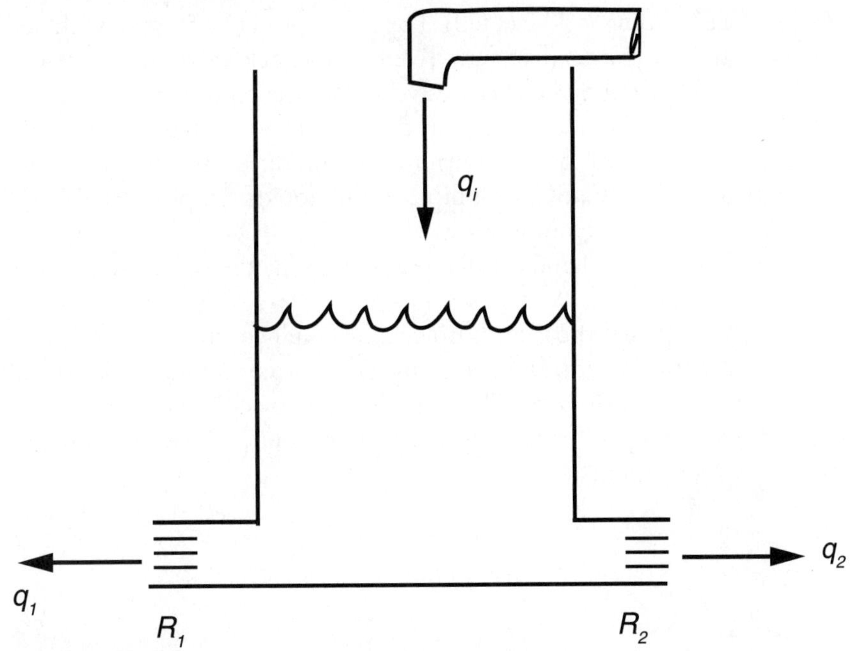

Figure 3–56

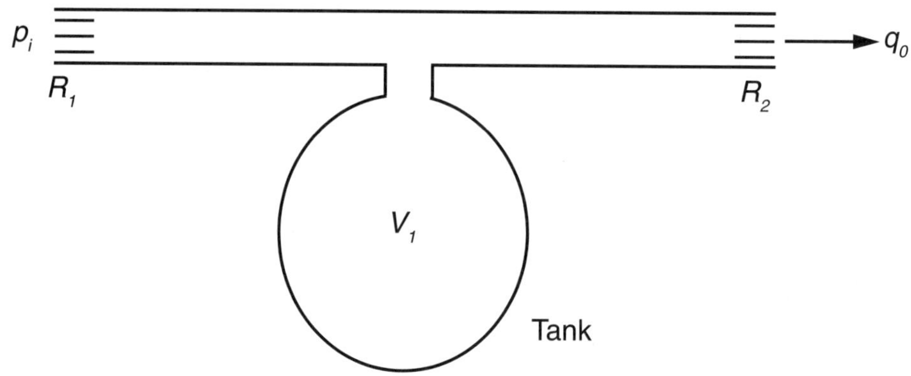

Figure 3–57

- Provide a model schematic representing the dynamic system immediately after the two marbles are dropped into the water.
- Produce a graph showing the expected temperature behaviors of the marbles (*T*1 and *T*2) and the water.

25. Consider a pump that sends water to a 10-foot tank on top of a 4-story building (the building is 40 feet tall). The pump produces 50 gpm with no back pressure and 30 psi dead-ended. There is a check valve just after the pump to prevent back flow. The check valve and all the piping offer a fluid resistance of 3 psi at 25 gal/min. Assume that the pressure flow characteristics of the piping system are linear (i.e., 50 gal/min would produce a pressure drop of 6 psi). The pipe is 1½″ PVC, SCH 40 pipe. At the top of the pipe is a valve. The pump motor is running continuously.
    - Using simple hand calculations, determine the steady-state flow of the system.
    - Assume that the valve is opened instantaneously. Using SIMULINK, simulate the system, and determine the time constant associated with this system (i.e., how long will it take the flow to achieve 63 percent of full flow).
    - How long will it take for the system to put 10 gal into the tank after the valve is opened?

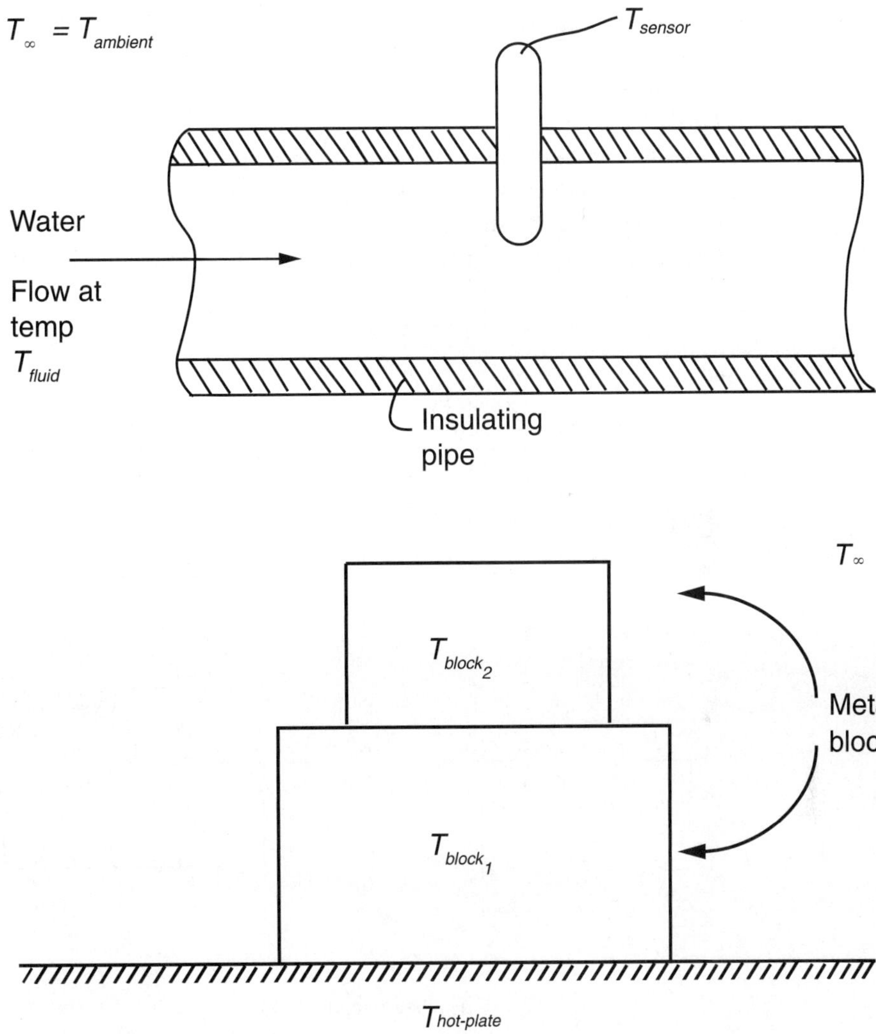

$T_\infty = T_{ambient}$
T_{sensor}
Water
Flow at temp T_{fluid}
Insulating pipe
T_∞
T_{block_2}
Metal blocks
T_{block_1}
$T_{hot\text{-}plate}$

Figure 3–58

26. For each of the physical systems in Figure 3–59, do the following:
    - Create a model schematic using lumped parameter components that approximate the real system.
    - Identify each element as:

      Energy dissipating

      T-type energy-storing element

      A-type energy-storing element
    - Determine the order of the differential equation that describes the system.

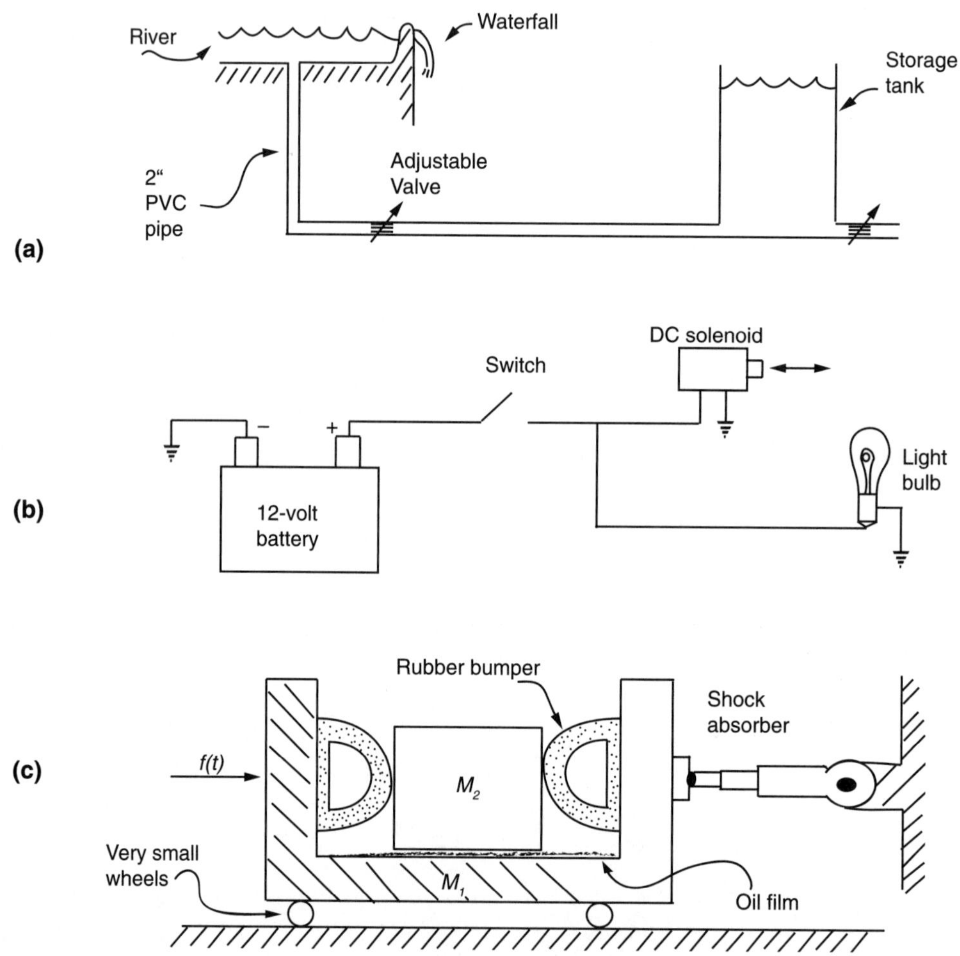

Figure 3–59

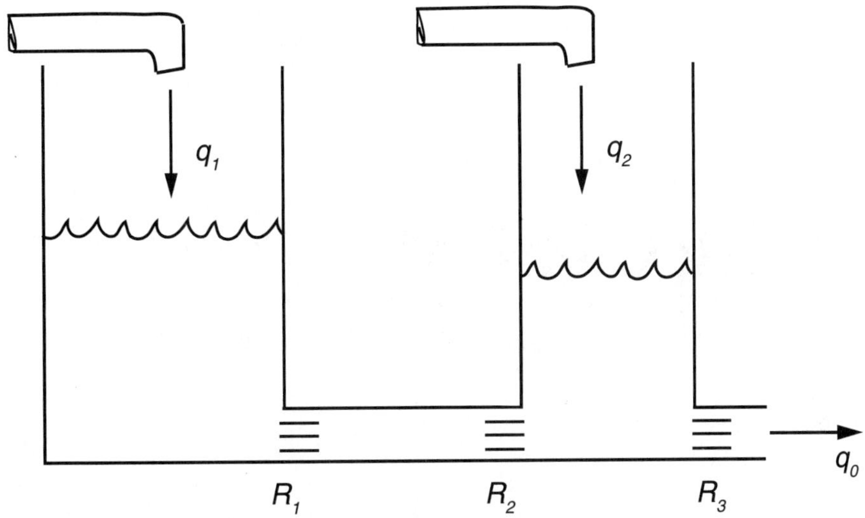

Figure 3–60

27. Show an equivalent electrical circuit for the hydraulic circuit in Figure 3–60.
28. Create a model schematic for the mechanical system in Figure 3–61.

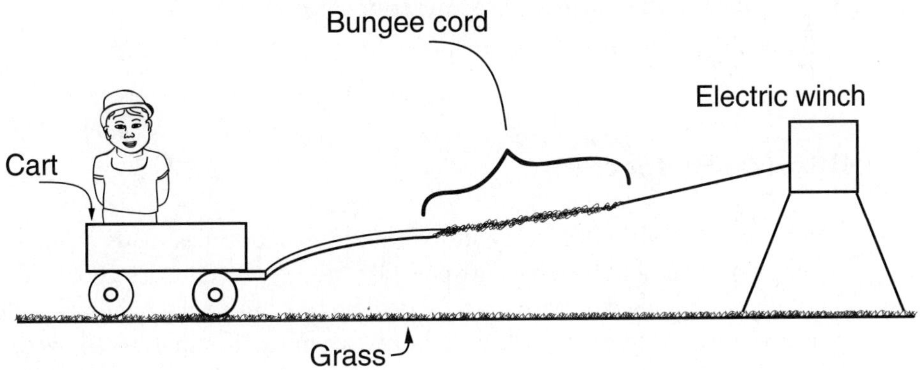

Figure 3–61

4

Feedback Control

Learning Objectives

In this chapter, you will learn how to:

- Identify strengths and weaknesses of open- and closed-loop systems.
- Discuss the dynamic characteristics of a DC motor.
- Describe the significance of the order of a system.
- Calculate the steady-state error.

4.1 Historical Perspective

Even a casual review of history shows that humankind has made tremendous technical progress in a relatively short period. Only a few centuries ago, land travel was limited—one walked or perhaps rode on a horse or in a vehicle horses pulled. Water travel was available, but it was useful only when the destination allowed. Today we depend on machines for transportation; we travel via bicycles, motorcycles, autos, buses, trains, aircraft, ferries, and cruise ships. Note that machines are dynamic devices. Chapter 3 covered some of the subsystems of such devices. We must be able to predict how machines work and be able to integrate machines' many dynamic devices into unified systems.

Humankind has depended heavily on machines to progress. In the Middle East, irrigation has always been necessary, and Babylonian records from several thousand years ago show that water flow was controlled by a type of float valve.[1]

Imagine trying to accurately measure time 2,000 years ago. In the Islamic Empire, water clocks using float valves were relatively common. Water clocks consisted of a device that delivered constant flow into a vessel of constant cross-section, as shown in Figure 4–1. Notice that the rate at which the level rises is constant. This basic idea can be used in two ways. One is by filling a given vessel to its top, which would represent a specific period. A second is marking the side of a constant cross-section vessel to indicate equal time intervals.

Float-valve regulators were used until about 1200 AD, when the Mongol invasion apparently destroyed the technology—at least no records have been found concerning them until about 1600 AD, when the float-valve regulator reappeared.

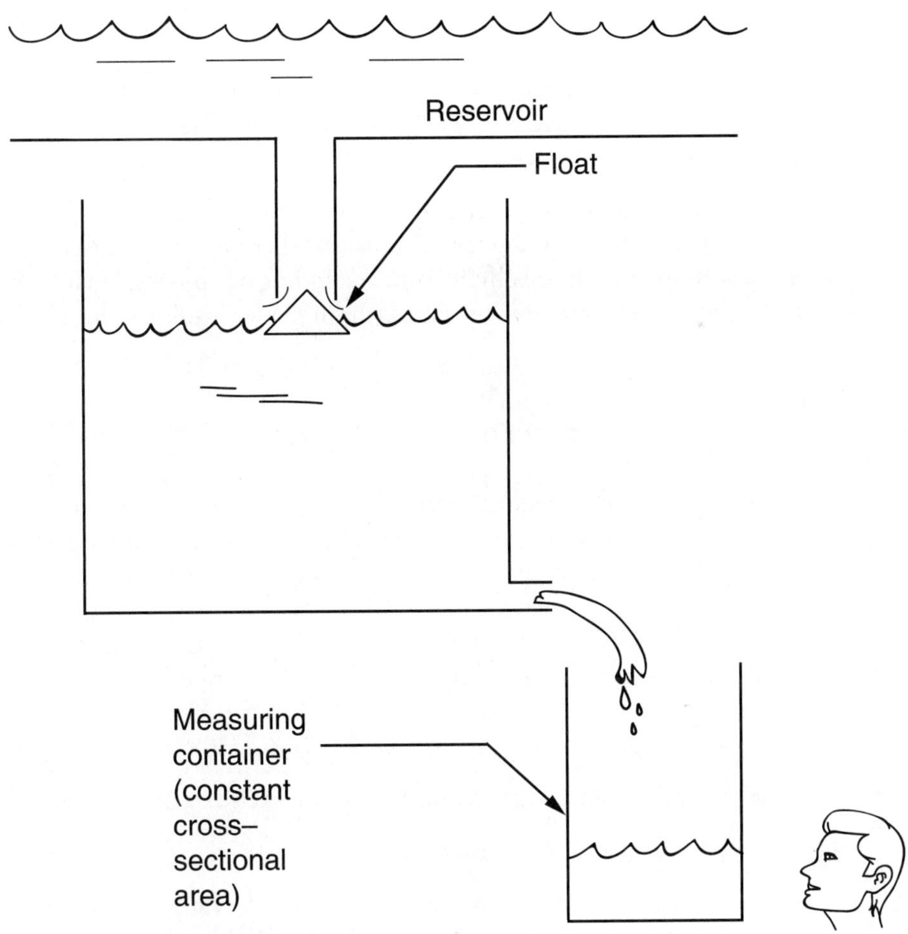

Figure 4–1 Water clock from antiquity

You probably have a float-valve regulator in your house, because many toilets use float-valve regulators. Today, some toilets do not use simple float-type water levelers, instead using small floats that activate hydraulic servos.

The Industrial Revolution introduced many practical undertakings that required regulation of some critical variable in a dynamic system, including things like:

- Liquid levels in process tanks.
- Chicken hatcheries, which require close temperature control. Many use incubators, but Conrad Drebbel invented an oven temperature regulator he used in chemistry studies.
- Flour-grinding mills, which often used wind power. For maximum efficiency, the mills were pointed into the wind by some type of device, such as the tail fan Andrew Meikle devised in 1750.[2] Such mills also required speed regulation, adjustment of the angle of attack of the sails, and adjustment of grindstone separation.
- Steam engines. Steam engines' speed was adjusted manually until James Watt developed the flyball governor in 1788. Steam engines also required pressure regulators and safety relief valves.

All these early dynamic systems, whether controlled automatically or by human intervention, were developed without mathematical theory. Inventions resulted from familiarity with systems. Basic physical understanding of a process led to improvements in the hardware and, ultimately, to a breakthrough idea leading to feedback control.

4.2 Definitions, Models, and Feedback Control

Chapter 1 noted that a feedback control system measures the output of a dynamic system and feeds this information back to the input. The output signal is then subtracted from the input signal. This was illustrated in the home heating system in Figure 1–2.

The feedback concept is illustrated in general terms in Figure 4–2a. Using the house heating system, the corresponding diagram appears in Figure 4–2b. The plant/controller model is represented by its transfer function, $G(s)$, and the feedback device/sensor is represented by $H(s)$. Some definitions associated with this system are:

R = the dynamic input command to the system, such as a step

C = the output signal in response to R

$E = R - C$ (or $R + C$ as will be explained in the following) for the signal obtained when the output is fed back and compared with R, commonly called the error signal

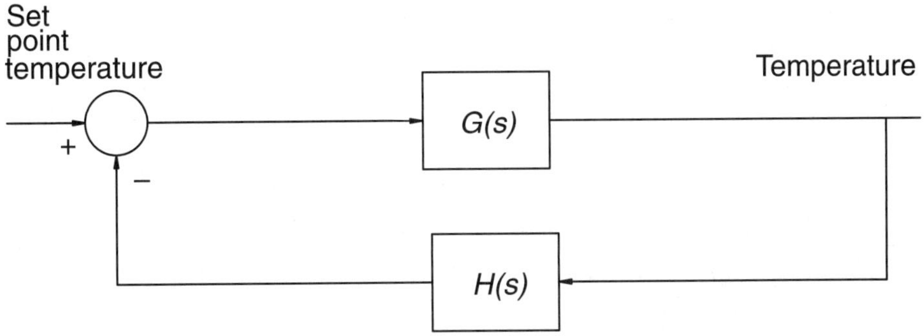

Figure 4–2a General feedback block diagram

When C is subtracted from R (as shown in Figure 4–2), the resulting system is a negative feedback system. On the other hand, when C and R are added, the resulting system is a positive feedback system.

When positive feedback is used, E is increased; for negative feedback, E is decreased. Positive feedback has been used in amplifiers to increase sensitivity, but it has the disadvantage of reducing stability. Consequently, negative feedback is used in almost all control systems.

Two types of transfer functions are very important when analyzing and designing dynamic systems: (1) the open-loop transfer function (OLTF) and (2) the closed-loop transfer function (CLTF). The OLTF represents the ratio of output to input, if the feedback signal path was severed just before the summing junction, as

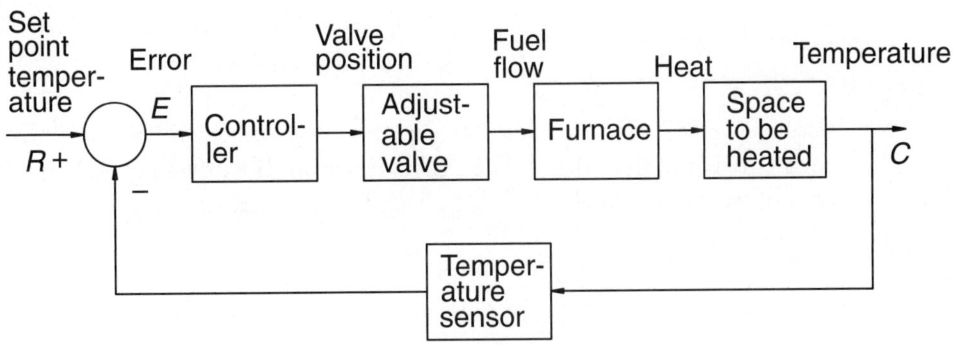

Figure 4–2b Specific closed-loop house heating system

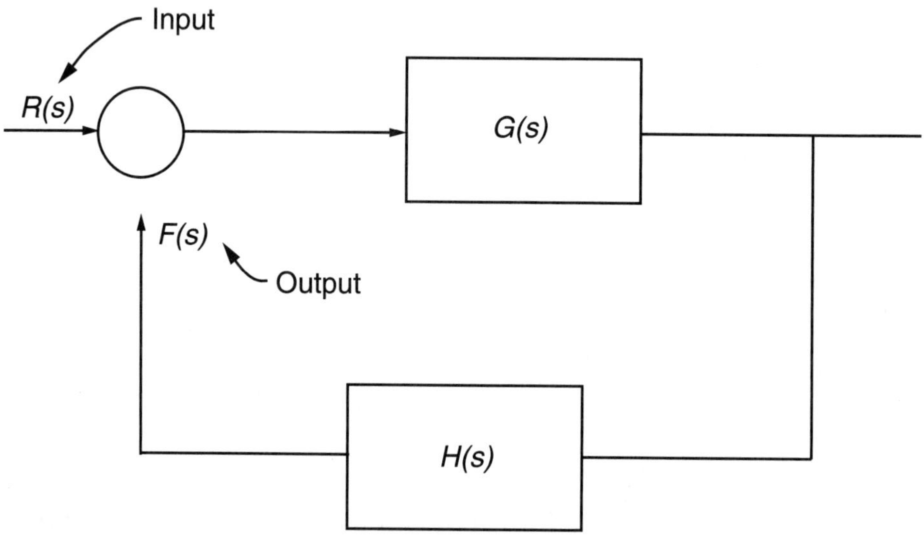

Figure 4–3 Open-loop representation

shown in Figure 4–3. Using the generic transfer functions $G(s)$ and $H(s)$, the OLTF can be expressed as:

$$OLTF = \frac{F(s)}{R(s)} = G(s)H(s) \quad \text{Eq. 4.1}$$

The CLTF, which represents the ratio of output to input of the complete and integrated system, is given by:

$$CLTF = \frac{C(s)}{R(s)} = \frac{G(s)}{1 + G(s)H(s)} \quad \text{Eq. 4.2}$$

Because we will be evaluating this expression often, we will present two shortcuts, one manual and the other a MATLAB function. If expressing $G(s)$ and $H(s)$ as the ratio of two polynomials in s using the following format:

$$G(s) = \frac{Gnum}{Gden} \quad \text{and} \quad H(s) = \frac{Hnum}{Hden} \quad \text{Eq. 4.3}$$

then the CLTF can be expressed as:

$$CLTF = \frac{CLTFnum}{CLTFden} = \frac{C(s)}{R(s)} = \frac{Gnum\ Hden}{Gden\ Hnum + Gnum\ Hden} \qquad \textbf{Eq. 4.4}$$

The *CLTFden,* a polynomial in terms of s that is very important, contains essential information about the system. We can also obtain these two polynomials using the following MATLAB function:

```
>>[CLTFnum,CLTFden] = feedback(Gnum,Gden,Hnum,Hden)
```

A very common type of feedback system is the unity feedback control system, which has a gain of 1 in the feedback path. This is shown in Figure 4–4. For this case:

$$OLTF = G(s)$$
$$CLTF = \frac{G(s)}{1 + G(s)} \qquad \textbf{Eq. 4.5}$$

Why devote any attention to the OLTF when we want to know the behavior of the system represented by the CLTF? The following illustration might shed some light on the motivation for dedicating a considerable amount of time and effort to the OLTF. Suppose that you are building a hoist that uses a pulley, as shown in Figure 4–5. Assume that the rollers in the pulley are frictionless and that a factor of safety

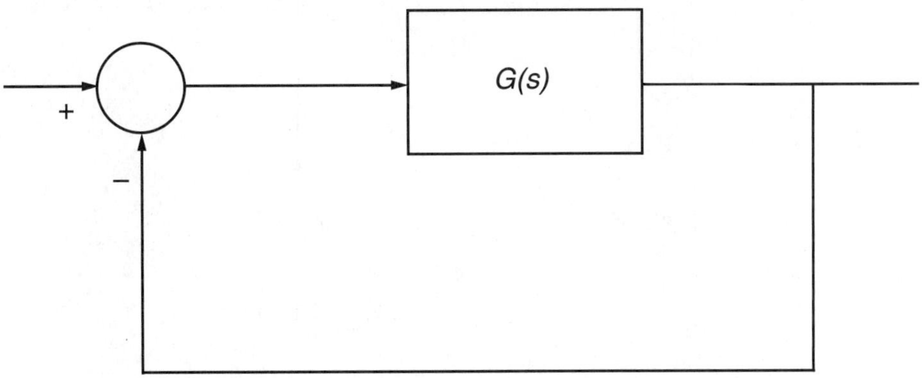

Figure 4–4 Unity feedback control system

has been incorporated into the 300 lbf load specification. Also assume that you want to ensure that the rope in your hoist can lift a 300 lbf load. Assume further that you have access to a hydraulic press, which will allow you to impose a predetermined tension on the rope, as shown in Figure 4–5.

Using the principles of statics, you realize that you will only have to apply a 150 lbf tensile load on the rope. If your test succeeds, and the rope can withstand the 150 lbf load, then you can conclude that the rope should be able to perform the 300 lbf lifting operation successfully. Even though this example/analogy has some severe limitations, it shows that working with one set of conditions (OLTF/hydraulic press) provides useful information on another (CLTF/hoist).

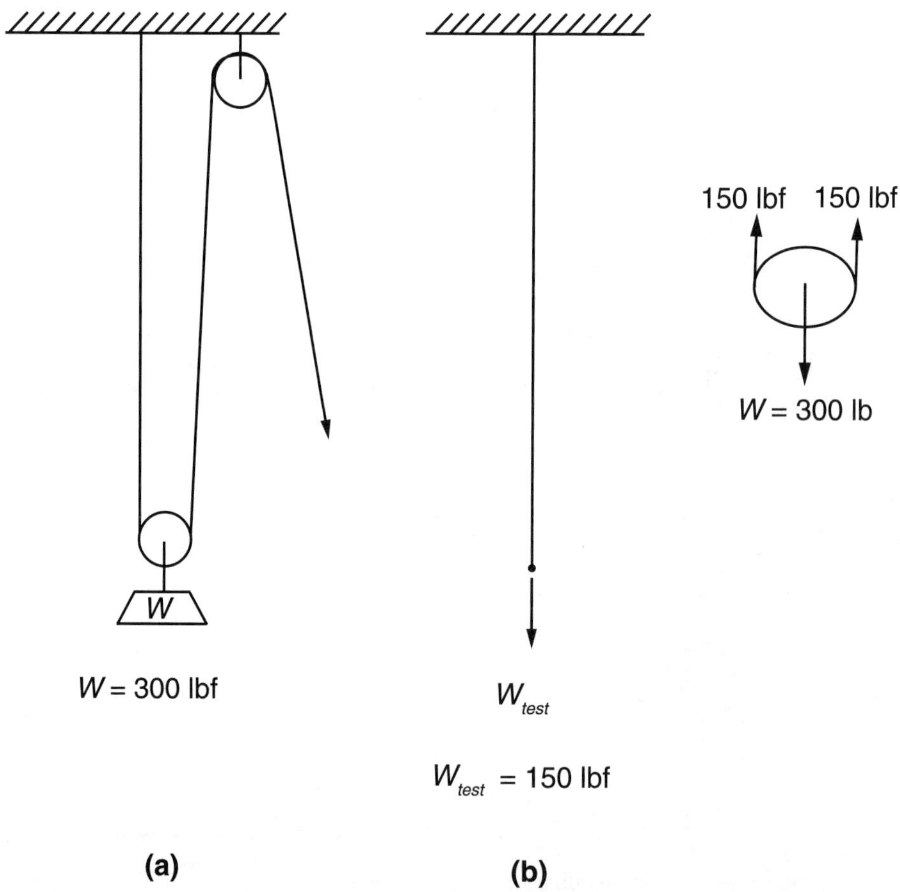

Figure 4–5 Analogy of OLTF-CLTF testing: (a) intended application and (b) test setup used to confirm the safety of intended application

Let us return to the unity feedback system. For the unity feedback system in Figure 4–4, the CLTF, which represents the complete feedback control system, is:

$$CLTF = \frac{C(s)}{R(s)} = \frac{G(s)}{1 + G(s)} \qquad \textbf{Eq. 4.6}$$

which is quite different from the plant transfer function $G(s)$. Both $G(s)$ and $1 + G(s)$ are ratios of polynomials in s.

The characteristic equation of the closed-loop system is obtained by setting the denominator of the CLTF (i.e., the ratio C/R) to 0. Then, if one desires the time response due to an input, such as a step, the characteristic polynomial must be factored. This task is very difficult, if not impossible, for any polynomial of fourth or higher order. It can be done with numerical methods using the computer. Because computer factoring of the characteristic polynomial is simple and accurate, it should be done in every dynamic analysis, because it provides insight regarding the dynamic characteristics of the system. The exponent of the highest power term in the characteristic polynomial (the denominator of the CLTF) defines the system order. Notice that this "order" is the same as the order of the OLTF denominator's polynomial.

4.3 System Order

By definition, the order of a differential equation is the highest derivative of the dependent variable. Similarly, the order of a polynomial is the highest power of the dependent variable. This number defines the number of roots of the characteristic equation, thereby indicating the complexity of the dynamic system and implicitly emphasizing that the analysis and design problems will be more difficult. Thus one can expect more work for high-order systems than for low-order systems.

The order of the system, which has several physical implications, indicates the number of:

- Energy storage devices
- Essential integrators in the system

It also indicates the minimum number of integrators in a SIMULINK/MATLAB environment required to solve the corresponding differential equation.

Another term that must be defined is "system type," which is the exponent on a factorable s term of the OLTF denominator. When the OLTF's denominator lacks a factorable s term, it is said that the OLTF is "type 0." When it has an s^2, it is said that it is "type 2." The physical meaning of "system type" is the number of integrators (or those devices acting like integrators) in the dynamic system.

For example, for the unity feedback system, $G(s)$ might be:

$$G(s) = \frac{C}{e} = \frac{As + Z}{Ps^4 + Qs^3 + Rs^2 + Ss + T} \qquad \text{Eq. 4.7}$$

The transfer function of the closed loop would be:

$$\frac{C}{R} = \frac{As + Z}{Ps^4 + Qs^3 + Rs^2 + (S + A)s + (T + Z)} \qquad \text{Eq. 4.8}$$

Notice that this example is a fourth-order system and type 0 system.

In short, both OLTFs and CLTFs are useful in analysis and design. The OLTF provides a simpler vehicle for predicting the dynamic behavior of the complete, closed-loop dynamic system. The approach is simpler because the mathematical expression is simpler, and the physical constants usually appear in isolated form.

Because of the challenges associated with a complex CLTF, analysts usually check other important features before attempting to solve the equation. They instead work with the OLTF on two things:

1. Stability
2. Steady-state accuracy

Analysts can check stability using Routh-Hurwitz methods (presented in Chapter 5) or frequency response methods using the Nyquist Criterion. They can calculate steady-state accuracy using the Final Value Theorem of the Laplace transform and the OLTF (i.e., the steady-state error coefficients), as demonstrated later in this chapter.

It helps, as shown later, to factor the polynomials of the OLTF so that the roots of both the numerator and the denominator are easily visible. The roots of the numerator and denominator polynomial of the OLTF are called zeros and poles of the OLTF, respectively. The "zero" name is given because the transfer function becomes 0 when evaluated at the "zeros." The "pole" name corresponds to the transfer function approaching infinity when evaluated at the "pole." In an absolute value of the transfer function versus s plot, this looks like a very steep spike or "flagpole." Consider the algebraic function:

$$y = \frac{x - 6}{x + 3} \qquad \text{Eq. 4.9}$$

and its corresponding plot in Figure 4–6. Notice that there is a 0-crossing at $x = 6$ and a "flag pole" at $x = -3$. As a result, we say that the function has a "0" at $x = 6$ and a "pole" at $x = -3$.

When the roots of the polynomials are complex, it is undesirable to show these roots in their rectangular form (i.e., $s - \sigma \pm j\omega$). Instead, we want to show the complex pair $(s - \sigma + j\omega)(s - \sigma - j\omega)$ in terms of its corresponding quadratic. The desired format of presentation, and the motivation for this, are discussed in the next chapter.

The OLTF can be expressed in two forms: (1) the Root Locus Form and (2) the Bode Form. Which form is preferred depends on the type of analysis to be performed. Before proceeding, we will present a specific notation that may be new to you: the Π notation. The Π symbol denotes the product of a series of terms, much like Σ denotes the summation of a series of terms. The two OLTF general forms are

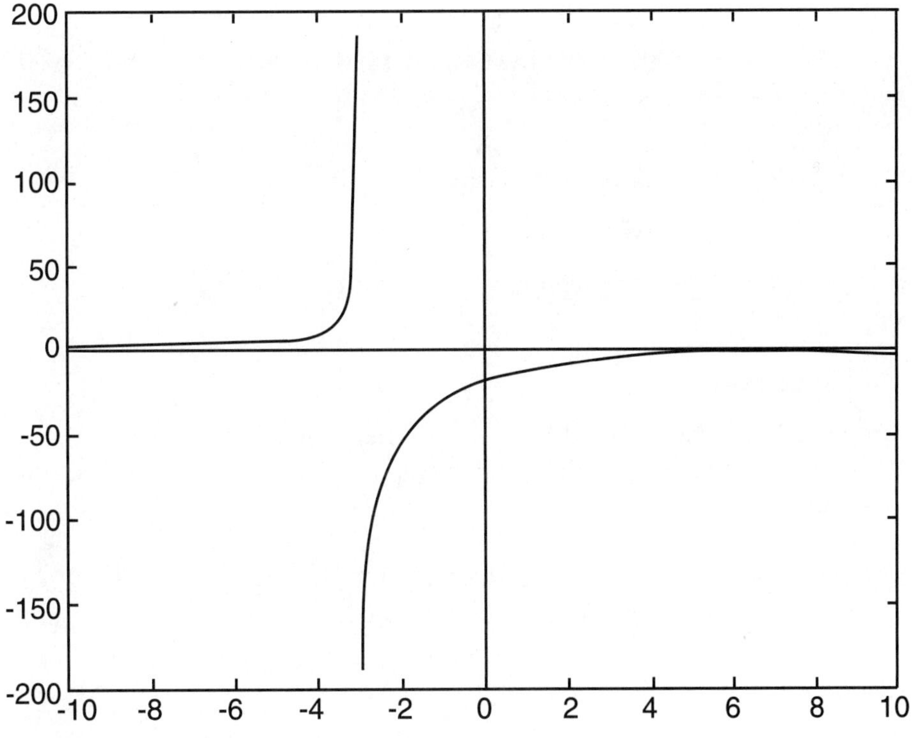

Figure 4–6 Plot of $y = x - 6/x + 3$

shown with the products of the quadratic terms (due to the complex pair of roots). As far as this chapter is concerned, you can ignore quadratic roots. Following we present the two forms.

Root Locus Form

The Root Locus Form is used to plot the Root Locus. Notice that the roots are readily visible.

$$G(s)H(s) = \frac{K_R \prod_{i=1}^{M}(s + z_i) \prod_{j=1}^{Q}(s^2 + 2\varsigma_j \omega_j s + \omega^2_{nj})}{s^N \prod_{k=1}^{T}(s + p_k) \prod_{l=1}^{U}(s^2 + 2\varsigma_l \omega_l s + \omega^2_{nl})} \quad \textbf{Eq. 4.10}$$

or:

$$G(s)H(s) = \frac{K_R \prod_{i=1}^{M}(s + z_i) \prod_{j=1}^{Q}(s + \{d_j + e_j j\})(s + \{d_j - e_j j\})}{s^N \prod_{k=1}^{T}(s + p_k) \prod_{l=1}^{U}(s + \{d_l + e_l j\})(s + \{d_l - e_l j\})}$$

where:

K_R = Root Locus Gain

N = system type

$N + T + 2U$ = system order

Bode Form

The Bode Form is used for, among other functions, calculating the steady-state error. Notice that, as the s variable approaches 0, all the factors, except the s^N, approach 1.

$$G(s)H(s) = \frac{K_B \prod_{i=1}^{M}(\tau_i s + 1) \prod_{j=1}^{Q}(a_j s^2 + b_j s + 1)}{s^N \prod_{k=1}^{T}(\tau_k s + 1) \prod_{l=1}^{U}(a_l s^2 + b_l s + 1)} \quad \textbf{Eq. 4.11}$$

or:

$$G(s)H(s) = \frac{K_B \prod_{i=1}^{M}\left(\frac{s}{w_i}+1\right)\prod_{j=1}^{Q}\left((\frac{s}{\omega_j})^2 + 2\frac{\varsigma_j}{\omega_j}s+1\right)}{s^N \prod_{k=1}^{T}\left(\frac{s}{w_k}+1\right)\prod_{l=1}^{U}\left((\frac{s}{\omega_l})^2 + 2\frac{\varsigma_l}{\omega_l}s+1\right)} \quad \text{Eq. 4.12}$$

where:

$K_B =$ Bode Gain

$N =$ system type

$N + T + 2U =$ system order

Just as the OLTF has poles and zeros, the CLTF has poles and zeros. The roots of the characteristic polynomial (CLTF denominator roots or poles) are called the eigenvalues or characteristic values, because they embody the essence or characteristics of the dynamic system. A given model schematic system can be represented mathematically in many ways, but those ways will share the same eigenvalues. The word *eigenvalue* is a mixture of German and English. The German prefix *eigen* can be translated as "proper" or "characteristic." Hence, eigenvalues are also called "proper values" or "characteristic values." In older literature, they are also called latent roots.

Now let us look at the benefits that are realized when the system is transformed from an open-loop system to a closed-loop one. As a first example, we will use a DC motor, which is to start at rest and accelerate to a constant speed.

4.4 DC Motor: Comparison of Open- and Closed-Loop Systems

An electric motor is a dynamic system composed of electrical, magnetic, and mechanical components. Imagine it as an electromagnetic subsystem that drives a mechanical subsystem, as shown in Figure 4–7a.

Notice that the magnitude of the torque produced by the electromagnetic subsystem of the motor, which is applied to the mechanical portion, is the same as the resisting torque the mechanical subsystem imposes on the electromagnetic portion.

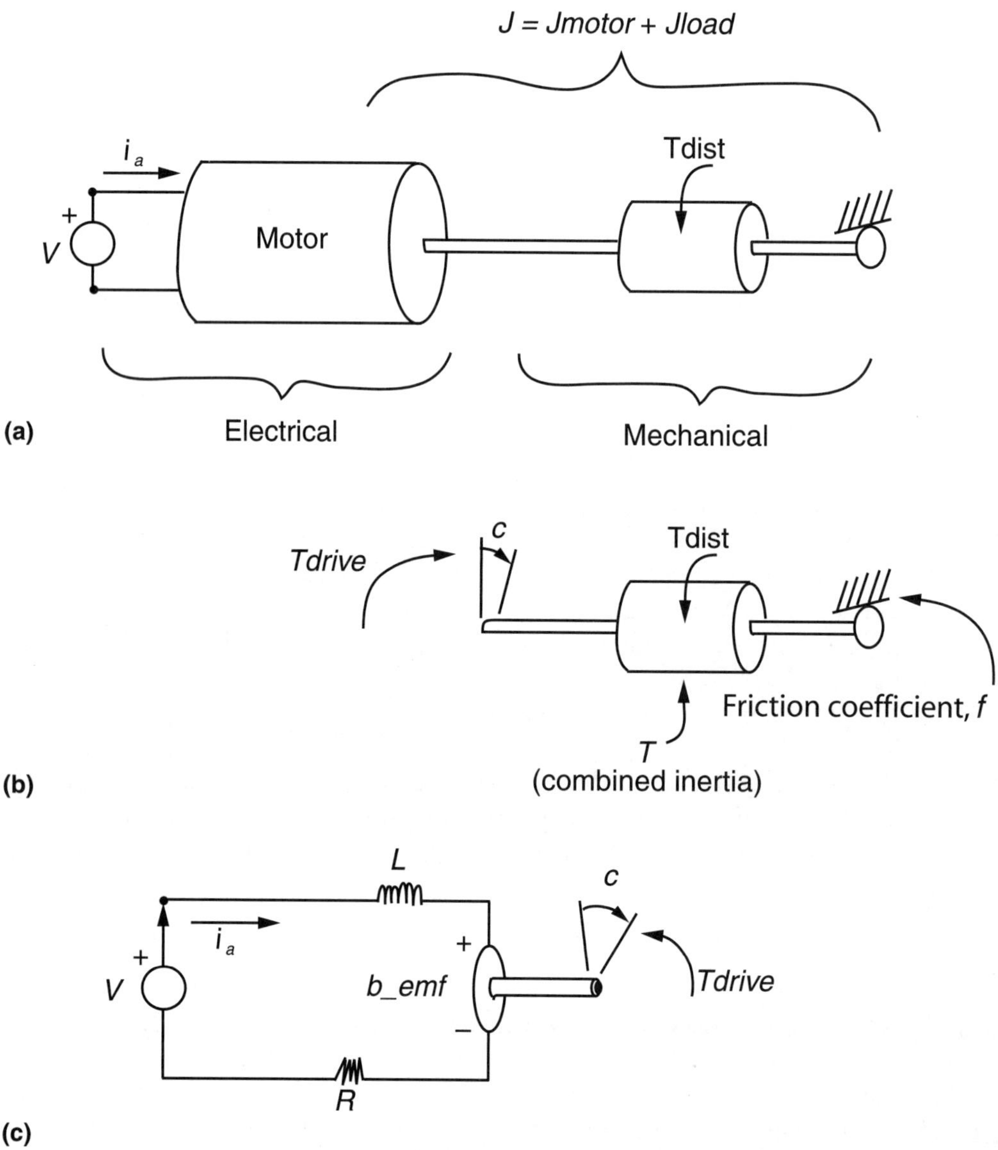

Figure 4–7 Electrical motor driving a mechanical load: (a) complete assembly, (b) mechanical subsystem, and (c) electrical motor equivalent

We can obtain the differential equation corresponding to the rotational mechanical system by applying Newton's Second Law:

$$\sum torques = J\ddot{C}$$
$$T_{drive} - T_{dist} - f\dot{C} = J\ddot{C}$$ **Eq. 4.13**

where:

J = moment of inertia of motor and load

f = viscous friction coefficient

$C, \frac{dC}{dt}(\text{or } \dot{C}), \frac{d^2C}{dt^2}(\text{or } \ddot{C})$ = motor shaft position, velocity, acceleration

T_{drive} = torque produced by the motor

T_{dist} = external disturbance torque being applied from outside the motor (i.e., a load imposed on the motor shaft)

The electromagnetic portion can be represented as shown in Figure 4–7c, an inductor (result of the winding of the armature), a resistor (from the ohmnic resistance of the wire), and a voltage source, *b_emf*(*t*), called back emf (electromotive force). This voltage is, under ideal conditions, proportional to the speed of the motor, or:

$$b\_emf = K_b\dot{C}$$ **Eq. 4.14**

Also under ideal conditions, the torque produced by the motor is strictly proportional to the current through the armature, or:

$$T_{drive=} = K_t\ i_a$$ **Eq. 4.15**

The two coefficients are:

K_t = motor torque constant

K_b = motor back emf constant

The derivation of these two equations includes an analysis of the electromagnetic phenomenon occurring inside the motor, a topic outside the scope of this book.

Applying KVL around the electrical circuit equivalent of Figure 4–7c, one obtains:

$$-V + L\frac{di_a}{dt} + Ri_a + b\_emf(t) = 0 \qquad \textbf{Eq. 4.16}$$

Substituting the corresponding expressions for T_{drive} and *b_emf* results in:

$$K_t i_a - T_{dist} = f\dot{C} + J\ddot{C} \qquad \textbf{Eq. 4.17}$$

and:

$$V = L\frac{di_a}{dt} + Ri_a + K_b\dot{C}$$

Because at this time we are primarily interested in the motor's speed, $\dot{C}$, rather than in position, C, we can use the variable ω for the speed of the motor:

$$\omega = \frac{dC}{dt} = \dot{C} \qquad \textbf{Eq. 4.18}$$

Redefining the two differential equations in terms of the newly defined variable ω:

$$K_t i_a - T_{dist} = f\omega + J\dot{\omega} \qquad \textbf{Eq. 4.19}$$

and:

$$V = L\frac{di_a}{dt} + Ri_a + K_b\omega$$

Now we take the Laplace transform of both equations, assuming that all the initial conditions are 0. (Note: We will forgo the conventional convention of changing the variable name from lowercase to uppercase as we go from the time domain to the *s*-domain.) The result is:

$$\begin{aligned} K_t i_a - T_{dist} &= f\omega + J\omega s \\ V &= Li_a s + Ri_a + K_b\omega \end{aligned} \qquad \textbf{Eq. 4.20}$$

Because i_a is not a visible variable, we will choose to eliminate it by solving the first equation for i_a and subsequently substituting it into the second equation. The result is:

$$V = \left\{\frac{LJs^2}{K_t} + \left(\frac{Lf + RJ}{K_t}\right)s + \left(\frac{Rf}{K_t} + K_b\right)\right\}\omega + \left\{\frac{L}{K_t}s + \frac{R}{K_t}\right\}T_{dist} \qquad \textbf{Eq. 4.21}$$

This expression can be represented as a transfer function for the independent variable ω or, more accurately, $\omega(s)$, in terms of the two inputs (independent variables) V and T_{dist}. Solving for the output (dependent) variable ω, we obtain:

$$\omega = \frac{1}{\left\{\frac{LJs^2}{K_t} + \left(\frac{Lf + RJ}{K_t}\right)s + \left(\frac{Rf}{K_t} + K_b\right)\right\}}V - \frac{\frac{L}{K_t}s + \frac{R}{K_t}}{\left\{\frac{LJs^2}{K_t} + \left(\frac{Lf + RJ}{K_t}\right)s + \left(\frac{Rf}{K_t} + K_b\right)\right\}}T_{dist} \quad \text{Eq. 4.22}$$

Multiplying the top and bottom of the right side by K_t:

$$\omega = \frac{K_t}{\left\{LJs^2 + (Lf + RJ)s + (Rf + K_bK_t)\right\}}V - \frac{LS + R}{\left\{LJs^2 + (Lf + RJ)s + (Rf + K_bK_t)\right\}}T_{dist} \quad \text{Eq. 4.23}$$

This can be represented in block diagram form as shown in Figure 4–8.

When obtaining a typical motor response, motor parameter values corresponding to a specific motor are used: Bodine N4022 (Bodine Electric Company, Chicago, Illinois). This manufacturer-provided data appears in Appendix F. The procedure for transforming the published parameters into appropriate units is left as an exercise. The result is:

$R = 0.16$ ohm

$L = 4.5\text{e-}4$ Henrys

$Kb = 0.086$ V/(rad/s)

$Kt = 0.0625$ lbf*ft/A

$J = 6.25\text{e-}4$ lbf*ft*s^2

Now assume that a 12 V excitation signal:

$$V(t) = 12u(t - 1) \quad \text{(volts)} \quad \text{Eq. 4.24}$$

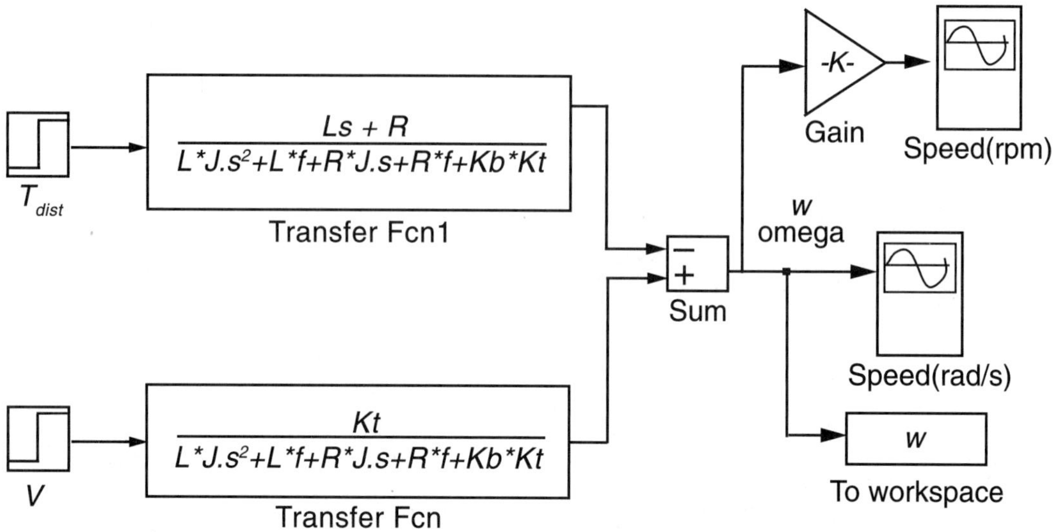

Figure 4–8 Open-loop transfer function block diagram of DC motor

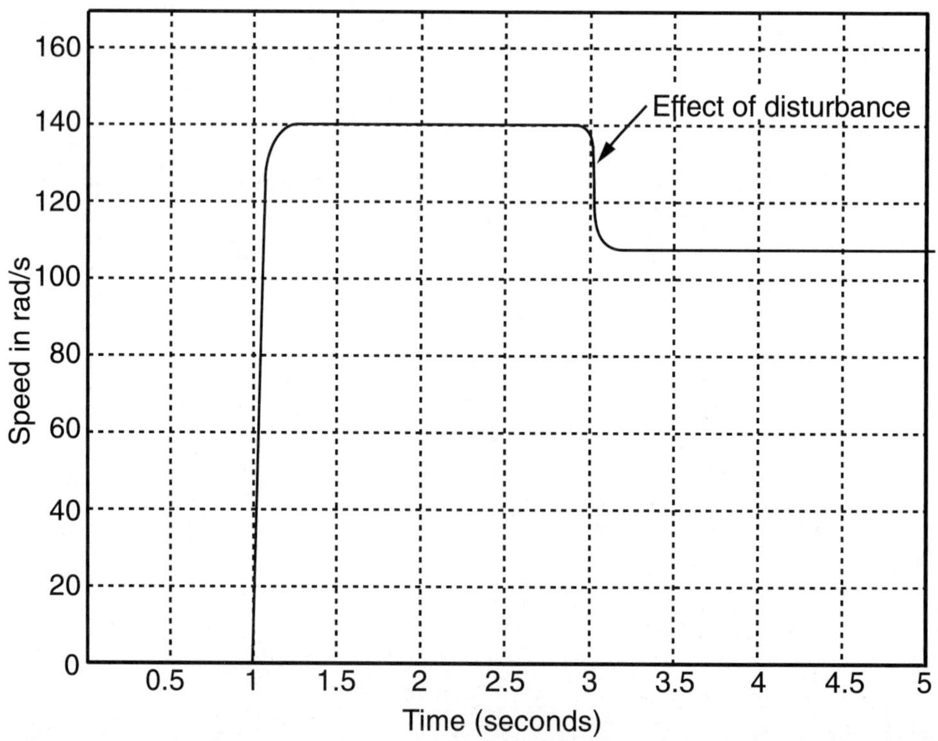

Figure 4–9 DC motor step input response and step disturbance response (open loop)

is applied to the motor and that once it achieves its final speed, a constant resisting torque is imposed on the motor:

$$T_{dist}(t) = 1.0\ u(t-3) \quad (\text{lbf} * \text{ft}) \qquad \textbf{Eq. 4.25}$$

SIMULINK is used to obtain the time response, as shown in Figure 4–9. Notice that the resulting steady-state speed is close to 140 rad/s and that once the disturbance torque, T_{dist}, is applied, it causes a drop in speed of 21 percent.

Now let us convert the open-loop system we just investigated into a closed-loop system and explore the potential benefits. All we must do is to feedback the output back into the input portion of the control system, as shown in Figure 4–10. We assume that we have a speed sensor that produces 1 volt for every rad/s produced by the motor. This conversion allows us to use the unity feedback type of representation, because the feedback block has a gain of 1. Our command signal is named *Vw,* a voltage that indicates the desired speed. The other block corresponds to *Kp,* the controller's gain. This block amplifies the error signal and transforms it

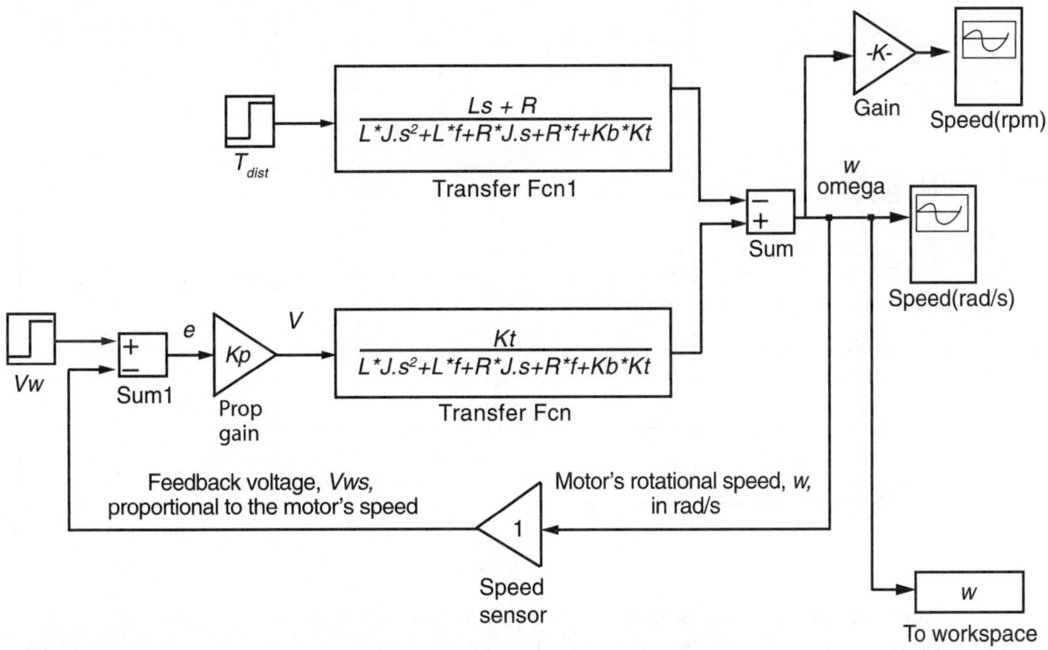

Figure 4–10 Closed-loop transfer function block diagram of DC motor

into an excitation voltage, *V*. For our simulation, we have chosen $Kp = 1$ *V*/(rad/s). Do not be tricked into thinking that this gain offers no amplification.

In this type of feedback scheme, formally known as proportional feedback, there will be a small error, *e*. *Vw* and *Vws* will be close, but not the same. This small error is multiplied by *Kp* to produce the driving voltage. If there was no error, *Vw* and *Vws* would be equal, and the driving voltage would be 0. We have selected $Vw = 152$V so that we obtain the steady-state speed as the open-loop system, namely 140 rad/s. This closed-loop system also experiences the same disturbances as the open-loop system. The resulting response appears in Figure 4–11. Compare this system behavior with that of the open-loop system. Notice that the torque disturbance only caused a 2 percent drop in speed, a significant improvement over the open-loop system (21 percent). However, notice that the system is more jerky (unstable), as observed in the initial overshoot of the speed.

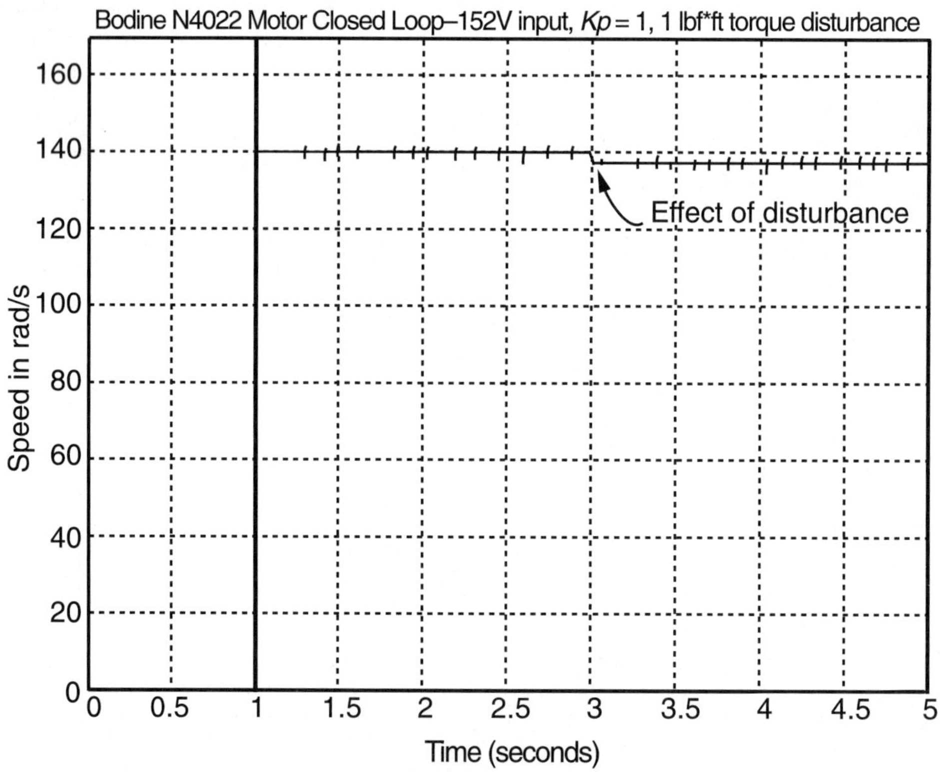

Figure 4–11 DC motor step input response and step disturbance response (closed loop)

4.5 Steady-State Error Analysis

You have seen that the closed-loop system is much better at disturbance suppression than its open-loop counterpart. Steady-state error analysis is an essential consideration in the design of a control system.

Engineers must also analyze the system for accuracy, in particular for steady-state accuracy. We will use the work done with the DC motor, but we will focus on the motor position instead of speed. We will start with the open-loop transfer function. If we wanted to obtain motor position, all we would need to do is integrate motor speed, as shown in Figure 4–12a. Because we are no longer interested in the motor's speed, we can collapse the integrator into the transfer function, as was done in Figure 4–12b.

Now we create a DC motor, position control, closed-loop system as shown in Figure 4–13. This system has an OLTF that equals $G(s)$, because the feedback $H(s) = 1$. Thus:

$$G = \frac{C}{E} = \frac{K_p K_t}{s[LJs^2 + (Lf + RJ)s + (Rf + K_b K_t)]} \quad \textbf{Eq. 4.26}$$

This OLTF is seen to have a factor of s in the denominator (i.e., a pole at the origin of the s-plane). Thus, the system is type 1. Many dynamic systems have open-loop poles at the origin of the s-plane. Typically, regulators (for temperature,

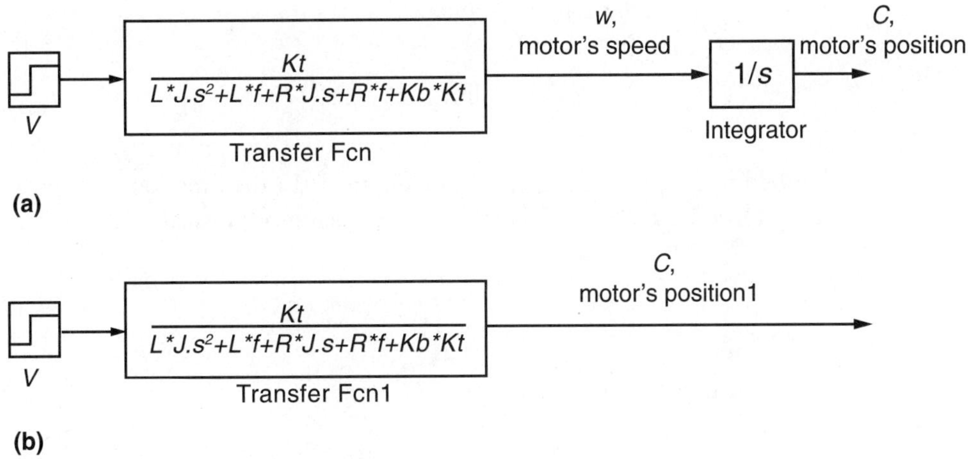

Figure 4–12 DC motor in terms of motor position: (a) external transformation and (b) incorporated into transfer function

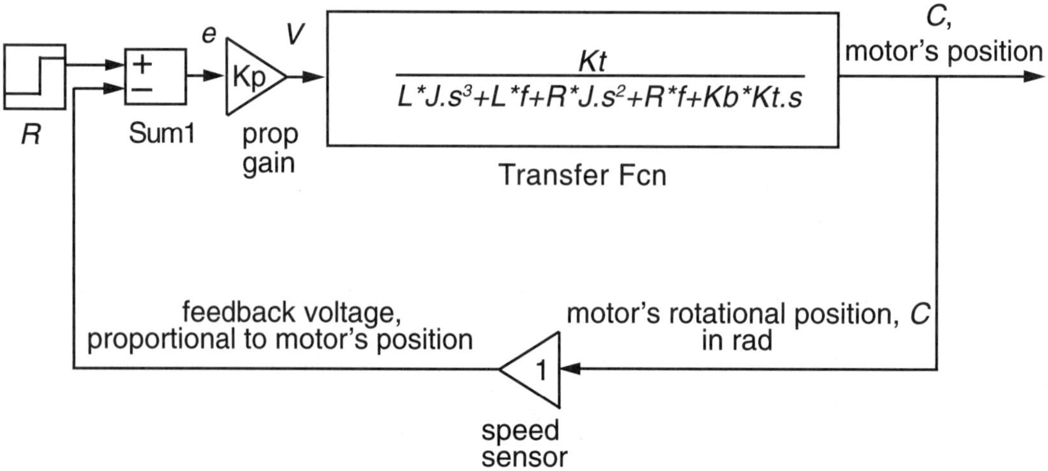

Figure 4–13 DC motor position control system

speed, etc.) have no poles at the origin. Position controls typically have one pole at the origin; tracking systems may have two or more transfer function poles at the origin.

The system type number is especially important, because such poles at the origin define steady-state accuracy (as will be shown) and thus permit us to determine system steady-state accuracy regardless of the system dynamics. The final value theorem of the Laplace transform states that:

$$\lim_{t \to \infty} f(t) = \lim_{s \to 0} s\,F(s) \qquad \textbf{Eq. 4.27}$$

Applied to the positioning system of Figure 4–13 the function of interest is $e(t)$, the error. Thus, Equation 4.27 is reformulated using $e(t)$ instead of $f(t)$:

$$\lim_{t \to \infty} e(t) = \lim_{s \to 0} s\,F(s) \qquad \textbf{Eq. 4.28}$$

The Laplace transform of the error, $E(s)$, is:

$$E(s) = \frac{R(s)}{1 + G(s)} \qquad \textbf{Eq. 4.29}$$

Then:

$$e_{ss} = \lim_{t\to\infty} e(t) = \lim_{s\to 0} s\ E(s) = \lim_{s\to 0} s\ \frac{R(s)}{1 + G(s)} \qquad \textbf{Eq. 4.30}$$

$$= \lim_{s\to 0} s \left[\frac{R(s)}{1 + \dfrac{K_p K_t}{s[LJs^2 + (Lf + RJ)s + (Rf + K_b K_t)]}} \right]$$

To simplify the expression for G, we let $s \to 0$ in G to eliminate the two s-terms of the polynomial. However, we keep the s-term corresponding to the pole at the origin. Thus:

$$E(s) = \frac{R(s)}{1 + \dfrac{K_p K_t}{s[(Rf + K_b K_t)]}} = \frac{R(s)}{1 + \dfrac{K_B}{s}} \qquad \textbf{Eq. 4.31}$$

where:

$$K_B = \frac{K_p K_t}{(Rf + K_b K_t)}$$

Notice that this simplified representation of $E(s)$ is only useful for determining the final value of the error, e_{ss}, but not the characteristics of the dynamic response.

Referring to the previous equation, you will be able to recognize K_B as the Bode gain. Then:

$$e_{ss} = \lim_{t\to\infty} e(f) = \lim_{s\to 0} s\ E(s) = \lim_{s\to 0} s \left[\frac{R(s)}{1 + \dfrac{K_B}{s}} \right] = \lim_{s\to 0} \left[\frac{s^2 R(s)}{s + K_B} \right] \qquad \textbf{Eq. 4.32}$$

To illustrate the process, we will evaluate the steady-state error for a specific type of input: a unity step change in position at $t = 0$. Let $R(s)$ be the Laplace transform of the input:

$$R(s) = \frac{1}{s} \qquad \textbf{Eq. 4.33}$$

Then:

$$e_{ss} = \lim_{t\to\infty} e(t) = \lim_{s\to 0} \frac{s}{s + K_B} = 0 \quad \textbf{Eq. 4.34}$$

That is, there is no steady-state error when the DC motor, closed-loop position control sysem is commanded a step input.

Now, let's try to generalize the results to any unity-feedback system. Because the Final Value Theorem points out that:

$$e_{ss} = \lim_{t\to\infty} e(t) = \lim_{s\to 0} s\left(\frac{R(s)}{1 + G(s)}\right) \quad \textbf{Eq. 4.35}$$

and by incorporating that the limit of $G(s)$ as $s \to 0$:

$$\lim_{s\to 0} (G)s = \lim_{s\to 0} \frac{K_B}{s^N} \quad \textbf{Eq. 4.36}$$

Then

$$e_{ss} = \lim_{s\to 0} s\,E(s) = \lim_{s\to 0} \frac{sR(s)}{1 + \dfrac{K_B}{s^N}} \quad \textbf{Eq. 4.37}$$

All that is needed to evaluate steady-state accuracy are specific values for $R(s)$, K_B, and N, the system type. The implications of the previous formula for different system types and different inputs are presented in Table 4–1. Note that stability and dynamic response issues are not considered in this type of analysis. The system must be stable if it is to achieve steady state, but if it is not stable, it may be stabilized without changing the accuracy.

Example 4.1

Determine the steady-state error for the DC motor, position control system for a unity step input. Use the previously derived equation.

↓ System Type	Input Type →	Step $A\,u(t)$	Ramp $A\,t\,u(t)$	Parabola $A\,t^2/2\,u(t)$
type 0		$\frac{A}{1+K_B}$*	∞	∞
type 1		0	$\frac{A}{K_B}$*	∞
type 2		0	0	$\frac{A}{K_B}$*

*K_B is Bode gain. This process assumes that the system is stable.

Table 4–1 Steady-state error, e_{ss}, for various system types and inputs (applies only to unity feedback systems)

Solution This simple procedure for evaluating steady-state accuracy may be applied for various input commands and for any type system provided that they are unity feedback.

For our positioning system, letting $s \to 0$ in $G(s)$ [Equation 4.25] except for the pole at the origin yields:

$$\frac{K_p K_t}{s[(Rf + K_b K_t)]} = \frac{K_B}{s} \qquad \text{Eq. 4.38}$$

Then, for a step input $R(s) = 1/s$:

$$e_{ss} = \lim_{s\to 0} s \left[\frac{\frac{1}{s}}{1 + \frac{K_B}{s}} \right] = \lim_{s\to 0} s \cdot \frac{1}{s + K_B} = 0 \qquad \text{Eq. 4.39}$$

which is the same result previously obtained.

For nonunity feedback systems, a slightly different procedure will be presented later in the chapter.

Example 4.2

Let the command to DC motor, closed-loop, position control system be a constant velocity, (i.e., a ramp of position) for which:

$$R(s) = \frac{A}{s^2} \quad \textbf{Eq. 4.40}$$

Solution With this input to the positioning system, the steady-state error is:

$$e_{ss} = \lim_{s \to 0} s \left(\frac{\frac{A}{s^2}}{1 + \frac{K_B}{s}} \right) = \frac{A}{K_B} \quad \textbf{Eq. 4.41}$$

At steady state there is a constant error $\frac{A}{K_B}$. This means that the output velocity is exactly the commanded velocity but that there is a displacement between reference positions on the input and output. This approach to evaluating steady-state accuracy permits a simple explanation based on the physical nature of the hardware. The positioning system provides an error signal when a command, R, is applied. The amplified error signal drives the rotor, rotating the output shaft, C, which reduces the error. When the error is reduced to 0, the motor stops.

Let us recap the result of the previous examples. The operation for a step change in position is:

1. The input, R, provides an error in position, which is the difference between R and C.
2. The motor rotates to reduce this difference.
3. There is a transient, but if the system is stable, the motor eventually stops at $E = 0$, and the desired position has been obtained.

When the command is a ramp of position (i.e., we want the output position to change at a specified rate), then:

1. The input, R, starts at 0 and changes at the desired rate.
2. The displacement of R provides an error signal that drives the motor in a direction to reduce this error.
3. After a transient, the output velocity duplicates the command rate.

4. Because the motor runs only when a voltage is applied, it follows that the error in position is not 0, but exactly the amount required to run the motor at the commanded rate. For the simple positioning system, this is $e_{ss} = \dfrac{A}{K_B}$.

The evaluation of steady-state error is relatively simple but sometimes tedious. It is easy to summarize the results in table form, as shown in Table 4–1:

4.6 Procedure for Determining the Steady-State Error in a Unity Feedback Control

Step 1

Put the open-loop transfer function in Bode form. In a unity feedback system, the OLTF is simply the transfer function, $G(s)$, or:

$$G(s) = \frac{K_B \prod_{i=1}^{M} (\tau_i s + 1) \prod_{j=1}^{Q} (a_j s^2 + b_j s + 1)}{s^N \prod_{k=1}^{T} (\tau_k s + 1) \prod_{l=1}^{U} (a_l s^2 + b_l s + 1)} \qquad \text{Eq. 4.42}$$

Step 2

Extract the Bode gain, K_B, and the system type, N, from the OLTF.

Step 3

Use Table 4–1 to determine e_{ss}. Use the table entry corresponding to the system type and the input used (e.g., step, ramp, parabola).

Example 4.3

Consider a unity feedback system whose transfer function is:

$$G = \frac{96}{(s+2)(s+3)} \qquad \text{Eq. 4.43}$$

The system is excited with a step function with a weight of 3. Determine the steady-state error and the final value of the output.

Solution The first step is to put the transfer function in Bode form:

$$G = \frac{96/(2 \cdot 3)}{(s/2+1)(s/3+1)} = \frac{16}{(s/2+1)(s/3+1)} \qquad \textbf{Eq. 4.44}$$

It is clear that the Bode gain, K_B, is 16 and that the system is type 0. Next we use Table 4–1. The steady-state error is given by:

$$e_{ss} = \frac{A}{1+K_B} = \frac{3}{1+16} = 0.18 \qquad \textbf{Eq. 4.45}$$

and the steady-state output is:

$$c_{ss} = r_{ss} - e_{ss} = 3 - 0.18 = 2.82 \qquad \textbf{Eq. 4.46}$$

Implement this block diagram in SIMULINK, and confirm that these results are true.

4.7 Error Due to Disturbances

Most control systems experience external disturbances that increase the steady-state error. Consider the unity feedback control system in Figure 4–14a, which shows both the input $c(t)$, and the disturbance, $d(t)$. Let us assume that the disturbance manifests itself in the form of a bias (i.e., a step). In obtaining the error due to the disturbance, assume that the input, $r(t)$, is 0. This will allow us to calculate the error due to the disturbance only. We can subsequently calculate the inherent error using the procedure previously presented. The total error will be the sum of the two. Also assume that the system is type 0. The first step is to redraw the block diagram, as shown in Figure 4–14b. Notice that because the input, $r(t)$ is 0, the error, $e(t)$, is simply the negative of the output, $c(t)$. To simplify the block diagram, we drop the input, $r(t)$ because it is 0, and move the inversion to the other summing junction, as shown in Figure 4–14c. Now, because this block diagram is not a unity feedback system, we cannot directly use the previous formulas unless we make some adjustments. Specifically, we would like to move the block G1 out of the feedback loop so

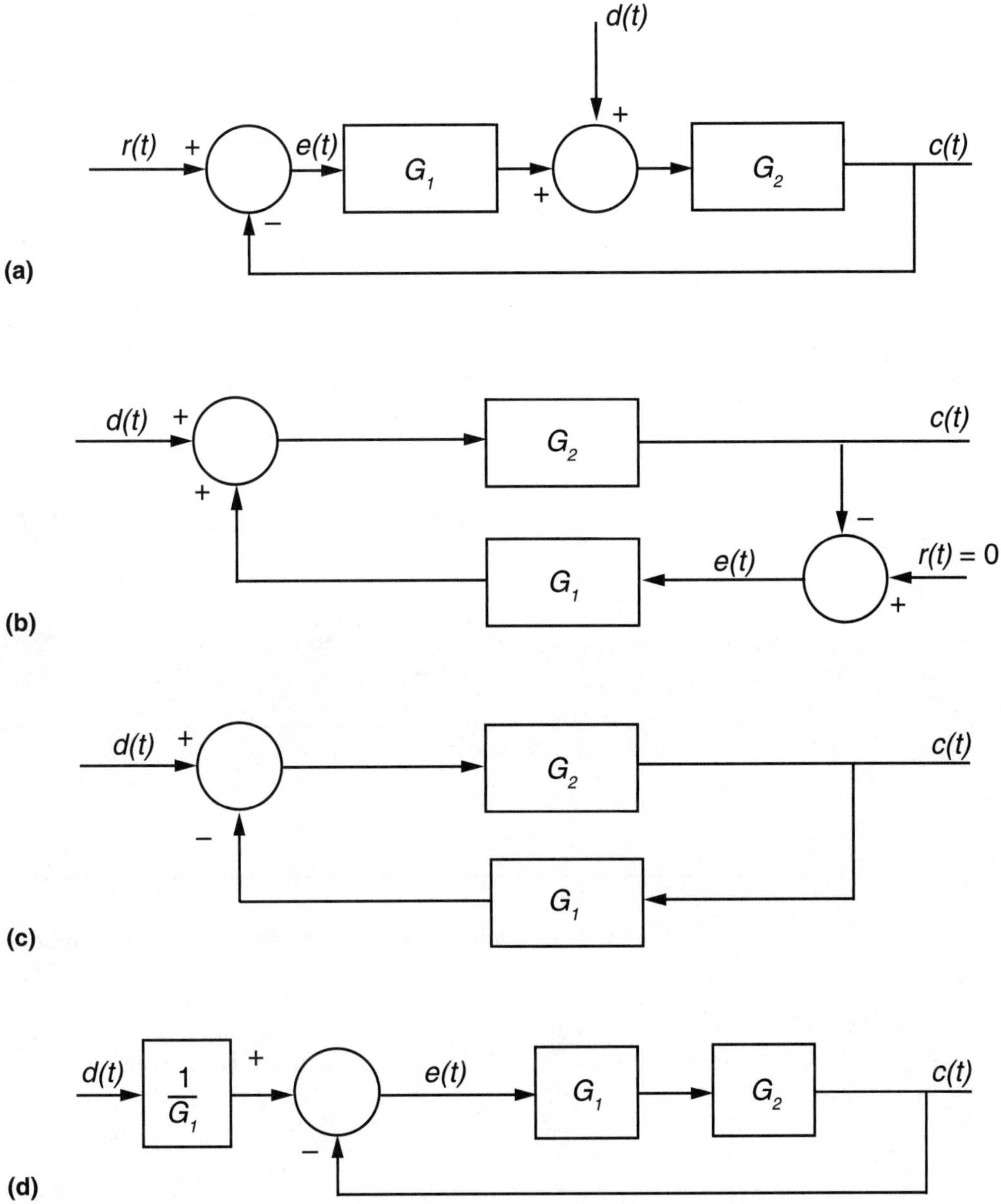

Figure 4–14 Block diagram used to obtain error due to disturbance

that we would have a unity feedback system. This has been done in Figure 4–14d. This final block diagram will be used as a basis for arriving at an expression to determine the error due to the disturbance. To facilitate the development, let us define a new variable, the intermediate error, e_i, which is given by:

$$e_i = \frac{\frac{A}{K_{B1}}}{1 + K_{B1}K_{B2}} \qquad \textbf{Eq. 4.47}$$

where A is the amplitude of the disturbance and K_{B1} and K_{B2} are the Bode gains of $G1(s)$ and $G2(s)$. Now, using Figure 4–14d as a reference, and remembering that the error is the negative of the output of the error due to the disturbance, $e_{ss-dist}$, is given by:

$$e_{ss-dist} = -e_i K_{B1}K_{B2} \qquad \textbf{Eq. 4.48}$$

or:

$$e_{ss-dist} = 1\frac{\frac{A}{K_{B1}}}{1 + K_{B1}K_{B2}}K_{B1}K_{B2} = -\frac{AK_{B2}}{1 + K_{B1}K_{B2}}$$

Example 4.4

Consider the system in Figure 4–15. Determine the error due to the disturbance.

Solution This system is similar to the previous example, but G has been separated into two transfer functions, $G1$ and $G2$. Using the formula previously derived, the error due to the disturbance is:

$$e_{ss-dist} = -\frac{AK_{B2}}{1+K_{B1}K_{B2}} = -\frac{2 \cdot 2}{1 + 8 \cdot 2} = -\frac{4}{17} = -0.235 \qquad \textbf{Eq. 4.49}$$

Again, simulate the system to confirm the results. Make sure that $r(t) = 0$ so that only the effect of the disturbance is obtained.

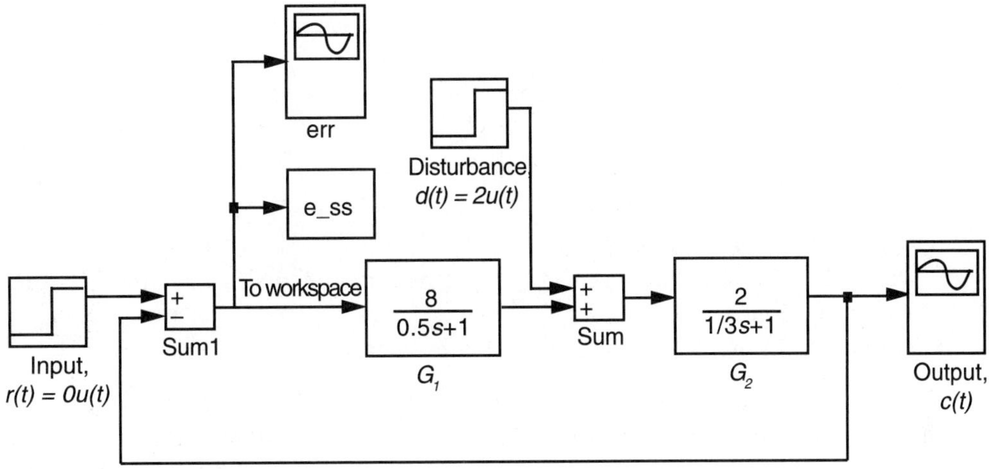

Figure 4–15 Sample problem: system with disturbance input

4.8 Total Error

Because the system is linear, we can use the Superposition Theorem. The total error is the sum of both errors: the inherent error in the system plus the error resulting from the disturbance. To arrive at the total error, first consider the error due to input by itself by following the procedure in the first example. Then, make the input equal 0, and calculate the error caused by the disturbance, as shown in the previous section.

4.9 Steady-State Error in Nonunity Feedback Systems

Often, the system in question is not a unity feedback system. This occurs because it might be problematic to perform the comparison operation (i.e., at summing junction) with the output's physical variable. Consequently, the physical output variable is typically converted to a variable that can easily be processed (e.g., a voltage, a current, a digital number). In such cases, manipulate the block diagram into unity feedback form as was done for the disturbance input development.

4.10 Nonlinearities

Nonlinearities exist in many kinds of dynamic components. Often, they can be ignored. Other times, the nonlinear system can be linearized (approximated with a linear system about its operating point).

Examples of existing, nondesired nonlinearities are:

- Saturation (as in amplifiers)
- Dead zone (for small disturbances; e.g., small input signals to an amplifier produce no response results)
- Coulomb friction and sticking of mechanical parts (e.g., sticking valves)

Perhaps the best way to analyze the impact of these nonlinearities is to incorporate them into the model in the SIMULINK environment and simulate the system. Even though the transfer function representation is only valid for linear components, SIMULINK allows the incorporation of nonlinear blocks. This is possible because SIMULINK uses numerical integration to propagate the solution of nonlinear components.

One can use nonlinearities to modify or improve system performance and/or counteract the effects of other nonlinearities. Some of these nonlinearities include:

- Switches
- Relays
- Square law or logarithmic electronic devices

Summary

1. Transformation of an open-loop system to a closed-loop system improves performance but also tends to make the system more unstable.
2. The order of a system indicates the number of energy-storing components and integrators in that system.
3. The system type indicates the number of integrators, or integrator-acting components, in a system.
4. The number of integrators in a controller determines the ability of the controller to follow a certain type of input (e.g., step, ramp, or parabola).
5. Steady-state error, error remaining after all the transients have died out, can be easily predicted using the tools of error analysis without having to solve the differential equation.
6. Eigenvalues (complex numbers) are the mathematical essence of a dynamic system. For a system to be stable, all its eigenvalues should have negative real parts.

Idea Portfolio

1. Does the system use negative feedback control? Can the system be tested open loop? How would you close the loop (i.e., subtract output for input)? What would happen if the comparison operation were inadvertently reversed? Would the system be damaged if the loop were opened?

Problems

1. For the systems in Figure 4.16 determine:
   - Order
   - System type
   - OLTF

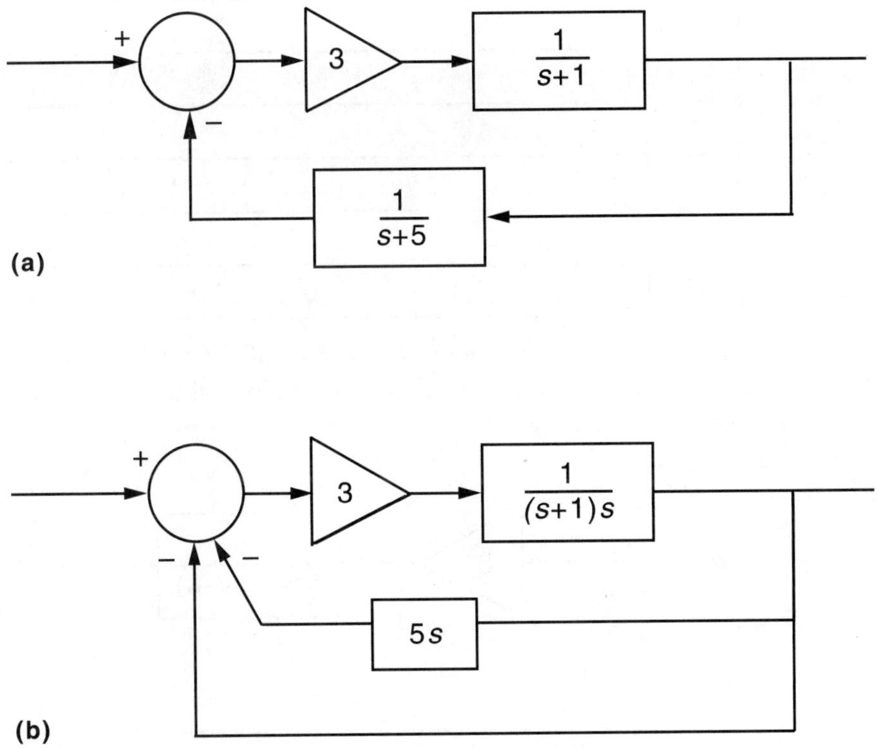

Figure 4–16

- OLTF zeros
- OLTF poles
- CLTF
- CLTF zeros
- CLTF poles
- Whether the system is stable

2. For the system in Figure 4–17, obtain:
   - Open-loop transfer function
   - OLTF zeros
   - OLTF poles
   - CLTF
   - Characteristic equation (characteristic polynomial). (Note: It will be a function of *K*.)
   - Solve the characteristic equation for the different values of *K* following. Complete the following table with the resulting roots:

K	Roots of Characteristic Equation (Eigenvalues)
1	
3	
4	
7	
10	
15	

3. Graph the migrating behavior of the roots of the CLTF from the previous problem using the results from the preceding table as *K* increases in the *s*-plane.

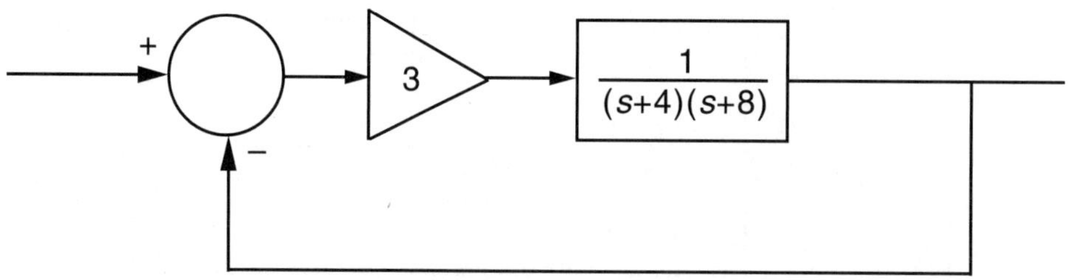

Figure 4–17

4. Express the following function in Bode form:

$$G_p = \frac{400}{s(2s+8)(s+10)}$$

5. A controller in Figure 4–18 is required to produce a steady-state output, $c(t)$, that is within 2 percent of the commanded input, $r(t)$. As the control engineer, you can introduce an amplifier with gain Kc. Determine the minimum value of Kc that would meet the requirement. Confirm your design using SIMULINK.
6. Determine the steady-state error for a unit step in a unity feedback system whose plant has the transfer function:

$$G_p = \frac{4}{(s+1)(0.1s+1)^2}$$

7. Assume that you are designing a system that must follow, for most of the trajectory without significant error, a saw-tooth waveform input. What should be the system's minimum type (i.e., number of integrators)?
8. Using MATLAB/SIMULINK, simulate the system, whose transfer function is given, to obtain its response to a unity step input. Determine from the response the effective time constant.

$$G = \frac{40}{(.2s+1)(5s+1)}$$

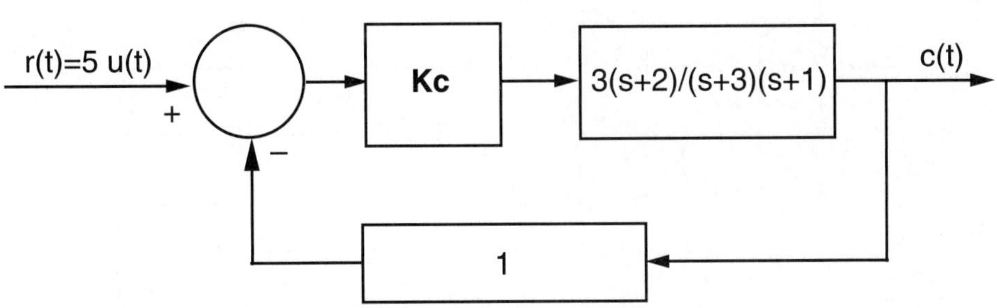

Figure 4–18

9. When the following unity feedback system is excited with a step of weight 5, what is the steady-state output for the system after the initial transients die out?

$$G = \frac{40}{(.2s+1)(5s+1)}$$

10. The transfer function relating input force and displacement of a mechanical system is:

$$\frac{X(s)}{F(s)} = \frac{2}{s^2+4s+9}$$

Determine the transfer function relating input force and velocity for the same mechanical system described by the above transfer function:

$$\frac{V(s)}{F(s)} = ?$$

11. Provide an example in which performing a test on a system, in its open-loop configuration, is not feasible.
12. The position control system in Figure 4–19 has a position sensor that converts position to voltage. The system also uses a prefilter with the same sensor gain. The sensor's sensitivity (i.e., *Ks*) is 50 mV/cm. If a command of 10 cm is given, determine the steady-state error. Modify its transfer function representation so that the steady-state error of Table 4–1 can be used. Afterward, simulate the system using SIMULINK to verify your answer.

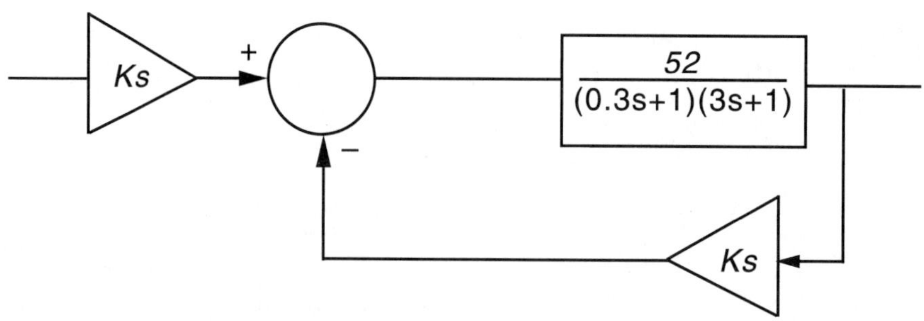

Figure 4–19

13. Explain back emf and its relationship to motor variables.
14. When a motor stalls, explain what happens to the current going through it and possible consequences on the motor.
15. Return to the racquetball problem of Chapter 2. Realize that the force of friction (drag) is not proportional to velocity but to velocity. In other words, it is a nonlinear relationship. Determine a coefficient, k_{DRAG} that produces the same terminal velocity as the linear coefficient. Formulate two models side by side in SIMULINK: (1) one for the linear case and (2) one for the nonlinear case. Simulate the two systems, and compare the resulting velocities.
16. Enter into SIMULINK the model in Figure 4–10, and resimulate. Do you obtain the same plot as shown in Figure 4–11?
17. Reformulate the block diagram in Figure 4–10 so that it can be used to estimate the steady-state operating speed with 1 lbf-ft torque disturbance.
18. What are some possible nonlinearities in a DC motor?
19. For the unity feedback system whose transfer function is shown below, determine the steady-state error for a:
    - Step input, $r(t) = u(t)$
    - Ramp input, $r(t) = t\,u(t)$

$$G = \frac{250}{s(s+2)(s+5)}$$

5

Stability and Dynamic Response

Learning Objectives

In this chapter, you will learn how to:

- Differentiate between the output characteristics of stable and unstable systems.
- Explain the implications of the location of the roots of the CLTF.
- Perform Routh analysis on the coefficients of the closed-loop transfer to determine stability.
- Use MATLAB to determine eigenvalues.

5.1 Introduction

As stated in Chapter 1, a dynamic system is one whose output varies with time when the system is disturbed by an input signal or a load disturbance. Because the dynamic system is used for some purpose, the nature and duration of the output response must be compatible with this purpose. The system features of importance may be classified as:

- Stability
- Nature of the dynamic response

A system is said to be stable if, after an input or a disturbance, the output eventually reaches a finite steady-state value. The actual steady-state value is not considered, nor is the time required to reach the steady-state value. The system is said to be unstable if the output oscillates with ever-increasing amplitude or if the output increases (decreases) unidirectionally and without limit.

After an input or a disturbance, the output undergoes a time variation, called a "transient response." This is true whether the system is stable or unstable, and this response must be acceptable (for the purpose of the system). When this dynamic response is not acceptable, further analysis and redesign are required. Thus, we need certain tools to analyze the dynamic system and still other tools to determine stability and acceptable dynamic response.

5.2 Stability Analysis

Because an unstable system is usually useless, a test for stability is needed, and we would like the analysis to be useful in designing for stability. A first step is to use the Laplace transform, converting the differential equation model to a more convenient, algebraic model. Appropriate manipulation can then provide the characteristic equation of the system. Once the characteristic equation is obtained, the following stability analyses are available:

- Use the computer to factor the characteristic polynomial for its roots, and use the root locations to predict stability. When any root is in the right half of the s-plane, the system is unstable. This method does not indicate the degree of instability, nor how to make the system stable.
- Use the characteristic polynomial to obtain the Routh array (presented in this chapter), and use the signs from this array to predict stability.
- Have MATLAB convert a large block diagram embedded in SIMULINK to a State Space representation (the topic of Chapter 10). Once it is in this format, the system can be transformed to transfer function format. In either format, the eigenvalues can be determined to ascertain stability.

5.3 The Routh Criterion

The stability of a system is determined by the roots of the characteristic polynomial of that system. When any root is in the right-half plane (is positive or has a positive real part), the system is unstable. A test for the existence of roots in the right-half plane was developed by E. J. Routh and is presented here without proof using examples to demonstrate the manipulations.

To create the Routh array, we need the characteristic polynomial:

$$A_7s^7 + A_6s^6 + A_5s^5 + A_4s^4 + A_3s^3 + A_2s^2 + A_1s + A_0 \quad \textbf{Eq. 5.1}$$

Rearrange the coefficients into two rows starting with the highest-order coefficient. Use odd-order coefficients in one row and even-order coefficients in the other:

$A_7 \quad A_5 \quad A_3 \quad A_1$

$A_6 \quad A_4 \quad A_2 \quad A_0$

The third row of the Routh array is formed using the first two rows and the algorithm:

$$B_1 = \frac{A_6A_5 - A_7A_4}{A_6}$$

$$B_2 = \frac{A_6A_3 - A_7A_2}{A_6} \quad \textbf{Eq. 5.2}$$

$$B_3 = \frac{A_6A_1 - A_7A_0}{A_6A}$$

Additional rows are formed using the two preceding rows until an all-0 row results.

Example 5.1

Check the stability using the Routh criterion for a simple, second-order positioning system, for which a block diagram model is given in Figure 5–1. Perform the analysis for two cases: (1) negative feedback and (2) positive feedback.

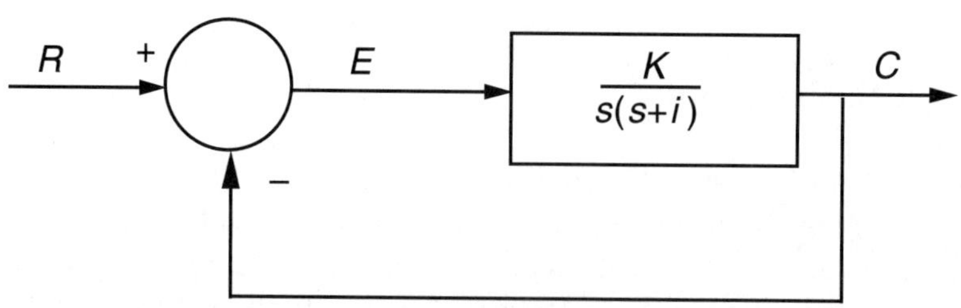

Figure 5–1 Block diagram for a simple positioning system

Solution—Negative Feedback For the system with negative feedback, the characteristic equation is:

$$CLTF = \frac{C(s)}{R(s)} = \frac{K}{(s^2 + 3s + K)} \qquad \textbf{Eq. 5.3}$$

The characteristic polynomial is:

$$s^2 + 3s + K \qquad \textbf{Eq. 5.4}$$

The Routh array is formed:

1	K
3	0
K	0

For $K > 0$, the Routh array indicates no changes in sign in the left-hand column, so we conclude that the system is always stable.

Solution—Positive Feedback For the system with positive feedback, the characteristic polynomial is:

$$s^2 + 3s - K \qquad \textbf{Eq. 5.5}$$

The Routh array is:

1	$-K$
3	0
$-K$	0

There is one change in sign in the left-hand column, so the system is unstable, with one root in the right half of the s-plane.

Example 5.2

Check the stability of the previous example by finding the roots of the characteristic equation. Perform the analysis for negative and positive feedback.

Solution—Negative Feedback Because the characteristic polynomial is of second order, we can use the quadratic formula:

$$r_1, r_2 = \frac{-1 \pm \sqrt{1-4K}}{2} = -\frac{1}{2} \pm \sqrt{\frac{1-4K}{4}} \qquad \textbf{Eq. 5.6}$$

Thus, the roots of the quadratic are complex conjugates, with real part at $-1/2$ (for $K > 1/4$) or are negative real (for $0 < K < 1/4$). In either case, there are no roots in the right half of the s-plane, so the system is stable.

Solution—Positive Feedback When the feedback is positive, the characteristic polynomial is:

$$s^2 + s - K \qquad \textbf{Eq. 5.7}$$

Applying the quadratic formula:

$$r_1, r_2 = \frac{-1 \pm \sqrt{1+4K}}{2} = \frac{-1}{2} \pm \sqrt{\frac{1+4K}{4}} \qquad \textbf{Eq. 5.8}$$

for which one root is in the right half plane.

Example 5.3

Use the Routh array to determine stability given the CLTF denominator polynomial:

$$s^5 + 5s^4 + 12s^3 + 15s^2 + 50s + 200 \qquad \textbf{Eq. 5.9}$$

Solution The Routh array is:

$$\begin{array}{ccc} 1 & 12 & 50 \\ 5 & 15 & 200 \\ \frac{5 \cdot 12 - 1 \cdot 15}{5} = 9 & \frac{5 \cdot 50 - 1 \cdot 200}{5} = 10 & \frac{5 \cdot 0 - 1 \cdot 0}{5} = 0 \end{array} \qquad \textbf{Eq. 5.10}$$

$\frac{9 \cdot 15 - 5 \cdot 10}{9} = 9.44$	$\frac{9 \cdot 200 - 5 \cdot 0}{9} = 200$	0
$\frac{9.44 \cdot 10 - 9 \cdot 200}{9.44} = -180.68$	0	0
$\frac{-180.68 \cdot 200 - 9.44 \cdot 0}{-180.68} = 200$	0	0

There are two changes in sign in the left-hand column, so there are two roots in the right-half plane, and the system is unstable. When we use the MATLAB root function:

```
>> roots([1 5 12 15 50 200])

Answer =
```

-3.1411

$-2.1760 + 2.5916i$

$-2.1760 - 2.5916i$

$1.2466 + 2.0015i$

$1.2466 - 2.0015i$

confirming that the system is unstable.

For certain types of stability problems, the Routh array is useful. When the gain of a dynamic system is adjustable, the limits on gain value can be established.

Example 5.4

Consider the third-order positioning system in Figure 5–2. Determine the range of gain, *K*, that makes the system stable.

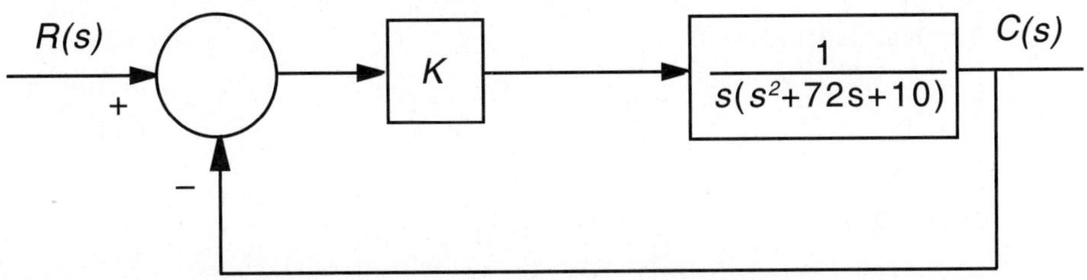

Figure 5–2 Position feedback control system

Solution The characteristic polynomial for this system is:

$$s^3 + 72s^2 + 10s + K \qquad \text{Eq. 5.11}$$

The Routh array is:

1	10
72	K
$(720 - K)/72$	0
K	0

From the left-hand column, K must be positive for the system to be stable, but it must be less than 720. Thus, with the bounds established ($0 < K < 720$), one may proceed to determine whether any value of K within these limits will provide an acceptable dynamic response.

The Routh array may also be applied to stability analysis when several system parameters are adjustable.

Example 5.5

Some systems may have two or more adjustable coefficients, so the stability limit is not unique for such cases but requires an appropriate value for each coefficient. A typical case is a control system for which the forward gain is adjustable to provide fast response but for which a velocity feedback path is provided (with adjustable gain, k) to adjust damping of the dynamic response. A block diagram for such a system is shown in Figure 5–3.

Solution The characteristic polynomial for such a system is:

$$s^3 + 15s^2 + (20 + k)s + K \qquad \text{Eq. 5.12}$$

The Routh array is:

1	$(20 + k)$
15	K
$(300 + 15k - K)/15$	0
K	0

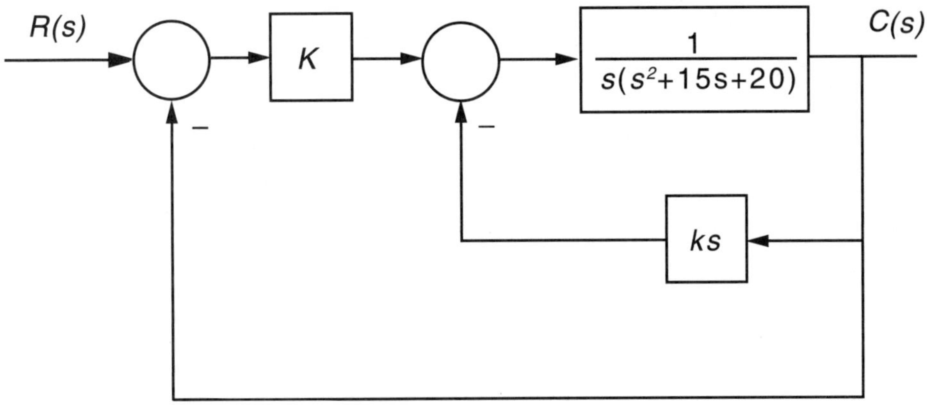

Figure 5–3 System with velocity feedback

Clearly, K must be positive, and the stability limit is defined by:

$$300 + 15k - K = 0 \quad \textbf{Eq. 5.13}$$

This is the equation of a straight line on the k versus K plane, as shown in Figure 5–4. It is clear that the stability of a dynamic system depends on the values of the parameters of the system. It is also clear that the parameters determine the values of the coefficients of the characteristic polynomial, and thus its roots, so that changing the value of a parameter causes the roots to move. This concept permits development of analysis and design methods that place roots where they are wanted. The best-known method uses the Root Locus method, which is developed in a later chapter.

Two especially interesting relationships are:

1. For a system to be stable, all coefficients must be greater than 0.
2. For a monic polynomial (a polynomial that has a coeficient associated with the highest-order term equal to 1, such as $s^3 + 2s^2 + 9s + 1$) of order n, the coefficient of the n-1 order term is the sum of all roots of the polynomial. For some systems, the sum of the roots remains unchanged for certain parameter changes, but locations of roots will change.

Example 5.6

Return to the DC motor problem of Chapter 4. Compute the Routh array for the DC motor.

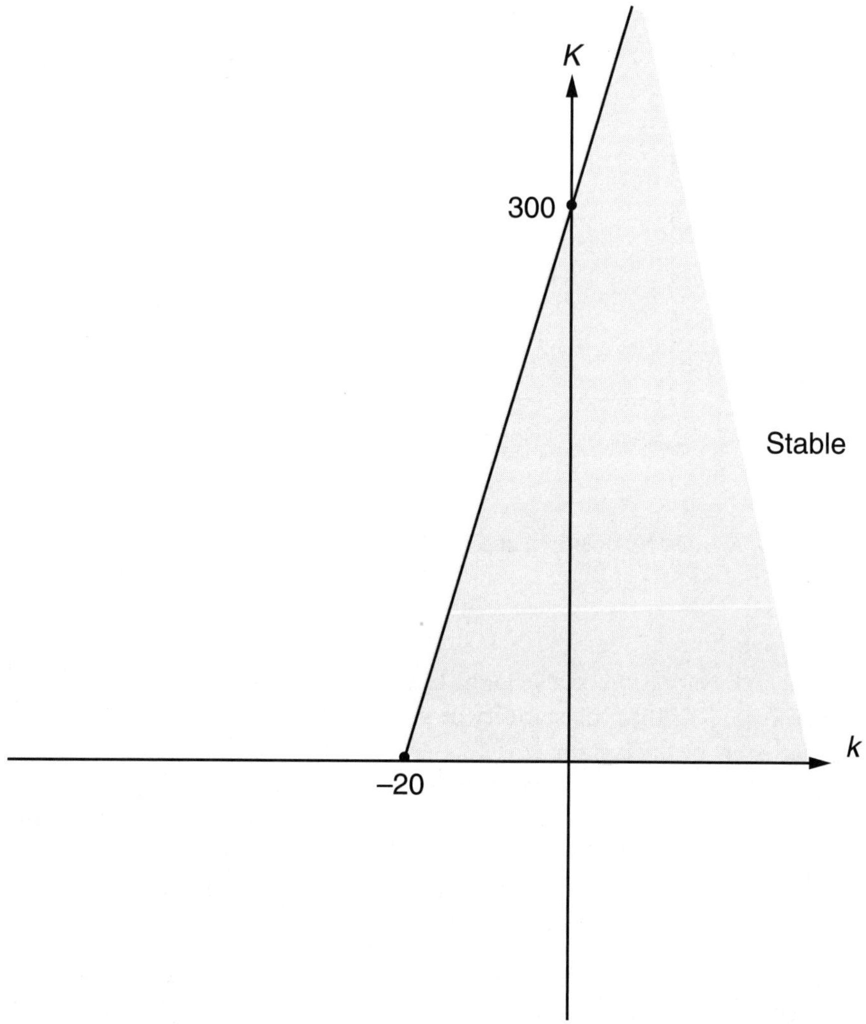

Figure 5–4 Stability limit on the *k* versus *K* plane

Solution The CLTF is:

$$\frac{C}{R} = \frac{Kk_t}{s[(Js + f)(ls + r) + k_b k_t] + Kk_t} \quad \text{Eq. 5.14}$$

The characteristic polynomial is:

$$JLs^3 + (JR + fL)s^2 + (fR + k_b k_t)s + Kk_t \quad \text{Eq. 5.15}$$

The Routh array is:

JL	$fR + k_b k_t$	0
$JR + fL$	Kk_t	0
$(JR + fL)(fR + k_b k_t) - (JL)(Kk_t)$	0	0

When K is large enough, the third row becomes negative, and the system is unstable. In practice, the engineer determines the numerical value of these parameters by measurement, experimental tests, or calculation, then determines the value of each coefficient, after which a computer may be used to determine stability.

Example 5.7

Let the characteristic polynomial for a dynamic system be:

$$s^3 + 2s^2 + 7s + K \quad \text{Eq. 5.16}$$

Determine the Routh array.

Solution The Routh array is:

1	7
2	K
$\dfrac{14 - K}{2}$	0

Stability is determined by the term 14-K; when:

- $K < 14$ stable, all roots in the left-half plane
- $K = 14$ two roots on the vertical axis
- $K > 14$ unstable, two roots in the right-half plane

For this example, the stability boundary is exactly $K = 14$, and there are no other adjustable variables. In problems where there are more than one adjustable variable, the stability boundary can be more complex. In many dynamic systems, even when they are stable, oscillations in the output may be reduced by introducing a damping device. One way to do this is to measure velocity and to subtract the velocity signal from the error signal as shown in Figure 5–5.

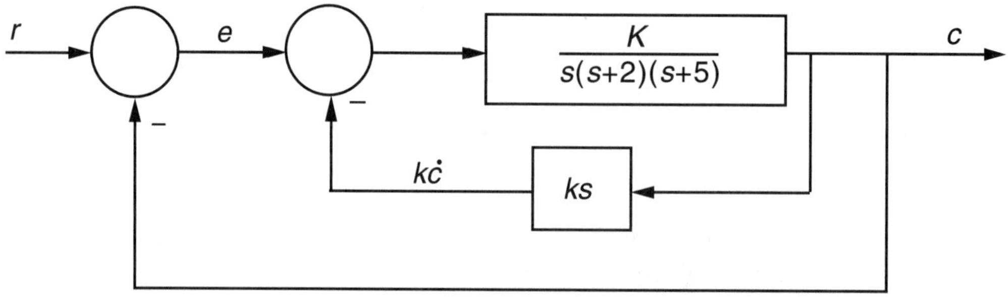

Figure 5–5 Positioning system with velocity feedback for damping

Example 5.8

Consider a dynamic system with:

$$G(s) = \frac{K}{s(s+2)(s+5)} \quad \text{Eq. 5.17}$$

and with velocity feedback, as shown in Figure 5–6a. Perform Routh analysis of this system.

Solution Reduction of the loop gives:

$$G_p = \frac{K}{s(s+2)(s+5) + Kks} \quad \text{Eq. 5.18}$$

and, introducing the position feedback (Figure 5–6b), gives:

$$\frac{C}{R} = \frac{K}{s(s+2)(s+5) + Kks + K} \quad \text{Eq. 5.19}$$

$$= \frac{K}{s^3 + 7s^2 + (10 + Kk)s + K}$$

as shown in Figure 5–6c.

The characteristic polynomial is:

$$s^3 + 7s^2 + (10 + Kk)s + K \quad \text{Eq. 5.20}$$

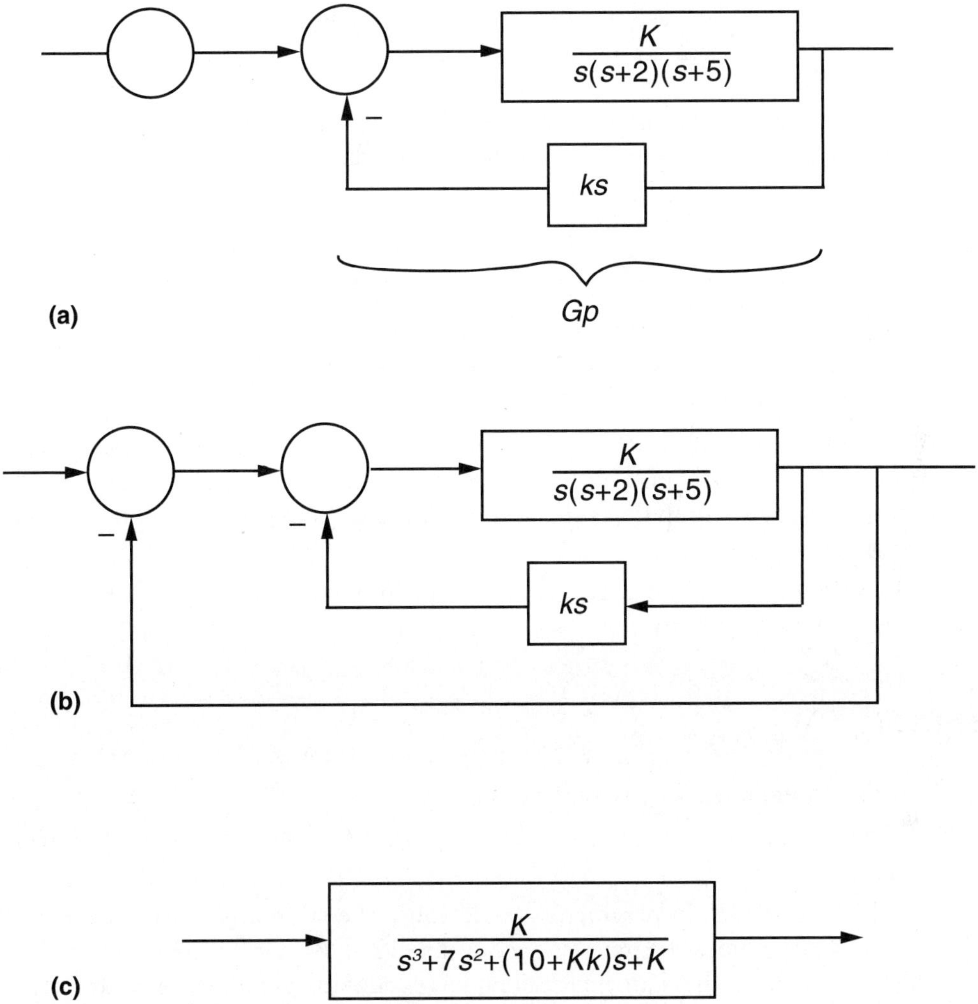

Figure 5–6 Stability analysis of dynamic system: (a) plant, (b) closed-loop block diagram (plant with feedback), and (c) reduced block diagram

and the Routh array is:

$$\begin{array}{lll}
1 & 10 + Kk & 0 \\
7 & K & 0 \\
\dfrac{70 + 7Kk}{7} - K & 0 & 0
\end{array}$$

Stability is determined as usual, but there are two adjustable parameters, K and k. These parameters define a stability boundary, which is a curve on the K versus k plane.

The Routh test is often convenient for doing stability analysis and can sometimes be used to improve stability and/or improve damping.

Summary

1. Stability is essential for a dynamic system to be useful.
2. One way to confirm stability is to find the roots (eigenvalues) of the characteristic system and ensure that they have negative real parts.
3. The Routh array allows the determination of the type of eigenvalues without actually solving for them. This technique was extremely useful in the early days of control, because computers capable of finding the roots of a polynomial were not typically available.
4. The Routh criterion can be used to establish boundaries on parameter values that guarantee stability.

Idea Portfolio

1. Is the system inherently stable?
2. Which safety device scheme should be included to ensure safety and equipment protection?
3. Which variations in parameters could cause the system to become unstable? Simulate the block diagram system obtained in Chapter 3 using SIMULINK. We will call this the mathematical model (or simply the model) for your system.

Problems

1. Which of the transfer function's denominator polynomial is used by the Routh criterion to determine stability?
   - OLTF
   - CLTF
2. Determine the characteristic equation for the closed-loop systems in Figure 5–7. Then, use the Routh Criterion to determine if the system is stable. Finally, use the `roots()` command in MATLAB to confirm your results.

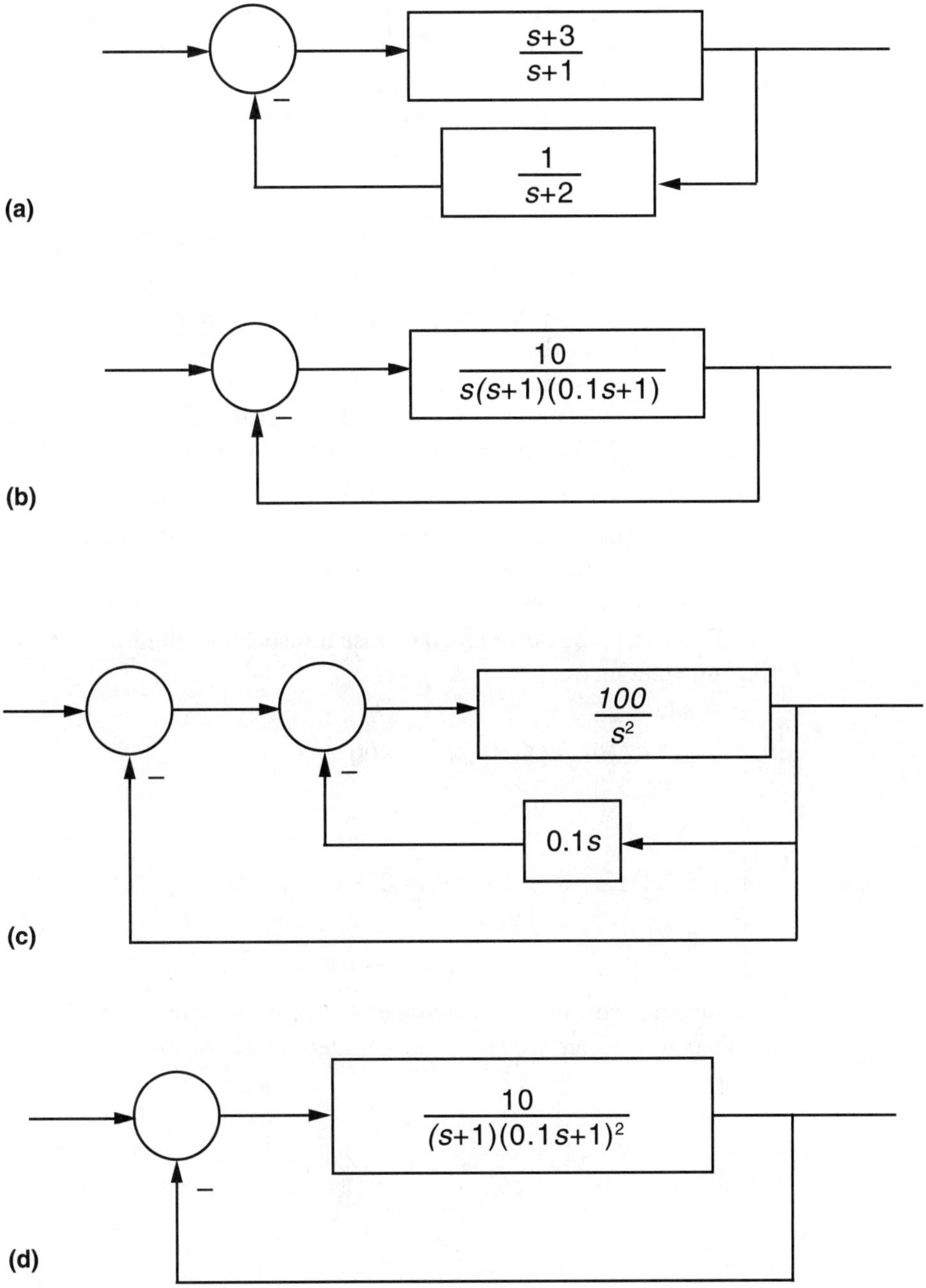

Figure 5–7

3. Use the Routh-Horwitz array to determine for the systems whose characteristic equations are given as follows: (a) whether the system is stable and (b) the number of roots in the right-half plane.
   - $s^7 + s^6 + s^5 + 2s^4 + 3s^3 + 4s^2 + 5s + 6 = 0$
   - $s^3 + 7s^2 + 2s + 3 = 0$
   - $s^4 + 5s^3 + 6s^2 + 4s + 6 = 0$
4. Using the motor coefficients in Chapter 4, and using the Routh array for the motor, determine which value of *K* would make the system unstable.
5. Stability for the system in Figure 5–6 depends on two parameters, *k* and *K*. Create a plot (*k* on the *y*-axis, *K* on the *x*-axis) showing the boundary that separates parameters that produce stable systems and those that do not. Afterward, pick two sets of *K*, *k* values and substitute them into the characteristic equation. Use MATLAB's `roots()` command to confirm your results.
6. Define a monic polynomial, and provide two examples.
7. Look at the following characteristic polynomials and determine if:
   - They have at least one unstable root
   - All roots are stable
   - This cannot be determined by visual inspection alone

   Explain your answer.

   $s^2 + 34s + 26 = 0$

   $s^5 + 5s^4 + 12s^3 + 15s^2 + 50s + 200 = 0$

   $-s - 200 = 0$

   $s + 34 = 0$

   $s^5 + 5s^4 + 15s^2 + 50s + 200 = 0$

   $s^3 + 15s^2 - 50s + 200 = 0$

   $s^5 + 5s^4 + 12s^3 + 15s^2 + 50s + 200 = 0$
8. Assume that you have constructed the Routh array and that the following numbers result from the first column. Determine the number of unstable roots.

   43

   22

   −33

   56

   −12

   31

   15

9. Suggest a control system that, even though unstable, can provide a useful purpose.
10. Indicate a present use of Routh's method, realizing that most control engineers can access other computer tools to determine system stability.
11. Given the following characteristic equation, determine the maximum value of K that will make the system stable:

$$s^3 + 10s^2 + 50s + K = 0$$

12. The characteristic polynomial (denominator of the CLTF) is:

$$s^3 + 4s^2 + s - 6$$

Use the Routh Criterion (i.e., set up the Routh array) to determine if the system is stable. Use the `roots()` command in MATLAB to confirm your results.

13. The Routh array indicates:
    - Number of roots in the right-half plane
    - Values of the roots

6

Time Domain Performance Characteristics

Learning Objectives

In this chapter, you will learn how to:

- Recognize the response of first- and second-order systems
- Calculate the response of first- and second-order systems to a step input
- Identify the various parameters and specifications that describe dynamic behavior, such as percent overshoot, rise time, and setting time

6.1 Introduction

Stability is essential for any practical control system. In Chapter 5, you learned techniques for confirming that a system is stable. However, the control engineer must know more than a system's stability. In Chapter 4, we explored methods for determining the steady-state error and making design changes to improve it. In addition to these aspects, a customer might specify how fast the system must respond to an excitation or how much the system must initially overshoot the steady-state output. These topics are the subject of this chapter.

For simple systems, it is possible to arrive at an expression that predicts the system's behavior in closed form. Further, it is often possible to represent complex systems with simplified ones. The procedure for evaluating this possibility and carrying it out will be presented here. For higher-order systems that cannot be simplified, the only way to predict behavior is through simulation using such tools as MATLAB/SIMULINK. Little can be generalized about such systems.

There are two main models for which dynamic behavior can be predicted in closed form: (1) first-order systems and (2) second-order systems. Fortunately, many complex systems can be approximated as one of these.

6.2 Pole Locations

The poles, or eigenvalues, the roots of the *s*-polynomial contained in the denominator of the CLTF, have a very strong effect on system behavior. For this reason, a control engineer often uses a graphical representation of these poles to understand and modify the behavior of the dynamic system. In a subsequent chapter, you will use a graphical design approach called the Root Locus method to position the system poles in the *s*-plane and thereby achieve a desired behavior (root locus). In the previous chapter, you learned that, for the system to be stable, all the CLTF poles must be in the left-half *s*-plane.

6.3 Disturbing the System to Investigate Its Characteristics

Often, the best way to inquire about system characteristics is to excite the system with a disturbance and observe its response. Special features of the resulting response can be used to describe its characteristics objectively.

Common types of excitation include step, square wave, sinusoidal input, and impulse. Because the response depends on the type of system being excited, we will discuss the responses to these inputs for the most important systems, namely first- and second-order systems.

6.4 First-Order System

A first-order system is one whose characteristic equation has only one root. Even though this type of system might appear too simplistic for the real world, it is an excellent descriptor of many real-world systems. (A process for reducing higher-order systems to simpler systems is treated later in the chapter.)

The transfer function of a first-order system can be represented in several ways, including:

$$\frac{K}{\tau s + 1} \qquad \text{Eq. 6.1}$$

where K is the DC gain and τ is the time constant.

The time constant, τ, is the time it takes system output to reach 63 percent of its final value when subjected to a step input. Notice that if the system's output changed from 20 to 35, the time constant of the system is the time it takes to go from 20 to 29.45, (20 + 0.63 (35 − 20)).

The preceding transfer function can be represented in a slightly different way:

$$\frac{K}{\frac{s}{\omega_0} + 1} \qquad \text{Eq. 6.2}$$

where ω_0 is called the corner or break frequency (the units of angular frequency are rad/s). It has two meanings: (1) the distance of the root from the origin in the s-plane and (2) the frequency of a sinusoid at the input that would produce the output's amplitude that is 70 percent of the maximum (also called the −3 dB frequency). When this corner frequency, ω_0 (rad/s), is expressed in Hz, it is called the bandwidth, BW, and it is given by:

$$BW = \frac{\omega_0}{2\pi} \qquad \text{Eq. 6.3}$$

6.5 First-Order System Response of a Step Input

Consider exciting a first-order system with a unit step where $r(t)$ is the input and $c(t)$ is the output:

$$r(t) = u(t)$$
$$R(s) = \frac{1}{s} \qquad \text{Eq. 6.4}$$

The transfer function is:

$$T(s) = \text{transfer function of system} = \frac{K}{\tau s + 1} \qquad \textbf{Eq. 6.5}$$

Thus:

$$\begin{aligned} C(s) &= T(s)R(s) \\ &= \frac{K}{\tau s + 1}\frac{1}{s} \end{aligned} \qquad \textbf{Eq. 6.6}$$

and:

$$c(t) = \mathcal{L}^{-1}\{C(s)\} = \mathcal{L}^{-1}\left[\frac{K}{\tau s + 1}\frac{1}{s}\right] \qquad \textbf{Eq. 6.7}$$

By means of partial-fraction expansion, we can rewrite the preceding equation as:

$$c(t) = K\left[\frac{A}{s} + \frac{B}{s + 1/\tau}\right] \qquad \textbf{Eq. 6.8}$$

Using a partial-fraction expansion, you obtain $A = 1$ and $B = -1$. Now, if you take the inverse Laplace transform, the result is:

$$c(t) = K\left[1 - e^{-t/\tau}\right] \qquad \textbf{Eq. 6.9}$$

To be mathematically accurate, we should multiply the right side of the equation by a unit step function to emphasize that the function is only defined for $t \geq 0$.

Experimentally, it is relatively easy to determine the time constant for a system whose dynamics are to be identified. The input is changed instantaneously, or nearly so, and the output is observed. The time it takes for the system to reach 63 percent of its final value is the time constant, τ, as shown in Figure 6–1a.

We can also use the initial slope of the response to determine the time constant, τ. Notice that K is the total change experienced by $c(t)$. If we differentiate $c(t)$ and evaluate it at $t = 0$:

$$\dot{c}(t)_{at\ t=0} = \frac{K}{\tau} \qquad \textbf{Eq. 6.10}$$

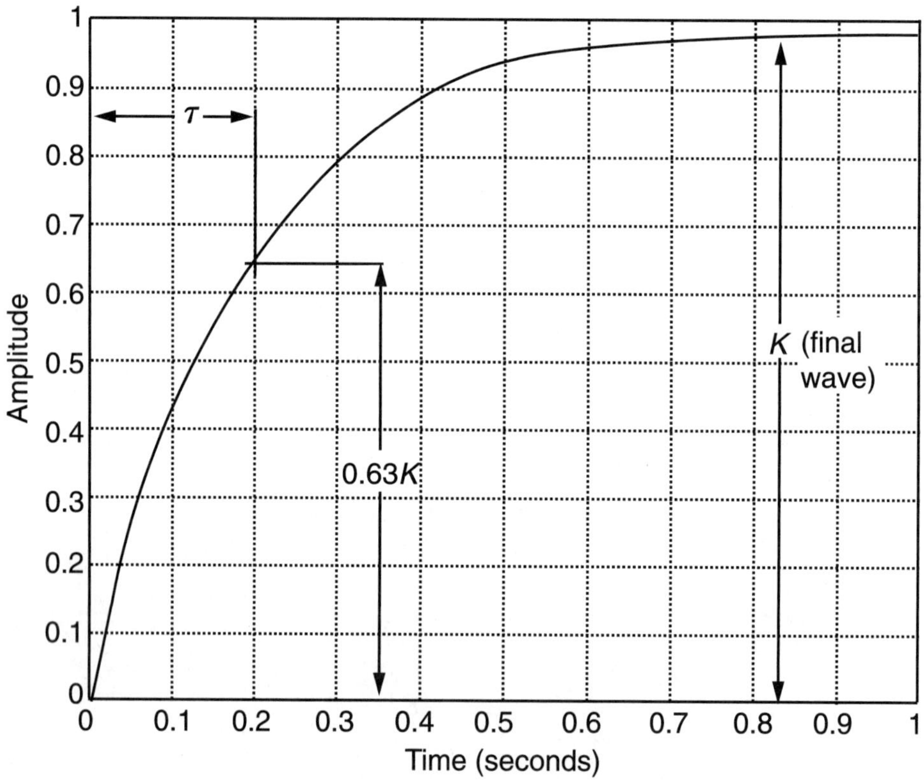

Figure 6–1a Step response to a first-order system and definition q time constant

Solving for τ:

$$\tau = \frac{K}{\dot{c}(t)_{at\ t=0}} \qquad \text{Eq. 6.11}$$

Figure 6–1b shows this alternate method for obtaining the time constant.

6.6 Excitation of a First-Order System with a Square Wave

Often, it is convenient to excite the system with a series of steps (i.e., a square wave) instead of one step, as shown in Figure 6–2. This is particularly true in the lab, because frequency generators with a square wave output are readily available.

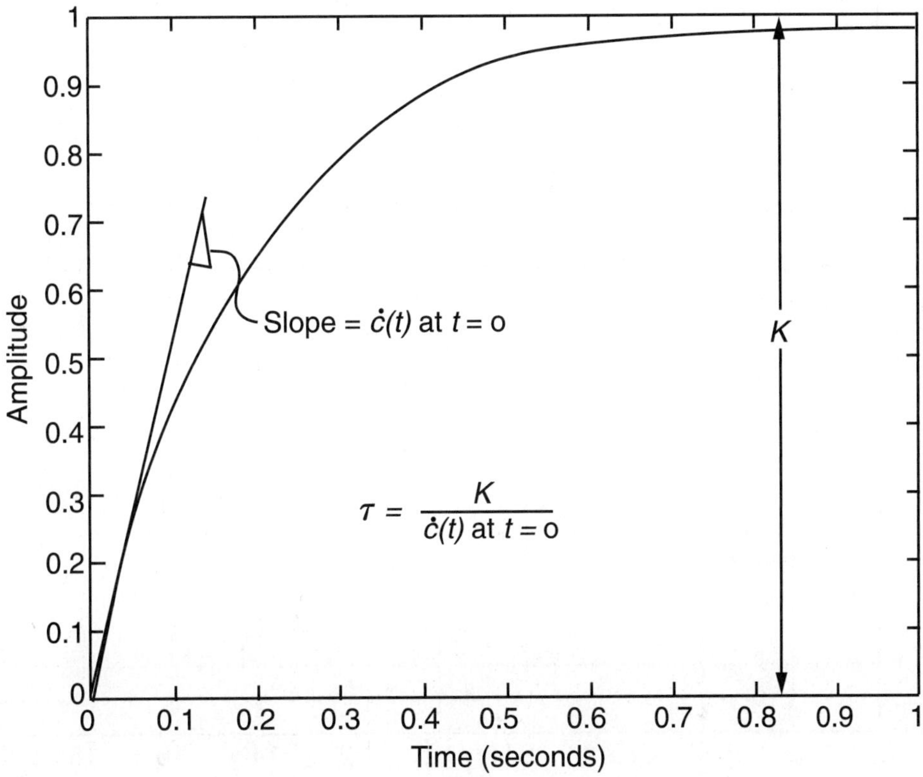

Figure 6–1b Alternate method for obtaining the time constant, τ

The response offers some advantages not present when the excitation is a simple step. Another descriptor of the speed of response is the rise time, *Tr*, the time it takes the output to go from 10 to 90 percent of its final value. With this purpose in mind, oscilloscopes have grids. To perform the measurement, the sensitivity and bias are adjusted so that output signal spans from the 0 to 100 percent lines, typically five full divisions on the vertical axis. Note that the oscilloscope has two dotted, horizontal lines at the 10 and 90 percent positions. The horizontal bias is then adjusted so the wave form begins at a vertical line, allowing *Tr* to be measured directly from the oscilloscope screen, as seen in Figure 6–2. For first-order systems, there is a fixed relation between time constant and rise time. This relationship, derived in Appendix A, is presented in the next section.

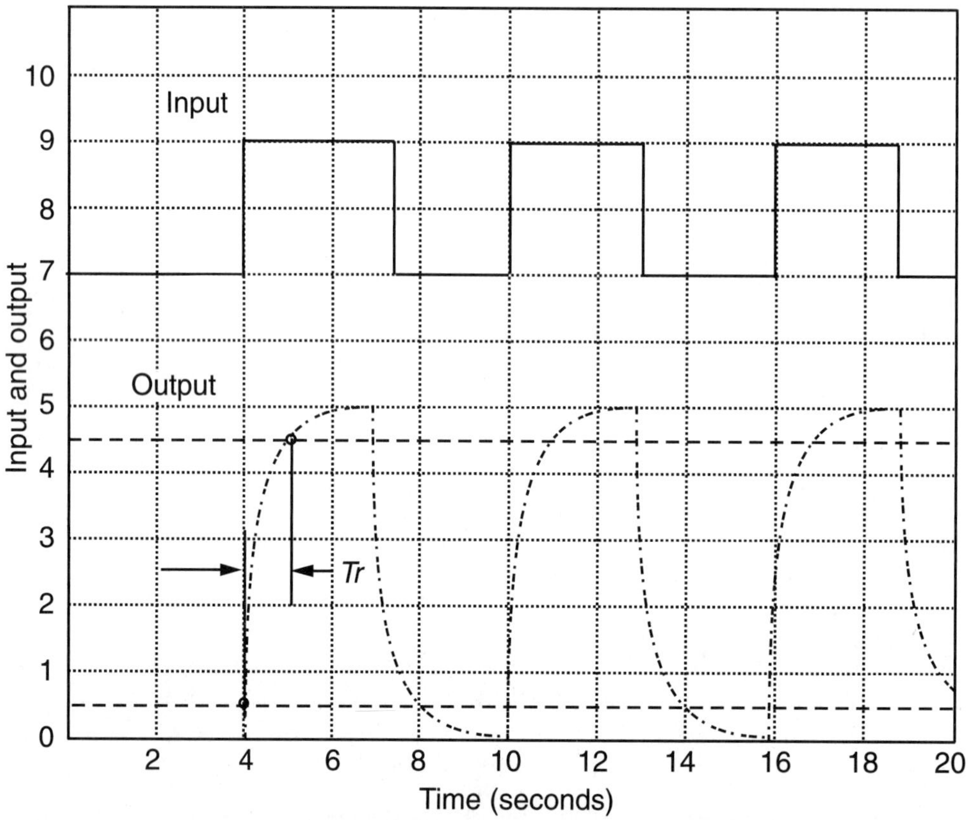

Figure 6–2 Excitation of a first-order system with a square wave for *Tr* determination

6.7 Relationship between *Tr* and *BW* and τ

A first-order system's response to a step input previously presented can be expressed as:

$$v(t) = V_F\left(1 - e^{-t/\tau}\right) \qquad \textbf{Eq. 6.12}$$

This equation can be evaluated at different time values (corresponding 10 and 90 percent of the final output value), then normalized (see derivation in Appendix A) to yield:

$$T_R = \frac{0.35}{BW} \qquad \textbf{Eq. 6.13}$$

This relationship allows a frequency parameter, *BW*, to be determined from a time-domain test. Using the preceding equation, with the relationship between τ and *BW*, we obtain:

$$T_R = 2.2\,\tau \qquad \textbf{Eq. 6.14}$$

This relationship is true only for first-order systems.

6.8 Second-Order System Classifications

Second-order systems can be classified into one of three categories:

1. **Underdamped**—having two complex poles (a complex conjugate pair)
2. **Critically damped**—having two equal, real poles
3. **Overdamped**—having two different, real poles (such a system can be represented as two separate first-order systems)

The locations of these roots dictate the behavior of the system when it is excited. Figure 6–3 shows the root arrangement and the corresponding response to a step input.

Of these responses, the one with the most identifiable characteristics is the first. Because this type of response occurs so often, control engineers have attached to it performance indicators associated with the system's response to a step input. Figure 6–4 shows an expanded step response plot with some of the most common performance indices, including:

- Rise time, T_r
- Peak time, T_p
- Settling time, T_s
- Maximum percent overshoot, *PO*
- Damped natural frequency, ω_d

Other specifications related to these values are:

- Natural frequency, ω_n
- Resonant frequency, ω_r

These parameters can be linked directly to the locations of the roots in the *s*-plane, as shown in Figure 6–5.

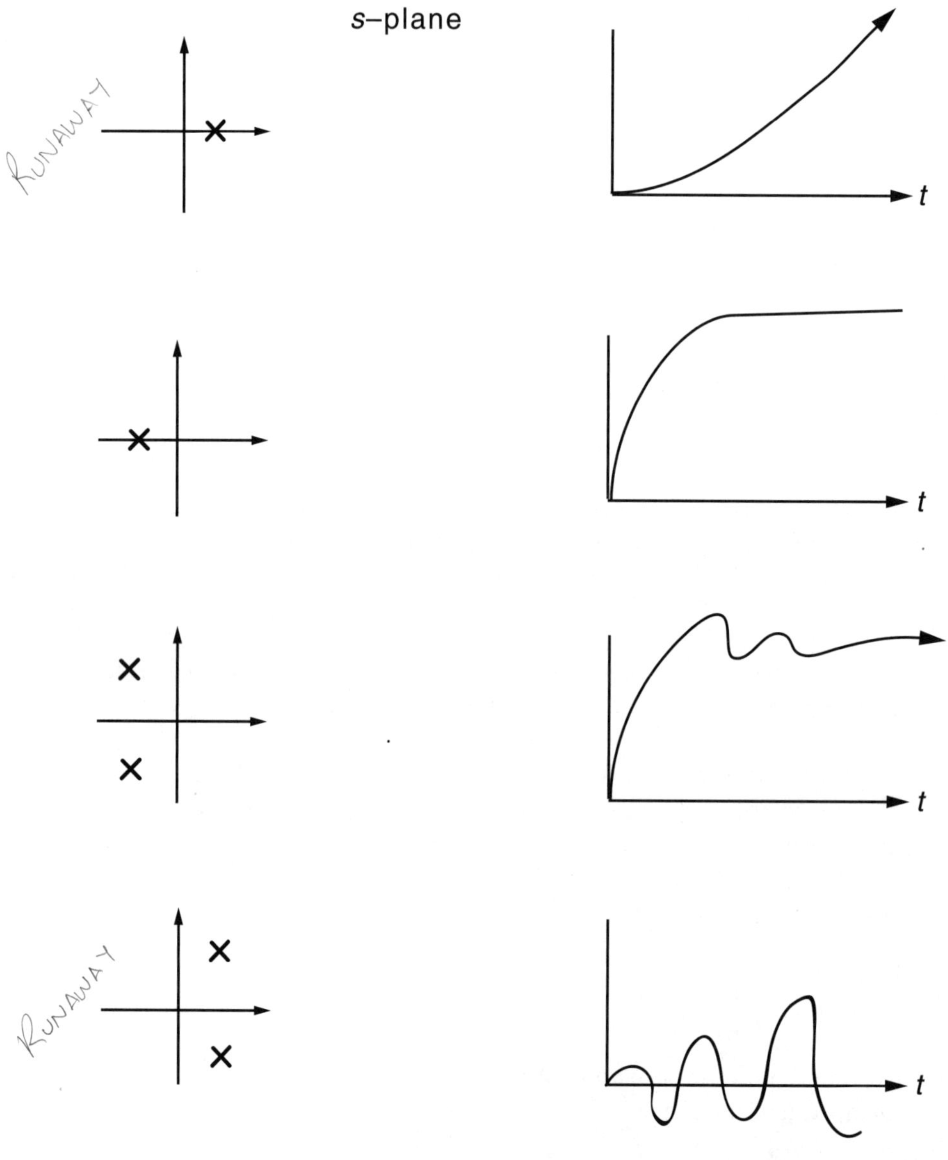

Figure 6–3 Relationship between root location and time response

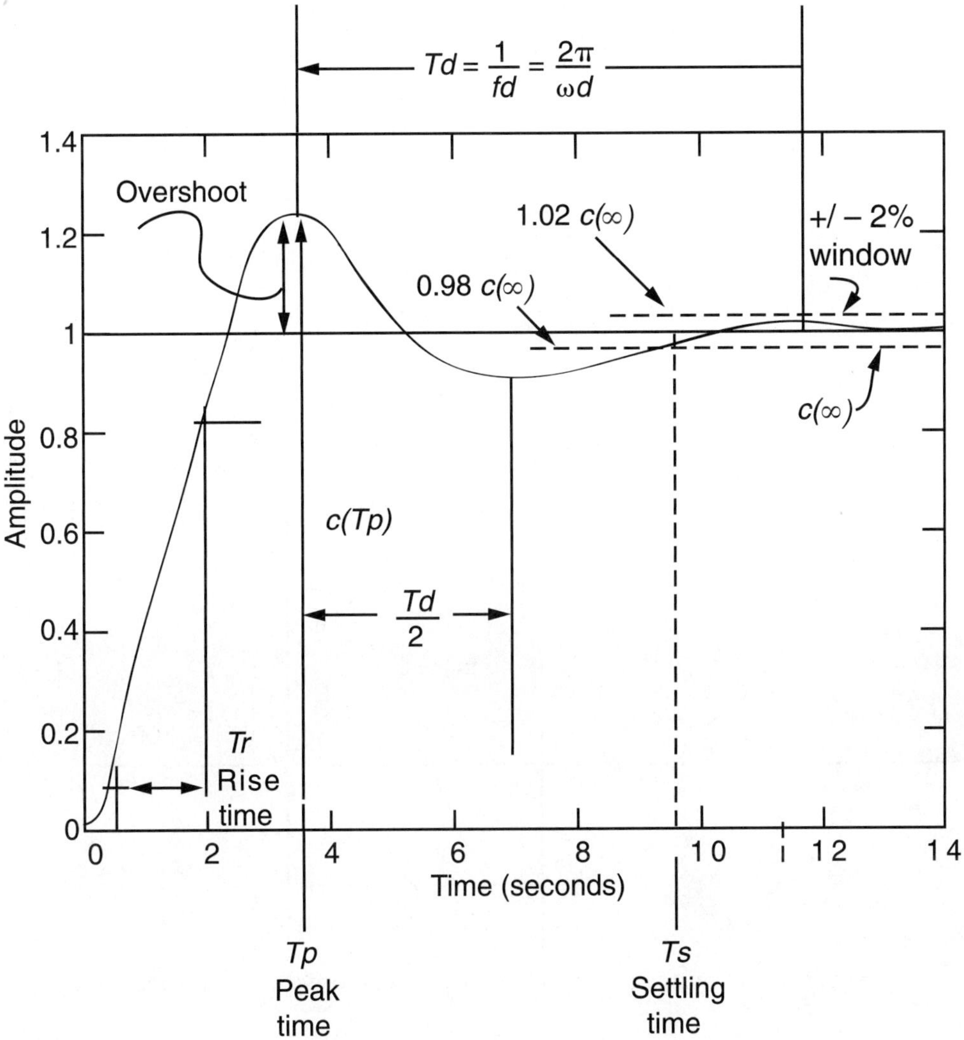

Figure 6–4 Performance indicators of a second-order system

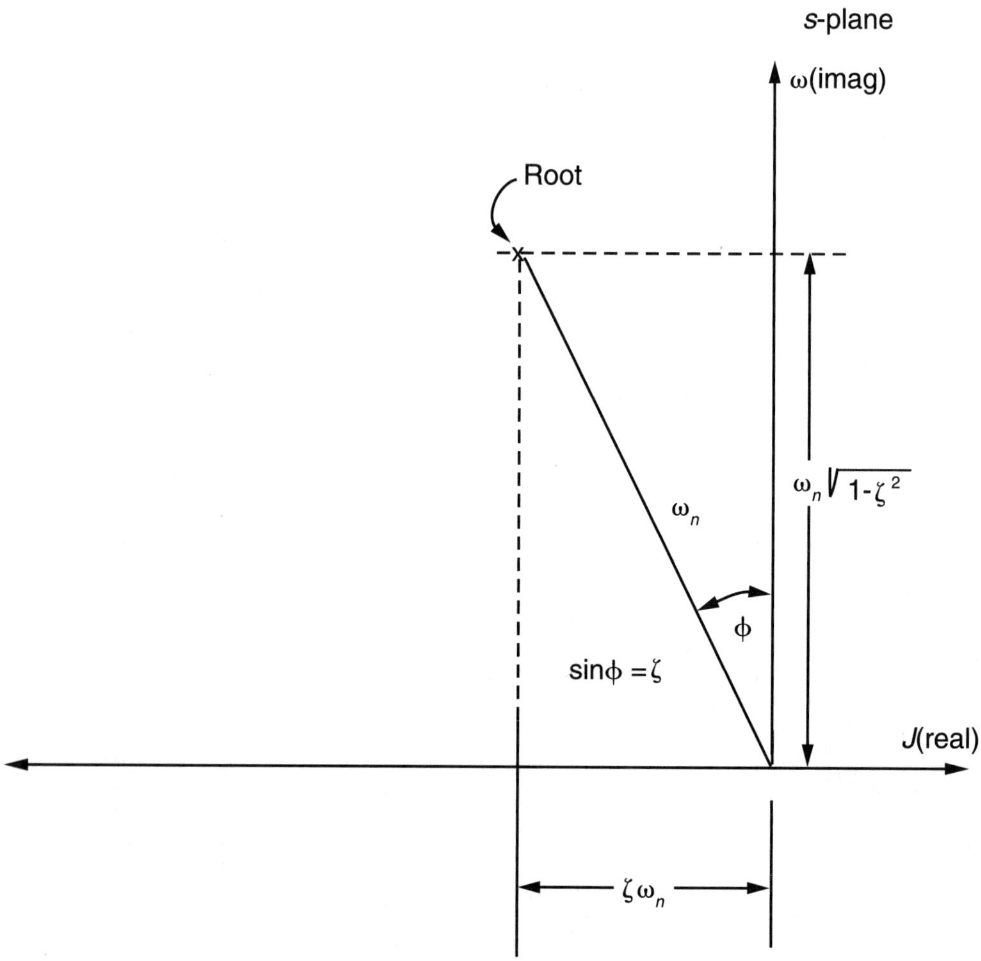

Figure 6–5 Generalized coordinates on the s-plane for a second-order, underdamped system

6.9 Second-Order System Response of a Step Input

The transfer function of a second-order system with a DC gain of 1 is:

$$\frac{C}{R} = \frac{\omega_n^{\ 2}}{s^2 + 2\varsigma\omega_n s + \omega_n^{\ 2}} \qquad \text{Eq. 6.15}$$

or, in another representation:

$$\frac{C}{R} = \frac{1}{\left(\frac{s}{\omega_n}\right)^2 + \frac{2\varsigma}{\omega_n}s + 1} \qquad \textbf{Eq. 6.16}$$

Using the first form, the output for a unit step input is:

$$C(s) = \frac{\omega_n^2}{s(s^2 + 2\varsigma\omega_n s + \omega_n^2)} \qquad \textbf{Eq. 6.17}$$

Factoring the quadratic:

$$C(s) = \frac{\omega_n^2}{s(s + \varsigma\omega_n + j\omega_n\sqrt{1-\varsigma^2})(s + \varsigma\omega_n - j\omega_n\sqrt{1-\varsigma^2})} \qquad \textbf{Eq. 6.18}$$

Taking the inverse transform:

$$C(\omega_n t) = 1 - \frac{e^{-\varsigma\omega_n t}}{\sqrt{1-\varsigma^2}}\sin\left(\omega_n\sqrt{1-\varsigma^2}\,t + \tan^{-1}\frac{\sqrt{1-\varsigma^2}}{\varsigma}\right) \qquad \textbf{Eq. 6.19}$$

Using $\omega_n t$ as abscissa and C as ordinate, a family of plots of the unit step responses is obtained for selected values of ς, as shown in Figure 6–6. Inspecting this family of curves, one observes that all start at $\omega_n t = 0$ and $C = 0$ and that all approach $C = 1$ as $\omega_n t$ approaches infinity.

For purposes of analysis and design, we need numerical values for the parameters that affect the particular details of the response:

ς = damping ratio, defined by the locations of the roots

ω_n = natural frequency, defined by the locations of the roots

$T_p = \dfrac{\pi}{\omega_n\sqrt{1-\varsigma^2}}$ peak time (time to reach the first maximum)

$M_{pt} = 1 + \exp\left(\dfrac{-\pi\varsigma}{\sqrt{1-\varsigma^2}}\right)$ height of first maximum of C (assumes $\lim_{t\to\infty} C(t) = 1$)

$PO = 100\exp\left(\dfrac{-\pi\varsigma}{\sqrt{1-\varsigma^2}}\right)$, percent overshoot

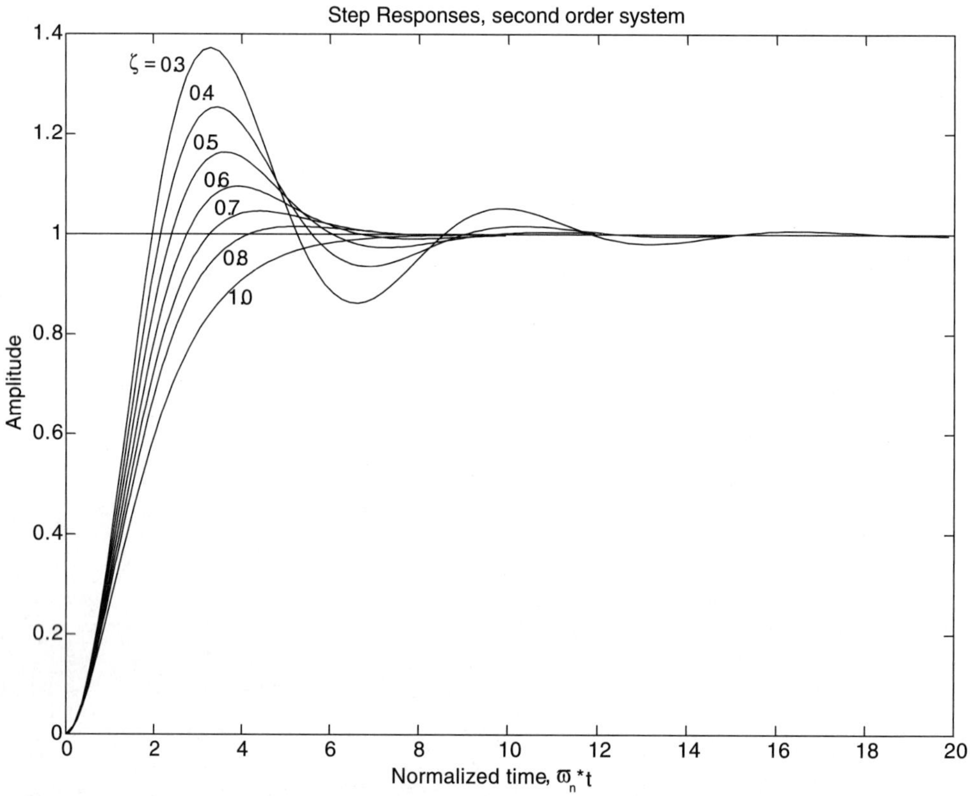

Figure 6–6 Step response of second-order systems with different damping ratios

which is defined experimentally as:

$$PO = \frac{C_{peak}(C(t) \text{ at } t = T_p) - C(\infty)}{C(\infty)} \times 100 \qquad \textbf{Eq. 6.20}$$

$\tau = \dfrac{1}{\varsigma\omega_n}$, time constant of the response envelope

$T_s = \dfrac{4}{\varsigma\omega_n}$, settling time

$\omega_d = \omega_n\sqrt{1 - \varsigma^2}$, damped natural frequency

$\omega_r = \omega_n\sqrt{1 - 2\varsigma^2}$, resonant frequency (applicable only for $\varsigma \leq 0.707$)

Two other relationships that relate some of these parameters are:[1]

$$T_r = \frac{2.16\varsigma + 0.60}{\omega_n}$$ **Eq. 6.21**

and:

$$\omega_{BW} = (-1.196\varsigma + 1.85)\omega_n$$
$$(\text{valid for } 0.3 \leq \varsigma \leq 0.8)$$

Plots of some of these characteristics can be useful in aiding analysis and design where one can move the poles. As a result, the overshoot, for instance, can be increased or decreased depending on the movement of the poles. Figure 6–7 shows *PO* as a function of ς.

Notice that these equations are exact only when the transfer function lacks numerator dynamics. In other words, the numerator is a constant. When the transfer function has numerator dynamics, these equations can be used to get approximations. The actual characteristics of the response must be obtained through simulation and trial and error.

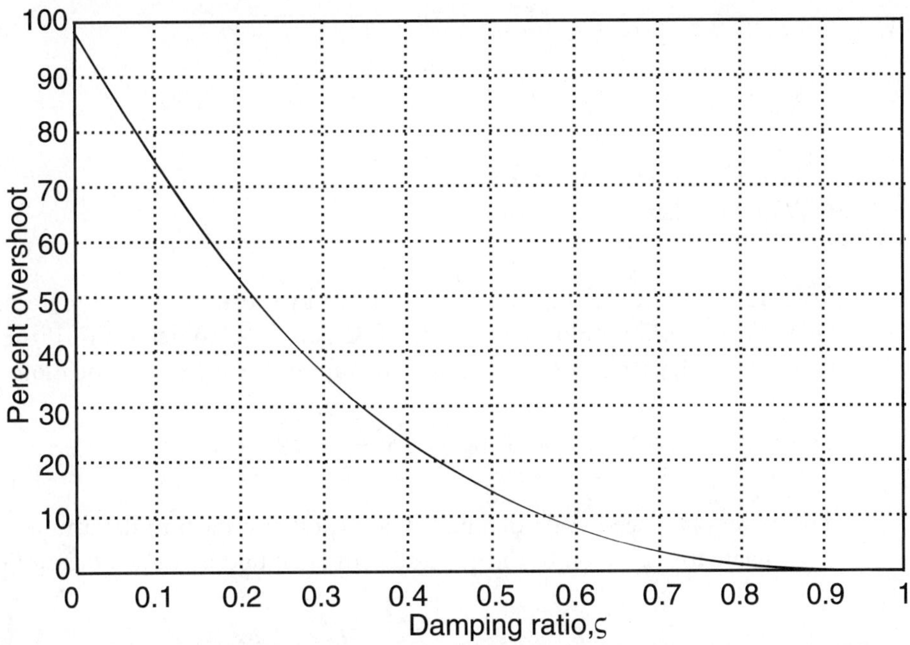

Figure 6–7 Percent overshoot (PO) as a function of damping ratio, ς

Specifications usually require bounds on the maximum overshoot and on the duration of the transient (settling time). One can enter the curve with the bound on the maximum overshoot and evaluate the corresponding damping ratio, ς. Using the curves again with the specific settling time, T_s, and the value of ς, one can obtain the value of ω_n. The value of T_p becomes readily available, as do other response values.

These values are especially useful when the actual higher-order system is available in hardware. Step responses of the second-order system can be compared with those of the hardware system. The next section presents some techniques for dealing with higher-order systems.

The detailed analysis of the step response of a second-order system is most valuable. This is especially true when used with a frequency response study of the same second-order system, which will be developed in a later chapter.

Example 6.1

Determine the parameters specified below for the following transfer function:

$$T(s) = \frac{72}{s^2 + 5s + 9} \quad \textbf{Eq. 6.22}$$

- What is the DC gain?
- Is it underdamped?
- If so, what is its damping ratio?
- What is its natural frequency?

Solution The system gain can be obtained by setting $s = 0$ and determining the ratio. This procedure indicates that the DC gain is 8. To determine if the system is underdamped, we can look at the discriminant of the quadratic equation:

$$\sqrt{b^2 - 4ac} = \sqrt{5^2 - 4(1)(9)} = \sqrt{-11} \quad \textbf{Eq. 6.23}$$

The system is indeed underdamped. The easiest approach to determine ω_n and ς is to correlate the equation to the general form of an underdamped quadratic, namely:

$$\frac{C}{R} = \frac{\omega_n^{\,2}}{s^2 + 2\varsigma\omega_n s + \omega_n^{\,2}} \quad \textbf{Eq. 6.24}$$

we can see that:

$$\omega_n^2 = 9$$

thus:

$$\omega = 3$$

also:

$$2\varsigma\omega_n = 5$$

$$\varsigma = \frac{5}{2\omega_n} = \frac{5}{2 \cdot 3} = 0.833 \quad \text{Eq. 6.25}$$

6.10 Dominant Roots

When a dynamic system is activated, as by an input command or by some disturbance, the dynamic response of the system is due to contributions from all the roots and in fact is the sum of all these contributions. For higher-order systems, the response is due mostly to those roots that are close to the origin of the *s*-plane. The reason for this is easily seen: For roots near the origin, the real parts of the roots are small so that the exponential decay of its response is slow and of relatively long duration. In addition, the magnitude of this response depends on the magnitude of the residue at the root, which is usually larger than that at more remote roots. Conversely, for roots farther from the origin, both the duration and magnitude of the response are smaller. Thus, one can say that the small roots are responsible for most of the dynamic response, and we say that such roots are dominant.

How many dominant roots are there for a given system? The answer differs with different systems. Not all systems are dominated by clearly identifiable roots. Most may be modeled with one real or two complex dominant roots. Occasionally, it is convenient to use only one root for a system. Some systems use three roots.

What is the value of the dominance concept?

- It permits us to study (analyze and design) a simple system rather than a complex system.
- It lets us define characteristics that are of interest for both simple and complex systems.

Most of these characteristics can be graphed on the *s*-plane or on a system step response or frequency response, which in turn can be plotted in dimensionless time and thus easily applied to a real system. Using such results (obtained for the second-order model), one can use the parameters of an actual system to determine the salient features of its response, as well as predict the design problems needed to meet the specifications of the desired system.

As a rule of thumb, if the real part of a root(s) is ten times larger than the real part of the root closest to the origin, it can be neglected. Two simple procedures for simplifying a transfer function are presented next. You might find one more intuitive than the other.

1. Put the transfer function in Root Locus form (see Chapter 4, Equation 4.12).
2. Calculate the DC gain of the transfer function.
3. Examine the roots of the numerator and denominator paying specific attention to their distance from the imaging axis.
4. Eliminate roots that fulfill the 10x rule of thumb (for both numerator and denominator).
5. Adjust the DC gain so that the reduced transfer function has the same DC gain as the original transfer function.

Example 6.2

Simplify the system whose transfer function follows:

$$T(s) = \frac{420}{s^2 + 38s + 105} \quad \text{Eq. 6.26}$$

The system has a DC gain of 4.

Solution The denominator can be factored:

$$s^2 + 38s + 105 = (s + 3)(s + 35) \quad \text{Eq. 6.27}$$

It is obvious that one root is 10x farther from the origin than the other. Thus, it can be eliminated. The reduced transfer function should be:

$$T(s) \cong \frac{?}{s + 3} \quad \text{Eq. 6.28}$$

Because the DC gains of the original and reduced transfer functions must match, or:

$$\frac{?}{3} = \frac{420}{105}$$

$$? = 12$$

Then completed, the reduced transfer function is:

$$T(s) \cong \frac{12}{s+3} \quad \text{Eq. 6.29}$$

Another approach is to put the transfer function in modified Bode form:

$$T(s) = \frac{K_B}{\left(\frac{s}{\omega_1}+1\right)\left(\frac{s}{\omega_2}+1\right)\left(\frac{s}{\omega_3}+1\right)\cdots} \quad \text{Eq. 6.30}$$

and eliminate terms that follow the 10x rule.

Example 6.3

Simplify the system whose transfer function is provided. Use the approach that uses the Bode form.

$$T(s) = \frac{420}{s^2+38s+105} \quad \text{Eq. 6.31}$$

Solution From the previous example, we know that the denominator can be factored as:

$$s^2+38s+105 = (s+3)(s+35) \quad \text{Eq. 6.32}$$

Thus the transfer function can be expressed as:

$$T(s) = \frac{\frac{420}{(3)(35)}}{\left(\frac{s}{3}+1\right)\left(\frac{s}{35}+1\right)} \quad \text{Eq. 6.33}$$

It is apparent that the second denominator term can be dropped. Thus the simplified transfer function becomes:

$$T(s) = \frac{\frac{420}{(3)(35)}}{\left(\frac{s}{3}+1\right)} = \frac{4}{\left(\frac{s}{3}+1\right)} \quad \text{Eq. 6.34}$$

Summary

1. Most dynamic subsystems can be approximated as first-order or second-order systems.
2. The time constant is a key parameter used to describe dynamic system behavior. It conveys the time it takes a system to reach 63 percent of its final value when subjected to a step input.
3. The time constant and bandwidth have an inverse relationship.
4. The response of first-order and second-order systems can be described analytically in general terms. This is not the case for systems of higher order than 2.
5. Second-order systems can be classified as overdamped, critically damped, or underdamped.
6. Underdamped second-order systems are characterized by their natural frequency and damping ratio.
7. Increasing the damping decreases the overshoot in most systems.

Idea Portfolio

1. If you were to change the command value instantaneously, how would the output change? Will the system act in an overdamped or underdamped manner?
2. For the existing system, estimate the:
    - Time constant
    - Setting time
    - Period of oscillation
3. Simulate your mathematical model and:
    - Determine typical input values.
    - Increase the input value after the initial transient dies out.
4. For the mathematical model, estimate the:
    - Time constant
    - Setting time
    - Period of oscillation

Problems

1. What is the DC gain of the following transfer function?

$$\frac{C}{R} = \frac{\omega_n^2}{s^2 + 2\varsigma\omega_n s + \omega_n^2}$$

2. If a second-order system is described as having the following parameters:

$$\varsigma = 0.5$$
$$\omega_n = 10$$

the system is:
- Overdamped
- Critically damped
- Underdamped

3. Determine if the system described by the following transfer function is:
   - Overdamped or underdamped
   - Calculate the parameters that apply:
     - Poles (if overdamped or critically damped)
     - ω_n and ς (if underdamped)

$$T(s) = \frac{80}{4s^2 + 12s + 40}$$

4. Indicate what will happen to the following parameters as the dominant poles are moved from 1 to 2 in Figure 6–8:
   ς ____, PO ____, ω_n ____, T_S ____, T_P ____
   Indicate your answer with the symbols NC = no change, ↑ = increases, ↓ = decreases.
5. Indicate what will happen to the following parameters as the dominant poles are moved from 1 to 2 in Figure 6–9:
   ς ____, PO ____, ω_n ____, T_S ____, T_P ____
   Indicate your answer with the symbols NC = no change, ↑ = increases, ↓ = decreases.

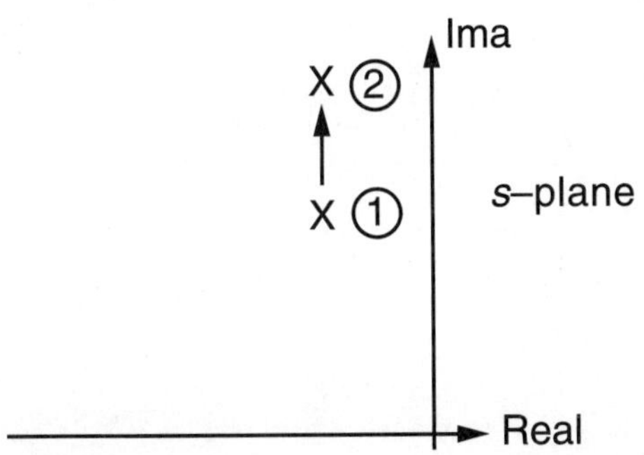

Figure 6–8

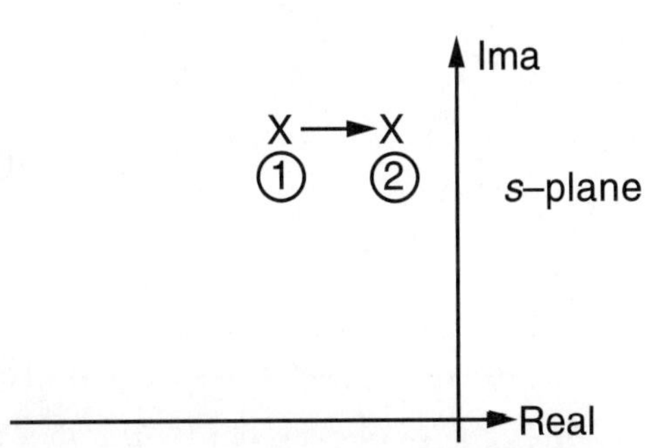

Figure 6–9

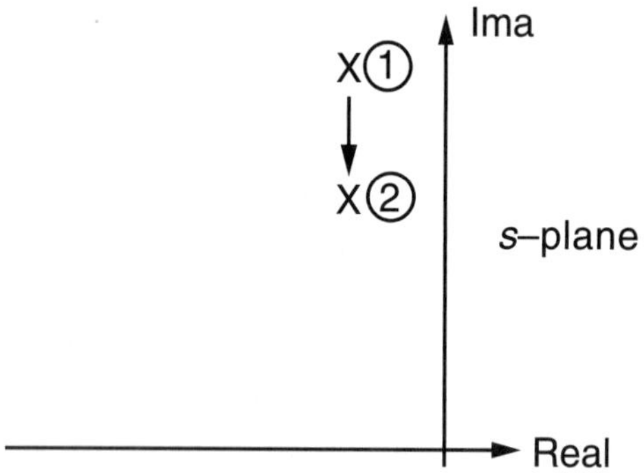

Figure 6–10

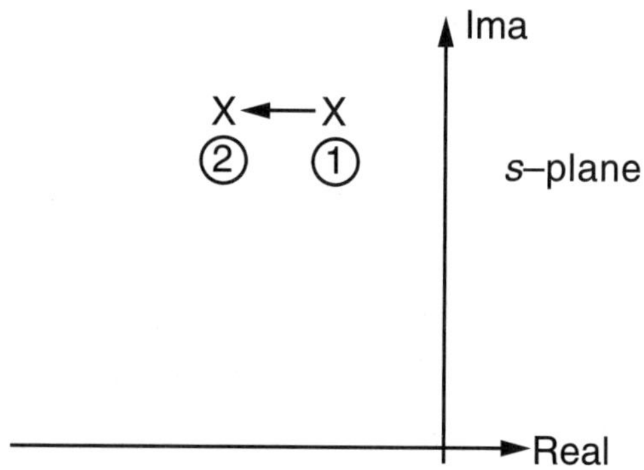

Figure 6–11

6. Indicate what will happen to the following parameters as the dominant poles are moved from 1 to 2 in Figure 6–10:
ς ____, PO ____, ω_n ____, T_S ____, T_P ____
Indicate your answer with the symbols NC = no change, ↑ = increases, ↓ = decreases.
7. Indicate what will happen to the following parameters as the dominant poles are moved from 1 to 2 in Figure 6–11:
ς ____, PO ____, ω_n ____, T_S ____, T_P ____
Indicate your answer with the symbols NC = no change, ↑ = increases, ↓ = decreases.
8. Indicate what will happen to the following parameters as the dominant poles are moved from 1 to 2 in Figure 6–12:
ς ____, PO ____, ω_n ____, T_S ____, T_P ____
Indicate your answer with the symbols NC = no change, ↑ = increases, ↓ = decreases.

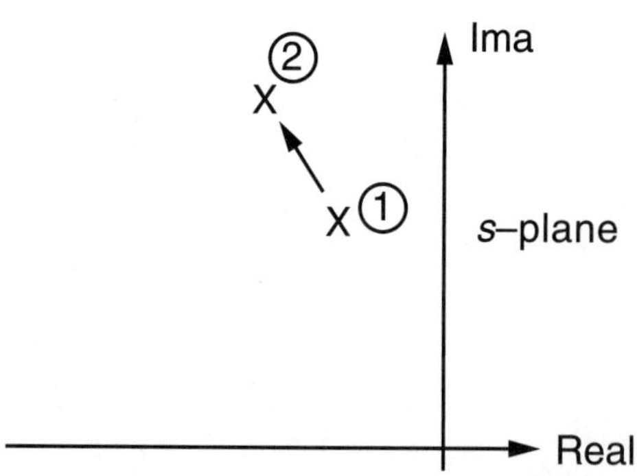

Figure 6–12

9. Various closed-loop systems have the pole arrangements shown in Figure 6–13. Sketch the unit step response for each.
10. Define the term *time constant* for a dynamic system.

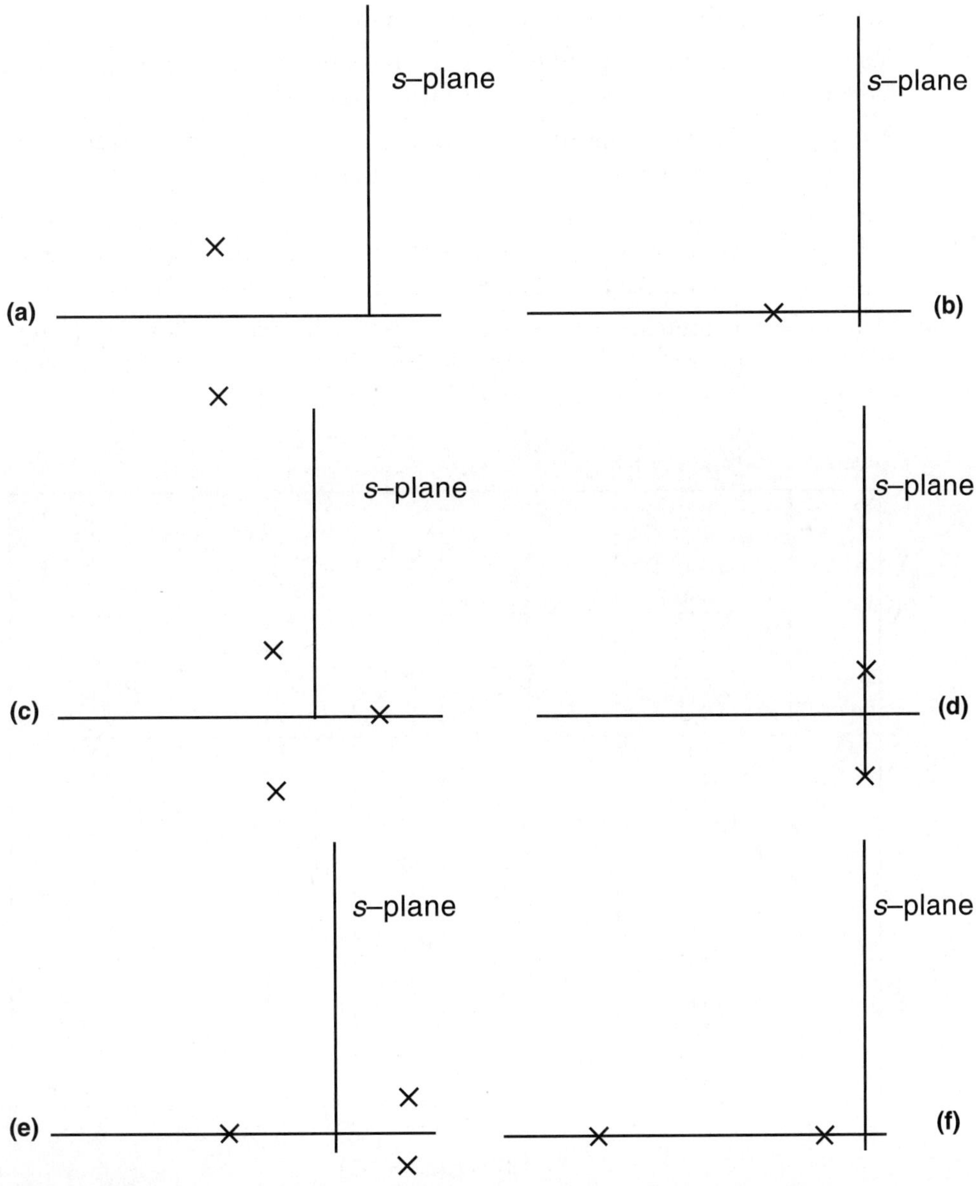

Figure 6–13

11. Explain two procedures for determining the time constant given its step response.
12. Must one need to know the magnitude of the step being applied to the system to calculate the system's time constant? Under what conditions is this true?
13. A step input change is applied to a system. Its output is shown in Figure 6–14. Determine the time constant associated with the system.
14. A step input change is applied to a system. Its output is shown in Figure 6–15. Determine the time constant associated with the system.
15. A step input change is applied to a system. Its output is shown in Figure 6–16. Determine the time constant associated with the system.
16. Determine whether the system whose pole distribution is shown in Figure 6–17 can be approximated with one complex pair. Explain. Create an example, and confirm your expectations by means of a SIMULINK simulation.
17. Determine whether the system whose pole distribution is shown in Figure 6–18 can be approximated with one pole. Explain. Create an example and confirm your expectations by means of a SIMULINK simulation.

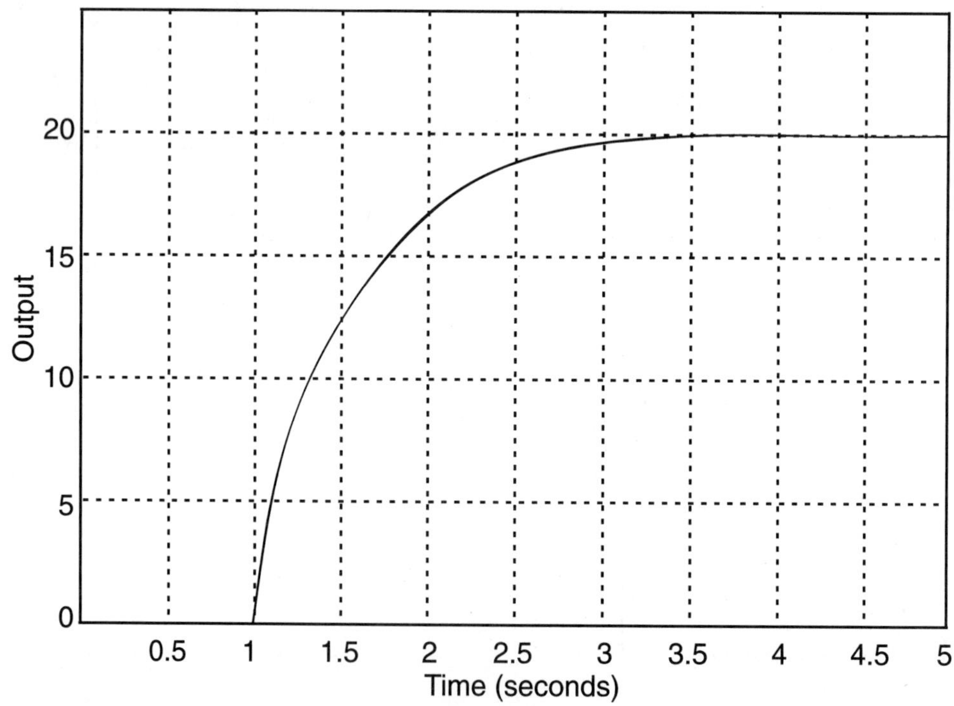

Figure 6–14

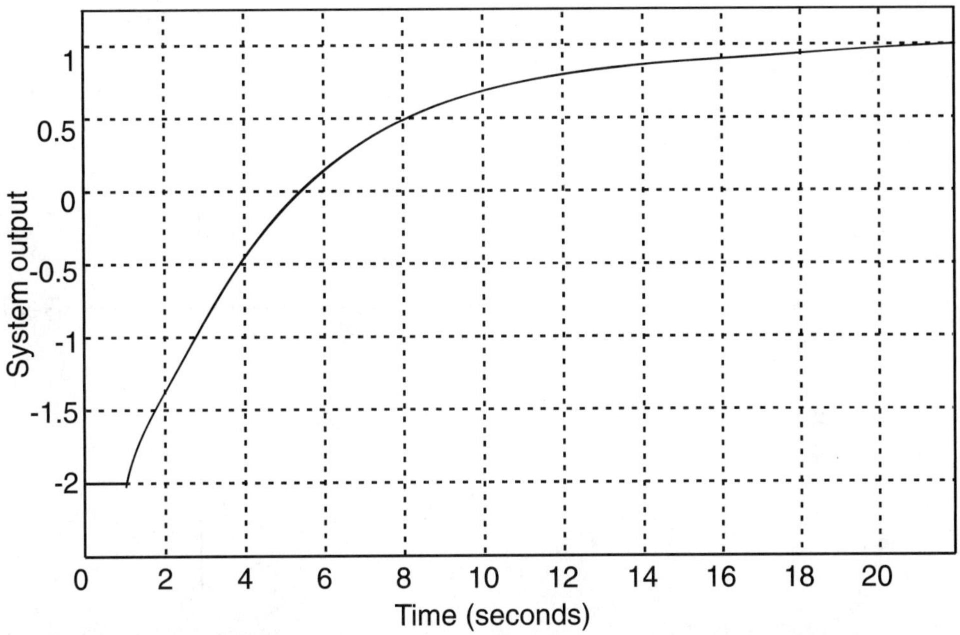

Figure 6–15

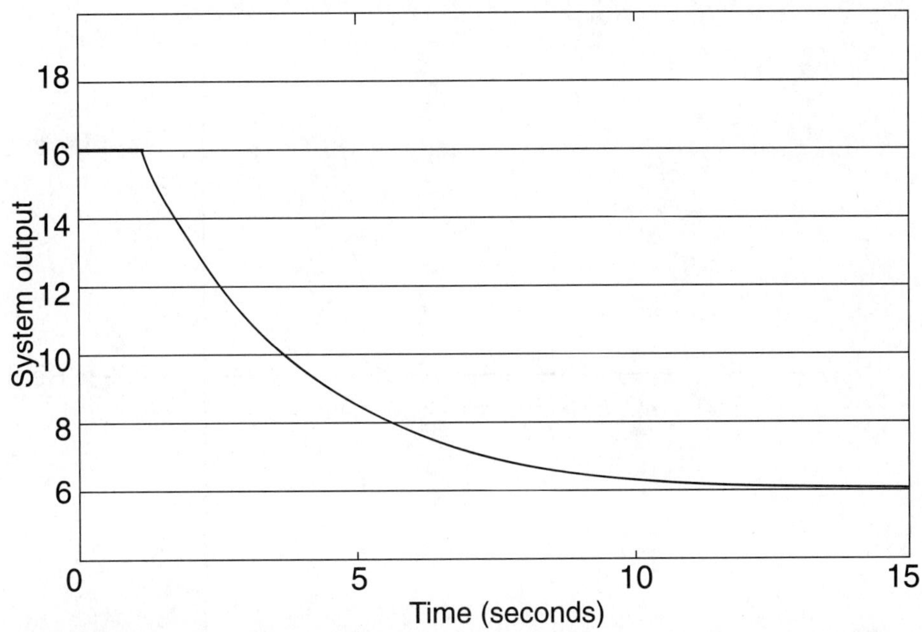

Figure 6–16

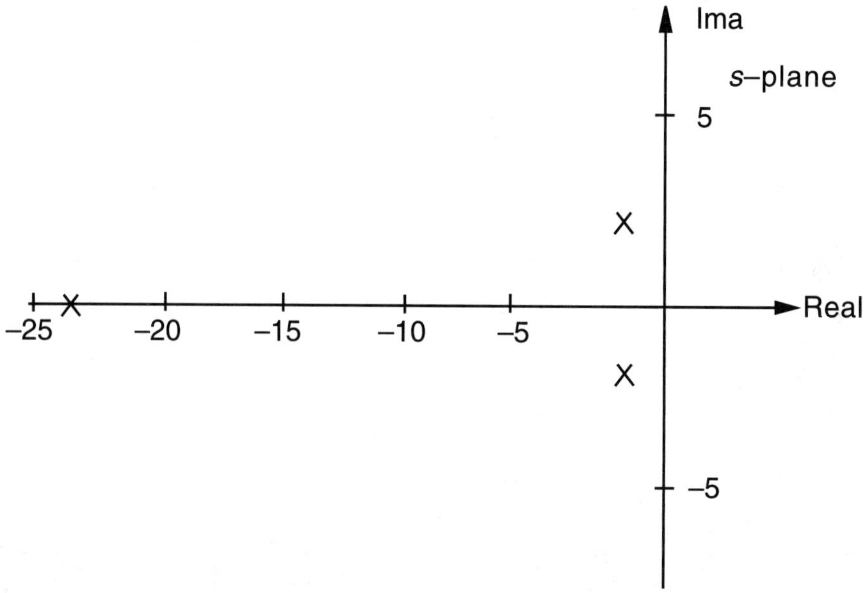

Figure 6–17

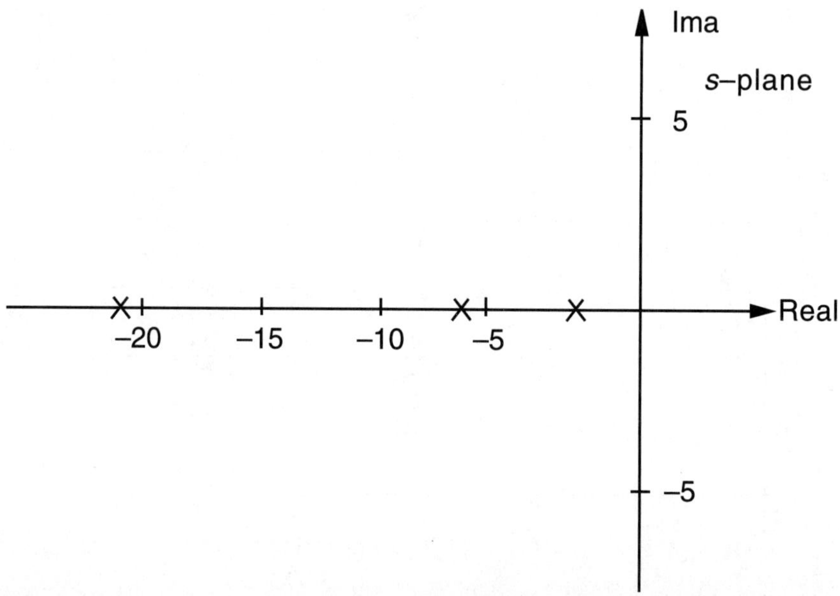

Figure 6–18

18. Consider the system with $G(s)$ and $H(s)$ specified as follows:

$$G(s) = \frac{7}{s + 2}$$

$$H(s) = \frac{8}{s + 10}$$

Determine:

- ω_n
- ς

19. If the system from the previous problem was excited with a step input, calculate the following:
    - *PO*
    - Frequency at which this system will oscillate as the output decays
20. Figure 6–19 shows the step response to a system under investigation. Determine from the response:
    - *PO*
    - ω_d

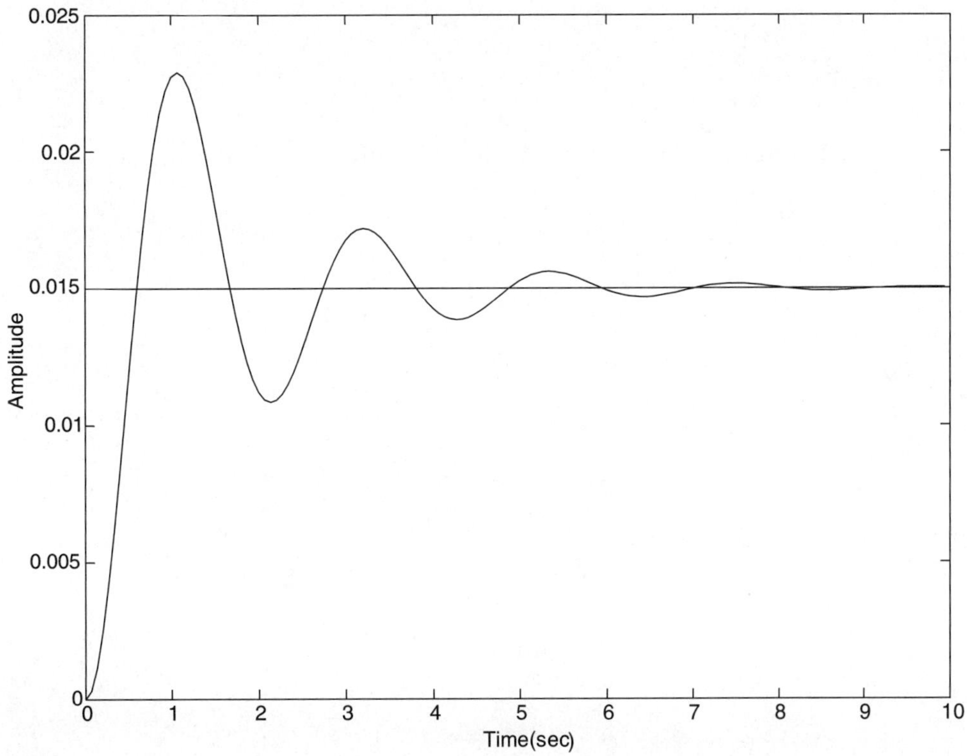

Figure 6–19

- ω_n
- ς
- Peak time

21. Consider applying a step input of the form:

$$r(t) = A\,u(t)$$

to a system whose transfer function is:

$$\frac{K}{\tau s + 1}$$

- Confirm, using the final value theorem, that the final value is *KA*.
- Confirm, using the Laplace methods of Chapter 2, that the response is given by:

$$c(t) = K\,A\left[1 - e^{-t/\tau}\right]$$

22. For the mechanical system in Figure 3–52a, $m = 5$ kg, $b = 25$ N(m/s), $k_2 = 65$ N/m, and $k_3 = 60$N/m, determine the following parameters using the formulas in the book:
   - ω_n
   - ς
   - ω_d
   - T_d
23. Simulate the previous problem to confirm your answer.

7

Root Locus Analysis

Learning Objectives

In this chapter, you will learn how to:

- Name the underlying principles for creating a root locus.
- Interpret root locus diagrams.
- Describe how the compensator affects the root arrangement.
- Use MATLAB to sketch the root locus.

7.1 History

Imagine trying to design a control system back in the 1940s, when computers were unavailable. Imagine trying to determine whether your system was stable or, if you were to increase the gain, whether the overshoot would increase or decrease. Our predecessors could make these determinations primarily due to an analysis and design method developed by Walter R. Evans (1920–2001).[1,2] This method, called root locus, is still useful today, because it allows control system designers to determine the general positions of the CLTF poles (eigenvalues). A set of rules allows control engineers to predict how the eigenvalues will move (migrate) as a specific parameter, typically the gain, changes. In other words, this method allows designers to quickly sketch the path roots take as the gain changes. In addition, with this method designers can predict how changing a parameter will affect system performance.

He was a remarkable man, as indicated in the following quotations[3]:

"He had a way of taking complex problems, finding answers to them, and communicate the process to people of more ordinary intellectual ability."

"The main key to learning, in my opinion, is to treat the problem as a game using all the simplifications possible to get the approximate answer."

His daughter Nancy recalls, "He did things differently and would think 'outside the box'; he encouraged us kids to think for ourselves." His son Greg said, "I remember some of [Dad's] ideas didn't catch on at the time: spray painting tennis balls bright yellow and orange to improve their visibility, mounting wheels on luggage to enhance their transportability, and staggering work hours to ease traffic congestion."

"Mr. Evans was not only a genius, but also a humanitarian."

7.2 Introduction

Consider a polynomial in terms of s. A nth order polynomial has n roots. When all the coefficients of the polynomial are known numerically, the polynomial can be factored and the n roots determined. When any coefficient changes value, all the roots change their values and, on the s-plane, every root changes location. Repeated changes in the value of a coefficient result in each of the roots tracing out a curve on the s-plane. These curves are the root loci, or the paths along which the roots move as the value of a coefficient (or parameter) changes. Once such curves have been obtained, inspection provides insight useful for analysis and aids in design.

7.3 Basic Theory of the Root Loci

The original techniques for determining the root loci were graphical. Computers let us factor a polynomial repeatedly for a number of values of a coefficient or parameter and show the results graphically. Thus, the computer does not use the original root locus method for determining the root loci. There are many root locus programs, and many are useful. In addition to providing the graphical results, they provide auxiliary information, which expands the benefits derived from analysis and aids in design. For instance, the computer program shows the value of the coefficient (usually the gain) that yields a particular root.

The characteristic equation of a dynamic system, in particular a feedback control system, may be written as:

$$1 + G(s)H(s) = 0 \qquad \textbf{Eq. 7.1}$$

in condensed notation:

$$GH(s) = -1$$

In general, s is a complex number, so $GH(s)$ is also complex. This means there are two types of information that can be obtained from the previous equation. The first, the magnitude condition, is obtained by taking the magnitude of both sides of the equation:

$$\begin{aligned} |GH(s)| &= |-1| \\ |GH(s)| &= 1 \end{aligned} \qquad \textbf{Eq. 7.2}$$

The second type of information leads to the angle condition:

$$\begin{aligned} \angle GH(s) &= \angle -1 \\ \angle GH(s) &= (2k - 1)180^\circ \quad \text{(expressed in degrees)} \end{aligned} \qquad \textbf{Eq. 7.3}$$

or:

$$\angle GH(s) = (2k - 1)\pi \qquad \text{(expressed in radians)}$$

where k is any integer.

A select point on the s-plane is a root of $1 + GH(s)$ if it satisfies both the preceding conditions (i.e., the magnitude and angle conditions). Each of these equations provides a curve, and the coordinates of the intersections of these curves are the roots of the polynomial. If we can find a way to construct the curve for $\angle GH(s)$ or $|GH(s)|$, the problem of finding the roots is simplified. A test point is selected on the available curve and substituted into the second equation. When this equation is also satisfied, the selected point is a root. This procedure is repeated as needed.

7.4 Proper Presentation of *GH*(*s*)

To perform root locus analysis and design by hand, it helps to represent the system in a way that facilitates the calculation of the desired system gain and the associated manipulation procedure. Consider the system in Figure 7–1a, where $G'(s)$ and

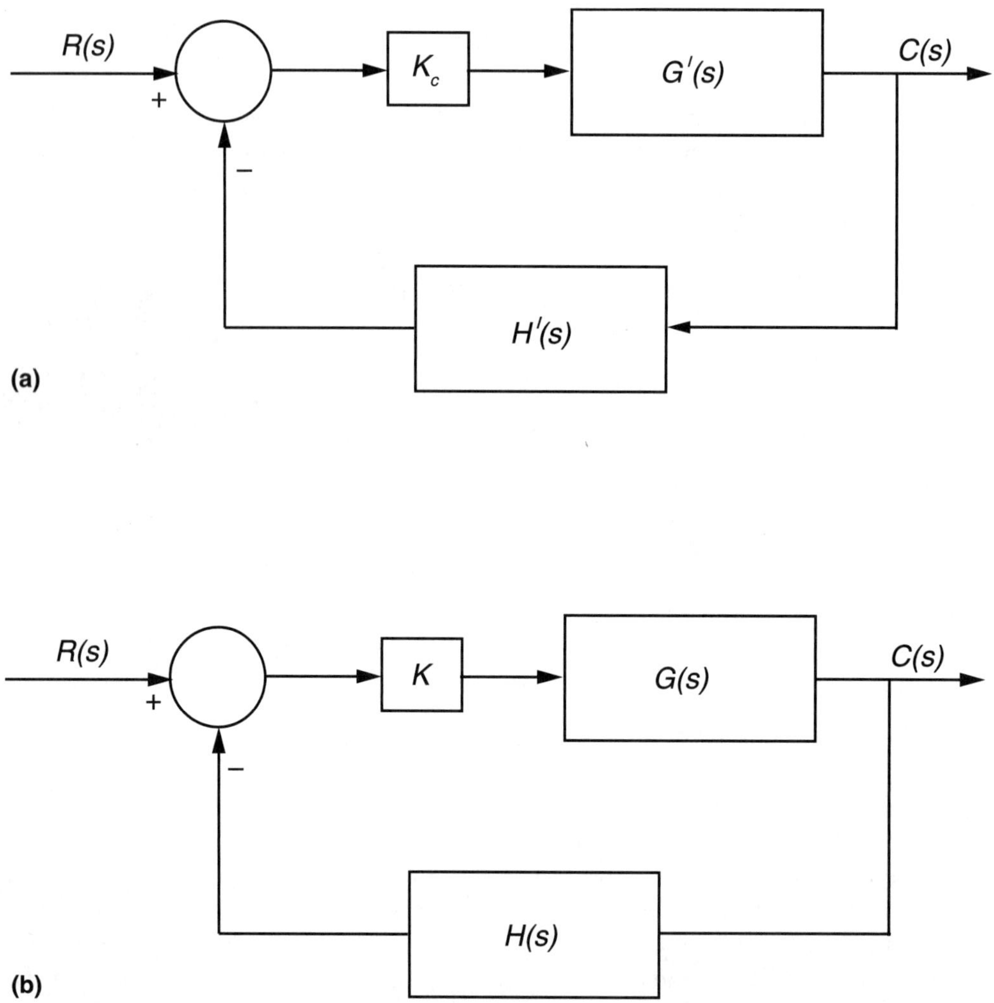

Figure 7–1 Representation of a closed-loop control system: (a)original representation and (b) working model, a modification of the original system so that $G(s)$ and $H(s)$ have unity root locus gains

$H'(s)$ represent the feedforward (plant) and feedback transfer functions, respectively. The controller in this presentation is a simple amplifier, represented with the gain K_C. To facilitate the process, it is desirable to make $G'(s)$ and $H'(s)$ have unity root locus gains by collapsing their respective gains into a newly defined controller gain, K (see Section 7.18). Thus, the system is transformed into the form in Figure 7–1b, where the newly defined $G(s)$ and $H(s)$ have unity root locus gains. Once an appropriate value of K is determined, the gain for the controller, K_C, can be calculated.

Using the representation of Figure 7–1b, the characteristic equation can be rewritten as:

$$1 + KGH(s) = 0 \qquad \textbf{Eq. 7.4}$$

which yields a revised magnitude condition:

$$|KGH(s)| = 1$$

and an unchanged angle condition because:

$$\angle KGH(s) = \angle K + \angle GH(s) = \angle GH(s) = (2k - 1)180^\circ \text{ where } k \text{ is an integer.}$$

7.5 Graphical Representation of Transfer Function Factors

Consider a point in the s-plane, s, as shown in Figure 7–2a. The vector from the origin to the point s is the vector s. What is the vector from the point -4 on the real axis to the point s? Because we do not at this time know what this vector is, let us call it x. A vector from the origin is again the vector -4. We can see by applying the laws of summation of vectors that:

$$-4 + x = s \qquad \textbf{Eq. 7.5}$$

thus:

$$x = s + 4$$

Consequently, the vector starting at -4 and ending at s is $s + 4$, its length is $|s + 4|$, and its angle is $\angle s + 4$, as shown in Figure 7–2b. We will use this result when we try to use the actual distances in calculations.

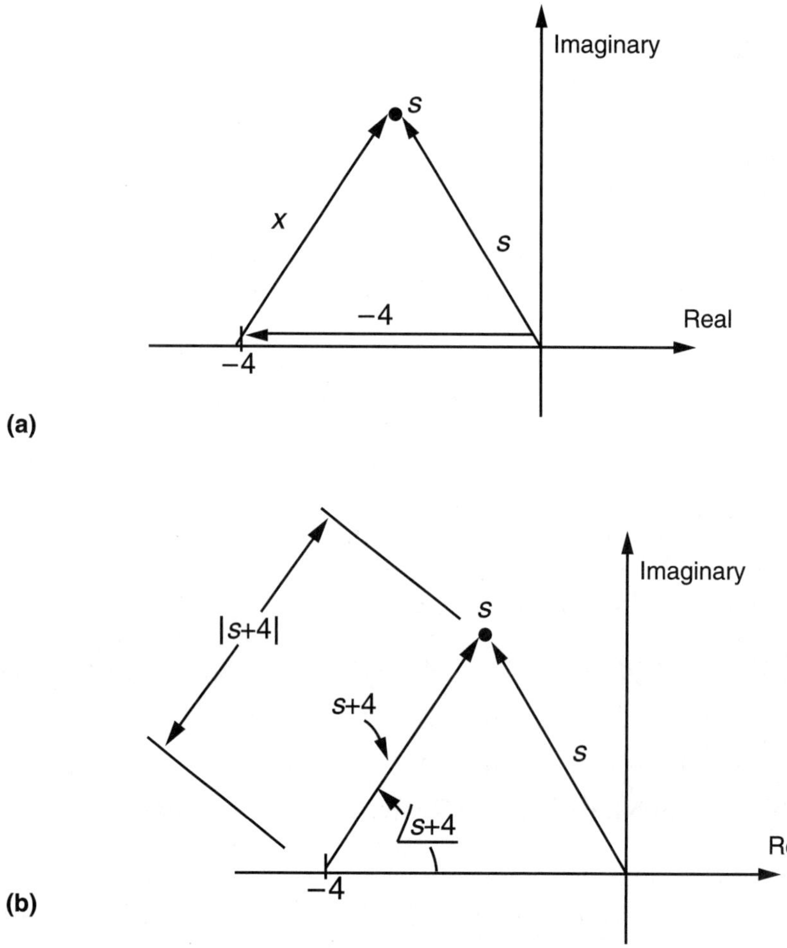

Figure 7–2 Determination of a point s in the s-plane: (a) set up and (b) solution

Number of Roots

To start the construction procedures, $GH(s)$ must be in factored form (i.e., its poles and zeros must be known and plotted on the s-plane). One would tend to think that the next step is to use trial and error: Select a test point on the s-plane, and see if it satisfies the angle condition. Repeat until a number of such points are found, and connect them with smooth curves. While not practical, this procedure is needed occasionally, as will be seen. To demonstrate, consider the following example.

Example 7.1

Explore the presence of loci by using the angle condition.

Let:

$$KGH(s) = \frac{K}{s(s+1)(s+5)} \qquad \text{Eq. 7.6}$$

Solution Plot the poles on the s-plane, select a test point (TP), and draw straight lines (vectors) from each pole to TP, as shown in Figure 7–3.

The angles defined at the poles are added to determine $\angle KGH(s)$. When the angle requirement is not satisfied, another TP is chosen and tested. The result at the first TP guides the choice of the second TP. In Figure 7–3, the angle at TP is less than the required -180 degrees. When the TP is moved to the left, all angles

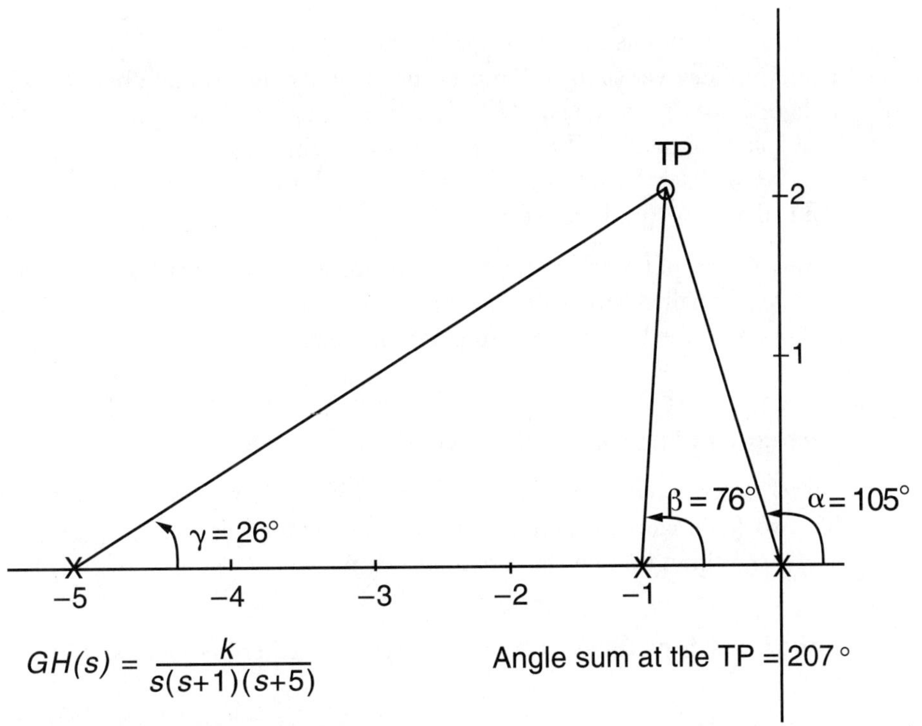

Figure 7–3 Sum of the angles at the TP

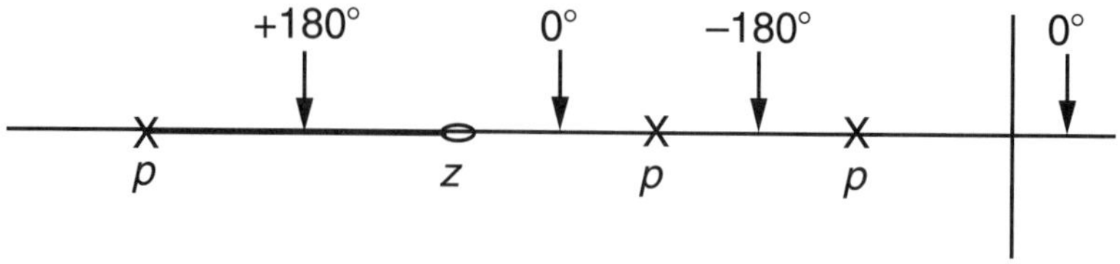

Figure 7–4 Root loci on the real axis

increase, and $\angle KGH(s)$ may approach -180 degrees. In this manner, points on the angle curve are found without excessive trial and error. With this background information, much simpler construction methods can be developed and used.

Root Locus Segments on the Real Axis

Most polynomials have some real roots. Certainly, an odd-order polynomial has at least one real root. To locate such roots, some segments of the real axis on the s-plane must satisfy the angle equation. Consider a TP on the real axis and to the right of all real poles and zeros. $\angle KGH(s)$ at such a TP must be 0 degrees and remains at 0 degrees as the TP moves in a negative direction. When the TP reaches a pole or 0 on the real axis, $\angle KGH(s)$ changes by 180 degrees, and the TP is on a root locus segment until another pole or 0 on the real axis is reached.

Consider the system of Figure 7–4, and note the locations of the root loci on the real axis. A simple rule for finding such segments is:

> When poles and zeros (singular points) on the real axis are numbered sequentially, with the rightmost point 1, segments of root loci on the real axis are to the left of odd-numbered singular points.

Terminal Points on the Root Loci

An nth order polynomial provides n root locus segments on the s-plane. Each curve has two terminal points, one for $K = 0$ and one for $K = \infty$. Evaluating these points provides help in finding the root locus segments.

Root Locations for $K = 0$ If the system transfer function is:

$$KGH(s) = K\frac{(s+z_1)(s+z_2)}{s^N(s+p_1)(s+p_2)} \quad \textbf{Eq. 7.7}$$

The characteristic polynomial is:

$$s^N(s+p_1)(s+p_2)\ldots + K(s+z_1)(s+z_2)\ldots \quad \textbf{Eq. 7.8}$$

If $K = 0$, this reduces to:

$$s^N(s+p_1)(s+p_2) = 0$$

The polynomial is in factored form, so the roots are at $s = 0, -p_1, -p_2 \ldots$

Thus, the root loci start when $K = 0$ at the poles of the OLTF, and each pole satisfies the angle condition.

Root Locations for K = ∞ Consider an all-pole transfer function, such as that in the previous example:

$$KGH(s) = \frac{K}{s(s+1)(s+5)} \quad \textbf{Eq. 7.9}$$

Plot the poles on the s-plane, select a TP, and draw lines from each pole to the TP. These lines define the angle contribution at TP for each of the poles of $GH(s)$. Add these angles. When the sum is an odd multiple of 180°, the TP satisfies the $\angle KGH(s)$ requirement and the TP is a potential root point.

To find the angle at which the root locus goes to ∞, consider a TP at ∞ on the s-plane. For the three-pole system of Figure 7–3, three angles are defined by straight lines from the poles to the TP. With TP at an infinite distance, these straight lines are essentially parallel, so each defines the same angle. As we let the TP traverse a circle of infinite radius, there will be N-points at which the sum of the angles will be $(2N\text{-}1)(180°)$. These are the terminal points at ∞. As a final illustration to convey this point, think of the case in which there is only one pole. Only one point (at an infinite distance) has an angle of 180°.

When $KGH(s)$ has zeros as well as poles, the angle technique can be used to show that each 0 is a terminal point for the root locus segment when $K = \infty$. The TP procedure readily locates all points that satisfy the $\angle KGH(s)$ requirement. This is easily seen for cases in which $KGH(s)$ is an all-pole function. One concludes that, for all such pole functions, the root loci go to ∞ as the variable gain becomes infinite. It has also been shown that the root loci terminate on finite zeros when the 0 is on the real axis. It is also true that if $KGH(s)$ has complex zeros, root loci terminate on complex zeros as the gain approaches infinity. This is shown in Figure 7–5.

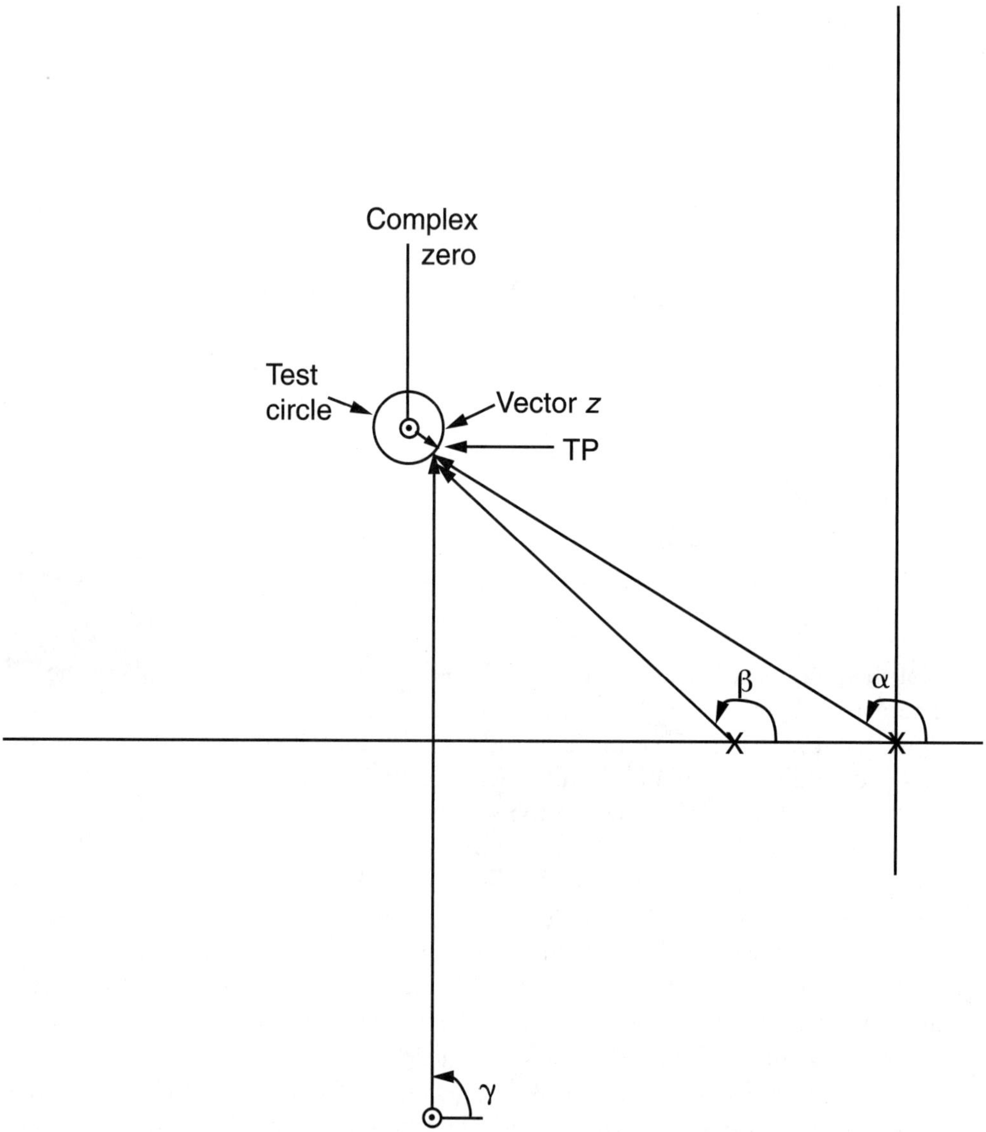

Figure 7–5 Root loci terminate at a complex zero

Choose a test point on a very small circle whose center is at a complex zero. The angle contribution of the 0 is the angle of the vector from the zero to the test point ($\angle z$), and this angle has a range of 360 degrees as the TP moves around the small circle. The sum of all other angles from other poles and zeros remains essentially constant, because the circle is small. Thus, $\angle KGH(s)$ must be an odd multiple of 180 degrees for some location of the TP and, as the radius of the circle approaches 0, there must be points near and at z that are on the root locus. Thus, the root loci are seen to terminate on zeros as the gain becomes infinite.

From the preceding analysis, one can say:

- Root loci start at the poles of $KGH(s)$ and, as the variable parameter (usually the gain) increases, the roots leave the OLTF poles and trace out the root loci.
- The number of root locus segments equals the number of poles of $KGH(s)$, providing $n \geq m$ (n is the order of the $KGH(s)$ numerator; m is the order of the $KGH(s)$ denominator).
- Root loci terminate on zeros of $KGH(s)$, and excess poles provide loci that terminate at infinity.

The root loci are planar curves, which can be found with the angle summation rule. This is occasionally of practical value. In the following paragraphs, other rules are developed to permit rapid evaluation of the loci.

7.6 Simple Example

Consider a simple example that brings together some fundamental ideas. For the system with:

$$GH(s) = \frac{K(s+3)}{(s+6)} \qquad \textbf{Eq. 7.10}$$

the root locus looks like the sketch in Figure 7–6. This means that the roots of the characteristic equation (poles of the CLTF or eigenvalues) migrate from $s = -6$ (when $K = 0$) to $s = -3$ (when $K = \infty$). When you check the angle condition, you will see that this is true. The next section allows you to determine which value of gain K produces any of the intermediate roots.

Let us apply the magnitude condition using the geometric condition previously derived. We want to know what value of K will produce a root of -5. We can

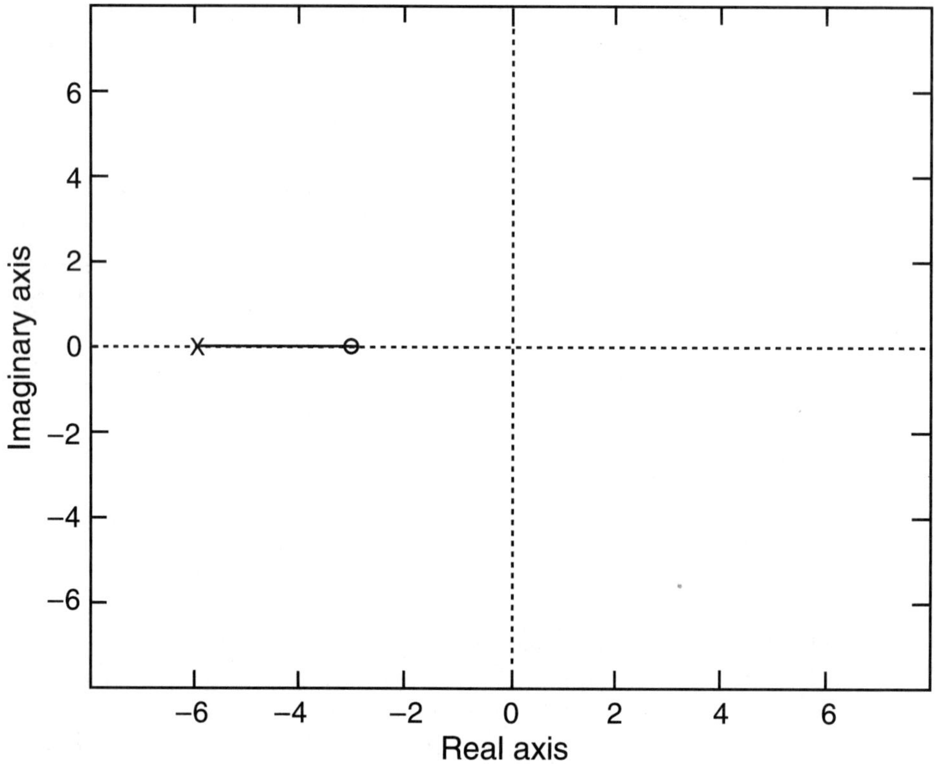

Figure 7–6 Root locus for $KGH(s) = K(s + 3)/(s + 6)$

apply the magnitude condition at point s, as shown in Figure 7–7. Applying the magnitude condition:

$$\begin{aligned} |KG(s)H(s)| &= 1 \\ \frac{K|s+3|}{|s+6|} &= 1 \\ \frac{K|2|}{|1|} &= 1 \\ K &= 1/2 \end{aligned} \qquad \textbf{Eq. 7.11}$$

This calculation can also be made by measuring the actual distance of the vectors. However, the scale (i.e., unit/inches) for both axes must be the same.

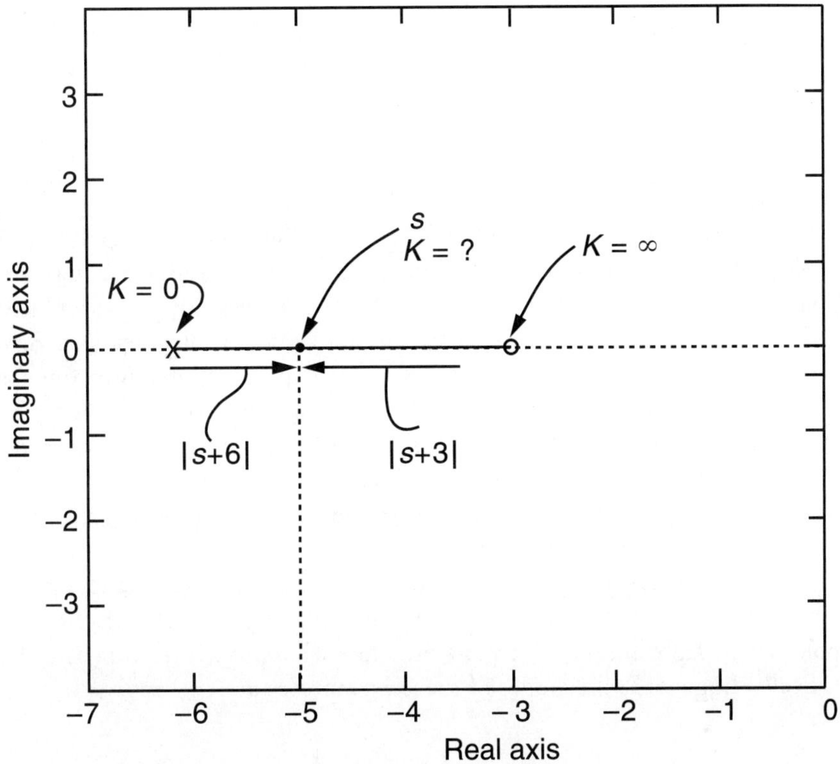

Figure 7–7 Application of the magnitude condition

7.7 Asymptotes

When $KGH(s)$ has more poles than zeros, the number of loci going to infinity equals the number of excess poles. Consider the case of two excess poles, remembering that root loci can exist only at points where:

$$\angle KGH(s) = (2k - 1)180^{\circ} \qquad \textbf{Eq. 7.12}$$

Then, as the locus approaches infinity:

$$\alpha = \alpha_1 + \alpha_2 \qquad \textbf{Eq. 7.13}$$

where α_1 and α_2 are the angles due to the two excess poles.

As the locus approaches infinity, only the two excess poles contribute to $\angle KGH(s)$. All others are canceled, because there remain equal numbers of poles and zeros. The contributions of the two excess poles must add to 180 degrees, because, for a TP at infinity, the angles from all poles and zeros are the same. It follows that each of the two excess poles contributes one-half of the required 180 degrees. For two excess poles, only two locations can satisfy the requirements (i.e., one root locus at +90 degrees, the other at −90 degrees). Remember that in this development locus is examined one at a time. For three excess poles, three loci go to infinity at +60 degrees, −60 degrees, and 180 degrees.

When there are four excess poles, at what angles do the loci go to infinity? At great distances from the origin of the s-plane, the root loci due to the excess poles are straight lines at fixed angles. The loci approach the straight lines asymptotically. In general, the angles of these asymptotes are:

$$\frac{(2k-1)}{n-m}180^\circ \qquad \textbf{Eq. 7.14}$$

One can easily solve for different $n - m$ (i.e., excess poles). The result can then be tabulated and is shown in Table 7–1.

$n - m$ (Excess Poles)	Angle of Asymptotes
1	180°
2	+90°, −90°
3	+60°, −60°, +180°
4	+45°, −45°, +135°, −135°

Table 7.1 Asymptote angles for various excess poles ($n - m$)

Example 7.2

Sketch the root locus for a system whose OLTF $KGH(s)$ is:

$$\frac{K}{(s+1)(s+3)} \quad \text{Eq. 7.15}$$

Solution The segment of the root locus on the real axis lies between the poles at -1 and -3. The root loci start at the poles but cannot terminate on the real axis segment; both these loci must terminate at infinity. There are two excess poles, so there will be two branches of the root locus going to infinity at angles of ± 90 degrees, and the angles of the asymptotes are also ± 90 degrees. Root loci that start at the poles must leave the real axis and terminate at infinity. The root locus of this system appears in Figure 7–8. Where do these root locus branches leave the real axis?

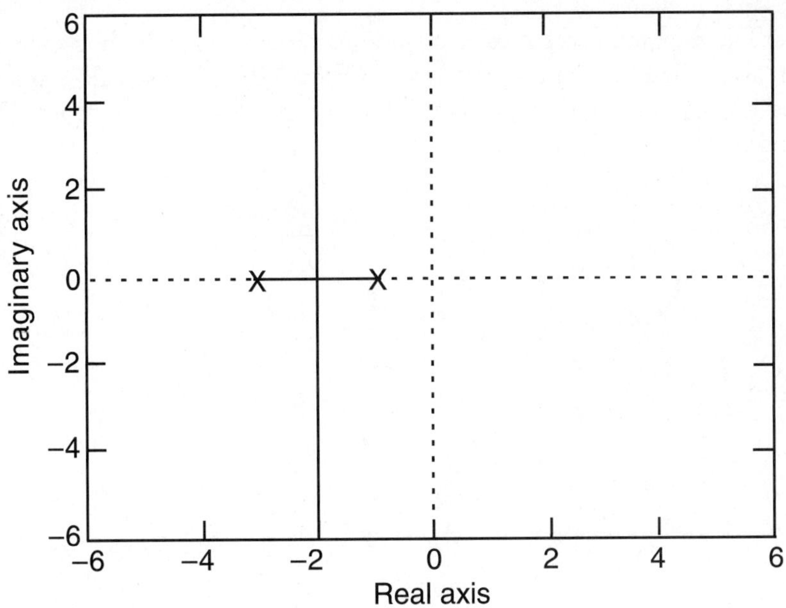

Figure 7–8 Root locus for $KGH(s) = K/(s + 1)(s + 3)$

7.8 Root Locus Intersections with the Real Axis

The locations of the roots on the s-plane depend on the angle condition and the magnitude condition. As the gain is raised, the roots move away from the poles, but the angle requirement forces the roots to stay on the root loci. For the second-order example, clearly the two roots must meet at some point between the poles. In addition, when the gain is again raised, the roots must emerge from the real axis. The point of emergence is not necessarily midway between the poles, but we can determine the emergence point easily using the magnitude condition. To do this, note that:

$$|KGH(s)| = \left|\frac{K}{|s+1||s+3|}\right| = 1 \quad \textbf{Eq. 7.16}$$

From this:

$$|K| = |s+1||s+3|$$

Choose points on the real axis, evaluate K, and plot s versus K. The locus emerges at the value of s that provides the maximum K. For different values of K, see Table 7–2.

The maximum value of K is obtained at $s = -2$, which is the emergence point, also called the point of departure. When $GH(s)$ has several zeros as well as the usual poles, the root locus may exhibit emergence and immergence. These conditions are also called points of arrival. For example, let:

$$KGH(s) = \frac{K(s+4)(s+5)}{(s+1)(s+2)} \quad \textbf{Eq. 7.17}$$

s	K
−1	0
−1.5	0.75
−2	1.0
−2.5	0.75
−3	0

Table 7–2 Value of K for different points

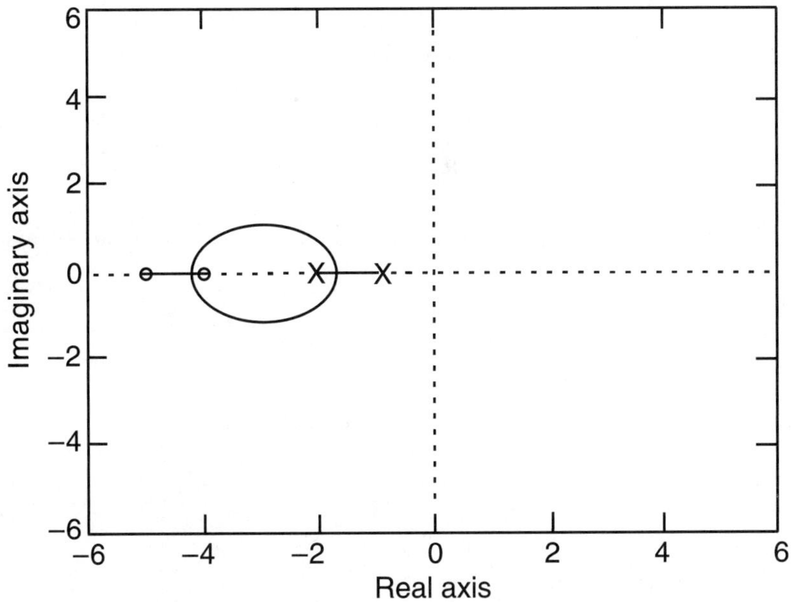

Figure 7–9 Root locus for $KGH(s) = K(s + 4)(s + 5)/(s + 1)(s + 2)$

There are two root locus segments on the real axis, as shown in Figure 7–9. The emergence and immergence points can be found using the magnitude condition for points in the real axis segments.

7.9 The Asymptote Centroid

The asymptotes, when extended, cross the real axis of the s-plane at some location we will call the asymptote centroid. It is clear that each pair of asymptotes must cross the real axis at the same point, and it should be clear that, for higher-order systems, all asymptotes cross the real axis at the same point. There are several ways to derive the location of the centroid, which is given by:

$$\sigma_x = \frac{\sum_{j=1}^{n} \text{poles}_j - \sum_{b=1}^{m} \text{zeros}_b}{n - m} \qquad \textbf{Eq. 7.18}$$

For the second-order example (Example 7.2), the point of departure appears to be exactly midway between the two poles. Note that the asymptote centroid is:

$$\sigma_x = \frac{\sum \text{poles} - \sum \text{zeros}}{n - m} = \frac{(-3-1)-0}{2-0} = -2 \qquad \text{Eq. 7.19}$$

which is exactly the same as the locus emergence point. This is not always the case.

Example 7.3

Sketch the root locus for a system where OLTF $KGH(s)$ is:

$$KGH(s) = \frac{K}{s(s+1)(s+3)} \qquad \text{Eq. 7.20}$$

Solution The poles are at 0, −1, −3. The real axis root loci are as shown in Figure 7–10.

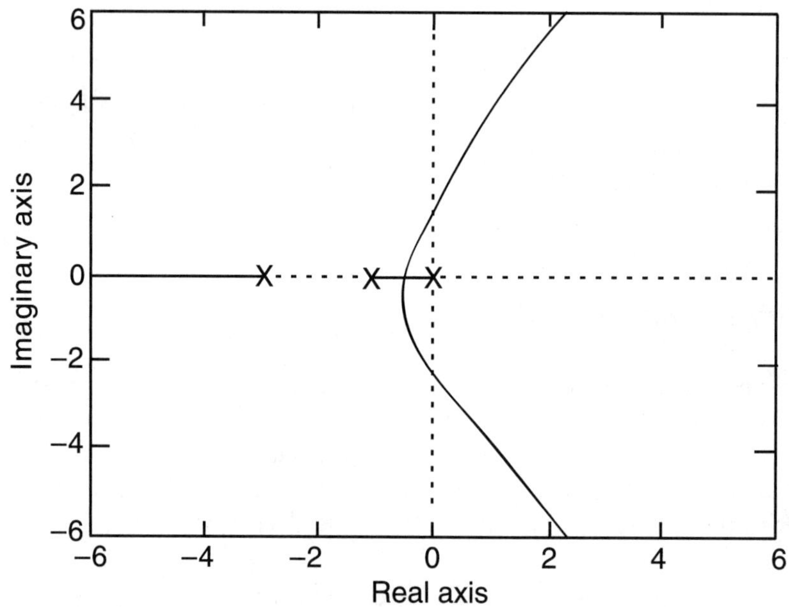

Figure 7–10 Root locus for $KGH(s) = K/s(s + 1)(s + 3)$

Three loci terminate at infinity. The asymptote centroid is at $\sigma_x = -\frac{4}{3}$, the asymptote angles are +60 degrees, −60 degrees, and 180 degrees.
The point of emergence is found using:

$$|K| = |s||s+1||s+3| \quad \text{Eq. 7.21}$$

The locus emerges from the real axis between $s = 0$ and $s = -1$, so we need not calculate outside that range. From the table of K values (Table 7–3), the locus emerges at $s = -0.45$, which is not the midpoint of the segment on the real axis. However, the difference is slight, and a root locus sketch using the midpoint would be sufficiently accurate for most analysis and design problems.

This information helps us to sketch the root loci with reasonable accuracy:

1. Plot the poles and zeros.
2. Determine and plot the asymptote centroid.
3. Determine the asymptote angles.
4. Draw the asymptotes.
5. Because loci start on poles, terminate on zeros or at infinity, and go to infinity asymptotic to the asymptotes, a reasonably good set of root loci can be obtained by sketching.
6. To improve accuracy, apply the TP and angle summation rule to locate points on the loci.

s	$\|s\|$	$\|s+1\|$	$\|s+3\|$	K
−0.1	0.1	0.9	2.9	0.261
−0.2	0.2	0.8	2.8	0.448
−0.3	0.3	0.7	2.7	0.567
−0.4	0.4	0.6	2.6	0.624
−0.45	0.45	0.55	2.55	0.63
−0.5	0.5	0.5	2.5	0.625
−0.6	0.6	0.4	2.4	0.566
−0.8	0.8	0.2	2.2	0.48
−1	1	0	2.0	0

Table 7–3 Values of K for different locations of the test point

7.10 Intersection of the Locus with the Imaginary Axis

When a root locus segment intersects the imaginary axis, the point of intersection satisfies $\angle KGH(s) = (2k-1)\pi$ and, for some value of gain, the point is a root of the characteristic polynomial, as well as a stability limit for the system. There are several ways to locate this intersection point:

- Choose TPs on the vertical axis, and evaluate $\angle KGH(s)$ at each TP until the angle condition is satisfied.
- Use the Routh criterion to find the gain at the stability limit. Then, choose TPs on the vertical axis using the condition to evaluate the gain at each point. When the value matches that obtained by Routh's, the intersection point has been located.
- Substitute $j\omega$ for s in the characteristic polynomial, then require that real and imaginary parts go to 0 independently. Next, solve the two resulting equations simultaneously.

7.11 Direction of the Locus at a Complex Pole or Zero

Using the TP concept, choose the TP to be at the point of interest. The angles contributed by the singular points are:

- α due to the complex pole or zero (this is the direction of the locus)
- β, γ, δ—angles contributed by other singular points
- -180 degrees due to the angle requirement $\angle KGH(s) = -180^\circ$

Solving for α:

$$-180^\circ \pm \beta \pm \gamma \pm \delta \ldots \qquad \textbf{Eq. 7.22}$$

7.12 Sketching the Loci

Before starting the graphical procedures, one must know *KGH*(s), including numerical values for the numerator and denominator. Both the numerator and denominator of *KGH*(s) must be in factored form so that the zeros and poles are known (root locus form). The first step is to choose the scale of the plot. Both abscissa and ordinate scales must be the same, because the angles at poles and zeros must be represented. The second step is to evaluate and plot:

- Open-loop poles and zeros
- Asymptote centroid and asymptotes

Example 7.4

Sketch the root locus for a system whose OLTF $KGH(s)$ is:

$$\frac{K}{(s+2)(s+4)} \tag{Eq. 7.23}$$

Solution This is a two-pole system, and the sketched plot (Figure 7–11) is exactly the root locus.

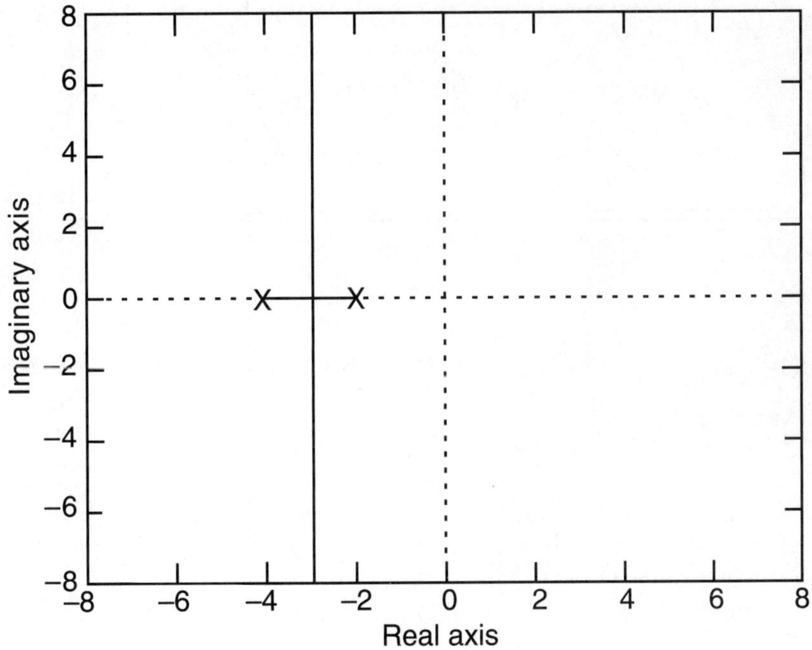

Figure 7–11 Root locus for $KGH(s) = K/(s+2)(s+4)$

Example 7.5

Sketch the root locus for a system whose OLTF $KGH(s)$ is:

$$\frac{K}{s(s+2)(s+4)} \qquad \textbf{Eq. 7.24}$$

Solution This is a three-pole system, and while the sketched plot (Figure 7–12) is not accurate in the vicinity of the origin, it permits interpretation of stability and dynamic response. When more accurate analysis is needed, the accuracy of the root locus curve can be improved using the TP angle summation procedure.

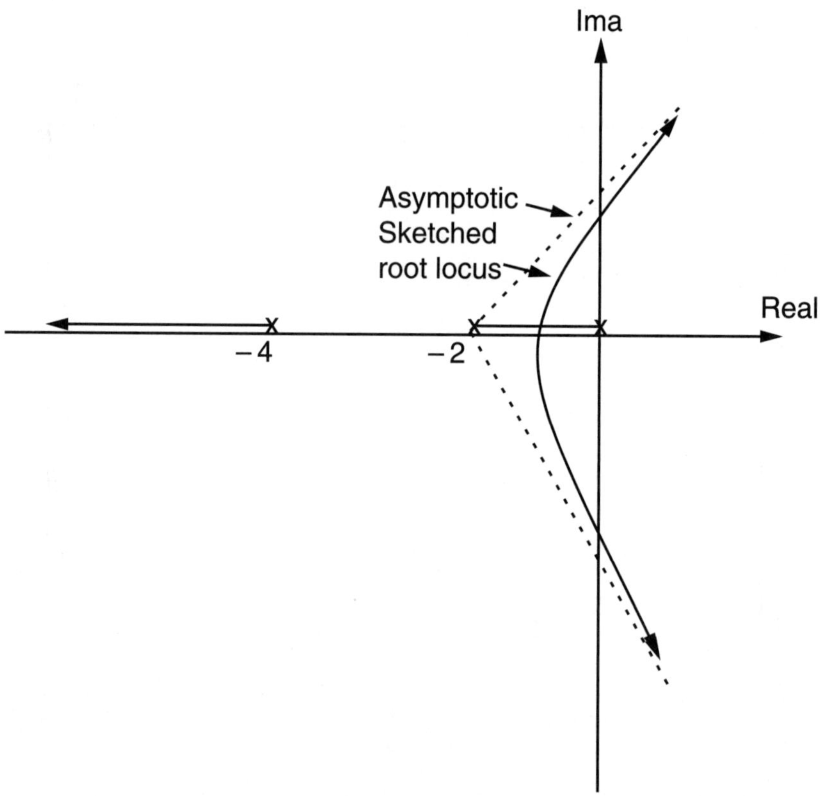

Figure 7–12 Root locus for $KG(s)H(s) = K/s(s+2)(s+4)$

Example 7.6

Sketch the root locus for a system whose OLTF $KGH(s)$ is:

$$\frac{K}{s(s+1)(s+3)} \tag{Eq. 7.25}$$

Asymptote centroid is:

$$\sigma_x = \frac{-6-0}{3-0} = -2 \tag{Eq. 7.26}$$

Angles of asymptotes:

$$+60°, -60°, 180°$$

Solution Figure 7–13 shows a sketched locus for a three-pole system.

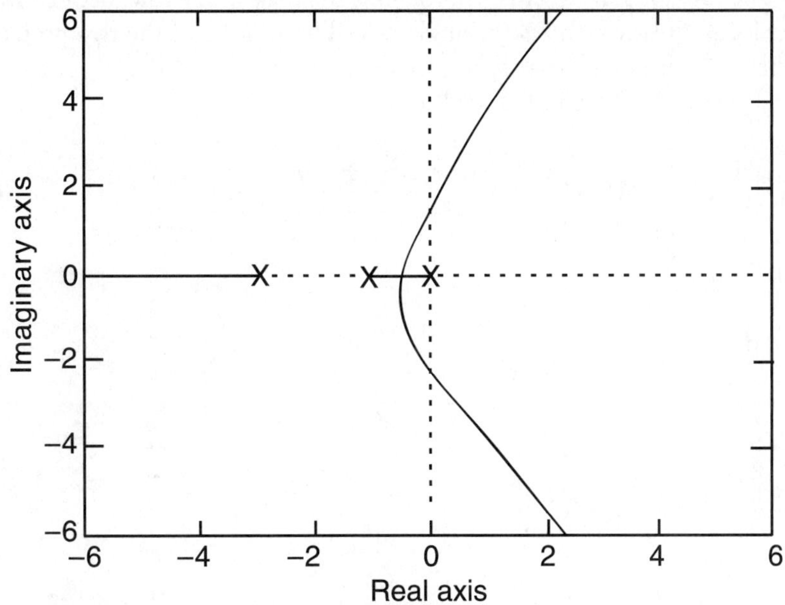

Figure 7–13 Root locus for $KG(s)H(s) = K/s(s+1)(s+3)$

7.13 Locating the Roots on the Loci

Once the root loci have been drawn to acceptable accuracy, the next step is to evaluate the locations of roots on the loci. To do this, one uses the magnitude rule:

$$|KGH(s)| = 1 \qquad \textbf{Eq. 7.27}$$

where $KGH(s)$ is a ratio of two polynomials, such as:

$$\frac{K(s^m + z_{m-1}s^{m-1} \ldots + z_1 s + z_0}{s^n + p_{n-1}s^{n-1} \ldots + p_1 s + p_0} \qquad \textbf{Eq. 7.28}$$

and K is the variable parameter, usually the gain.

Remember that for each value of K, there will be n roots, one on each branch of the locus. There are two types of problems:

1. For a chosen point on one locus, find the value of K that satisfies the magnitude condition.
2. For a chosen value of K, find the points on the loci that satisfy the magnitude condition.

The basic tool for solving either problem is the magnitude rule, because every point on the locus already satisfies the angle rule. The simpler of the two problems is the first, because repeated trials are not needed.

Rewriting $GH(s)$ in factored form:

$$KGH(s) = K\frac{(s + z_1)(s + z_2)\ldots(s + z_m)}{(s + p_1)(s + p_2)\ldots(s + p_n)} = 1 \qquad \textbf{Eq. 7.29}$$

It follows that:

$$|KGH(s)| = K\frac{|s + z_1|\ldots|s + z_m|}{|s + p_1|\ldots|s + p_n|} = 1 \qquad \textbf{Eq. 7.30}$$

Thus:

$$K = \frac{|s + p_1|\ldots|s + p_n|}{|s + z_1|\ldots|s + z_m|} \qquad \textbf{Eq. 7.31}$$

and the quantities on the right-hand side of the equation for K are the distances from the poles and zeros to the desired root point. Once a length has been determined for each factor, simple multiplication and division gives the required value of K.

Solution of the second problem requires more work but no new techniques. One chooses a point on the locus that hopefully is a root, uses the preceding technique to evaluate K, and compares this with the specified value of K. One wants the two values of K to be the same, but this is not likely on a first trial. Therefore, be prepared to choose another point and repeat the calculations.

For this second trial, the result of the first trial provides guidance for choice of the second TP: Knowing that a root starts on a pole and moves away as K increases, and knowing the value of K at the first TP, see which is larger. When the first TP requires a value of K larger than the desired K, the next TP should be moved toward the pole from which the root locus segment started and K reevaluated. Again, comparing values of K, the second choice should be better than the first. A third trial will usually be close enough for the purposes of your problem.

7.14 Root Loci for Parameters Other Than Gain

It is important to realize that the root locus method can be used to factor any polynomial. The variable parameter need not be the gain; it may be any parameter we wish. As a result, root locus studies become more valuable in both analysis and design. One new technique, partitioning, is needed. In this technique, the terms in the polynomial are rearranged into two groups, one of which contains all the terms, with the variable parameter as a factor.

As a simple example, consider the polynomial in which B is a constant:

$$s^3 + (A + 1)s^2 + As + B = 0 \quad \text{Eq. 7.32}$$

To study the effect of A as a parameter, group the terms as:

$$s^3 + s^2 + B + As^2 + As = 0 \quad \text{Eq. 7.33}$$

Divide by $s^3 + s^2 + B$ to obtain:

$$1 + \frac{A(s^2 + s)}{s^3 + s^2 + B} = 0 \quad \text{Eq. 7.34}$$

This puts the polynomial in the desired algebraic form, and all the root locus techniques may be used for a given value of the constant B.

Example 7.7

The methods can be applied to find the roots of a given polynomial (no specified variable parameter). Consider the polynomial:

$$s^4 + 20s^3 + 50s^2 + 100s + 300 \qquad \textbf{Eq. 7.35}$$

Solution This can be partitioned in various ways, but most are undesirable because of the time and labor they require. A simple partition is:

$$(s^4 + 20s^3) + (50s^2 + 100s + 300) \qquad \textbf{Eq. 7.36}$$

Dividing by $s^4 + 20s^3$ results in:

$$1 + \frac{50(s + 1 + j5)(s + 1 - j5)}{s^4 + 20s^3} = 0 \qquad \textbf{Eq. 7.37}$$

with three poles at the origin and one at -20.

The pole-zero plot is given in Figure 7–14. Use the magnitude condition to find the roots at $K = 50$. These roots will be located on each of the loci emanating from the poles. We can use MATLAB to confirm the location of the roots:

```
>> roots([1  20  50  100  300])
yields
ans=
    -17.39
    - 2.96
      0.18 + 2.41i
      0.18 - 2.41i
```

Example 7.8

A control system has high gain, and velocity feedback is used for damping. How does the feedback gain change the root loci and the roots?

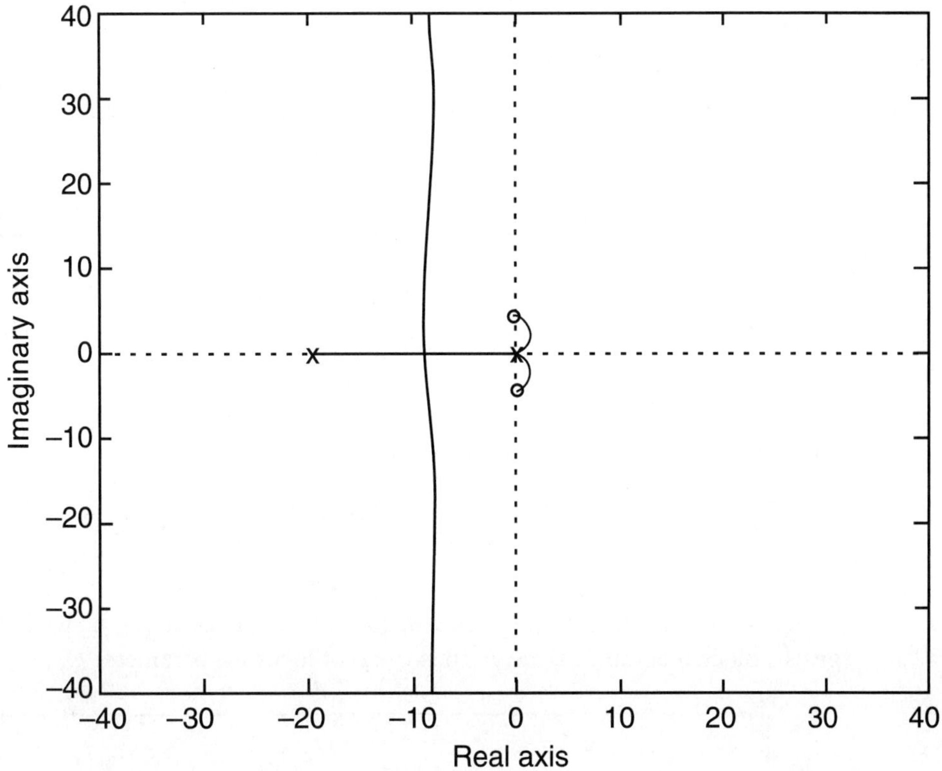

Figure 7–14 Application of root locus for solving for roots

Solution For this system, a transfer function block diagram appears in Figure 7–15. For this system:

$$G(s) = \frac{100}{s^2 + 100ks} \qquad \textbf{Eq. 7.38}$$

from which the characteristic polynomial is:

$$s^2 + 100ks + 100$$

Partitioning:

$$(s^2 + 100) + 100ks$$

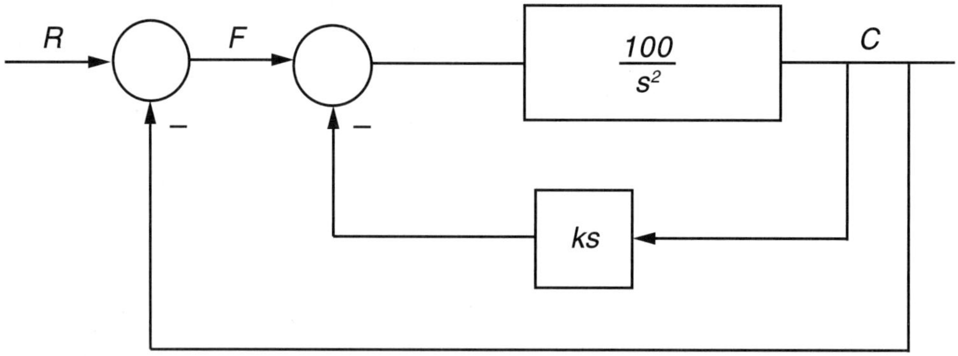

Figure 7–15 Block diagram for positioning system with tachometer feedback

and the desired algebraic form is:

$$1 + \frac{100ks}{s^2 + 100}$$

There is one 0 at the origin of the s-plane and two poles at $s = \pm j10$. Figure 7–15 gives a block diagram; Figure 7–16 is the root locus for parameter k.

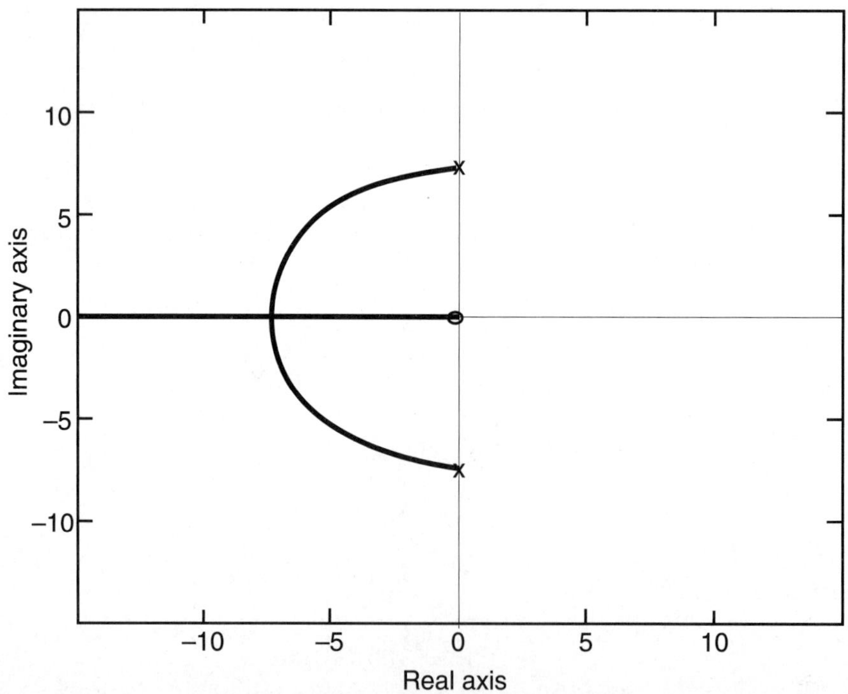

Figure 7–16 Root locus for parameter k in Figure 7–15

7.15 System Analysis from the Root Locus

Consider the root locus for the $KGH(s)$ presented earlier:

$$KGH(s) = \frac{K}{s(s+1)(s+5)} \qquad \text{Eq. 7.39}$$

The root locus is shown in Figure 7–17.

The locations of the roots on the loci depend on the value of the gain, K, which has not been specified. However, because two segments of the loci cross into the right-half plane from the left-half plane, it is clear that the system will be stable for low gains but unstable for high gains. The value of gain at the stability limit can be found using the magnitude rule or by using the Routh criterion. Within the stable region, one can choose a root location with damping ratio as desired, between 0 and 1.0.

There is one pole at the origin, so the open-loop system is type 1. There will be no steady-state error in response to a step input.

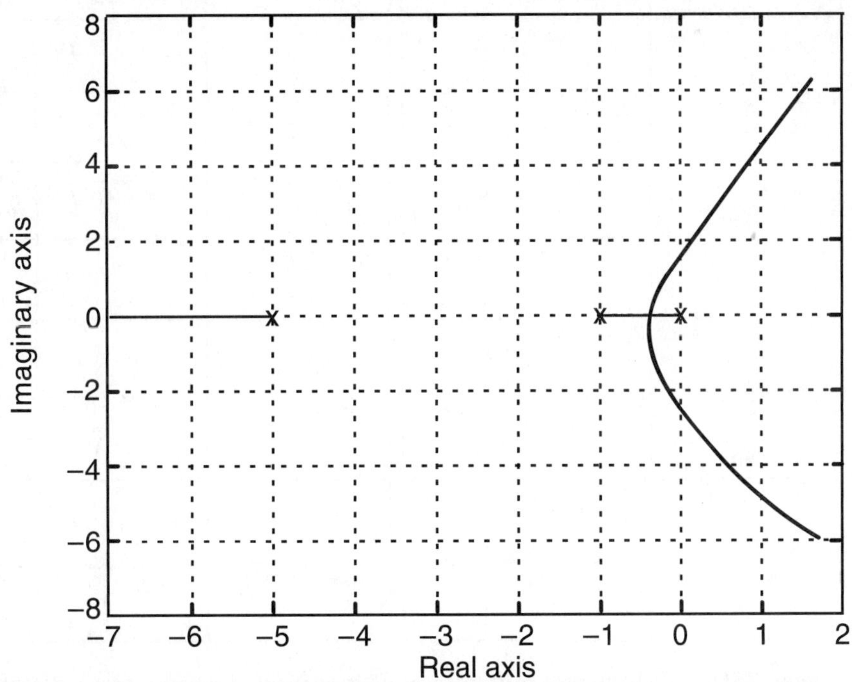

Figure 7–17 Root locus for $KGH(s) = K/s(s + 1)(s + 5)$

The third root must be real and more negative than −5. The two complex roots are clearly dominant and may be used to define an equivalent second-order system, if desired. Using such a reduced-order model and the known characteristics of a second-order system, the response may be predicted and closely approximated.

7.16 Collection of Typical Root Locus for Various Pole-Zero Arrangements

Once you have practiced creating various root locus, it helps to identify their common underlying characteristics. Ogata has produced such a plot, shown in Figure 7–18.[4]

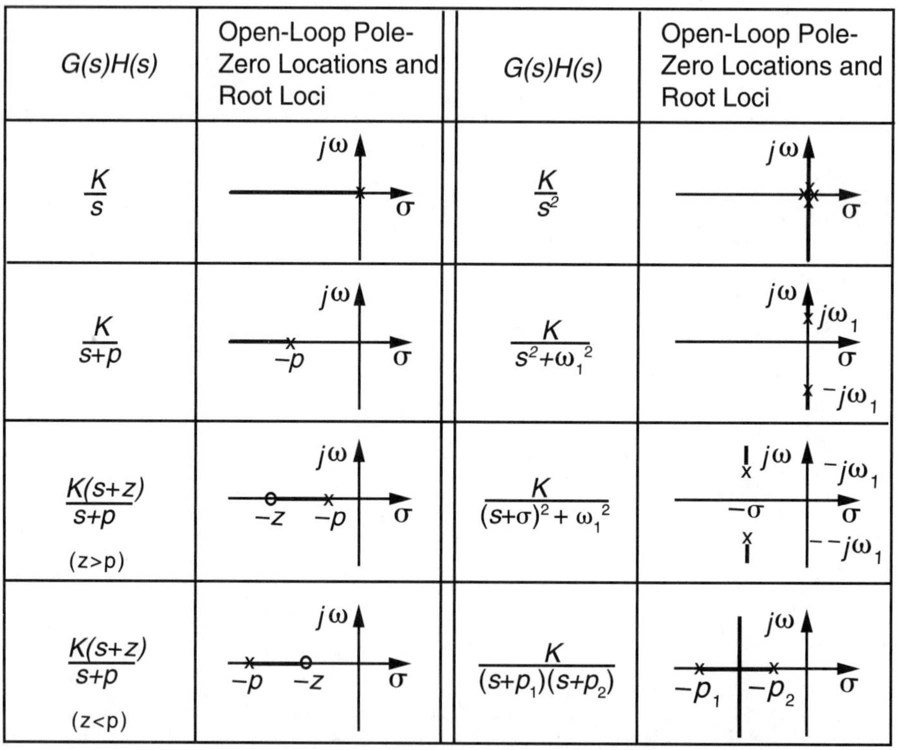

Figure 7–18 Collection of root locus plots for various pole-zero arrangements

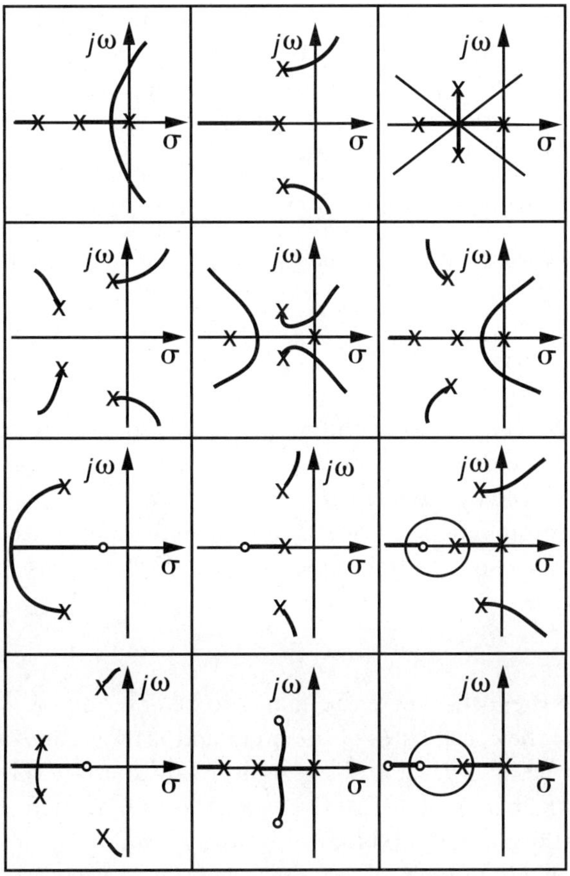

Figure 7–18 *Continued*

7.17 Using MATLAB to Draw the Root Locus

MATLAB facilitates plotting of the root loci by solving for the roots of the characteristic equation. It uses none of the rules we have just developed.

To invoke the root locus command, simply type:

```
rlocus(num,den)
```

where num is the OLTF numerator and den is the OLTF denominator. This assumes that $K = 1$ in $KGH(s)$.

To use the plot to obtain a root locus plot produced by MATLAB that can be used at a later time in order to get information, a couple of steps are necessary.

An example would be the determination of the gain value that produces a certain root. These steps ensure that the scale spacing is uniform in both the x (real) and y (imaginary) directions. The first step is to select a window that has an equal number of units in both directions. To do this after calling the `rlocus` function, invoke the `axis()` command with the size of the window (i.e., $[x_{min}, x_{max}, y_{min}, y_{max}]$):

```
>> axis([-10  0  -5  5])
```

Because MATLAB does not produce a square plot, it is necessary to invoke the `axis` function once more:

```
>> axis('square')
```

Now you can use this plot to calculate gains manually using the magnitude condition. Newer versions of MATLAB do not allow the use of "square." In such cases, one must choose axis parameters that produce the same x and y units on the standard rectangular plot produced by `rlocus`).

Another useful command is `rlockfind`, which determines the gain at a particular point. To use it, after generating the root locus plot with `rlocus(num,den)`, type

```
rlocfind(num,den)
```

and point to the place in the locus where the gain is to be determined. Two numbers will be produced: (1) the coordinates of the point and (2) the value of the gain at that point. MATLAB has also developed a graphical user interface (GUI) in the control toolbox, called `rltool`. This GUI allows the user to define each of the transfer functions in the closed-loop block diagram and, following, it shows the root locus. This is an excellent design tool.

7.18 Summary of Rules for Drawing the Root Locus

The following rules summarize the root locus procedure, a graphical technique that yields an approximate representation of the migration of the CLTF poles (eigenvalues) as the controller gain changes from 0 to ∞.

Referring to Figure 7–1a, this procedure can be used to determine the value of the gain, K_C, that provides a specific transient response (in terms of stability, speed to response, and steady error coefficient). Recall the definitions:

K_C = compensator gain (to be determined)

K_G = feedforward gain (DC gain of $G'(s)$) (transfer function in root locus form)

K_F = feedback gain (transfer function in root locus form)

It is easier to work with transfer functions with unity root locus gains. For that reason, we collapse the gains into the controller's gain. We redefine $G(s)$ and $H(s)$ to have unity root locus gains. Then K is defined as:

$$K = K_C K_G K_F \quad \textbf{Eq. 7.40}$$

This representation appears in Figure 7–1b. Once the design procedure produces a specific value of K, the controller gain can be computed from:

$$K_C = (K_G K_F)/K \quad \textbf{Eq. 7.41}$$

Nomenclature

OL = open loop

CL = closed loop

m = number of open-loop zeros = order of OL numerator, $N(s)$

n = number of open-loop poles = order of OL denominator, $D(s)$

$$KGH(s) = \frac{N(s)}{D(s)} \quad \textbf{Eq. 7.42}$$

The root locus approach deals with the denominator of the CLTF (characteristic equation):

$$CLTF = \frac{C(s)}{R(s)} = \frac{KG(s)}{1 + KG(s)H(s)} \quad \textbf{Eq. 7.43}$$

or:

$$1 + KG(s)H(s) = 0$$

which leads to the two conditions:

magnitude condition $= \left|KG(s)H(s)\right| = 1$

angle condition $= \angle KG(s)H(s) = (2k - 1)180^\circ$ where k is an integer.

Number of Roots

Each value of K produces n roots (CL poles or eigenvalues). As K changes, these roots move, creating a locus, which is what we are trying to sketch.

Migration

CL poles move from the OL poles toward OL zeros as K goes from 0 to ∞. If there are more CL poles than CL zeros, $n - m$ loci will migrate toward infinity (i.e., the CL zeros reside at infinity).

Real Axis

Choose a TP on the real axis. When the total number of OL real poles and OL real zeros to the right of this TP is odd, then this point lies on the root locus.

Number of Asymptotes

The number of asymptotes is $n - m$.

Angle of Asymptotes

$$\text{Angle of asymptotes} = \frac{(2k-1)180^\circ}{n-m} \quad k = 0, \pm 1, \pm 2, \ldots \qquad \textbf{Eq. 7.44}$$

If this formula were evaluated for different values of k, the following angle of asymptotes would result (Table 7–4):

$n - m$ (Excess Poles)	Angle of Asymptotes
1	180°
2	+90°, −90°
3	+60°, −60°, +180°
4	+45°, −45°, +135°, −135°

Table 7–4 Asymptote angles for various excess poles ($n - m$)

Origin of Asymptotes (Centroid)

Origin of asymptotes on real axis:

$$\sigma_x = \frac{\sum \text{poles} - \sum \text{zeros}}{n - m} \qquad \textbf{Eq. 7.45}$$

Break-Away and Break-In Points

Solve the characteristic equation for *K*, then perform *dK*/*ds* and set the result equal to 0:

$$D(s)\frac{d}{ds}[N(s)] - N(s)\frac{d}{ds}[D(s)] = 0 \qquad \textbf{Eq. 7.46}$$

Values of *K* at Break-In and Break-Away Points

Substitute values of *s* obtained in previous step into:

$$K = \frac{1}{G(s)H(s)} \qquad \textbf{Eq. 7.47}$$

Crossing of Imaginary Axis by Loci

Using trial and error, choose a TP, s_1, and determine if it meets magnitude and angle conditions.

Routh Criterion

Substitute $s = j\omega$ into a characteristic equation, and equate real and imaginary parts to 0. Solve for ω and *K*.

Angles of Departure (in Complex Zeros)

Angle of departure = 180

$-\Sigma$—angle of vectors from other poles to complex zero in question

$+\Sigma$—angle of vectors from other zeros to complex zero in question

Angles of Arrival (in Complex Poles)

Angle of arrival = 180

$-\Sigma$—angle of vectors from other zeros to complex pole in question

$+\Sigma$—angle of vectors from other poles to complex pole in question

Summary

1. Root locus provides a graphical representation of the behavior of the roots of the characteristic equation. It shows the path of migration of the roots as the system gain increases.
2. The method can also be used to investigate parameters besides system gain.
3. The root locus method provides some help in obtaining the actual roots of the characteristic equation.
4. The root locus method continues to be useful today in providing the control engineer insight into how parameters should be changed to improve performance or guarantee stability.
5. MATLAB facilitates the construction of the root locus, even though it does not use the root locus technique.

Idea Portfolio

1. Perform a root locus analysis on your system in which you change the controllers gain, K. Reformulate the root locus problem in terms of another critical parameter, and evaluate its input on system stability.

Problems

1. Construct a root locus plot for a unity feedback system whose transfer function follows. Show the value of K at the pole and the 0, as well as the direction of increasing K.

$$G(s) = \frac{K(s+4)}{(s+2)}$$

2. For Problem 1, pick a point between $s = -4$ and $s = -2$, $s = -3$.
   - On your root locus sketch, show the vectors going from the 0 to the TP (i.e., $(s + 4)$) and the pole to the TP (i.e., $(s + 2)$).
   - Confirm that the angle condition is met.
   - Use the magnitude condition to determine the value of K that would place the root at this point.
   - Enter this value of K into the characteristic equation, and confirm that it produces a root corresponding to the chosen point.
3. Sketch the root locus for a unity feedback system whose transfer function is:

$$G(s) = \frac{K(s+5)}{(s+1)(s+2)}$$

4. For Problem 3, answer the following with underdamped, critically damped, or overdamped:
   - Very small values of K produce roots that are ______.
   - Intermediate values of K produce roots that are ______.
   - Large values of K produce roots that are ______.
5. Sketch the root locus for a unity feedback system whose transfer function is:

$$G(s) = \frac{K(s+7)}{s(s+1)(s+2)}$$

6. Obtain a root locus for the systems whose block diagram is shown in Figure 7–19.
7. Sketch the root locus of a unity feedback system whose transfer function is:

$$G(s) = \frac{K(s+2)}{(s+1)(s^2+6s+10)}$$

8. Sketch the root locus of a unity feedback system whose transfer function is:

$$G(s) = \frac{K}{s(s+3)(s^2+2s+2)}$$

9. For Problem 8, it is desired to have a system with a PO of 17 percent. Using MATLAB, determine the value of K that will produce it. Hint: After drawing

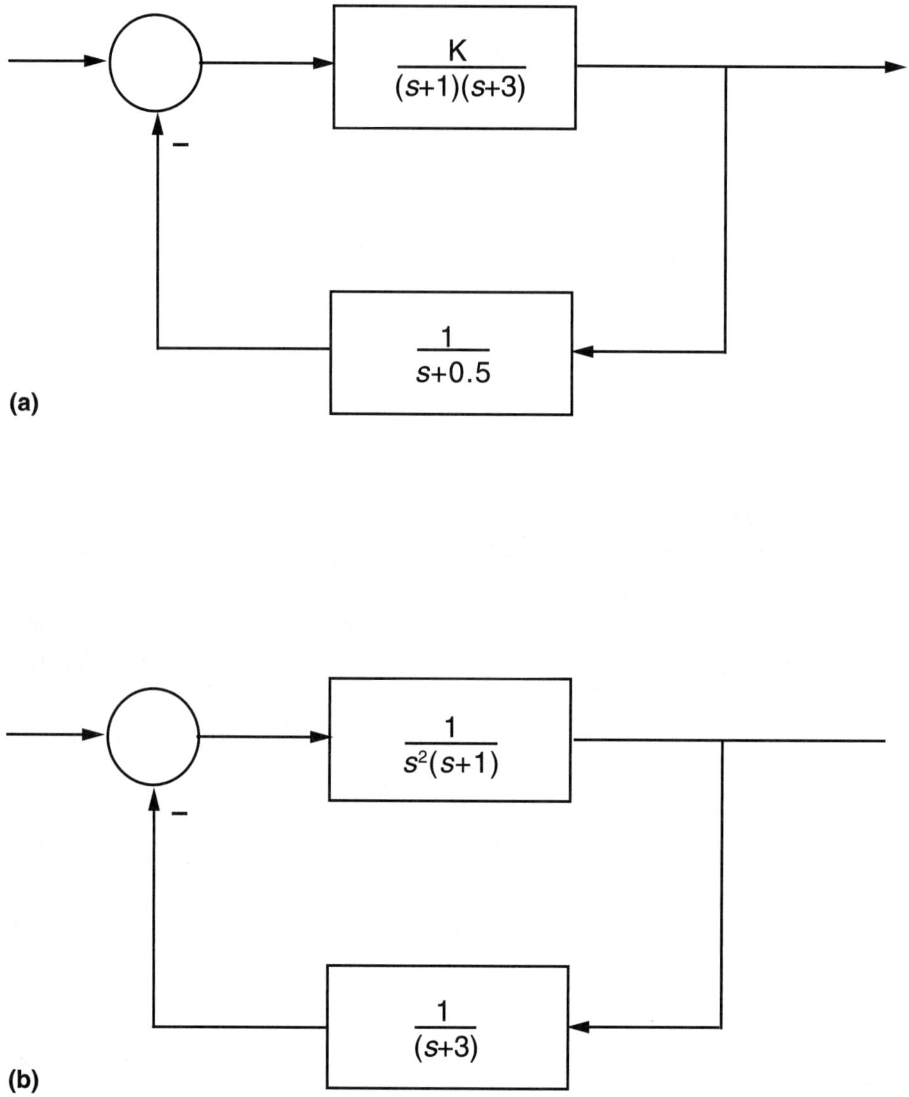

Figure 7–19

the root locus using MATLAB, and ensuring that the plot is square, hold the plot and draw a line of constant ς, which would have a slope of:

$$\frac{-1}{\tan(\sin^{-1}(\varsigma))}$$

Next, determine the value of K at that intersection. You can use `rlocfind (Gnum, Gden)`, then click at the intersection of the line and the locus to obtain the value of K.

10. Simulate the system using the value of K obtained in Problem 9. Apply a step input, and observe the response. Using the `roots()` command, determine all the eigenvalues corresponding to this value of K. Does the PO correspond to the design value? Explain.
11. Perform the exercise in Problem 10 using the control toolbox GUI `rktool`.
12. Using root locus analysis, determine the value of K for a unity feedback system with the trasfer function:

$$G(s) = \frac{K}{(s+1)(s+3)}$$

such that the resulting damping ratio, $\varsigma = 0.7$. Simulate your results to confirm your design.

13. A unity feedback control system has a plant with the following transfer function:

$$G(s) = \frac{20(s^2 + As + 10)}{s^2(s+1)}$$

Obtain a root locus as a function of A. Confirm your results using MATLAB. Determine:

- Asymptote angles and centroid
- Break-away point and break-in point
- Departure angle
- Imaginary axis crossing

14. Sketch the root locus of a unity feedback system whose transfer function is:

$$G(s) = \frac{K(s+10)(s+6)}{(s^2+9)}$$

8

Frequency Response Analysis

Learning Objectives

In this chapter, you will learn how to:

- Interpret the information in a Bode plot.
- Predict the dynamic characteristics of a system by examining its Bode plot.
- Draw a Bode plot given the transfer function.
- Use MATLAB to obtain a Bode plot.
- Describe the implications of gain and phase margin.

8.1 Historical Perspective

As has been shown, a dynamic system is unstable when any root of its characteristic polynomial is in the right half of the *s*-plane. A test for the existence of such roots is a stability test and bears the name of its developer, Henry Nyquist.[1] At the time, Bell Laboratories was developing negative feedback amplifiers for use as telephone repeaters. The amplifiers were used in transoceanic telephone lines laid at the bottom of the ocean. The telephone signal was transmitted through wires but, after a certain distance, the signal lost some of its power. For that reason, "repeater" or amplifier stations were needed every few miles.

When testing these amplifiers, Harold S. Black found that sometimes the units became unstable and acted as oscillators. When an amplifier became an oscillator, it had to be brought to the surface of the ocean and replaced. Nyquist was asked to study the problem and, in doing so, developed his stability test.

Nyquist was concerned with feedback amplifiers operating with sinusoidal signals. In this text, we are concerned with dynamic systems that have the same structure as the feedback amplifier and produce the same equations. The knowledge gained by years of experience with communication systems applies readily to other dynamic systems described by similar equations.

The human ear is a filter with logarithmic frequency response characteristics. Hendrik W. Bode developed design methods for feedback amplifiers in telephone communication systems.[2,3] Bode used the basic relationships for a transfer function expressed in the time constant format, what we call the Bode format, which allows the dynamics associated with each term to be directly recognizable.

8.2 Introduction

For most dynamic systems, and especially feedback control systems, the characteristics of primary concern are time-domain characteristics. The engineer is most concerned with stability, with the nature and duration of transients, and with steady-state accuracy.

8.3 Mathematical Basis

Using frequency response methods we can determine stability, estimate the nature and duration of the transient response, and check steady-state accuracy. The justification for this lies in the Fourier transform:

$$f(t) = \frac{1}{2\pi}\int^{\infty} F(j\omega)e^{j\omega t}\,d\omega$$

$$F(\omega) = \int_{-\infty}^{\infty} f(t)e^{-j\omega t}\,dt \qquad \textbf{Eq. 8.1}$$

The time domain and frequency domain are related, so knowing one we can calculate the other. It is also possible to establish a correlation between the time domain and the frequency domain. The equations themselves are seldom used when calculations are needed, because they require time and labor.

Most mathematical work uses ω, radian frequency, which has units of rad/s. The equation relating the usual frequency, f, in Hz to w, or radian frequency, is:

$$f = \frac{\omega}{2\pi} \tag{Eq. 8.2}$$

8.4 Returning to the Original Notation for OLTF

At this point, it helps to simplify the feedback system in terms of $G(s)$ and $H(s)$. The Kc controller gain, introduced in Chapter 7, or any transfer function it represents, should be collapsed into $G(s)$. Thus, from this section on, the OLTF is simply $G(s)H(s)$ or $GH(s)$ in shorthand.

8.5 System Response to a Sinusoidal Input

When analyzing dynamic systems, it is sometimes useful to excite the system with a sinusoidal input and observe its output. When the system is linear, the output is also a sinusoidal function, but it will have a different amplitude and phase than the input. When the input sine wave is applied first, the output may not be a perfect sine wave initially, but eventually, after about five time constants, the output will become a sine wave. The initial part of the response is usually called the transient response; the remainder is called steady-state sinusoidal response.

8.6 Application of the Frequency Response Diagram

Many applications represent information in the frequency domain, rather than in the time domain. One special graphic application is the Bode plot, which we will use extensively in this chapter. Some other applications include a graphical method for repesenting:

- A transfer function
- The characteristics of a dynamic system (i.e., a sensor/device).

The following examples illustrate these applications.

Audio Amplifier

The Bode plot of the audio amplifier conveys whether the amplifier amplifies all the frequency components in a given signal with the same gain.

Flow Meter

The Bode plot of the flow meter conveys if a flow meter can detect quick changes in the measured flow.

Microphone

A frequency response plot reveals if a microphone can capture the high-frequency components of music.

Speaker

The Bode plot of a speaker reveals if the speaker can reproduce the signals received from the audio amplifier. If not, the music will be distorted.

Acoustics of a Room

A plot of acoustic characteristics conveys whether the sound from musical instruments and/or a singer will be muffled by the characteristics of the room. It also helps the sound staff adjust for it.

Such a plot can also be used as a control system design tool by providing a method for showing the energy characteristics of a given signal, such as the acoustic energy in an industrial noise or a musical instrument. Figure 8–1 shows the wave form and frequency content of a Piano C^{II} and a Clarinet C^{I}.[4]

The dynamic characteristics of an unknown device can help predict the device's dynamic behavior. By performing sinusoidal tests, you can experimentally obtain the transfer function of a system.

The frequency response of the OLTF can help determine the CLTF's behavior. For instance, the frequency response of the OLTF system is stable, as well as the corresponding margin of safety (gain margin, phase margin).

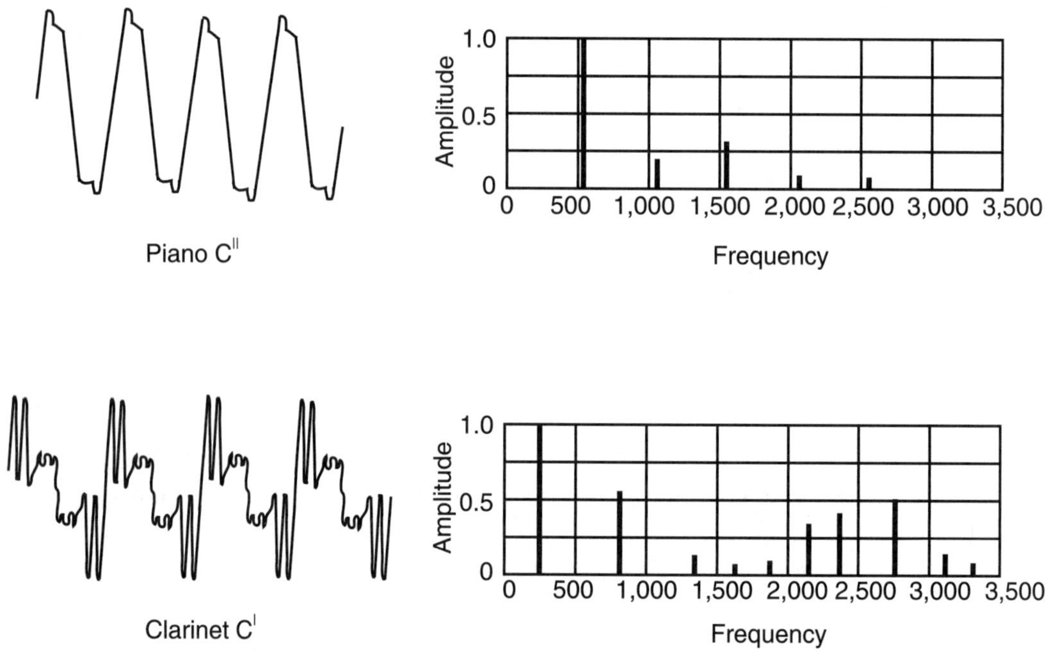

Figure 8–1 Wave forms and frequency distributions of two musical instruments (adapted from Weber, While, and Manning 1959)

8.7 First-Order System Response

To understand how a dynamic system responds to a sinusoidal input, consider the following simple, first-order system corresponding to a low-pass filler:

$$\frac{C(s)}{R(s)} = \frac{1}{\frac{s}{\omega_o} + 1} \qquad \text{Eq. 8.3}$$

Assume a sinusoidal input of adjustable frequency, ω, then $s = j\omega$, and:

$$\frac{C}{R}(j\omega) = \frac{1}{\frac{j\omega}{\omega_o} + 1} \qquad \text{Eq. 8.4}$$

Relative Value of ω	Magnitude Ratio	Phase
$\omega << \omega_o$	1	0
$\omega = \omega_o$	$1/\sqrt{2} = 0.707$	$-45°$
$\omega >> \omega_o$	0	$-90°$

Table 8–1 Magnitude and phase for a low-pass filler for different values of ω

Because this is a complex quantity, it embodies two types of information: (1) magnitude and (2) phase. For the magnitude:

$$\left|\frac{C}{R}(j\omega)\right| = \frac{1}{\left[\left(\frac{\omega}{\omega_o}\right)^2 + 1\right]^{1/2}} \qquad \textbf{Eq. 8.5}$$

and for the phase:

$$\angle\frac{C}{R}(j\omega) = -\tan\frac{\omega}{\omega_o}$$

we can perform some simple calculations to determine the value of the magnitude ratio and the phase for some relative values of ω, as shown in Table 8–1.

Notice that low frequencies (relative to ω_o) are passed through unchanged. Higher frequencies are attenuated and acquire some phase shift. For this reason, this transfer function is appropriately named a low-pass filter. A frequency response plot appears in Figure 8–2.

8.8 Second-Order System Response

We can analyze a second-order system with the transfer function:

$$\frac{C(s)}{R(s)} = \frac{\omega_n^2}{s^2 + 2\varsigma\omega_n s + \omega_n^2} \qquad \textbf{Eq. 8.6}$$

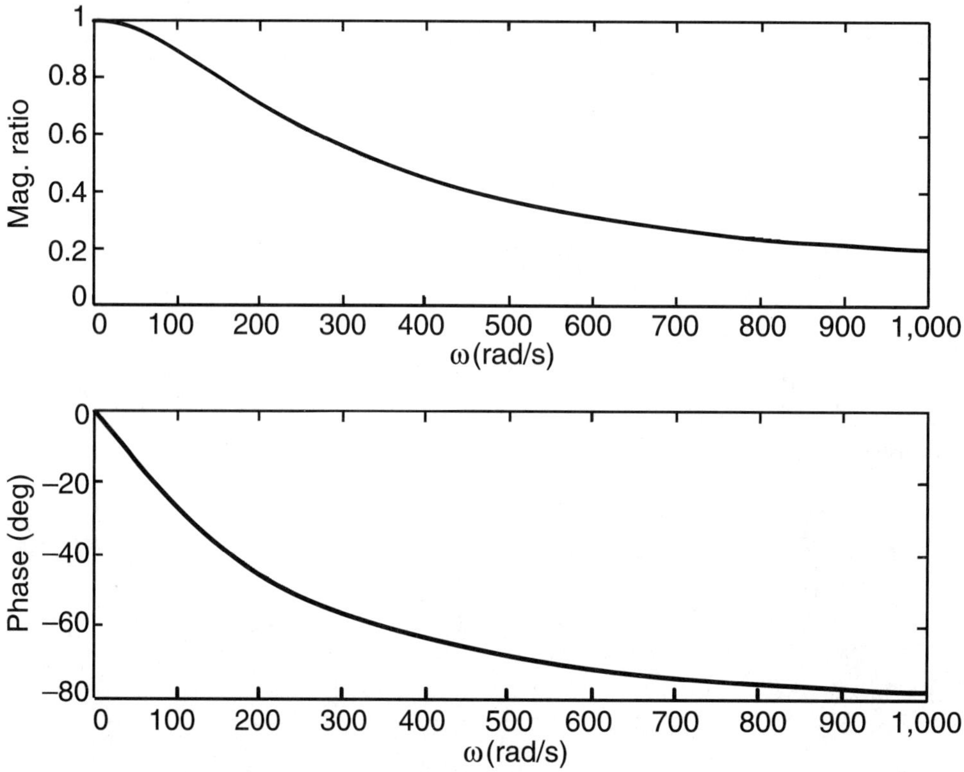

Figure 8–2 Magnitude and phase for a low-pass filter

Again, assuming a sinusoidal input of adjustable frequency, ω, then $s = j\omega$ and:

$$\frac{C}{R}(j\omega) = \frac{\omega_n^2}{(j\omega)^2 + 2\zeta\omega_n j\omega + \omega_n^2} = \frac{\omega_n^2}{(\omega_n^2 - \omega^2) + j2\zeta\omega_n\omega} \qquad \textbf{Eq. 8.7}$$

or:

$$\left|\frac{C}{R}(j\omega)\right| = \frac{\omega_n^2}{[(\omega_n^2 - \omega^2)^2 + (2\zeta\omega_n\omega)^2]^{1/2}}$$

We find the maximum value of $\left|\frac{C}{R}(j\omega)\right|$ by differentiating with respect to ω, setting to 0, then manipulating to provide the value of ω at which the maximum occurs (i.e., the resonant frequency). This provides:

$$\omega_R = \omega_n\sqrt{1 - 2\varsigma^2} \qquad \textbf{Eq. 8.8}$$

and:

$$\left|\frac{C}{R}(j\omega)\right|_{max} = M_{p\omega} = \frac{1}{2\varsigma\sqrt{1 - \varsigma^2}} \qquad \textbf{Eq. 8.9}$$

the height of the resonance peak, which is a function of ς only. Using this, we can also determine the bandwidth, which is the range of frequencies over which $\left|\frac{C}{R}\right|$ exceeds 70.7 percent.

$$\left|\frac{C}{R}(\omega = \omega_b)\right| = \frac{1}{\sqrt{2}} = \frac{\omega_n^2}{[(\omega_n^2 - \omega_b^2)^2 + 4\varsigma^2\omega_n^2\omega_b^2]^{1/2}} \qquad \textbf{Eq. 8.10}$$

from which:

$$\omega_b = \omega_n\sqrt{1 - 2\varsigma^2 + (2 - 4\varsigma^2 + 4\varsigma^4)^{1/2}} \qquad \textbf{Eq. 8.11}$$

We can also define phase margin (PM or ϕ_m), a measure of relative stability that will be developed and used later. It is defined as:

$$\phi_m = 180\,\text{deg} - \angle HG(j\omega)_{|GH(j\omega)=1|} \qquad \textbf{Eq. 8.12}$$

but:

$$GH(j\omega) = \frac{\omega_n^2}{j\omega(j\omega + 2\varsigma\omega_n)}$$

$$|GH(j\omega)| = 1.0 = \frac{\omega_n^2}{[(-\omega^2)^2 + (2\varsigma\omega\omega_n)^2]^{1/2}} \qquad \textbf{Eq. 8.13}$$

from which:

$$\omega_\phi = \omega_n\sqrt{-2\varsigma^2 + (4\varsigma^4 + 1)^{1/2}} \qquad \textbf{Eq. 8.14}$$

and:

$$\angle GH = -\frac{\pi}{2} - \tan^{-1}\frac{\omega}{2\varsigma\omega_n} = -\tan^{-1}\frac{2\varsigma\omega_n}{\omega} \qquad \textbf{Eq. 8.15}$$

Then the phase margin is:

$$\phi_m = \tan^{-1}\frac{2\varsigma}{\sqrt{-2\varsigma^2 + (4\varsigma^4 + 1)^{1/2}}} \qquad \textbf{Eq. 8.16}$$

Note that all these characteristics are functions of ς only. They are plotted on Figure 8–3 as ordinates, with ς as abscissa. Each of the curves corresponds to one of the axes on the left. Using the curves, we can evaluate ς from the open-loop frequency response using the phase margin or from the closed-loop frequency response using the height of the resonance peak, $M_{p\omega}$.

We would like to estimate the time response of higher-order, minimum-phase systems using the results of these second-order system studies. Note:

- The second-order system cannot be unstable unless the phase margin and/or gain margin is negative.
- When the phase and gain margins are small, the system will be barely stable and very oscillatory.
- If we assume the system has a pair of dominant complex roots, then its dynamic behavior will be approximately the same as for a second-order system with comparable phase and gain margins. Gain margin will be defined later.
- The settling time is then $T_s = \dfrac{4}{\varsigma\omega_n}$

To develop useful relationships, we need a starting point. For this we consider stability analysis.

8.9 Resonance

The frequency response tools in this chapter are useful for analyzing systems that could experience sinusoidal inputs. Many systems, especially mechanical and electrical systems, experience resonance. When a system is excited at the resonant

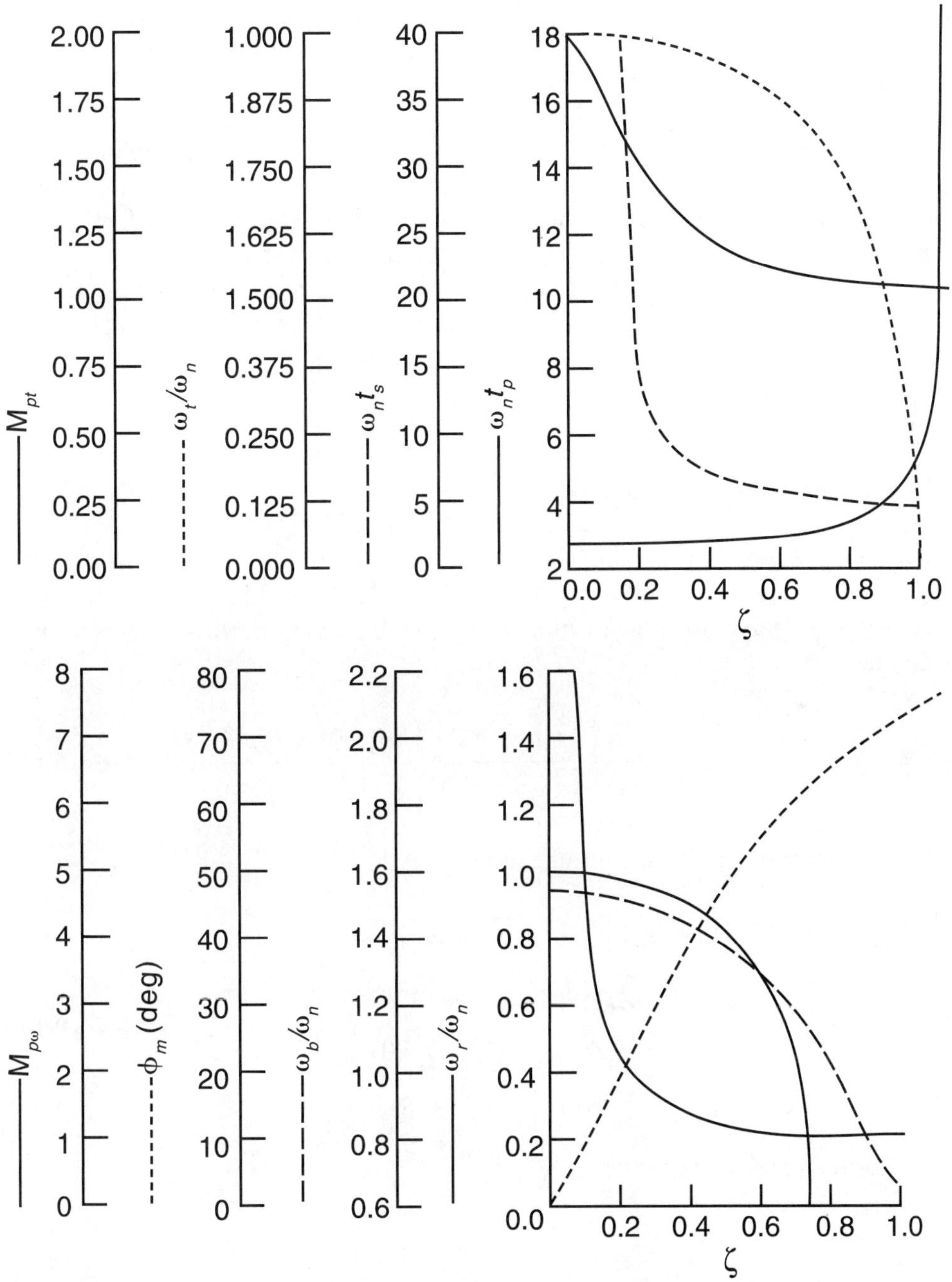

Figure 8–3 (a) Time-response characteristics of a second-order system and (b) frequency-response characteristics of a second-order system

frequency (ω_R), it will exhibit large oscillations. For example, aircraft designers ensure that the wings of an airplane or other control surfaces lack a natural frequency close to the excitation frequency caused by moving air.

Like aircraft, bridges and other structures experience oscillations. They must be able to withstand the effects of earthquakes or gusts of wind. In July 1940, a bridge over Tacoma Narrows at Puget Sound, Washington, was opened to traffic. Unfortunately, any time the wind blew, the bridge oscillated. Finally, on November 7, 1940, the oscillations became so large that the bridge disintegrated.

In contrast, some systems should operate at resonance, for example, a bandpass-filter with a narrow passband, or an aquarium air pump that is operated at the resonant frequency of the diaphragm-supporting structure. The equations for these systems were presented in Chapter 6.

8.10 Drawing the Bode Diagram

Bode recognized that plotting the transfer function in a form different than the classical frequency response would yield some benefits. He proposed to plot 20 $\log_{10}G(jw)$ and use a logarithmic axis for ω. Let our generalized transfer function be:

$$GH(j\omega) = \frac{K\prod(j\omega\tau_z + 1)\prod[A(j\omega)^2 + Bj\omega + 1]}{(j\omega)^N\prod(j\omega\tau_p + 1)\prod[x(j\omega)^2 + y(j\omega) + 1]} \quad \textbf{Eq. 8.17}$$

Using logarithms for the magnitude terms gives:

$$\begin{aligned} 20\log_{10}|G(j\omega)| &= 20\log_{10} K + \sum 20\log_{10}|j\omega\tau_z + 1| \\ &\quad + \sum 20\log_{10}|A(j\omega)^2 + Bj\omega + 1| \ldots. \\ &\quad -20\log_{10}|(j\omega)^N| - \sum 20\log_{10}|j\omega\tau_p + 1| \\ &\quad - \sum 20\log|A(j\omega)^2 + Bj\omega + 1| \end{aligned} \quad \textbf{Eq. 8.18}$$

and for the phase angle terms:

$$\begin{aligned} \angle G(j\omega) &= \sum \tan^{-1}\omega\tau_z + \sum \tan^{-1}\frac{\omega B}{1 - A\omega^2} \\ &\quad - N\frac{\pi}{2} - \sum \tan^{-1}\omega\tau_p - \sum \tan^{-1}\frac{\omega B}{1 - A\omega^2} \end{aligned} \quad \textbf{Eq. 8.19}$$

When plotting curves for these equations, semilog paper is used with ω as abscissa, using a logarithmic scale and magnitude in decibels (dB) for the ordinate using a linear scale. For the angle curve, the ordinate is in degrees.

Several advantages are realized when using the logarithmic Bode form of the equations. The obvious advantage is in the conversion of multiplication to addition. A less obvious, but perhaps more valuable, advantage lies in the fact that the magnitude curves for each term in the addition can be represented by their asymptotes. This simplifies the drawing of the curves, as well as analysis and design. The algebra of the transfer function allows only four types of terms:

1. $20 \log_{10} K$ (gain)
2. $\pm 20 \log_{10} \left|(j\omega)^N\right|$ (poles or zeros at the origin)
3. $\pm 20 \log_{10} \left|j\omega\tau + 1\right|$ (simple poles or zeros)
4. $\pm 20 \log_{10} \left|A(j\omega)^2 + Bj\omega + 1\right|$ (complex poles or zeros)

While curves can be obtained by substituting a series of values for ω into the expressions, it is more convenient to draw asymptotes and to calculate points on the curve for only a few, select values.

Consider the terms and their asymptotes. For $20 \log_{10} \left|GH(j\omega)\right| = 20 \log_{10} \left|K\right|$, the term K is not a function of ω and plots at all frequencies at the same value, yielding a horizontal, straight line.

For the term $20 \log_{10} \left|j\omega\right|$, the curve plots as a true straight line with slope of ± 20 dB/decade and crosses the horizontal axis at $\omega = 1$. The phase angle is 90° throughout.

The Bode plot for $GH(s) = s$ (magnitude and phase) appears in Figure 8–4. This corresponds to a differentiator. A differentiator has high gain at high frequencies, which causes concern when trying to differentiate a signal with noise.

An integrator $GH(s) = 1/s$ has a transfer function that is the reciprocal of the differentiator. Its Bode plot appears in Figure 8–5.

For $\pm 20 \log_{10} \left|j\omega\tau + 1\right|$, two asymptotic segments are needed. For $\omega\tau << 1$, we consider $\omega\tau \sim 0$ and $20 \log_{10} \left|j\omega\tau + 1\right|$, which plots as a straight, horizontal line with ordinate at 0 dB. When $\omega\tau >> 1$, the 1 can be neglected, and $20 \log_{10} \left|\omega\tau\right|$ plots as a straight line with a slope of ± 20 dB/decade. This line intersects the horizontal asymptote at $\omega = 1/\tau = \omega_o$, which is called a corner frequency, or a break frequency. Consider the transfer function $GH(s) = (s/200 + 1)$. Its Bode plot appears in Figure 8–6.

Now consider a transfer function that is the reciprocal of the previous one: $GH(s) = 1/(s/200 + 1)$. Its Bode plot appears in Figure 8–7. This transfer function, called a low-pass filter, is the same as what was plotted using a regular frequency plot in Figure 8–2.

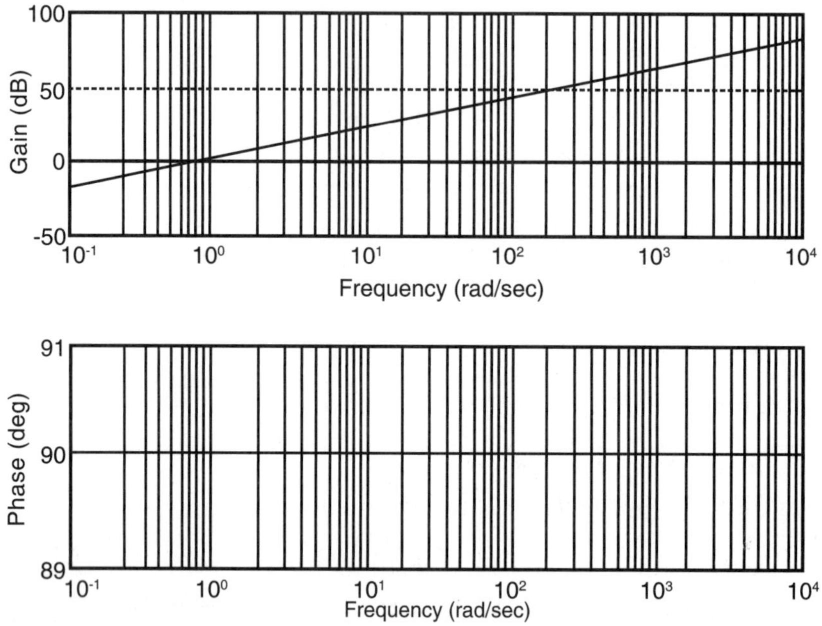

Figure 8–4 Bode plot of $GH(s) = s$

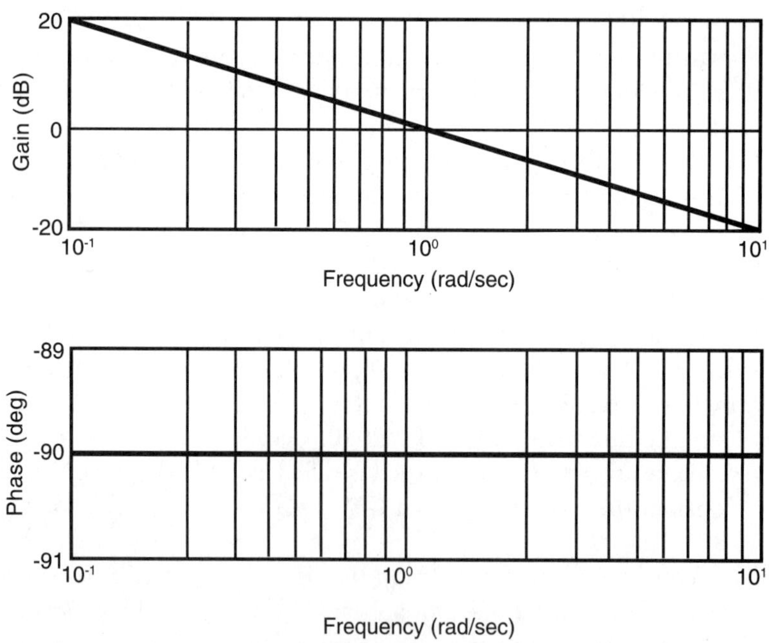

Figure 8–5 Bode plot of $GH(s) = 1/s$

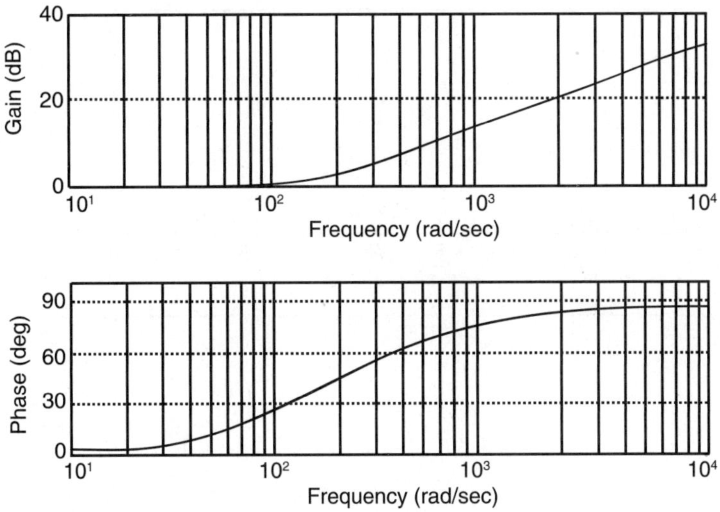

Figure 8–6 Bode plot of $GH(s) = (s/200 + 1)$

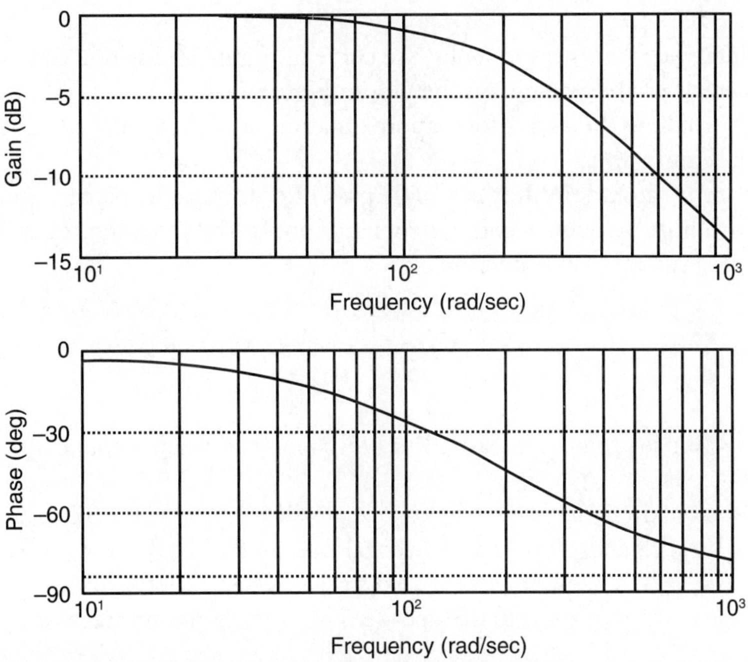

Figure 8–7 Bode plot of $GH(s) = 1/(s200 + 1)$

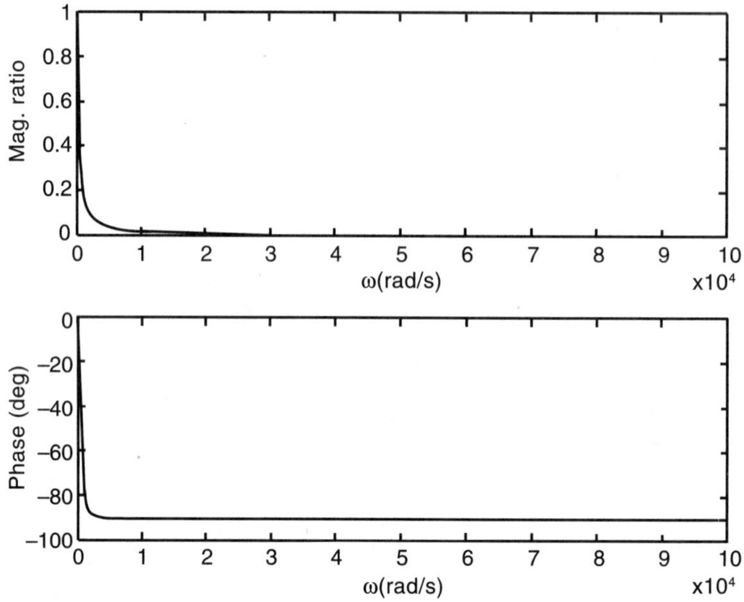

Figure 8–8 Frequency plot of a low-pass filter using similar values of *w*

Note that if we were to replot the curve in Figure 8–2 with a similar range in the frequency variable, we obtain the plot in Figure 8–8.

Let us draw the asymptotic approximation over the actual curve. The result appears in Figure 8–9.

Another common filter is a high-pass filter that, as its name implies, passes signals with high-frequency components but blocks low-frequency ones. The standard form for this transfer function is:

$$GH(s) = \frac{s/\omega_o}{s/\omega_o + 1} \qquad \text{Eq. 8.20}$$

The Bode plot for this high-pass filter (with $\omega_o = 200$ rad/sec) appears in Figure 8–10.

For a quadratic term like $20\log_{10}\left|Aj\omega^2 + Bj\omega + 1\right|$, when ω is small, this becomes $20\log_{10}\left|1\right|$ and plots as a horizontal line on the 0 dB ordinate. For large ω, the term $Bj\omega + 1$ is negligible compared to $A\omega^2$. Then, $20\log_{10}\left|A(j\omega)^2\right|$ plots as a straight line with slope of ± 40 dB/decade. This asymptote intersects the horizontal asymptote at $\omega = \dfrac{1}{\sqrt{A}}$, which is also called a corner or break frequency.

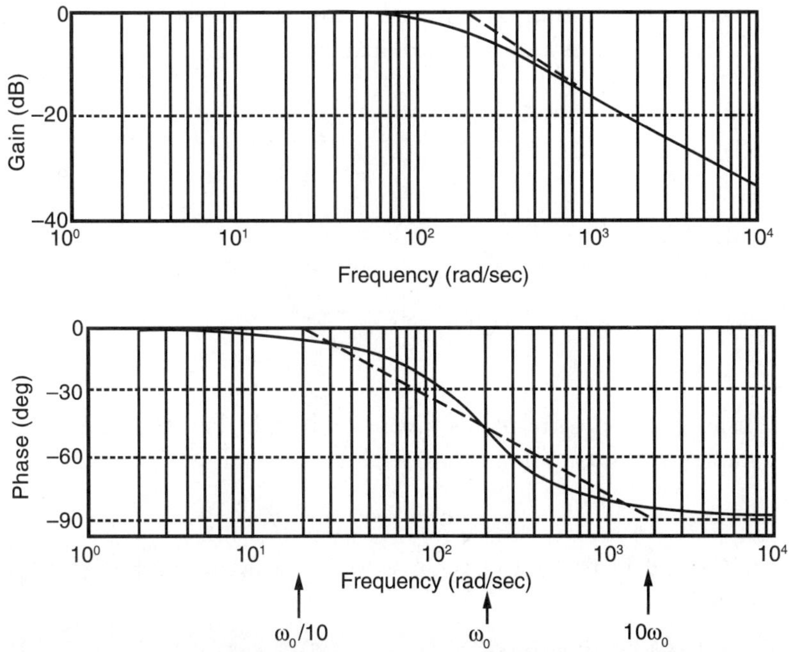

Figure 8–9 Bode plot of $GH(s) = 1/(s200 + 1)$ and the asymptotic approximation

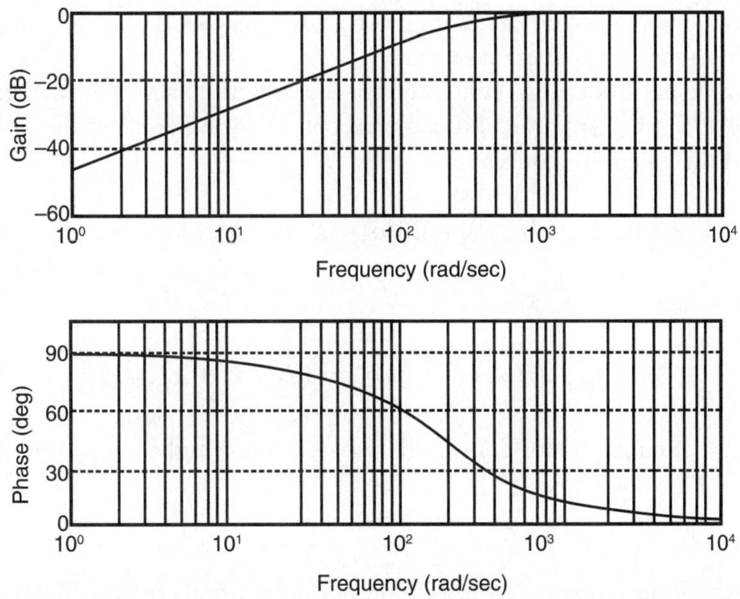

Figure 8–10 Bode plot for the high-pass filter

Next, consider a transfer function whose denominator has the same mathematical form as the term just discussed, namely:

$$\frac{C(s)}{R(s)} = \frac{\omega_n^2}{s^2 + 2\varsigma\omega_n s + \omega_n^2} \qquad \textbf{Eq. 8.21}$$

or, in terms of $j\omega$:

$$\frac{C}{R}(j\omega) = \frac{\omega_n^2}{(\omega_n^2 - \omega^2) + j2\varsigma\omega_n\omega}$$

This transfer function corresponds to a second-order, low-pass filter arising from a set of complex quadratic poles. Its Bode plot appears in Figure 8–11.

Recommended Plotting Procedure for Magnitude Curve (Longhand Calculation)

1. Choose abscissa and ordinate scales on semilog paper.
2. Determine the system type number, N, from your transfer function (i.e., the number of poles at the origin). This also determines the slope of the lowest frequency asymptote.
3. Determine and mark on the plot all corner frequencies. These are the poles and zeros of the transfer function.
4. Determine the Bode gain. When the transfer function is in Bode constant form, $(j\omega/\omega o + 1)$ or $(j\omega\tau + 1)$, all factors except poles and zeros at the origin go to 1.0 and the equation reduces to:

$$20\log_{10}|GH(j\omega)| = 20\log_{10}|K_B| - 20N\log_{10}|j\omega| \qquad \textbf{Eq. 8.22}$$

Rearranging:

$$20\log_{10}|GH(j\omega)| = -20N\log|(j\omega)| + 20\log_{10}|K_B| \qquad \textbf{Eq. 8.23}$$

which is the equation on semilog coordinates of a straight line of the form:

$$\mathrm{y = mx + b} \qquad \textbf{Eq. 8.24}$$

For a type 0 system $m = 0$, the asymptote is horizontal at:

$$|GH| = 20\log_{10}|K_B| \qquad \textbf{Eq. 8.25}$$

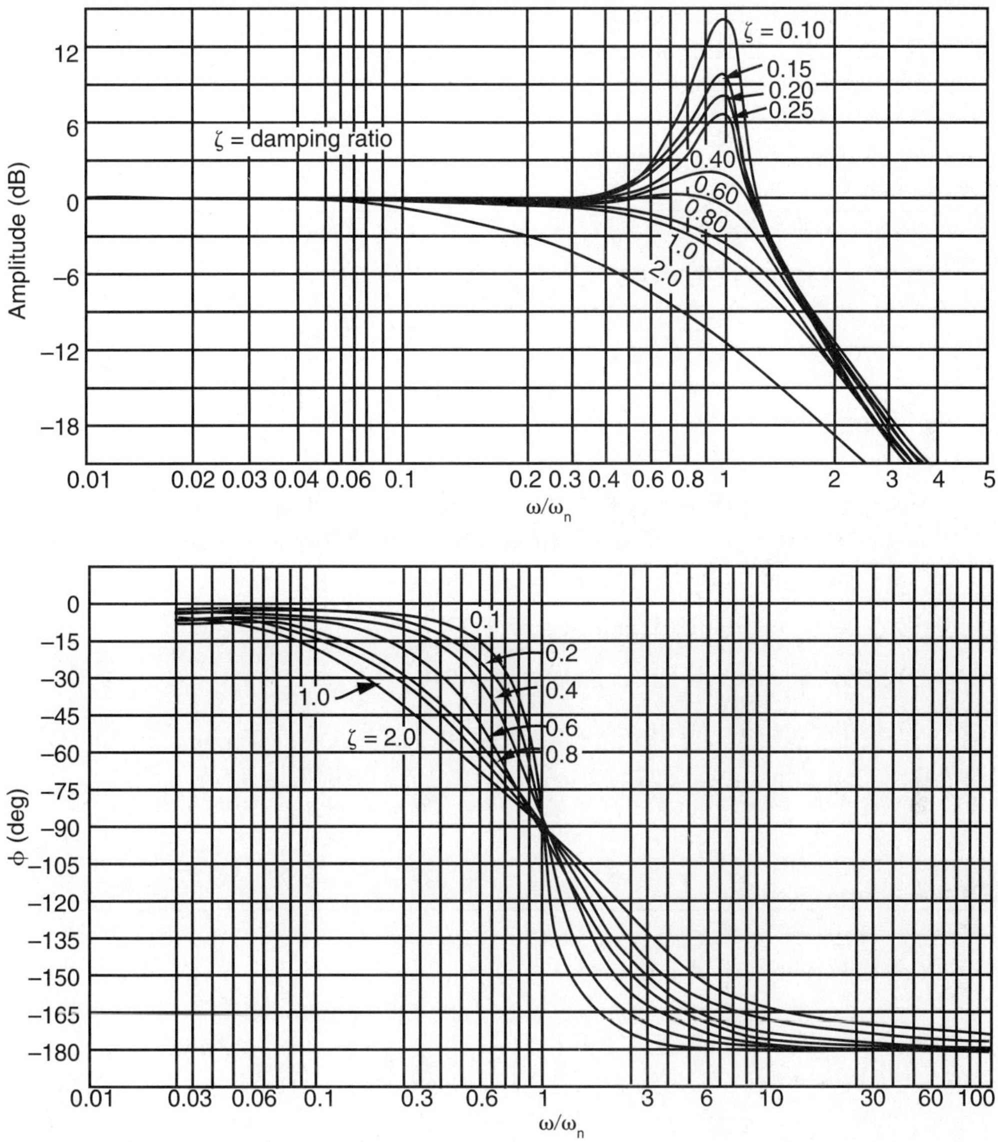

Figure 8–11 Bode plot of four quadratic poles

For a type 1 system, the asymptote has a slope of −20 dB/decade and intersects the 0 dB axis at $\omega = K_B$. Because this defines a point on the asymptote, the asymptote is drawn toward lower frequencies starting at $\omega = K_{B},\ \ |G(j\omega)| = 0$.

5. In similar manner, for a type 2 system, the asymptote has a slope of −40 dB/decade and intersects the axis at $\omega = \sqrt{K_B}$.

6. To complete the asymptotic magnitude curve, draw the low-frequency asymptote. Then, following it in the direction of increasing frequency, change the slope as you reach each corner frequency.

Example 8.1

Draw the asymptotic gain curve for:

$$G(j\omega) = \frac{100}{(s+1)\left(\frac{s}{5}+1\right)\left(\frac{s}{20}+1\right)} \qquad \text{Eq. 8.26}$$

Solution The system is type 0, the asymptote at low frequency is horizontal, and the gain is 100, so the ordinate is 40 dB. Starting at lowest frequency, change the asymptotic slope from:

0 dB/decade to -20 dB/decade at $\omega = 1.0$

-20 dB/decade to -40 dB/decade at $\omega = 5.0$

-40 dB/decade to -60 dB/decade at $\omega = 20.0$

The completed plot appears in Figure 8–12.

Example 8.2

Draw the asymptotic diagram for:

$$G(j\omega) = \frac{200\left(\frac{s}{20}+1\right)}{s\left(\frac{s}{5}+1\right)\left(\frac{s}{50}+1\right)} \qquad \text{Eq. 8.27}$$

Solution The system is type 1, and the lowest frequency asymptote has a slope of -20 dB/decade. The gain is 200, and the low-frequency asymptote (extended) intersects the 0 dB axis at $\omega = 200$. Follow this asymptote from low frequency in the direction of increasing frequency, changing the slope from:

-20 dB/decade to -40 dB/decade at $\omega = 5.0$

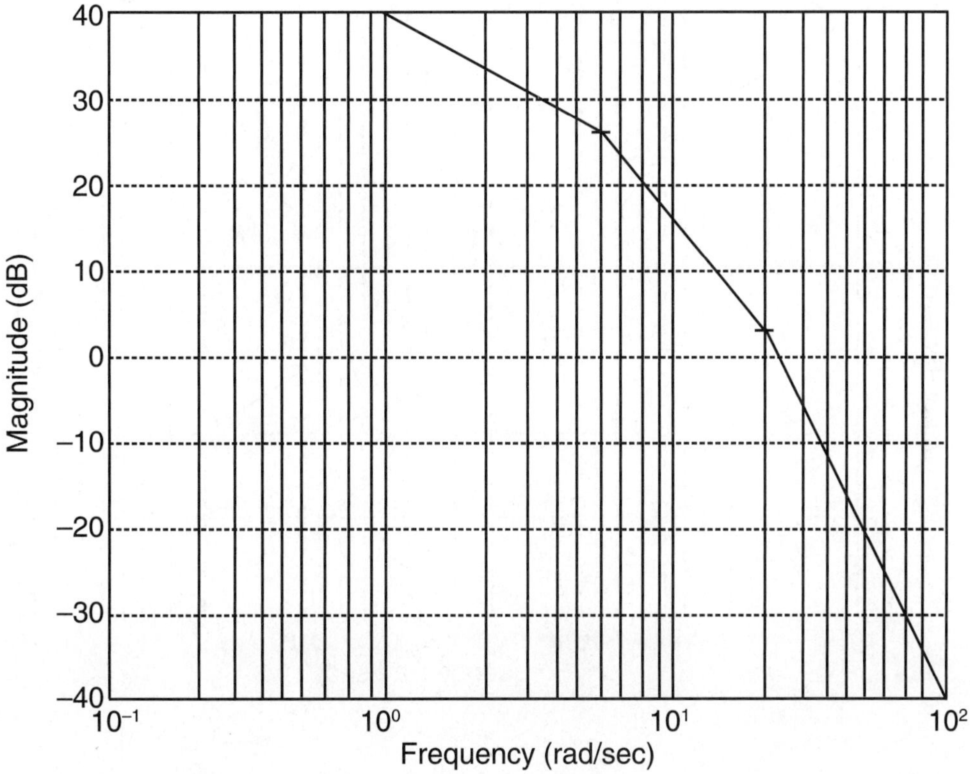

Figure 8–12 Asymptotic Bode magnitude plot for $GH(s) = \dfrac{100}{(s+1)\left(\dfrac{s}{5}+1\right)\left(\dfrac{s}{20}+1\right)}$

-40 dB/decade to -20 dB/decade at $\omega = 20$

-20 dB/decade to -40 dB/decade at $\omega = 50$

The completed plot appears in Figure 8–13.

Corrections to the Asymptotic Magnitude Curve

For many problems, the asymptotes are adequate descriptions of the magnitude curves. However, there are problems for which a more accurate curve is needed, especially at, or near, any corner frequencies. One of the advantages of the logarithmic plot is that the frequency range in which magnitude corrections are needed is obvious on the plot. Thus, the choice of frequencies at which magnitude corrections are

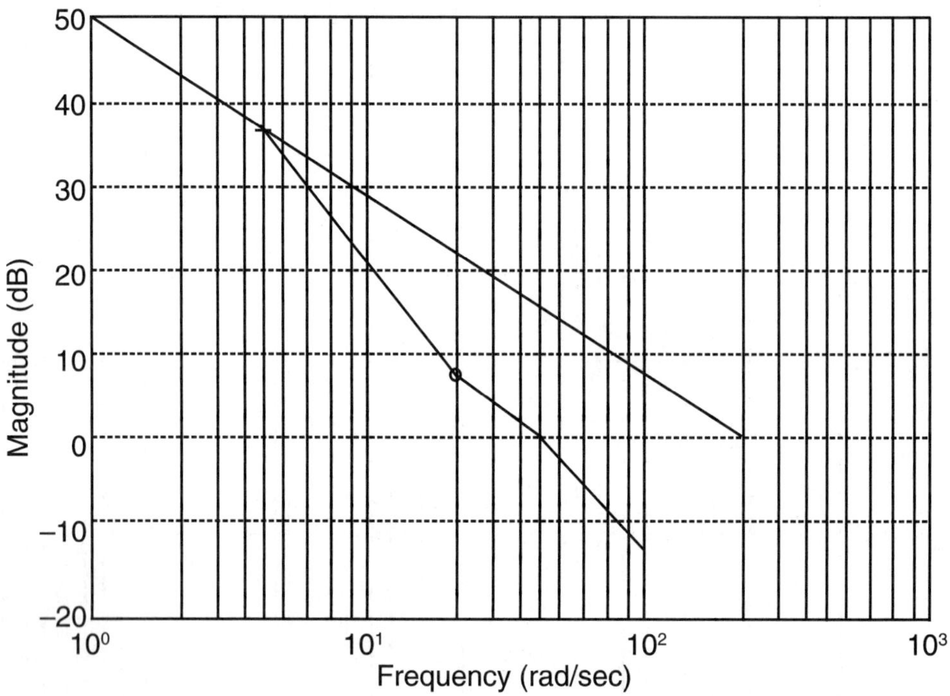

Figure 8–13 Asymptotic Bode magnitude plot for $GH(s) = \dfrac{200\left(\dfrac{s}{20}+1\right)}{s\left(\dfrac{s}{5}+1\right)\left(\dfrac{s}{50}+1\right)}$

needed is optimized. There is a simpler way, however. Corrections can be made by inspection because:

- At a corner frequency for which the slope change is 20 dB/decade, the true magnitude differs from the asymptotic value by 3 dB.
- At frequencies one octave (factor of 2) above and below the corner frequency, the magnitude difference is 1.0 dB.
- This gives three corrected points and two asymptotes to guide drawing the corrected curve.
- When the transfer function contains a quadratic term (complex poles or zeros), the low-frequency asymptote is a horizontal line at 0 dB. There is a double corner at $\omega = \dfrac{1}{\sqrt{A}}$, and the high-frequency slope is 40 dB/decade. The correction at frequencies near the corner frequency depend on the damping ratio of the

quadratic. When the magnitude curve in that frequency range is important (i.e., when the phase margin and gain margin are affected), select several frequencies near the corner frequency and calculate points on the true magnitude curve.

The Angle Curve

To complete the Bode diagram, the angle curve ($\angle GH(j\omega)$) must be added. The abscissa for the angle curve is ω, as it was for the magnitude curve. The ordinate is $\angle G(j\omega)$ in degrees. The scale may be chosen to suit the user, and the use of 30 deg/inch is common. This angle scale should be located such that -180 degrees on this scale is colinear with 0 dB on the magnitude scale. This alignment permits simple interpretation of phase margin and gain margin, which are discussed in Section 8.15.

There are several ways to plot the $\angle GH(j\omega)$ curve. The three following are presented here:

1. Select a value of ω, substitute all pertinent terms, add, and plot the point. Repeat for another value.
2. Make a template, and use it. Note that almost all the terms in the $\angle GH$ equation are arc-tan terms $\left(\text{i.e., } -\tan^{-1}\frac{\omega\tau}{1}\right)$ for a simple pole or 0. For many problems, this template is the only one needed, because quadratic terms are seldom encountered. (It was common to make an arc-tan template from semilog graph paper, plotting on it points taken from trig tables, then drawing the curve through the points and cutting along the plotted curve. This template was then backed with cardboard or made on a flat, plastic sheet.) Two indexing lines should be put on the template to aid in alignment: (1) a vertical line at $\omega\tau = 1$ (the corner frequency used in making the template) and (2) a horizontal line at $\angle GH(j\omega) = -45$ degrees so that the -45 degree line on the template is on the -45 degree line of the curve sheet.

   To use the template, align it on the horizontal axis, place the $\omega\tau = 1$ (or $\omega/\omega = 1$) line of the template) at each of the corner frequencies in turn, and draw an $\angle G$ curve for each position of the template. Then, add ordinates at select values to get points on the $\angle GH(\omega)$ curve.
3. Plot each term of $GH(s)$ then add all of them. For a 0 at the origin, the angle is 90 degrees; for a pole at the origin, the angle is -90 degrees. For other single poles and zeros, use the following asymptotic approximation. Consider one pole as an illustration of the asymptotic approximation:

$$\omega \leq \omega_0/10 \qquad \angle GH(s) \cong 0$$

$$\omega = \omega_0 \qquad \angle GH(s) = -45$$

$$\omega \geq \omega_0 * 10 \qquad \angle GH(s) \cong -90$$

Example 8.3

Sketch the following transfer function:

$$GH(s) = \frac{1}{s(s/50 + 1)} \qquad \text{Eq. 8.28}$$

Solution The preceding transfer function has two poles: (1) one at the origin and (2) another at 50 rad/s. The angle contribution of the first pole is -90 degrees. The second can be approximated as indicated previously. Each contribution appears in Figure 8–14. The complete asymptotic approximation is obtained by adding the two curves.

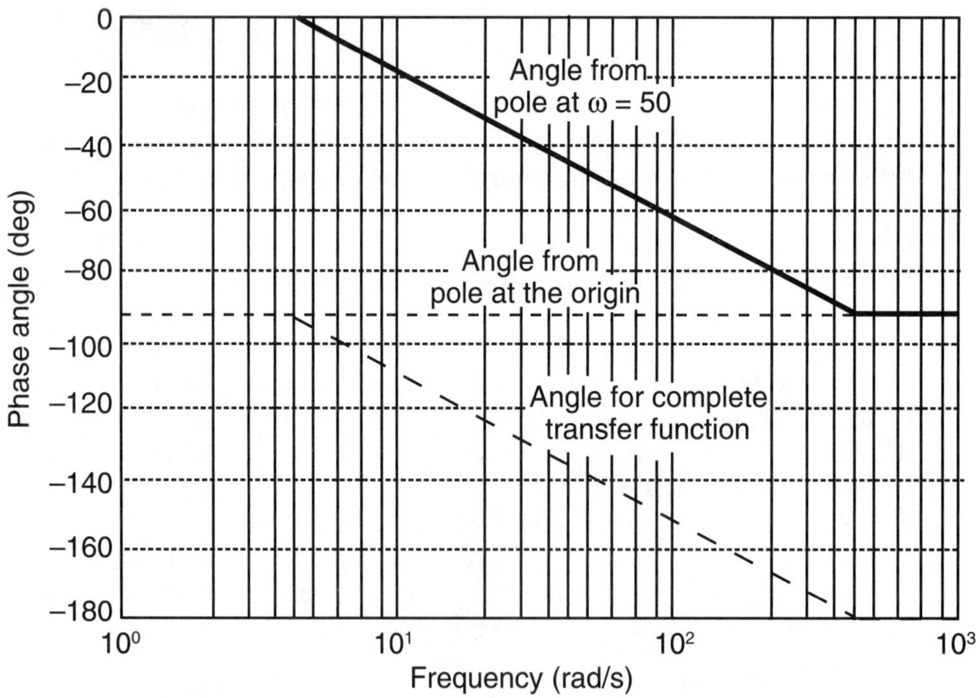

Figure 8–14 Asymptotic Bode phase plot for $GH(s) = \dfrac{1}{s\left(\dfrac{s}{50} + 1\right)}$

8.11 Using MATLAB to Obtain Bode Plots

MATLAB facilitates the creation of Bode plots, but you must define the numerator and denominator of the OLTF, assuming a controller gain of 1. Consider creating a Bode plot corresponding to the high-pass filter, shown previously in Figure 8–10. The information is entered into MATLAB as follows:

define the numerator polynomial:

```
>>num=[1/200 0]
```

define the denominator polynomial:

```
>>den= [1/200 1]
```

invoke the Bode function:

```
>>bode(num,den)
```

The function automatically picks the range for ω. If this is not adequate, however, the user can select a more appropriate range. For instance, if the desired frequency range is from 10 rad/s (= 10^1) to 1,000 rad/s (= 10^3), the user would invoke the following function:

```
>>w=logspace(1,3)
```

then:

```
>>bode(num,den,w)
```

One can also obtain the magnitude and phase at a particular frequency, $w1$, by:

```
>>[magnitude,phase]=bode(num,den,w1)
```

8.12 Cauchy's Principle of the Argument

The Nyquist stability criterion is developed from Cauchy's Principle of Argument, which may be paraphrased as follows:

> If we choose a closed contour on the s-plane and also a mapping function, when the closed contour is mapped using the mapping function, the change in the angle (argument) of the map indicates the number of singular points of the mapping function inside the contour.

Consider the following mapping function:

$$GH(s) = \frac{K}{(s+A)(s+B)(s+C)} \quad \text{Eq. 8.29}$$

and a contour on the s-plane, which is to be mapped using $GH(s)$. The s-plane and contour appear in Figure 8–15a, as do the poles of $GH(s)$.

Choosing a point x on the contour as the first point to be mapped, the angle of $GH(s)$ at x is $\angle GH(s) = \alpha + \beta + \gamma$. The magnitude of $GH(s)$ is unneeded. Let the mapping point move clockwise around the contour, and observe that, for this contour:

- The angle of $GH(s)$ first goes through a maximum value.
- The $\angle GH(s)$ then decreases to a minimum value.
- The mapping point finally returns to x.

The angle of each vector has returned to its starting value, so the $\angle GH(s)$ has made no net change.

Next, consider that $GH(s)$ has an additional pole that is inside the contour at $-D$ and thus $\angle GH(s) = \alpha + \beta + \gamma + \delta$ as shown in Figure 8–15. Now, let the mapping point start at x and progress clockwise until it returns to point x. Observe that the three vectors from poles $-A, -B, -C$, which are outside the contour, start and finish at point x with no net rotation, but angle δ makes a net rotation of 2π for the same traverse of the mapping point. This indicates that there is a pole inside the contour.

The contour in Figure 8–15 was chosen arbitrarily to demonstrate the interpretation of $\angle GH(s)$ as a contour is mapped. To use this information for stability studies, the mapping contour must be chosen so that it encloses the entire right half of the s-plane.

This closed contour, often called the $\mathcal{D}$ contour, in the limit, encloses the entire right half of the s-plane and appears in Figure 8–16.

1. The contour starts at the origin of the s-plane and lies on the imaginary axis from $0 \leq j\omega \leq +\infty$.
2. At $j\omega = \infty$, the contour changes to a circular arc of infinite radius, proceeding clockwise to $-j\infty$.
3. From $-j\infty$, the contour lies on the negative imaginary axis from $-\infty \leq j\omega \leq 0$.
4. Excludes the origin.

8.13 Stability Analysis

Having chosen a suitable mapping contour independent of $GH(s)$, an appropriate mapping function is needed. $GH(s)$ is not suitable, because it is the transfer function of the open-loop system, and it is the stability of the closed-loop system that is of interest. To make this transition, consider the geoemetric construct in Figure 8–17a.

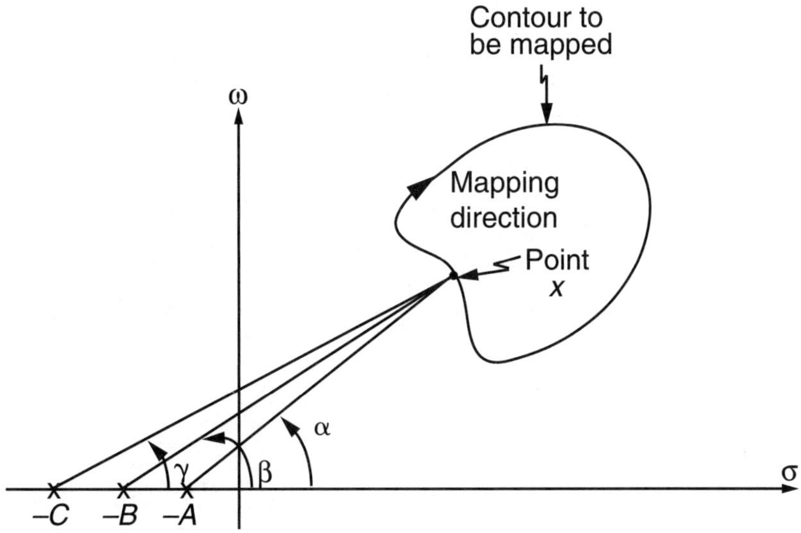

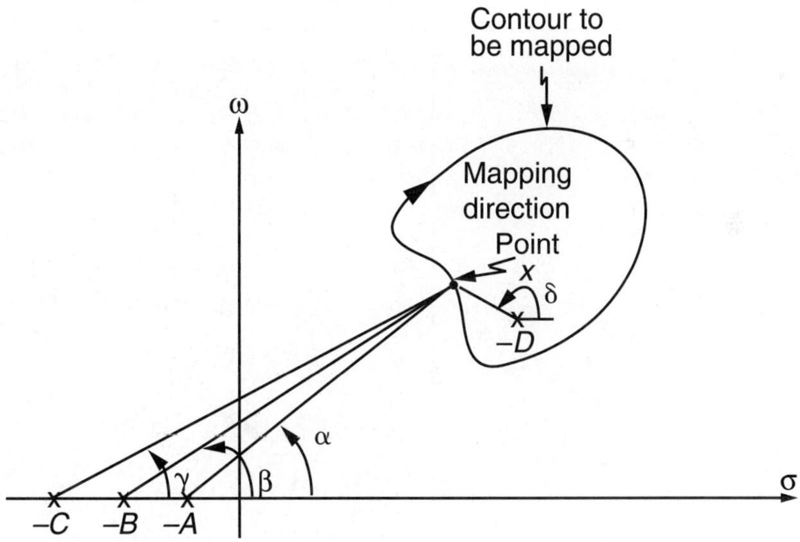

Figure 8–15 Polar plot illustrating Cauchy's Principle of the Argument with (a) no poles inside the contour and (b) one pole inside the contour

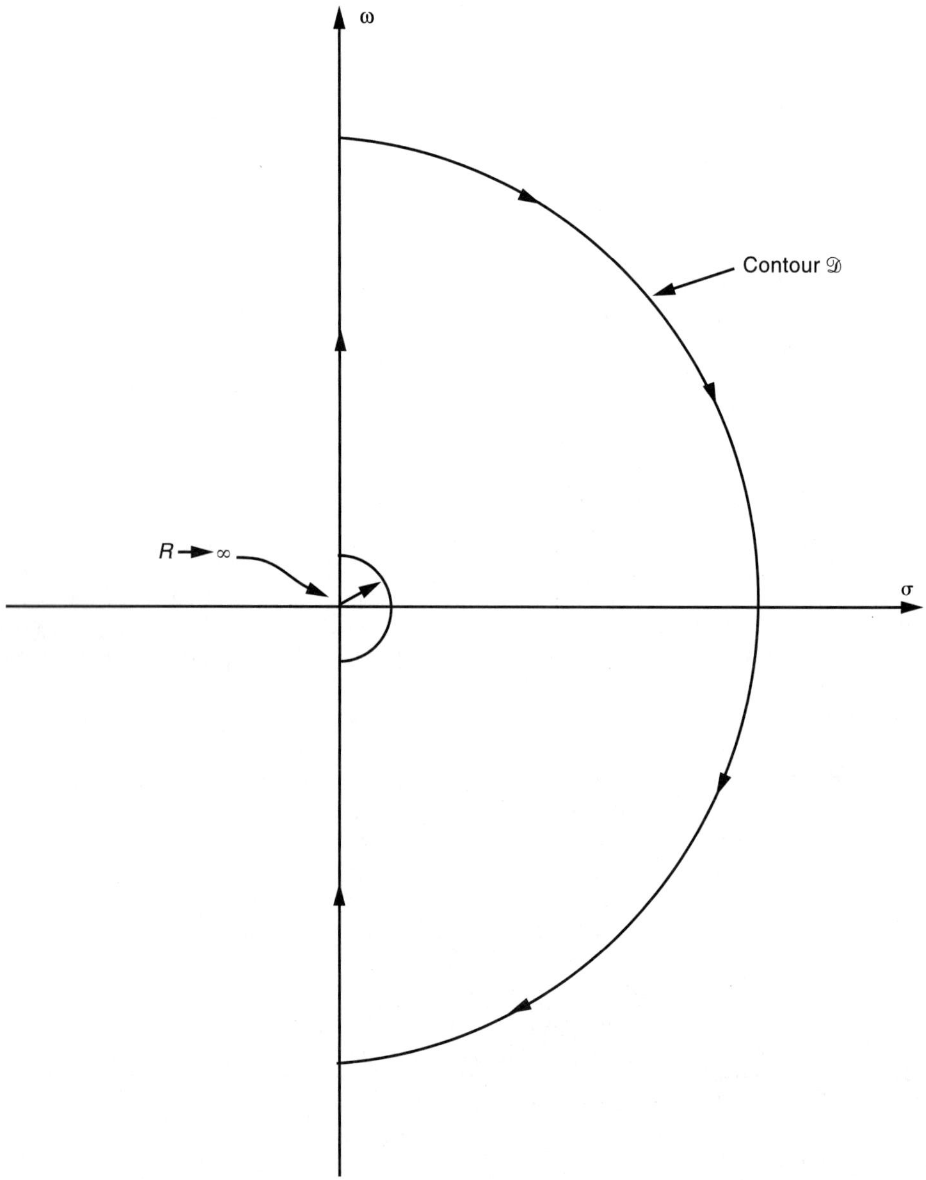

Figure 8–16 Mapping contour to apply Cauchy's Principle of the Argument

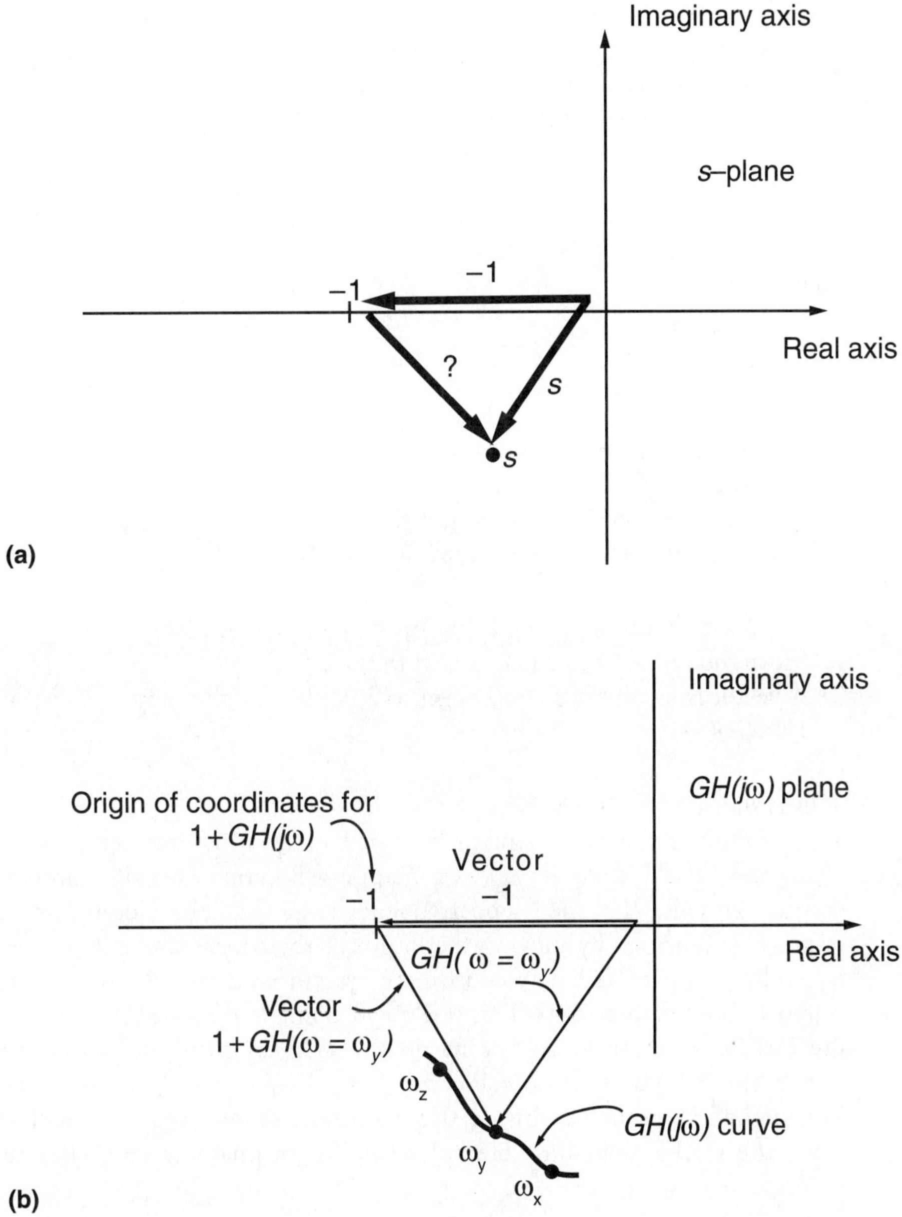

Figure 8–17 Defining the closed loop: (a) general representation and (b) representation specific to the $GH(j\omega)$ plane

It shows the vector -1 extending from the origin to the point -1. It also shows a vector s going from the origin to the point s. What is the vector from -1 to point s? To answer this question, set up a vector equation that adds the components, then solve for the unknown vector as follows:

$$-1 + ? = s \qquad \textbf{Eq. 8.30}$$

then,

$$? = s + 1 \qquad \textbf{Eq. 8.31}$$

The vector from the -1 point to point s is simply $s + 1$. We will use this realization to work with the CLTF.

The characteristic equation of the closed-loop system, $1 + GH(s) = 0$, is a suitable mapping function that is readily available as follows:

1. Using the chosen mapping contour, obtain a map of $GH(s)$.
2. Move the origin of this map 1 unit to the left.
3. A vector from this $-1 + j0$ point to a point x on the map of $GH(s)$ is a vector $1 + GH(x)$.

This is shown in Figure 8–17b.

Observe that the map of a closed contour is another closed contour. Using the OLTF $GH(s = j\omega)$ as a mapping function and the chosen mapping contour provides the open-loop frequency response. The closed-loop frequency response is obtained by moving the origin of the coordinate system 1 unit to the left. The origin of the new coordinate system is at the $-1 + j0$ point of the original coordinate system. This $-1 + j0$ point is the critical point. From Figure 8–17b, a vector to any point on the $GH(j\omega)$ curve defines $1 + GH(j\omega)$. When the -1 point lies outside the $1 + GH(j\omega)$ curve, the system is stable. When the -1 point lies inside the curve, the system is unstable. This is true when the OLTF lacks unstable poles (minimum phase system). The results can be presented as:

number of unstable closed-loop poles (poles in the right half of the s-plane)	=	number of clockwise encirclements of the -1 point in the $GH(s)$ plane as s traverses the contour $\mathcal{D}$

Example 8.4

Consider the following transfer function:

$$GH(s) = \frac{10(s + 10)}{(s + 1)(s + 3)} \quad \text{Eq. 8.32}$$

which has two real poles and one 0. Determine if the system is stable.

Solution The Nyquist plot of this transfer function appears in Figure 8–18. Notice that the −1 point is never encircled, confirming that the system is stable.

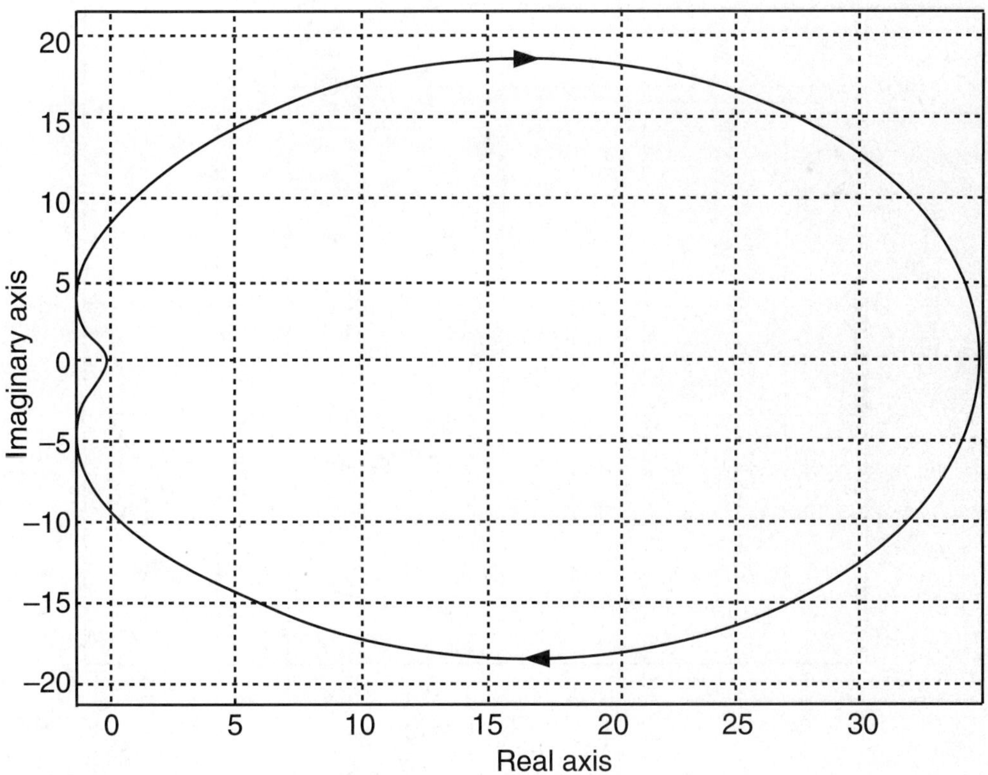

Figure 8–18 Nyquist diagram for $GH(s) = 10(s + 10)/(s + 1)(s + 3)$

Example 8.5

A regulator has the following OLTF:

$$GH(s) = \frac{1,000}{(s+1)(s+21)(s+120)(s+100)} \qquad \text{Eq. 8.33}$$

Determine if the system is stable.

Solution This system is a fourth-order type, with error coefficient $K_p = 252000$. The Nyquist diagram, given in Figure 8–19, does not encircle the $-1 + j0$ point, so

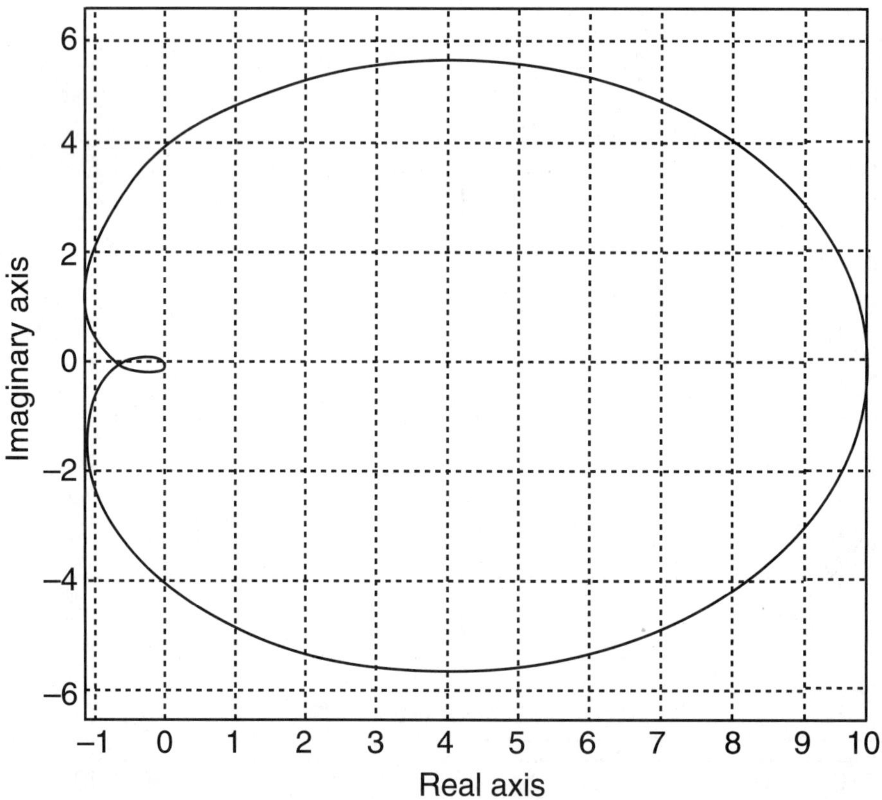

Figure 8–19 Nyquist diagram for $GH(s) = 1{,}000/(s + 1)(s + 21)(s + 120)(s + 100)$

the closed-loop system is stable. Following is the characteristic equation (obtained from the transfer function):

$$s^4 + 242s^3 + 16861s^2 + 268620s + 252000 = 0 \qquad \text{Eq. 8.34}$$

which has roots at:

$-1.2030 + j0.4513$

$-1.2030 - j0.4513$

$-0.0007 + j0.0385$

$-0.0007 - j0.0385$

Example 8.6

Determine if the transfer function for a fourth-order regulator with two zeros and four poles is stable.

$$GH(s) = \frac{10(s^2 + 12s + 72)}{(s + 1)(s + 3)(s + 5)(s + 10)} \qquad \text{Eq. 8.35}$$

Solution The Nyquist diagram for this transfer function appears in Figure 8–20. It does not encircle the $-1 + j0$ point, so the system is stable. The characteristic equation is:

$$s^4 + 19s^3 + 3013s^2 + 245s + 150 \qquad \text{Eq. 8.36}$$

and the roots are obtained using the `roots( )` MATLAB function. They are located at:

$-9.4595 + j54.0549$

$-9.4595 - j54.0549$

$-0.0405 + j0.2195$

$-0.0405 - j0.2195$

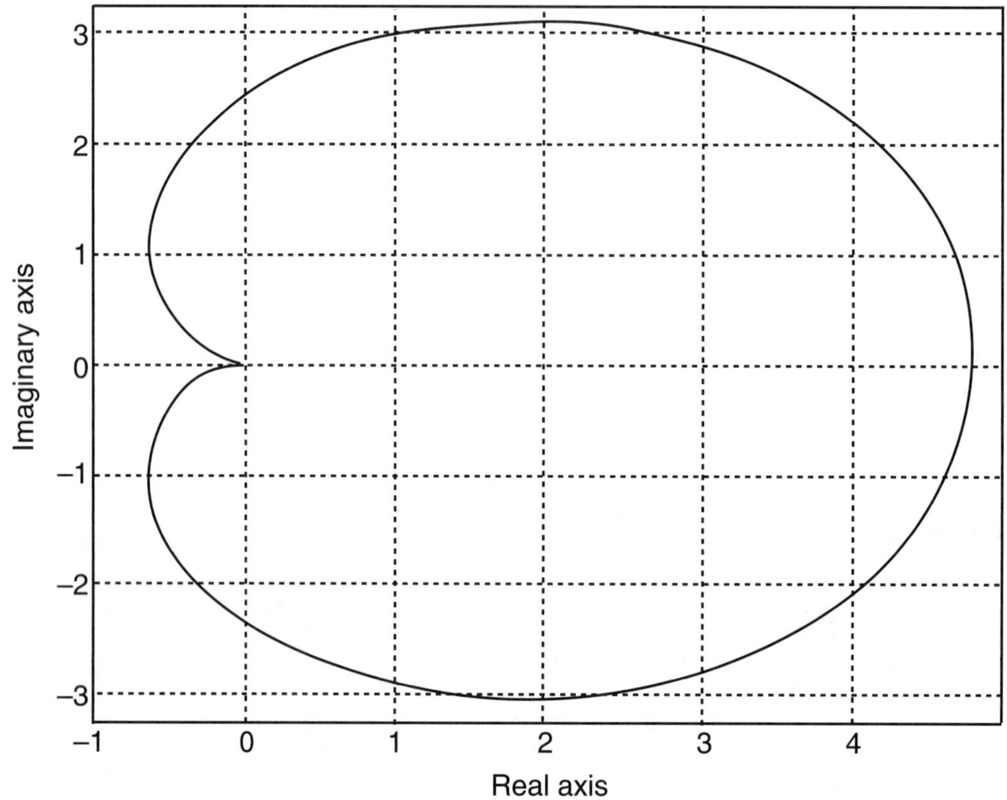

Figure 8–20 Nyquist plot for $GH(s) = 10(s^2 + 12s + 72)/(s + 1)(s + 3)(s + 5)(s + 10)$

The system is stable and should be well damped. This can be confirmed using the `damp()` MATLAB function. It is similar to the `roots()` function, but it also calculates the damping ratio. For example:

```
>>damp([1 19 3013 245 150])

   Eigenvalue              Damping    Freq. (rad/sec)

  -0.0405 +   0.2195i      0.1816           0.2232

  -0.0405 -   0.2195i      0.1816           0.2232

  -9.4595 + 54.0549i       0.1724          54.8764

  -9.4595 - 54.0549i       0.1724          54.8764
```

8.14 Nonminimum Phase Systems

The mapping function $1 + GH(s)$ is a ratio of two polynomials and thus has both poles and zeros. The numerator of $1 + GH(s)$ is the characteristic equation of the system, and its zeros are the roots that define stability. The denominator of $1 + GH(s)$ gives the poles of the OLTF (i.e., the poles of $GH(s)$). When all the zeros and poles are in the left half of the s-plane, the dynamic system is stable, as well as minimum phase. When any of the poles of $GH(s)$ are in the right half of the s-plane, the system is nonminimum phase. When any of the zeros of $1 + GH(s)$ (roots of the characteristic equation) are in the right half of the s-plane, the system is unstable. Systems with open-loop poles in the right half of the s-plane may still be stable.

When a nonminimum phase system has zeros and poles, the system is unstable when any zero of $1 + GH(s)$ is in the right half of the s-plane. To interpret stability using the Nyquist criterion (when the poles and zeros are unknown [i.e., the polynomials are not factored]), observe the angle of the $1 + GH(s)$ vector as the mapping point traces out the mapping contour:

- When there are any net clockwise rotations of the $1 + GH(s)$ vector, the system is unstable.
- When there are no net rotations of the $1 + GH(s)$ vector, or when there are some counterclockwise rotations, the situation requires further study.
    - The number of right-half plane poles of $GH(s)$ must be known to predict stability.
    - Because the poles of $1 + GH(s)$ are the roots of $1 + GH(s)$, the number of poles in the right half of the s-plane may be determined using the Routh array.

In summary, the number of closed-loop poles in the defined contour is related to the number of clockwise encirclements of the -1 point in the $GH(s)$ plane. The relationship can be expressed by the following equation:

number of unstable CL poles (pole in the right half of the s-plane) = number of unstable OL poles (poles in the right half of the s-plane) + number of clockwise encirclements of the -1 point in the $GH(s)$ plane as s traverses contour $\mathcal{D}$ **Eq. 8.37**

Example 8.7

A second-order system has the following OLTF:

$$G(s) = \frac{100}{(s+5)(s-10)} = \frac{2}{\left(\frac{s}{5}+1\right)\left(\frac{s}{10}-1\right)} \quad \textbf{Eq. 8.38}$$

Determine if the system is stable.

Solution There is one pole in the right half of the plane, so the system is non-minimum phase. The Nyquist diagram for this system appears in Figure 8–21. The Nyquist curve encircles the $-1 + j0$ point in a clockwise direction. There is one root in the right half of the plane, so the system is unstable. From the transfer function, the error coefficient is $Kp = 2.0$.

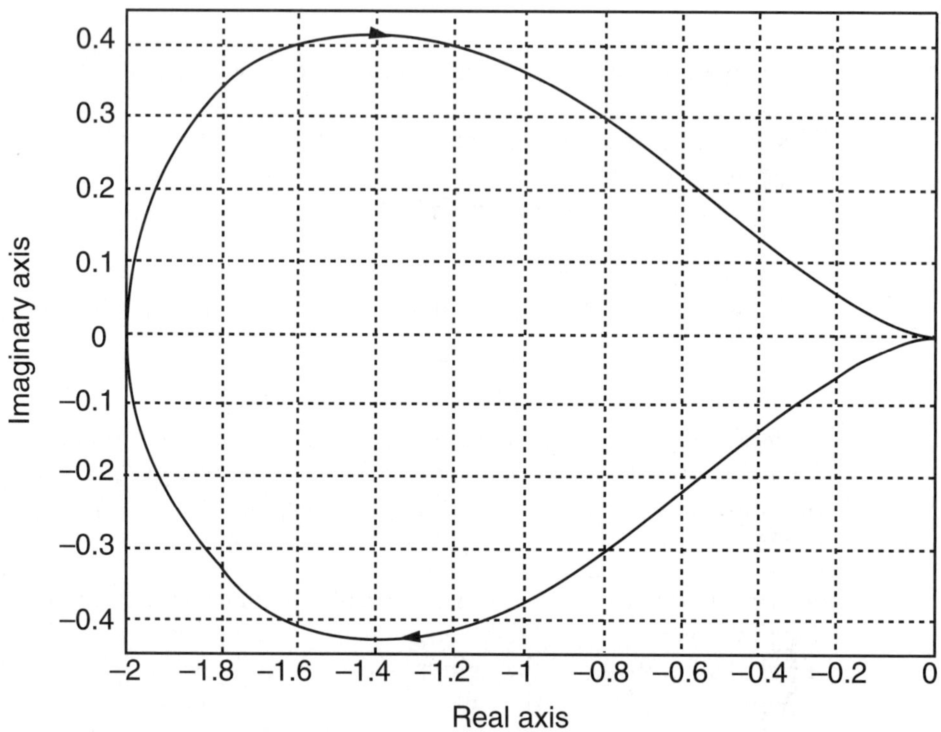

Figure 8–21 Nyquist plot for $GH(s) = 100/(s + 5)(s - 10)$

8.15 Relative Stability (Gain and Phase Margins)

The control system must be stable to be useful, but stability is not enough. A control system must also have acceptable dynamic and steady-state response. Does the control system respond quickly enough to a command or to a disturbance? How long does the transient last? Is the response oscillatory? Is it well damped? These matters, and others, constitute a system's relative stability.

The Nyquist criterion is simply a test for the absolute stability of the system, which is interpreted by inspection of a graph that defines the frequency response of the system. The relative stability can be measured by inspecting the curves on the graph.

Consider the Nyquist plot of a control system in Figure 8–22, which is for a stable system. The curve does not encircle the critical −1 point. It is clear that

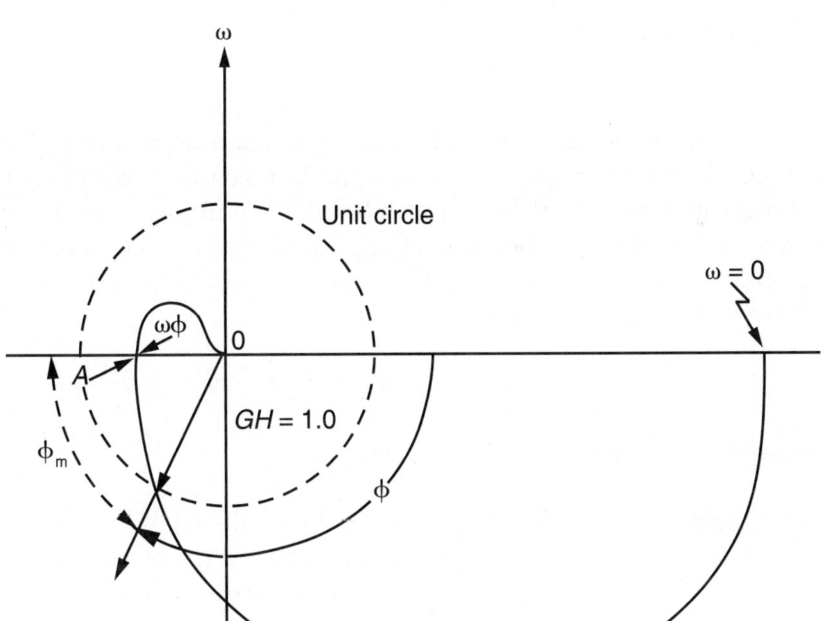

Figure 8–22 Nyquist plot of a stable system where

$$GH = \frac{K}{\left(\frac{j\omega}{P_1}+1\right)\left(\frac{j\omega}{P_2}+1\right)\left(\frac{j\omega}{P_3}+1\right)},\ \text{gain margin} = \frac{1}{OA},\ \text{and phase margin} = \phi_m$$

the distance from the -1 point to the Nyquist plot is important, but, the "distance" from a point to a curve is not easily defined. Two distances are defined and used for relative stability analysis. If the length of the 180 degree vector is increased or shortened, its tip may be placed on the $-1 + j0$ point. The amount of this change in length is $1/|GH|$ and is called the gain margin (GM). In similar fashion, one locates the vector $|GH| = 1.0$ and defines the phase margin (PM) to be the angle through which this vector must be rotated to place its tip on the $-1 + j0$ point. When the curve goes through the $-1 + j0$ point, the system is at the limit of stability. From the vector from the origin of the $GH(s)$ coordinate system to a point on the GH curve is the $GH(s)$ vector. When the vector has $\angle GH(s) = 180$ degrees, adjusting the gain can provide a vector length $|GH| = 1$, and the GH curve then passes through the $-1 + j0$ point. Raising the gain increases the length of the vectors, the -1 point then lies inside the map, and the system is unstable. Conversely, reducing the gain shortens the GH vectors and the -1 point is outside the map, so the system is stable. Thus, the value of gain at which the tip of the GH vector lies on the -1 point defines the stability limit.

In practice, it is usually possible to adjust the gain. While doing so may stabilize the system, the resulting stable system may not meet transient specifications. Rotating the unit length vector can also place the tip of the $GH(s)$ vector on the $-1 + j0$ point, and the angle through which this vector must be rotated is the PM. For a given system, adjusting the gain only is not likely to be completely satisfactory, and neither is adjusting the phase. However, evaluating PM and GM gives guidance as to what is needed to meet specifications.

8.16 Closed-Loop Frequency Response

The preceding discussion has been concerned with the stability of the closed-loop system, but has used the open-loop frequency response to predict it. There are various problems for which the closed-loop frequency response is a better tool. For example:

- When experimental testing of hardware is needed to define a dynamic model, measuring the closed-loop frequency response often gives better data.
- When the system is to be operated with a periodic command signal, the steady-state response may be more meaningful.

When needed, the open-loop characteristics may be calculated from the closed-loop data:

Let:

$$M = \left|\frac{C}{R}(j\omega)\right| = \left|\frac{GH(j\omega)}{1 + GH(j\omega)}\right| \qquad \textbf{Eq. 8.39}$$

and:

$$N = \angle\frac{C}{R}(j\omega) = \angle GH(j\omega) - \angle 1 + GH(j\omega) \qquad \textbf{Eq. 8.40}$$

Curves of constant M and N are circles. For the M circles, the coordinates of the center of the circle are:

$$x = -\frac{1}{2} \quad y = +\frac{1}{2N} \qquad \textbf{Eq. 8.41}$$

with radius:

$$R = \sqrt{\left[\frac{1}{4} + \left(\frac{1}{2N}\right)^2\right]} \qquad \textbf{Eq. 8.42}$$

Once the closed-loop frequency response has been obtained, the open-loop frequency response is available using the circles as a nomograph. Conversely, if the open-loop frequency response is known, the circles may be used to obtain the closed-loop response. In practice, the polar plot and the M and N circles are seldom used, because the Bode diagram and Nichols chart are more convenient.

8.17 Stability and Relative Stability from the Bode Diagram

The Nyquist criterion uses one polar plot of the frequency response to determine the stability of a dynamic system. A system is unstable when the polar plot encloses the $-1 + j0$ point and, in the process of analyzing stability, it was convenient to define PM and GM.

The Bode diagram, a parametric plot of the frequency response using logarithmic coordinates, requires two curves: (1) one for $|GH(j\omega)|$ and (2) one for $\angle GH(j\omega)$. On the Bode plot, the PM and GM are easily evaluated and are used to evaluate both stability and relative stability.

The abscissa for both curves is the frequency in radians per second. Also, both curves normally cross the horizontal axis. The 0 crossing on the magnitude curve, called the gain crossover, locates the frequency at which the length of the gain vector is 1.0 (0 dB). This frequency is defined as the gain crossover frequency, w_x. In like manner, the -180 degree crossing of the angle curve is called the phase crossover and locates the frequency at which the angle of $GH(j\omega)$ is -180 degrees. This frequency is defined as the phase crossover frequency, w_ϕ. When the two curves cross the horizontal axis at the same point (this corresponds to the $-1 + j0$ point), the system is at the stability limit, and both the PM and GM are 0. When the two curves do not cross the axis at the same point, it is still true that the PM is defined at the gain crossover, and the GM is defined at the phase crossover. With a positive phase margin, the system is stable. With a positive gain margin, the system is stable. When the phase curve is above the gain crossover, PM is positive; when the gain curve is below the phase crossover, the GM is positive.

The following are of interest in both analysis and design. When the gain and phase scales are aligned so that 0 dB on the magnitude scale and -180 degrees on the angle scale are the same line, then:

1. When the gain crossover lies below the phase curve, the PM, read from the angle scale, is $\phi_m = 180$ degrees $- \phi_x$, where ϕ_x is the phase angle at the gain crossover frequency.
2. At the phase crossover, read the GM
3. For most systems, $\varsigma = 0.01 * \phi_m$
4. $\omega_n \cong \omega_x$
5. $T_s = 4/\varsigma\omega_n$
6. $M_{p\omega} \cong M_{pt}$

8.18 Using MATLAB to Obtain GM and PM

To obtain the phase and margin, invoke:

```
>>[Gm,Pm,Wcg,Wcp] = margin(num,den)
```

Doing so returns the gain margin (GM) and phase margin (PM), the gain crossover frequency (*Wcg*), and the phase crossover frequency (*Wcp*) for a system in the *s*-domain, whose OLTF is defined in terms of numerator and denominator polynomials.

8.19 Using MATLAB to Obtain a Nyquist Plot

To have MATLAB produce a Nyquist plot, invoke:

```
>>nyquist(num,den)
```

where num and den are the OLTF polynomial coefficients.

8.20 Nichols Coordinates and the Nichols Chart

When using the frequency domain for analysis and design, the usual tool is the Bode diagram, and one obtains two curves for an open-loop system. These curves are concerned with $|GH(s)|$ in dB as ordinate versus ω as abscissa and $\angle GH(s)$ as ordinate versus ω as abscissa. There are problems for which this is not the most convenient representation. The parameter ω is common to both curves and can be eliminated, thereby providing a coordinate system with $|GH(s)|$ as ordinate and $\angle GH(s)$ as abscissa. The curve this provides is called the Nichols plot. Thus, the coordinate system is commonly called "Nichols coordinates." The Nichols curve has some very useful characteristics, as Figure 8–23 shows. The coordinate point (-180 deg, 0 dB) corresponds to the -1 point of the Nyquist plot and is the stability limit for the system. The PM and GM appear in Figure 8–23. This Nichols

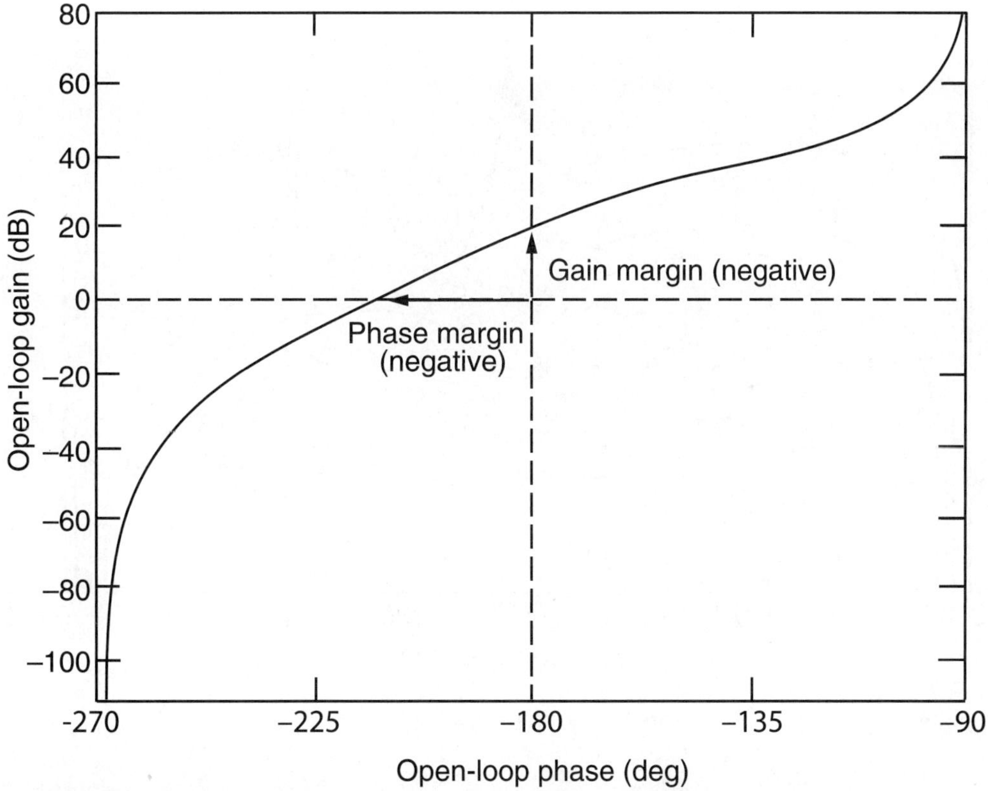

Figure 8–23 Nichols chart

chart corresponds to an unstable system. Values of w for points on the curve are available from the Bode diagram used to provide the coordinate pairs, which help estimate the bandwidth.

For some problems, the closed-loop frequency response is needed. It may be obtained using the Nichols chart in Figure 8–24. The curves on the chart define closed-loop loci for constant magnitude or constant angle. To convert the open-loop data to closed-loop data, the Nichols curve is plotted on the Nichols chart using the rectangular coordinates (each point is identified by the corresponding value of w). Then, at each value of w, the closed-loop magnitude and phase are determined from the intersection of the Nichols curve with the curves of the chart.

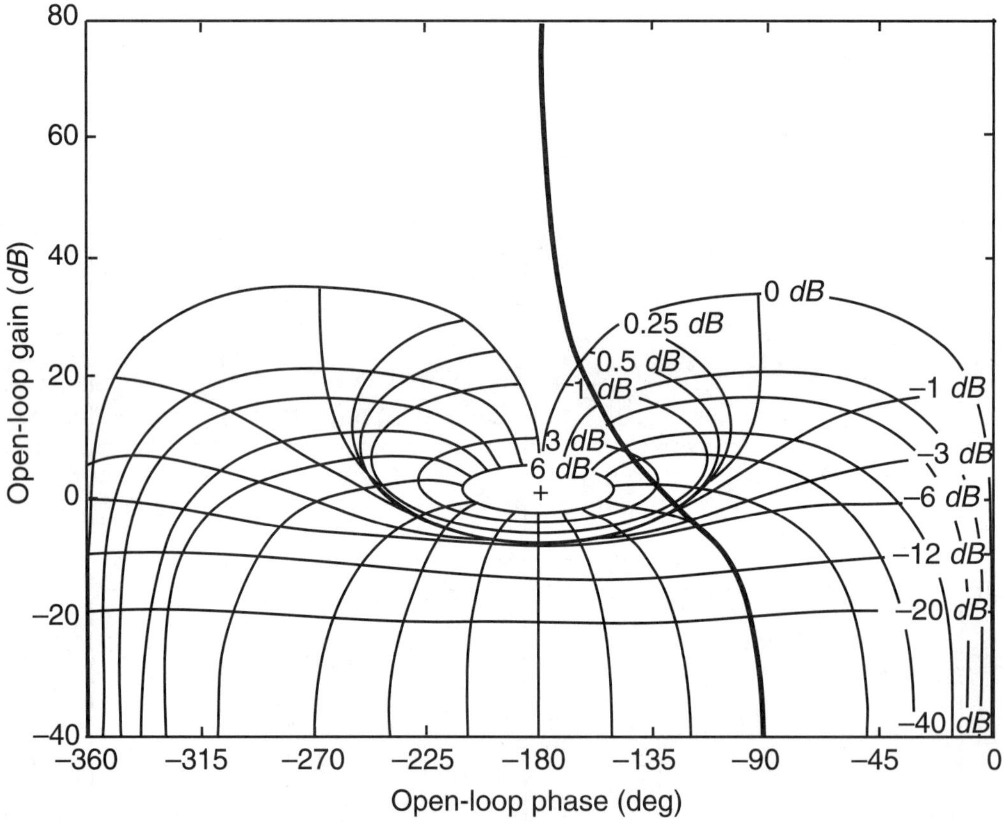

Figure 8–24 Nichols chart with closed-loop information

8.21 Using MATLAB to Produce a Nichols Chart

To have MATLAB produce a Nichols chart, invoke:

```
>>nichols(num,den)
```

where num and den are the OLTF polynomial coefficients.

8.22 Comparison of Various Methods

It helps to compare the several methods of representing relative stability. Compare the same information presented in three ways. See Figure 8–25.

Summary

1. When a linear system is excited with a sinusoid, its response will be a sinusoid of the same frequency but different amplitude and phase.
2. Application of a sinusoidal input and subsequent analysis of the resulting output can be used to determine the system's dynamic characteristics and its transfer function.
3. The Bode diagram captures the frequency response of dynamic components in a format that facilitates the prediction of response when several components are cascaded.
4. Stability of the entire system (i.e., in closed-loop mode) can be predicted by evaluating the Bode plot of the OLTF.
5. A Bode plot consists of two plots: (1) magnitude and (2) phase. There are asymptotic approximations for most transfer functions, making manual analysis relatively easy.
6. The Nyquist plot, another way of presenting frequency response information, also reveals whether a system is stable.
7. Both Bode plots and Nyquist plots provide numbers (PM and GM) that provide a "factor of safety" for the control system. In other words, these plots reveal how far or close the system is to instability.

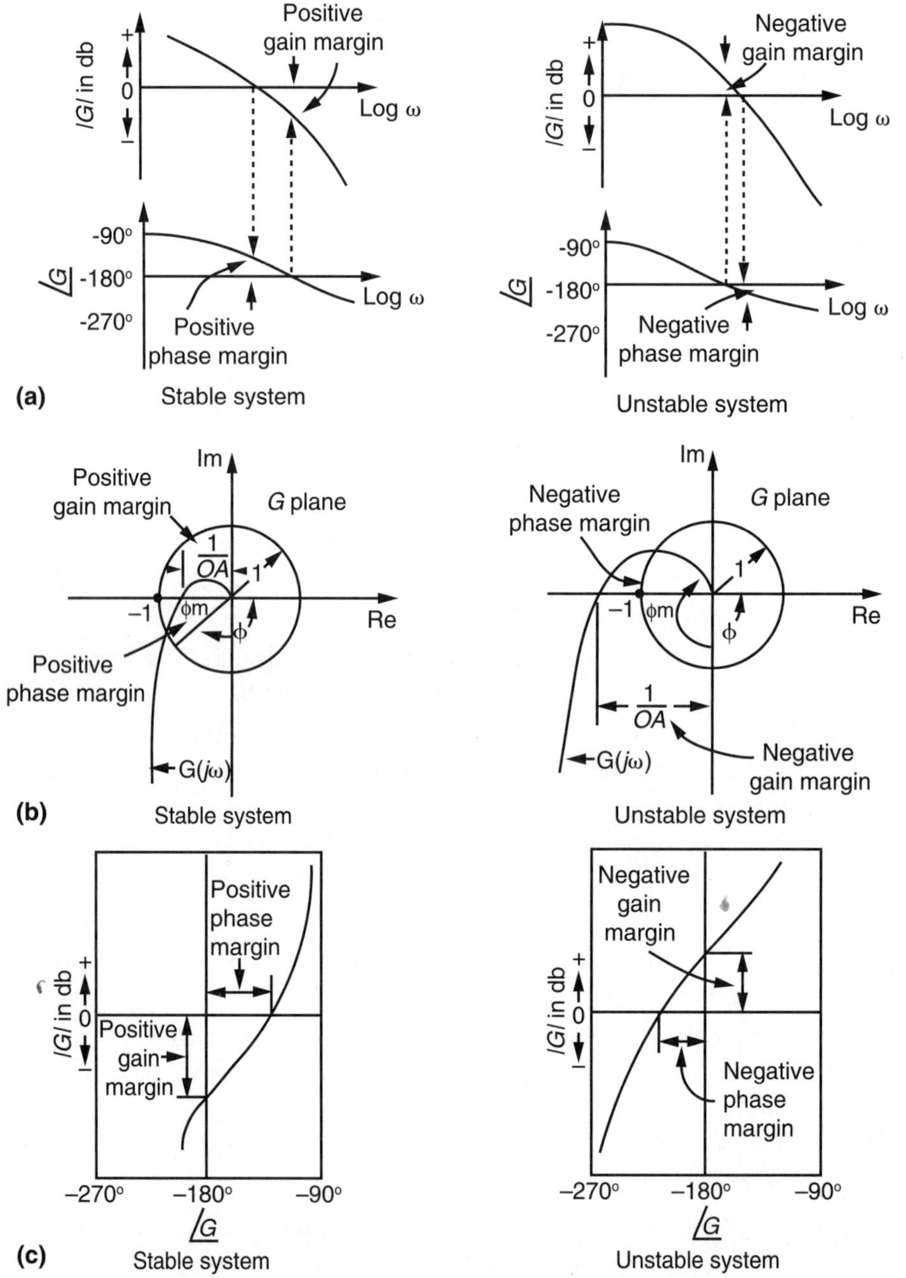

Figure 8–25 Comparison of phase and gain margins using several diagrams: (a) Bode plot, (b) Nyquist plot, and (c) Nichols chart (adapted from Ogata 1970)

Idea Portfolio

1. Would it be possible to perform a frequency response test on the system? What would be the consequences on the output product? For instance, if the controller controls the temperature of a yogurt batch, how would the taste and quality of the yogurt be affected?
2. Which physical device would you use to create the sinusoidal input?
3. How much noise is in the output, which would determine the minimum amplitude of the input?
4. Could the test be performed open loop, or would it have to be performed under closed-loop operation?

Problems

1. Sketch the asymptotic Bode plots (magnitude and angle) for the following transfer functions:

   - $T(s) = \dfrac{20,000s}{(s+100)(s+2,000)}$
   - $T(s) = \dfrac{21,000}{(s+30)(s+700)}$
   - $T(s) = \dfrac{15s}{(s+500)}$
   - $T(s) = \dfrac{120}{s(s+30)}$

2. Assume that the signal:

$$v_i(t) = 23\sin 200t$$

   is applied to a black box whose Bode plot is known. At 200 rad/s, the Bode plot indicates a gain of -13 dB and a phase of 32 degrees. Determine the output of the system (after the initial transients dies out) for the given input signal.

3. If the following transfer function is excited with an input signal of the form:

$$4.4\sin(400t)$$

Calculate at the following input conditions:

- Magnitude in dB
- Magnitude as a gain
- Phase (angle in degrees)
- Steady-state, sinusoidal output

$$T(s) = \frac{25}{\left(\frac{s}{200} + 1\right)}$$

4. You have performed a frequency response on a "black box," put the magnitude data in Bode format, and obtained the plot in Figure 8.26. Use the asymptotic approximation guidelines to suggest an appropriate transfer function that describes this black box.
5. A dynamic system has a controller gain of 3. If this gain is increased, the system becomes marginally stable (i.e., becomes oscillatory with a constant amplitude). The gain that produces this condition is a gain of 12. When the gain is set at 3, determine the following:
   - PM (in dB)
   - GM (in dB)
   - Whether the system has an adequate gain margin
6. Create a rough magnitude Bode plot for each of the following filters:
   - Low pass
   - High pass
   - Bandpass
   - Notch filters (or bandstop)

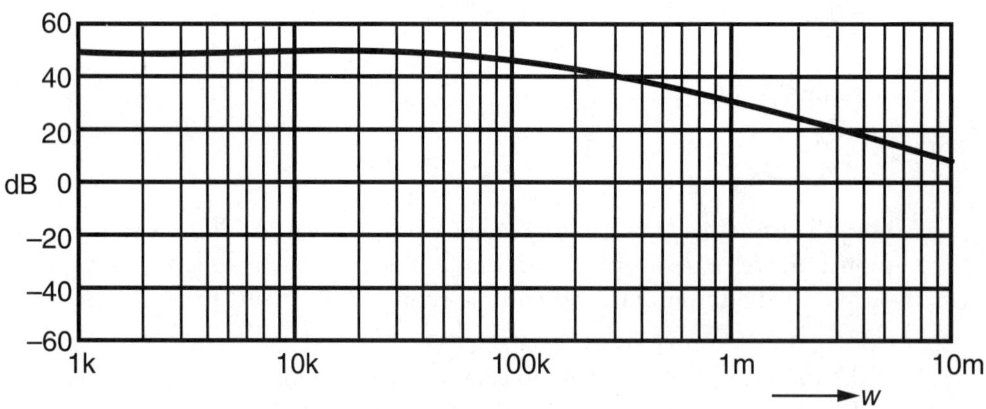

Figure 8–26

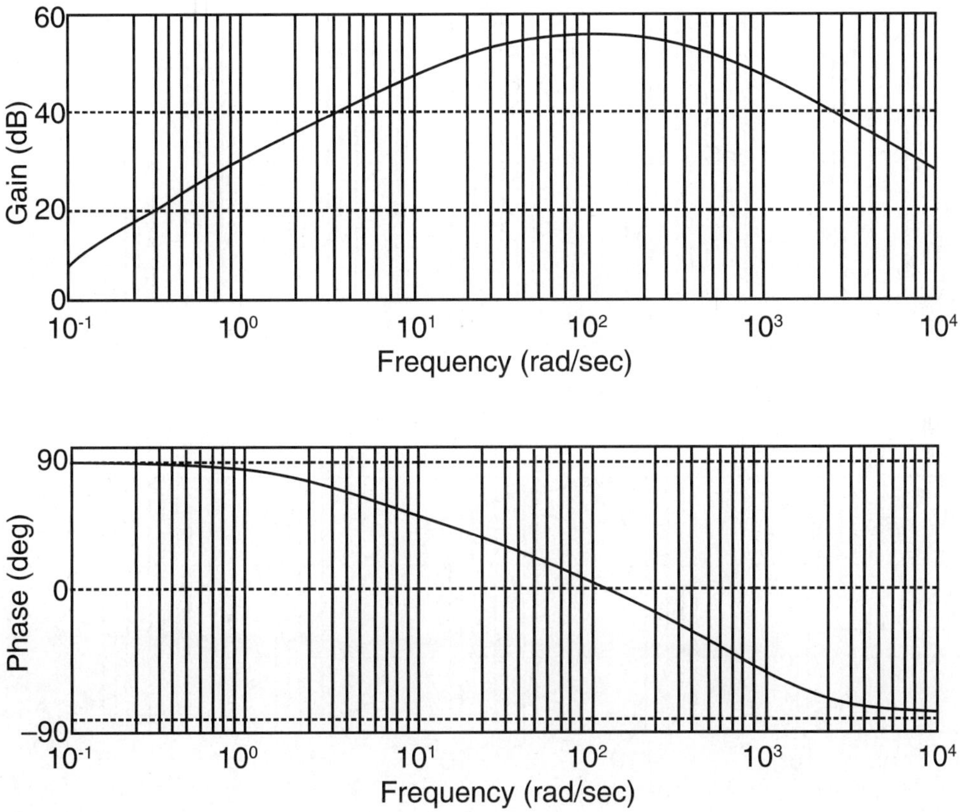

Figure 8–27

7. Propose a Bode plot for a transfer function whose magnitude Bode plot is shown in Figure 8.27.
8. Propose a Bode plot for a transfer function whose magnitude Bode plot is shown in Figure 8.28.
9. Propose a Bode plot for a transfer function whose magnitude Bode plot is shown in Figure 8.29.
10. Use MATLAB-created Bode plots to determine the PM and GM for a unity feedback system whose transfer function is:
    - $G(s) = \dfrac{4}{s^2(s+1)}$
    - $G(s) = \dfrac{100}{s^2(0.01s+1)}$

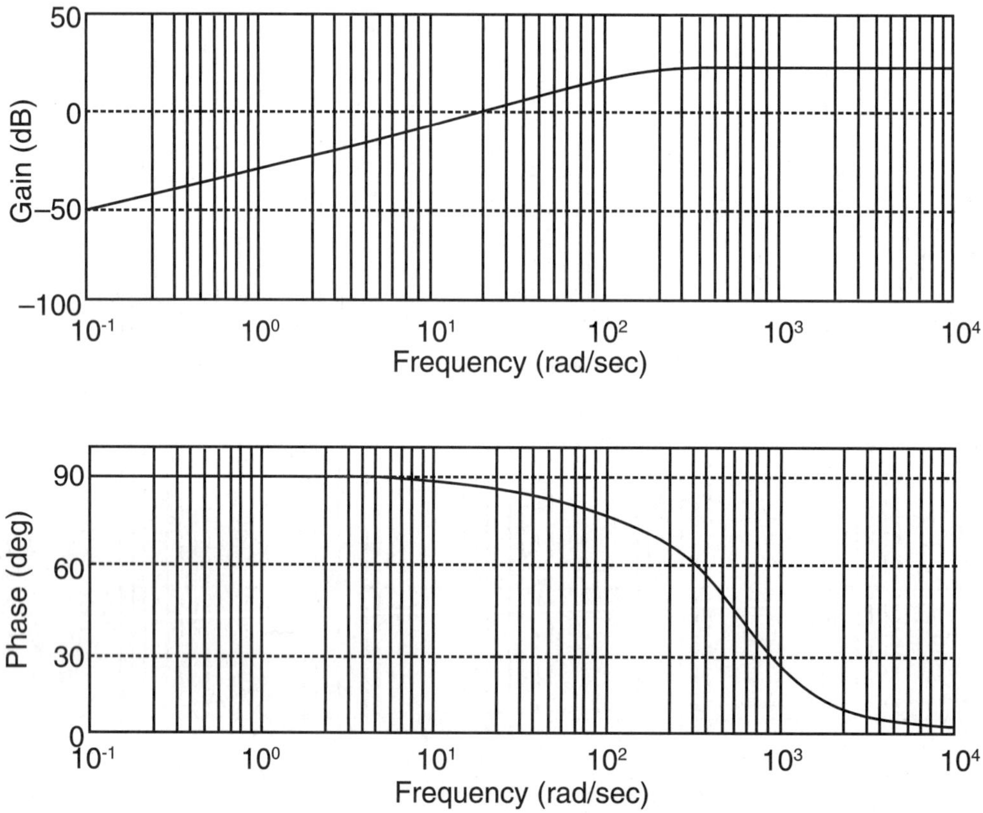

Figure 8–28

- $G(s) = \dfrac{500}{s(0.01s + 1)(0.1s + 1)}$

In each case, determine whether the system is stable.

11. Use the `margin()` function in MATLAB to determine the PM and GM for the systems of Problem 10. In each case, determine whether the system is stable.
12. A Nyquist plot for a system appears in Figure 8.30. Estimate the following:
    - GM
    - PM
    - System stability
13. Determine the transfer function whose Nyquist plot appears in Figure 8.31.

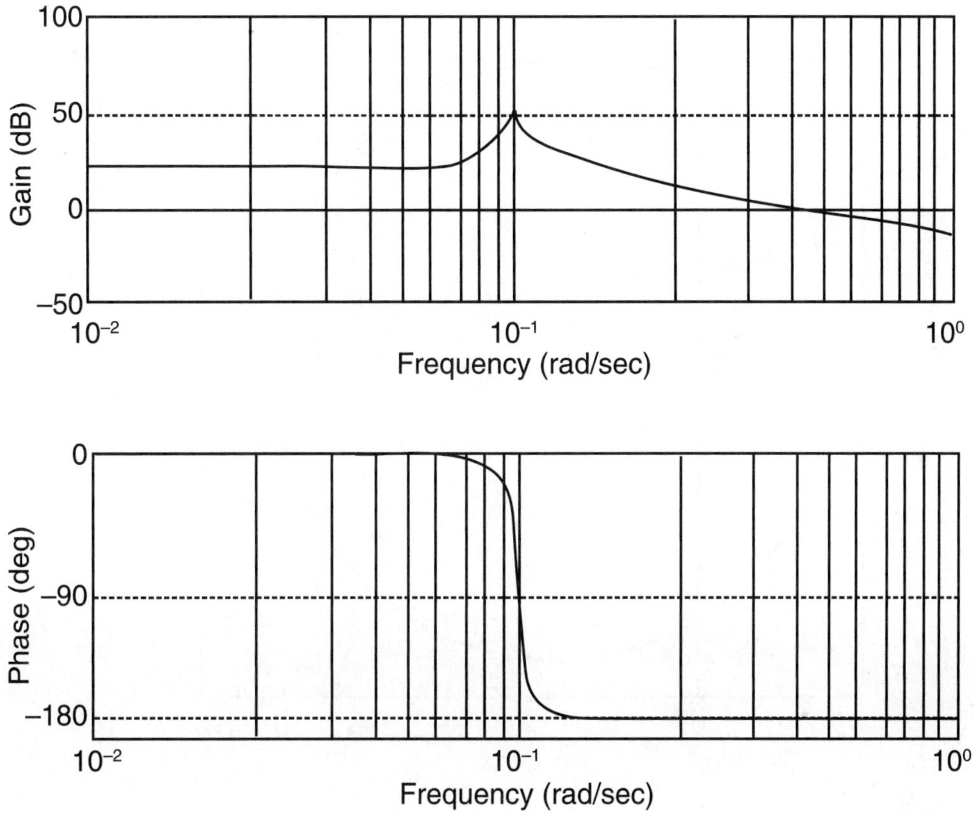

Figure 8–29

14. Sketch the Nyquist plot for the following transfer function:

$$T(s) = \frac{10}{s + 5}$$

15. Sketch the Nyquist plot for the following transfer function:

$$G(s) = \frac{100}{(s + 2)(s + 20)}$$

16. The response in an amplifier is given by the following:

$$G(s) = \frac{Ks}{(s + a)(s + b)} \quad \text{where } K = (10)^5,\ a = 10, \text{ and } b = (10)^4$$

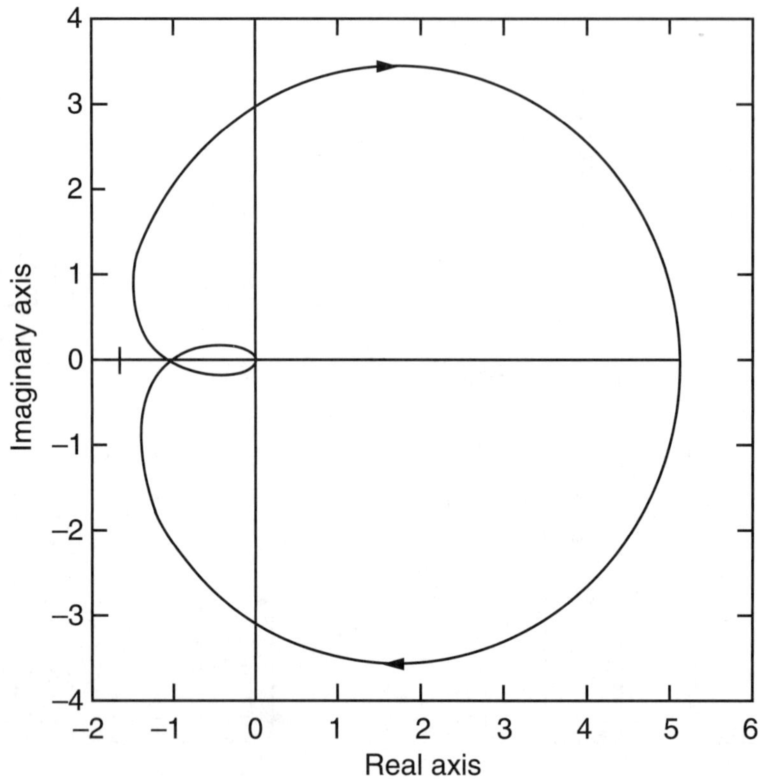

Figure 8–30

- Sketch the magnitude and the phase asymptotic (straight-line) Bode plots of $G(s)$. Clearly label all slopes, constant values, and corner frequencies.
- Determine the 3-dB bandwidth, ω_B.
- Determine the two gain crossover frequencies and the PM at each.
- Confirm your results with MATLAB.

17. A unity feedback system has the following plant:

$$G(s) = \frac{K(s+3)}{s(s+1)(s+5)}$$

- Find the gain crossover frequency needed to give a PM of 40 degrees. (Hint: This may require trial and error.)
- Determine the value of K required.
- For the preceding value of K, find the phase crossover frequency.
- Determine the GM.

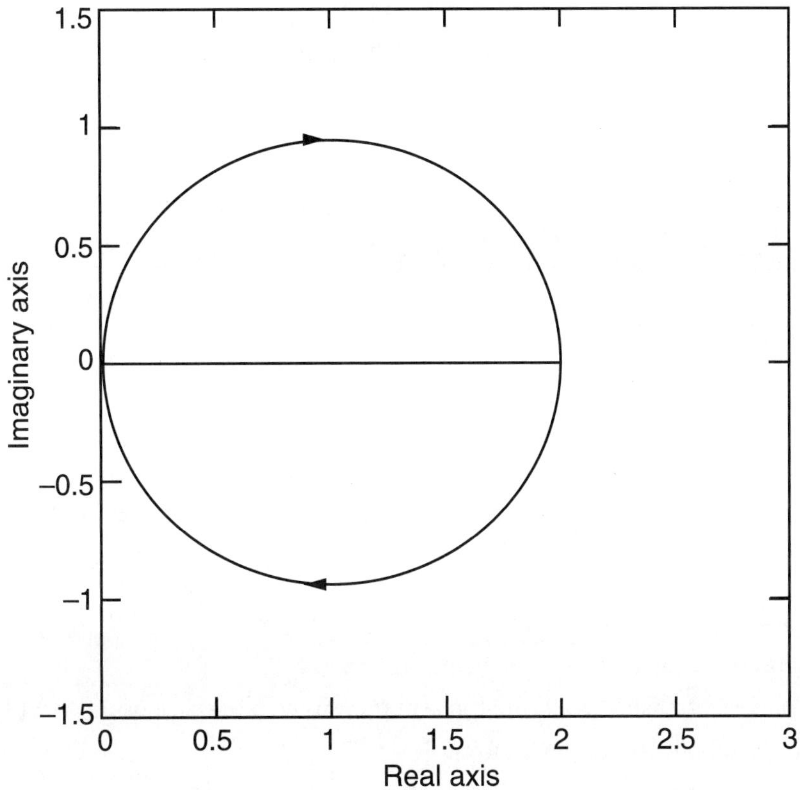

Figure 8–31

18. Explain the differences between a Nyquist plot and a Nichols chart.
19. Indicate the peak, in dB, of an underdamped second-order transfer function with $\varsigma = 0.26$.
20. What are the consequences of a system being excited at its resonant frequency?
21. Differentiate between the natural frequency, the resonant frequency, and the damped natural frequency in an underdamped, second-order system. What is the typical numerical difference between them?

9

Introduction to State Space Methods

Learning Objectives

In this chapter, you will learn how to:

- Represent a dynamic system using state space.
- Convert a transfer function representation into state space using MATLAB.
- Perform basic matrix operations.
- Interpret and use the information in the rank of a matrix.
- Render a SIMULINK block diagram into a single transfer function.

9.1 Historical Perspective

The state space discipline of control systems began to develop in the 1960s with the advent of the digital computer. Today, state space methods are the main design and analysis tools for complex systems, especially for multiple-input and/or multiple-output systems.

9.2 Introduction

State space is an excellent bookkeeping/book-handling method for dynamic systems, especially multiple input–multiple output (MIMO) systems. Even though the classical techniques, which were developed for single input–single output

(SISO) systems, can be applied iteratively to MIMO systems, they are unlikely to obtain an optimal design.

The state space method can handle very large systems with multiple inputs and outputs. It uses matrices to encapsulate the description of the dynamic system. It also provides a medium that is particularly amenable to computer manipulations or calculations.

Many contemporary closed-loop dynamic systems have their design roots in state space and are often called modern control systems.

9.3 Matrix Fundamentals

An $n \times m$ matrix is characterized by its number of rows, (n), and columns, (m). For instance, a sample 2×3 matrix, A, is:

$$A = \begin{bmatrix} 1 & 22 & 12 \\ 3 & 45 & 11 \end{bmatrix}$$ **Eq. 9.1**

A sample column vector is:

$$x = \begin{bmatrix} 2 \\ 12 \\ 13 \end{bmatrix}$$ **Eq. 9.2**

A sample row vector is:

$$y = \begin{bmatrix} 2 & 12 & 13 \end{bmatrix}$$ **Eq. 9.3**

By transposing the vectors (denoted with single quotes):

$$x = y' \quad \text{and} \quad y = x'$$ **Eq. 9.4**

Two matrices can be multiplied as long as their inner characteristics match. Consider the following:

$$C = A \cdot B$$ **Eq. 9.5**

where A is $i \times j$ and B is $j \times k$

Their product, C, will be $i \times k$.

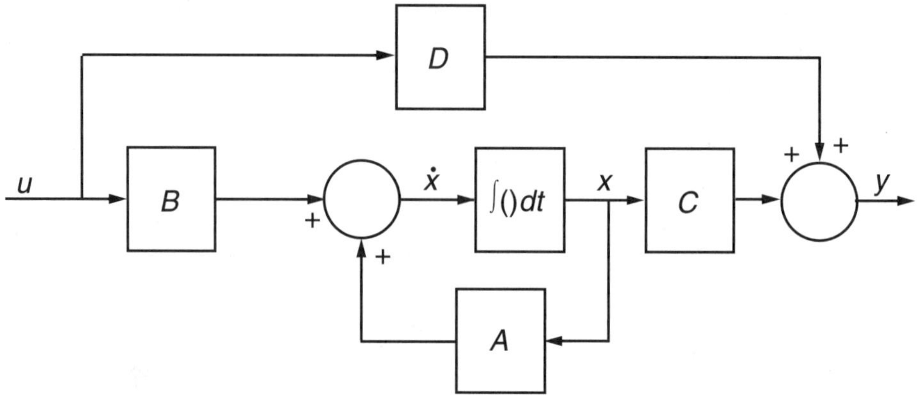

Figure 9–1 State space representation

9.4 Dynamic System Representation

A dynamic system is represented as:

$$\dot{x} = Ax + Bu$$
$$y = Cx + Du \qquad \textbf{Eq. 9.6}$$

where:

x is the state vector

u is the input

y is the output

A, B, C, and D are the matrices

This representation appears in Figure 9–1. The matrix A contains all the stability information about the systems, as does the CLTF's characteristic equation. MATLAB offers the command `eig()` to obtain the eigenvalues.

9.5 Convert a Differential Equation to State Space

To convert a differential equation to a state space representation, make the following substitutions. Notice that there are as many states as the order of the differential equation. Consider the following differential equation:

$$\dddot{x} + 5\ddot{x} + 4\dot{x} + 3x = 15r(t) \qquad \textbf{Eq. 9.7}$$

Using the variable z to denote the states, make the following substitution:

$$
\begin{aligned}
z_1 &= x \\
z_2 &= \dot{z}_1 = \dot{x} \\
z_3 &= \dot{z}_2 = \ddot{x} \quad \Rightarrow \quad \dot{z}_3 = \dddot{x}
\end{aligned}
$$

Eq. 9.8

The first two state equations result from the preceding second and third equations or:

$$
\begin{aligned}
\dot{z}_1 &= z_2 \\
\dot{z}_2 &= z_3
\end{aligned}
$$

Eq. 9.9

The third state equation comes from the preceding differential equation. The differential equation must be solved for the highest-order derivative. Next, each x variable is expressed in terms of state variables (z_1, z_2, z_3):

$$
\begin{aligned}
\dddot{x} &= -5\ddot{x} - 4\dot{x} - 3x + 15r(t) \\
\dot{z}_3 &= -5z_3 - 4z_2 - 3z_1 + 15u(t)
\end{aligned}
$$

Eq. 9.10

Notice that $u(t)$ is not the step function but a general variable denoting the input. The result can now be expressed in matrix form as:

$$
\begin{bmatrix} \dot{z}_1 \\ \dot{z}_2 \\ \dot{z}_3 \end{bmatrix} = \begin{bmatrix} 0 & 1 & 0 \\ 0 & 0 & 1 \\ -3 & -4 & -5 \end{bmatrix} \begin{bmatrix} z_1 \\ z_2 \\ z_3 \end{bmatrix} + \begin{bmatrix} 0 \\ 0 \\ 15 \end{bmatrix} u(u)
$$

$$
A = \begin{bmatrix} 0 & 1 & 0 \\ 0 & 0 & 1 \\ -3 & -4 & -5 \end{bmatrix}
$$

Eq. 9.11

$$
B = \begin{bmatrix} 0 \\ 0 \\ 15 \end{bmatrix}
$$

Because the output variable equals z_1, the output can be expressed as:

$$
y = z_1
$$

Eq. 9.12

Thus:

$$y = Cz + Du$$
$$C = \begin{bmatrix} 1 & 0 & 0 \end{bmatrix}$$
$$D = \begin{bmatrix} 0 \end{bmatrix}$$
Eq. 9.13

9.6 Convert a Transfer Function to State Space

Often, the mathematical model of a physical system is expressed in terms of a transfer function. There are several procedures for converting a transfer function representation to state space. One procedure suggests first converting the transfer function back into a differential equation, then applying the procedure outlined in the previous section.

Another approach capitalizes on the fact that, customarily, the numerator and denominator polynomials of the transfer function are known in factored form, because they represent specific physical components of the system.[1] Thus, when this system is represented in state space form, the choice of variables is dictated by the intermediate variables between the transfer functions, because they represent physical variables. However, when the transfer functions are expressed as a ratio of two polynomials, they must be factored, and there is no physical meaning associated with the intermediate variables. The grouping of factors is arbitrary as long as the order of the denominator polynomial is greater than or equal to the order of the numerator polynomial.

Four general forms can be used to decompose any transfer function:

1. Simple pole with gain
2. Single pole–single zero
3. Quadratic set of poles
4. Single zero and a quadratic set of poles

These transfer functions, and their corresponding block diagrams, appear in Figure 9–2.

There is another case that is relatively easy to represent in state space: a transfer function with no zeros and a denominator polynomial in an unfactored form. Consider the following:

$$\frac{X(s)}{R(s)} = \frac{K}{s^3 + as^2 + bs + c}$$
Eq. 9.14

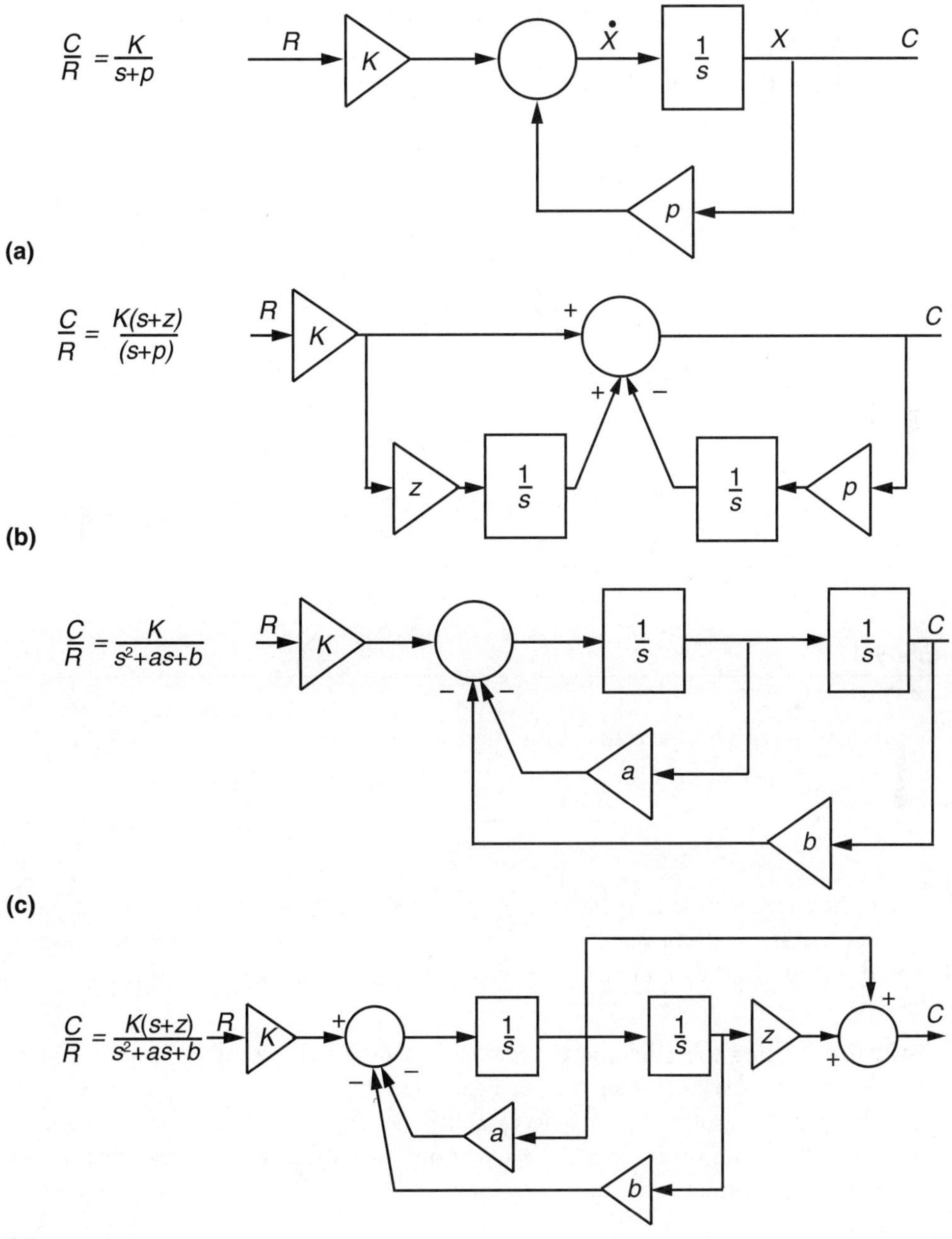

Figure 9–2 Block diagram realizations for various transfer functions (a) single pole, (b) pole-zero, (c) quadratic, (d) quadratic-zero

Notice that we have used the variable $X(s)$ instead of $C(s)$ to represent the Laplace transform of $x(t)$ to avoid confusion with the constant c.

When the preceding transfer function is expressed in the time domain, the result is:

$$\dddot{x}(t) + a\ddot{x}(t) + b\dot{x}(t) + cx(t) = Kr(t) \qquad \text{Eq. 9.15}$$

or:

$$\dddot{x}(t) = -a\ddot{x}(t) - b\dot{x}(t) - cx(t) + Kr(t)$$

Defining the state variables using the letter z, and using the procedure in Section 9.5, the result is:

$$\begin{bmatrix} \dot{z}_1 \\ \dot{z}_2 \\ \dot{z}_3 \end{bmatrix} = \begin{bmatrix} 0 & 1 & 0 \\ 0 & 0 & 1 \\ -c & -b & -a \end{bmatrix} \begin{bmatrix} z_1 \\ z_2 \\ z_3 \end{bmatrix} + \begin{bmatrix} 0 \\ 0 \\ K \end{bmatrix} r(t) \qquad \text{Eq. 9.16}$$

Example 9.1

Consider a system whose transfer function is:

$$\frac{X(s)}{R(s)} = \frac{K}{(s+p_1)} \frac{(s+z)}{(s+p_2)} \qquad \text{Eq. 9.17}$$

where the transfer functions have been separated according to the physical separation of subsystems. Represent these transfer functions using the general forms presented previously, then represent the system using state variables.

Solution Using the guidelines previously proposed, the result appears in Figure 9–3. State variables are associated with each variable appearing after each integrator. This representation allows us to establish w_1, w_2, w_3 and $\dot{w}_1, \dot{w}_2, \dot{w}_3$ (the variable preceding the integrator). We can now determine the equivalence of the derivatives of the state variable:

$$\dot{w}_1(t) = r(t)K - p_1 w_1(t)$$

$$\dot{w}_2 = z\,w_1(t)$$

$$\dot{w}_3 = p_2(w_1(t) + w_2(t) - w_3(t))$$

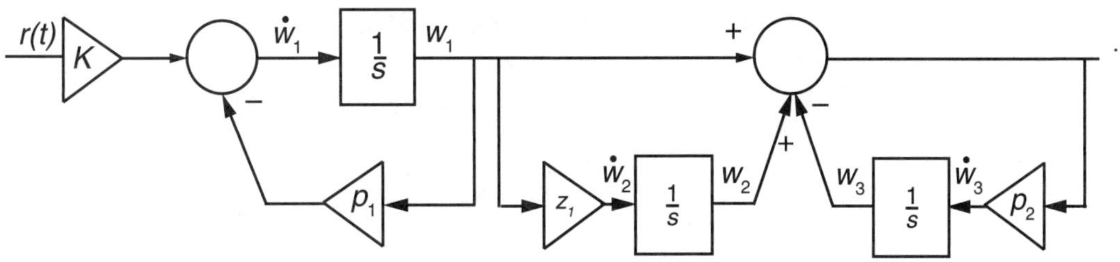

Figure 9–3 Block diagram realization for example

The output of the system is given by:

$$y(t) = w_1(t) + w_2(t) - w_3(t) \tag{Eq. 9.18}$$

$$\begin{bmatrix} \dot{w}_1 \\ \dot{w}_2 \\ \dot{w}_3 \end{bmatrix} = \begin{bmatrix} -p_1 & 0 & 0 \\ z & 0 & 0 \\ p_2 & p_2 & -p_2 \end{bmatrix} \begin{bmatrix} w_1 \\ w_2 \\ w_3 \end{bmatrix} + \begin{bmatrix} K \\ 0 \\ 0 \end{bmatrix} u(u)$$

and the output

$$y = \begin{bmatrix} 1 & 1 & -1 \end{bmatrix} \begin{bmatrix} w_1 \\ w_2 \\ w_3 \end{bmatrix}$$

9.7 Modeling Using State Space

The procedure for obtaining a state space dynamic representation of a dynamic system is similar to that presented in Chapter 3. Each elemental equation corresponding to energy-storing components produces a state. Each of the resulting equations is solved for the differentiated variable, and the rest of the equation is expressed in terms of nondifferentiated states and the input. Finally, the result is expressed in matrix notation.

Linear system theory indicates that the output(s) can always be expressed as a linear combination of the states and the output.

9.8 Convert a SIMULINK Block Diagram to State Space

To convert a SIMULINK model to state space, do the following:

1. Connect an "Import" icon at the input and an "Outport" icon at the output.
2. Save your SIMULINK file. For illustrative purposes, assume that you have named your SIMULINK file as `mymodel`.
3. In the MATLAB window, issue the following commands,

```
>>[A,B,C,D]=linmod('mymodel')
```

   The numerical value of the computed state space matrices *A*, *B*, *C*, and *D* are given.
4. Then:

```
>>[num,den]=ss2tf(A,B,C,D)
```

This stores the numerator and denominator polynomial into `num` and `den`, which are also displayed. Now you can use `[num,den]` for any other time analysis (e.g., Bode) that is desired.

Example 9.2

Consider the block diagram in Figure 9–4.

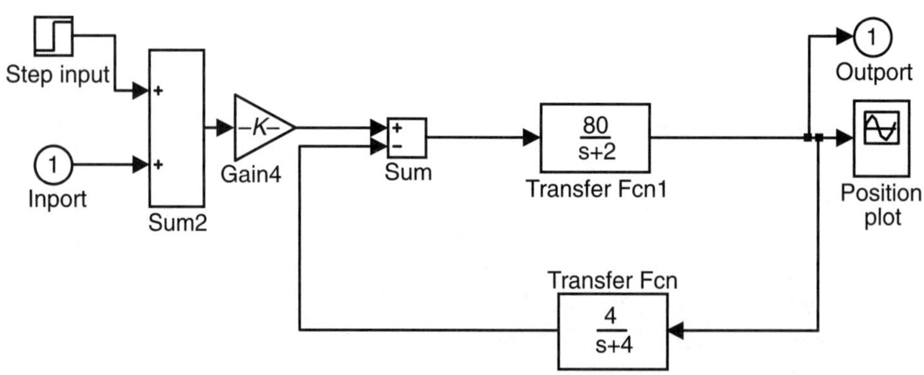

Figure 9–4 SIMULINK diagram with blocks needed for state-space conversion. Filename is `exo9_red.mdl`.

Solution Notice that it is necessary to place the Import and Outport icons at the input and output of the system. Here, a summer is used at the input so that the input block will not have to be reconnected. This SIMULINK file has been named `ex09_red`. The sequence of inputs follows:

```
>>[A,B,C,D]=linmod('ex09_red')

A =

     -2   -4

     80   -4

B =

      1

      0

C =

     80   0

D =

      0

>> [num,den]=ss2tf(A,B,C,D)

num =

               0   80.0000   320.0000

den =

      1.0000    6.0000   328.0000
```

The resulting numerator and denominator of the CLTF are the same as would be obtained using block-reduction techniques.

9.9 Rank of a Matrix

The rank of a matrix indicates the number of linearly independent rows or columns of the matrix. The rank of the matrix A, preceding, indicate the number of essential eigenvalues; that is, it can be used to determine the true order of a system. MATLAB determines the rank of a matrix A with the following function:

```
>>rank(A)
```

Example 9.3

Consider the mechanical system in Figure 9–5a.

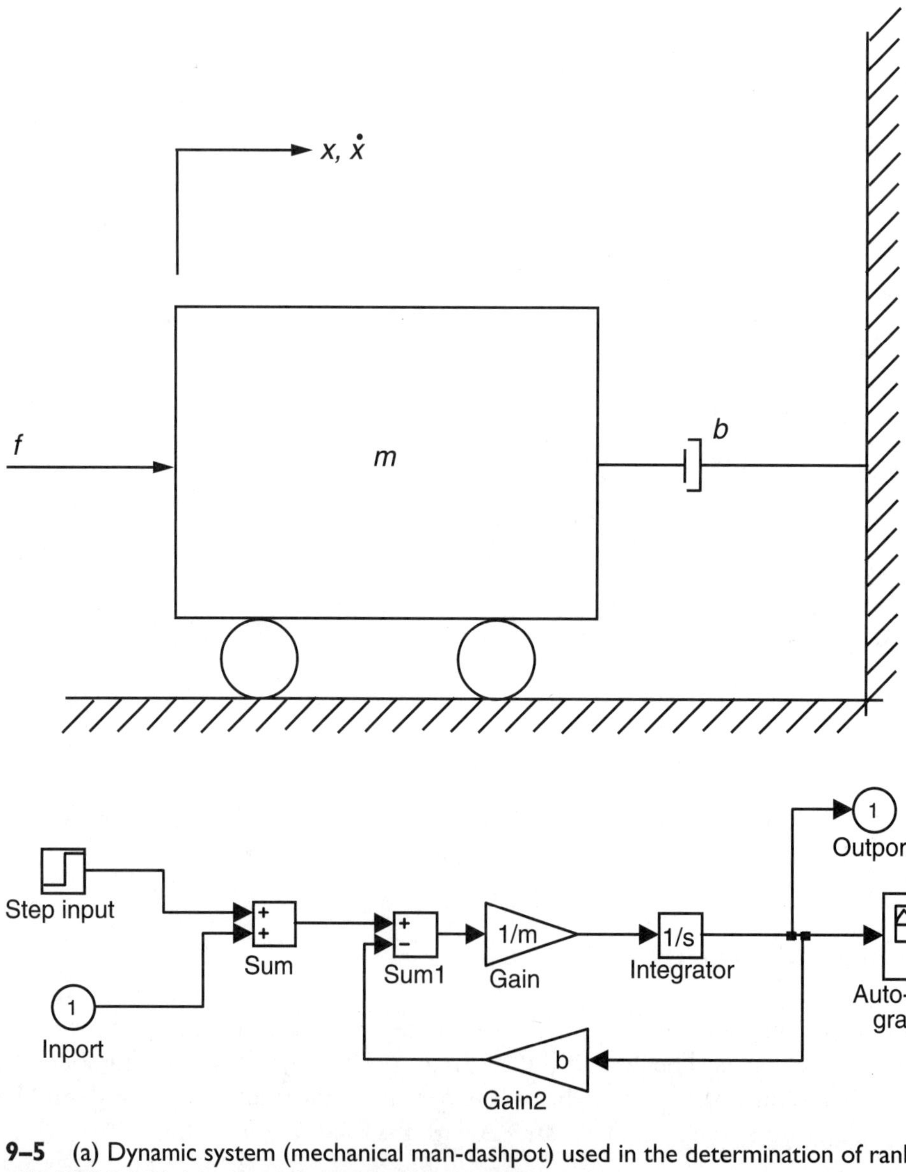

(a)

(b)

Figure 9–5 (a) Dynamic system (mechanical man-dashpot) used in the determination of rank example, (b) SIMULINK diagram with essential blocks, and

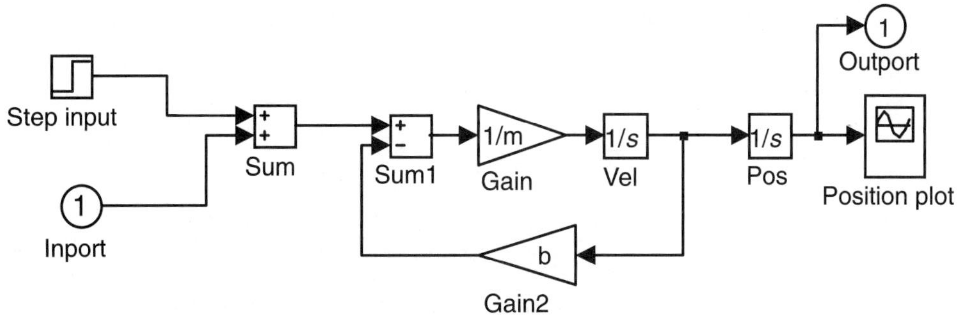

(c)

Figure 9–5 (continued) (c) SIMULINK diagram with additional blocks

Solution The system here is of order 1, because it only has one energy-storage component, namely the mass. A SIMULINK implementation to solve the differential equation appears in Figure 9–5b. If we were to perform the "linmod" procedure presented in the previous section, we would obtain an A matrix with rank of 1 (for this case, A is a scalar). Assume that we add another integrator, because we are also interested in position, as shown in Figure 9–5c. Performing "linmod" analysis again would produce an A matrix that is 2×2, but it would still have a rank of 1. Thus, the rank of the model is invariant with respect to superfluous integrators.

9.10 Consolidation of First-Order Differential Equations

A system can be described by several first-order differential equations, each with one or more system variables. This representation format is commonly called the state space form.

The state space representation is used widely, because it lends itself to computer manipulation and is often the result of a computer-based algorithm that takes as its input the elemental and topological equations and produces the system equations in state space form.

A procedure is presented for transforming a set of first-order differential equations into a single, higher-order differential equation in terms of one variable. When the system is linear, one can always arrive at the solution by taking the Laplace transform and using Cramer's rule. For a nonlinear system, however, Cramer's rule cannot be used. For this reason, the following method is presented.[2]

The procedure will be illustrated by means of an example. Consider the following system:

$$\dot{X}_1 = X_1 + 2\,X_2 + X_3 \qquad \text{Eq. 9.19}$$

$$\dot{X}_2 = 2\,X_1 + X_2 + X_3 \qquad \text{Eq. 9.20}$$

$$\dot{X}_3 = X_1 + X_2 + 2\,X_3 \qquad \text{Eq. 9.21}$$

It is desirable to have one differential equation in terms of X_2 only. The differential equation in which the derivative of the variable of interest appears is called the primary equation.

1. Differentiate the primary equation.

$$\ddot{X}_2 = 2\,\dot{X}_1 + \dot{X}_2 + \dot{X}_3 \qquad \text{Eq. 9.22}$$

2. Eliminate the derivatives of the other variables using remaining fundamental Equation 9.19 and equation 9.21. After similar terms are combined, the result is:

$$\ddot{X}_2 = 3\,X_1 + 5\,X_2 + 4\,X_3 + \dot{X}_2 \qquad \text{Eq. 9.23}$$

3. Decide which variable to eliminate next. Let us eliminate X_3 first.
4. Solve the primary equation 9.20 for X_3:

$$X_3 = \dot{X}_2 - 2\,X_1 - X_2 \qquad \text{Eq. 9.24}$$

5. Substitute into equation 9.23:

$$\ddot{X}_2 = 3\,X_1 + 5\,X_2 + 4\,(\dot{X}_2 - 2\,X_1 - X_2) + \dot{X}_2 \qquad \text{Eq. 9.25}$$

$$\ddot{X}_2 = -5\,X_1 + X_2 + 5\,\dot{X}_2 \qquad \text{Eq. 9.26}$$

6. Substitute into equation 9.19:

$$\dot{X}_1 = X_1 + 2\,X_2 + (\dot{X}_2 - 2\,X_1 - X_2) \qquad \text{Eq. 9.27}$$

$$\dot{X}_1 = -X_1 + X_2 + \dot{X}_2 \qquad \text{Eq. 9.28}$$

This is the end of one cycle. At this time we have two equations, 9.26 and 9.28, in which X_3 has been eliminated.

7. Start the next cycle by differentiating equation 9.26:

$$\dddot{X}_2 = -5\,\dot{X}_2 + \dot{X}_2 + 5\,\dot{X}_2 \qquad \textbf{Eq. 9.29}$$

8. Eliminate X_1 with equation 9.28:

$$\dddot{X}_2 = -5\,(-X_1 + X_2 + \dot{X}_2) + \dot{X}_2 + 5\,\ddot{X}_2 \qquad \textbf{Eq. 9.30}$$

$$\dddot{X}_2 = 5\,X_1 - 5\,X_2 - 4\,\dot{X}_2 + 5\,\ddot{X}_2 \qquad \textbf{Eq. 9.31}$$

9. Eliminate X_1 by solving equation 9.26 for X_1:

$$X_1 = \frac{1}{5}\,X_2 + \dot{X}_2 - \frac{1}{5}\,\ddot{X}_2 \qquad \textbf{Eq. 9.32}$$

10. Substitute into equation 9.31:

$$\dddot{X}_2 = 5\left(\frac{1}{5}\,X_2 + \dot{X}_2 - \frac{1}{5}\,\ddot{X}_2\right) - 5\,X_2 - 4\,\dot{X}_2 + 5\,\ddot{X}_2 \qquad \textbf{Eq. 9.32}$$

$$\dddot{X}_2 = -4\,X_2 + \dot{X}_2 + 4\,\ddot{X}_2$$

or:

$$\dddot{X}_2 - 4\,X_2 - \dot{X}_2 + 4\,\ddot{X}_2 = 0 \qquad \textbf{Eq. 9.33}$$

This procedure also works when there are forcing functions in the equations and when some of the terms are nonlinear.

9.11 Transformation Procedure Using MATLAB

The preceding procedure can be carried out using MATLAB. This will be done by reworking the previous example. On the MATLAB prompt, define:

```
>>A=[1 2 1 ; 2 1 1 ; 1 1 2]

A =

      1  2  1

      2  1  1

      1  1  2
```

```
>>B=[0;0;0]

B =

     0
     0
     0

>>C=[0 1 0]

C =

     0  1  0

>>D=[0]

D =

     0

>>[num,den]=ss2tf(A,B,C,D)

num =

     0  0  0  0

den =

    1.0000   -4.0000   -1.0000   4.0000
```

This is the result obtained previously.

9.12 Other Areas of Control

Many areas related to control systems use state space as a basis. Following are some of them.

Digital Control

The digital control applies when the controller is implemented with a digital computer, and the information (both input and output) is only available at finite points in time. The Z-transform is used to analyze this type of system. This process approaches the continuous analysis approach (Laplace operator s) if the sampling is done very quickly.

System Identification

Measure the input(s) and output(s) of a system for a period of time, and use the information to create a mathematical model of the system.

Observers

In the observer field, the observer performs calculations on the system's easily accessible variables to recreate or predict those variables that are inaccessible or difficult to measure.

Optimal Control

The optimal control field has to do with designing a controller that is optimal, according to some cost function.

Adaptive (Self-Tuning) Controllers

The adaptive controller modifies itself so that it is optimal under different operating conditions.

Summary

1. State space is a matrix-based approach for capturing dynamic system characteristics. It can be used for analysis and design. The method lends itself to computer implementation.
2. State space is the choice method for multiple input–multiple output (MIMO) systems.
3. MATLAB can represent dynamic systems using both transfer functions and state space representations. It can also convert from one to the other.
4. MATLAB/SIMULINK provide tools for producing an equivalent state space representation for a system, represented as many interconnected transfer functions (i.e., a SIMULINK block diagram).
5. It is possible to represent an nth differential equation as n first-order differential equations. The converse is also true.

Idea Portfolio

1. Transform the SIMULINK block diagram model to a state space model. Simulate this model, and confirm that it behaves as the block diagram representation behaves. What are the eigenvalues of the system?

Problems

1. If matrices A and B are 3x2, can they be:
   - Added?
   - Multiplied?
2. If matrices A and B are 3x2, are the following operations valid?
   - A+B'
   - A B'
3. Give examples of column and row vectors.
4. Convert the following differential equation to state space:

$$2\dddot{x} + 16\ddot{x} + 4\dot{x} + 10x = 150r(t)$$

5. Convert the following differential equation to state space:

$$\frac{d^4x}{dt^4} + 3\frac{d^3x}{dt^3} + 2\frac{d^2x}{dt^2} + \frac{dx}{dt} + 5x = 44f(t)$$

6. Manually convert the following transfer function to state space:

$$\frac{X(s)}{R(s)} = \frac{10}{s^3 + 5s^2 + 4s + 2}$$

7. Determine the poles (eigenvalues) for the above transfer function using the `roots()` function.
8. Enter the preceding transfer function into SIMULINK, then use `linmod()` to convert it to state space.
9. Compute the eigenvalues using the MATLAB function `eig()`.
10. Express the following transfer function in state space format by:
    - Using the template in the book

- Converting the transfer function to the time domain, implementing a SIMULINK solution (on paper), then obtaining a state space representation using 'linmod'

$$\frac{C(s)}{R(s)} = \frac{9}{s^4 + 14s^3 + 71s^2 + 154s + 120}$$

11. Calculate the eigenvalues of the preceding section by:
    - Finding the roots of the denominator polynomial
    - Using MATLAB to calculate the eigenvalues of A, (i.e., `eig(A)`).
12. Consider a system described by the following transfer function:

$$\frac{C(s)}{R(s)} = \frac{(s+2)}{(s+4)} \cdot \frac{5}{(s+1)}$$

Represent this system using the suggested block diagram representation. Remember, this format will include no transfer function blocks, only integrators.

13. Create a SIMULINK diagram for the system from the previous problem. The diagram should have two system representations in parallel:
    - Using the original transfer function given
    - The expanded block diagram obtained in the previous problem.

Connect the outputs of the two representations into a MUX, then into a Scope so that their outputs can be compared easily. Apply a unity step input to both systems, and confirm that the outputs are the same.

10

Design of Control Systems

Learning Objectives

In this chapter, you will learn how to:

- Identify the various design techniques of control systems.
- Describe the effect of lead and lag compensators on the dynamic characteristics of a system.
- Relate a set of specifications to the transfer function characteristics.
- Perform the experimental Ziegler-Nichols method for determining optimal PID coefficients.

10.1 Historical Perspective

Design is much more difficult than analysis. For example, a particular electrical circuit can be analyzed using several approaches to determine, let's say, the voltage at a particular node. All analysis approaches should yield similar results; many designs can meet the same set of specifications. In the early days of control, an implementation that was stable, and remained stable, was considered acceptable. The control scheme was improved further by tweaking the system gain until acceptable behavior was obtained. A system with acceptable behavior was neither too noisy nor exhibited excessive error.

As the body of knowledge grew, control engineers began to seek more powerful analytical approaches and "optimal designs" controllers. Two methods began to appear for determining the best controller gains:

1. **Analytical design of a compensator that would fulfill certain design requirements.** This approach required a mathematical description of the plant, the sensors, and the disturbance affecting system behavior. Many design approaches have been developed for single-input, single-output (SISO) systems, as well as for multiple-input, multiple-output (MIMO) systems. The latter uses modern control techniques, including state space.
2. **Experimental tuning of a compensator to produce an acceptable response.** This approach required some testing (either in open-loop or closed-loop mode) where results were used to calculate the "best" system gains. Often, some additional tweaking was required. Knowledge of the plant transfer function was not required.

During design task, the control engineer tames the system by introducing a compensator. Often, a compensator is the only means of affecting system behavior, because the plant already exists. Design is best done by someone who understands all the ramifications and implications of the controller and the plant.

A. Whitely[1] and D. Graham[2] developed performance indexes, mathematical tools that objectively compare systems' behaviors. These performance indexes can be applied when there is an accurate mathematical model of the system and controller. In some cases, however, there is no model for the system, and obtaining one could be extremely time consuming. This is particularly true in chemical processing plants. In cases like these, trial and error is all that is left.

John Ziegler and Nathaniel Nichols,[3] who worked for Taylor Instruments, a company that led the pneumatic control industry, attacked the trial-and-error dilemma by seeking a more objective method. In the early 1940s, Ziegler and Nichols developed test procedures for both open-loop and closed-loop configurations that produced optimal gains for the controller. After tuning using their guidelines, such "optimal" systems produced quarter-decay responses, indicating that each subsequent peak was one-quarter the amplitude of the preceding peak (relative to the final, steady-state value). The pair's method and its subsequent requirements, commonly called the Ziegler-Nichols method, became one of the main tools for tuning control systems, especially in the complex and difficult-to-model process control systems found in the chemicals manufacturing industry. In fact, in 1995, the editors of *In Tech,* the Instrument Society of America magazine, surveyed their readership to determine which was the "all-time classic control article." The survey revealed that, without question, Ziegler's and Nichols's work was the classic.

10.2 Introduction

Once a decision has been made to implement a feedback control system as a solution to some performance requirement, the first step is to select appropriate hardware. In most cases, the "plant," or the device or process to be controlled, already exists, and the sensors and other controller components must be selected. In some cases, the plant does not exist, and the engineer may be required to also design the plant. In either case, before design of the control system can be undertaken, the plant must be defined in terms of a reasonable mathematical model. Because the problem of designing the controller begins after the plant model is obtained, the design methods in this text assume that the mathematical model of the plant is known.

Most systems contain some nonlinearities. Saturation limits always exist, and other nonlinearities are encountered frequently. Why, then, emphasize linear control systems? The primary reason is that almost all control systems can be considered essentially linear provided that the range of operation is confined to a specific window. This is true for either operation as a regulator or as commands of small magnitude.

Because stable, accurate, and well-damped operation is necessary, a linear design is clearly indicated. This does not mean that nonlinear effects are neglected. They must eventually be considered, but the nonlinear part of the problem is usually deferred until an acceptable linear design has been completed. The design problem starts with a set of specifications that must be met if the system is to operate as desired. Of course, specifications are as varied as the systems to which they apply, but certain requirements, like those following, are always included:

- The system must be stable.
- Accuracy is required:
    - In steady state.
    - When responding to various commands.
- In the presence of noise or other random signals, the dynamic response must have acceptable characteristics (e.g., settling time).

Additional specifications depend on the application. Steady-state accuracy specifications are usually satisfied by choosing the proper system type number (number of open-loop poles at the origin, i.e., the number of integrators) and gain settings. Stability problems and other problems with the dynamic response are often caused by the design required for steady-state accuracy. One way to design is to satisfy dynamic requirements for a linear system by placing the roots of the characteristic equation at suitable locations. In practice, we can interpret specifications in terms of desirable locations for two or three roots. For systems of fourth and higher order, we can name locations for four or more roots, but we usually lack good reasons for any of these except a dominant two or three. Thus, the "classical" design tools—Bode, Nyquist, Nichols; root locus; parameter plane—focus on only the dominant roots. Any other roots are allowed to move in whatever fashion the mathematical relationship dictates. The design is checked to ensure stability and acceptable performance. When these are not achieved initially, the design is modified by trial and error, until the performance is acceptable.

Modern control methods use state variables and matrix manipulations as basic tools. Two kinds of design procedures are well established in the literature:

1. Pole placement
2. Optimal control

In pole placement, the roots of the characteristic polynomial are placed at specified locations using state variable feedback. For linear systems, all the roots can be placed at previously specified locations when all states are fed back, each with an adjusted gain. To use this method, all the root locations must be specified. This can be an advantage or a disadvantage. It is nice to have control of all the roots, but the physical system design may have undesirable features when unsuitable root locations are chosen. (Note: There are few good rules for choosing root locations that are not for dominant roots.)

Optimal control methods are concerned with finding control parameter values that minimize a cost function. A cost function provides an objective index of how well a particular criteria is satisfied.

10.3 Accuracy Requirements

One of the primary requirements of almost all control systems is that those systems be "sufficiently" accurate. For most control systems, accuracy is specified for steady-state conditions as follows:

- **Regulators.** Regulators are supposed to keep the controlled variable at or near a specified constant value. When we define error to be the difference between desired and actual variable values, the regulator may be required to keep the error at 0. When the system is subjected to disturbances, the regulator must maintain a steady-state value with error less than a specified amount, but it may also be required to keep the peak dynamic error (due to a disturbance) within acceptable limits.
- **Tracking systems.** Tracking systems are subject to time-varying command inputs, such as ramps. When a steady-state condition exists, an acceptable steady-state error may be specified and the system designed to meet the specification.

For both cases, desired steady-state accuracy may be achieved by some combination of system type, number, and gain adjustment. In some cases, a steady-state error may not exist or may not be of interest. For example, when a system is subject to "noise," the best possible design may only guarantee that the system error is bounded by some maximum deviation from the desired value. Another example would be that of tracking a target using evasive maneuvers. A steady-state condition does not exist, and the design, at best, can only guarantee that the error is bounded by some value that can be determined.

10.4 Requirements for Dynamic Behavior

Clearly, a control system must be stable. The dynamic characteristics of the transient response are usually described and specified in the time domain (e.g., for a step command) as follows:

- Amount of overshoot allowed, if any
- Number of overshoots and undershoots allowed
- Frequency of any transient oscillations
- Duration of the transient, or settling, time, and so on

Some specifications may be given in the frequency domain, such as bandwidth, phase margin (PM), and gain margin (GM).

To proceed, the engineer must interpret these specifications in terms of the analysis and design methods. Clearly, the requirement for stability of a linear system is that all roots of the characteristic polynomial be in the left half of the s-plane. In addition, the characteristics of the transient response are determined by the locations of these roots. When the engineer makes an intelligent choice for desired root locations, a dynamic response can be acceptable. Unfortunately, we lack good methods for interpreting specifications in terms of a unique set of desired root locations. We also lack general design methods to force the roots to such locations, except for full state output feedback.

Frequency response methods (e.g., Nyquist, Bode, Nichols) normally use PM, GM, and bandwidth to place the two dominant roots, usually complex conjugate, while occasionally considering a third root. Root locus and parameter plane methods normally consider dominant complex roots. It is well known that trial-and-error procedures are used when frequency domain methods are the design tools. For pole placement designs, the initial choice of root locations may not provide an acceptable response, leading to trial and error in the choice of desired roots. Accuracy requirements may also lead to design modifications. Optimal designs and function minimization methods with the computer both require a cost function, or performance index, to be minimized, but the cost function is just a mathematical expression that is supposed to describe both dynamic and accuracy requirements. These requirements must be verified in the resulting design. When they are not, the weighting factors in the cost function are adjusted.

The design procedure starts with a first trial that is not expected to be the final design but that provides information that guides successive trial designs and leads to a final system that satisfies all requirements. For complex systems, a large number of trials may be required, and the final design is not likely to be the best possible design, though most of the practical systems designed with these methods have proven satisfactory. The resulting design usually has a pair of dominant complex roots, but there are others. The locations of these roots have not been previously specified, and we accept our results as long as the system meets specifications.

10.5 Standard Forms

One reason for trial-and-error design methods is the lack of a precise definition of desired dynamic performance. In early attempts, Whitely and Graham and Lathrop formulated performance indexes and standard forms for the desired, or "optimal," step response of a closed-loop control system. One of the objectives of such an approach was to eliminate trial-and-error design and provide a means for direct synthesis. Because the primary function of the control system can be thought of as the control of the error, e, and because one usually wants to minimize the error, the dynamics are usually defined as some function of the error that is to be minimized. Early proposals included:

$$J = \int_0^T e\,dt \quad \text{(IE)} \qquad \text{Eq. 10.1}$$

$$J = \int_0^T e^2\,dt \quad \text{(ISE)} \qquad \text{Eq. 10.2}$$

$$J = \int_0^\infty |e|\,dt \quad \text{(IAE)} \qquad \text{Eq. 10.3}$$

$$J = \int_0^\infty t|e|\,dt \quad \text{(ITAE)} \qquad \text{Eq. 10.4}$$

$$J = \int_0^\infty te^2\,dt \quad \text{(ITSE)} \qquad \text{Eq. 10.5}$$

The "best" or "optimal" system was defined as the system that minimized the value of *J*. To truly minimize the value of the integral, every parameter in the system should be a variable. In other words, all the coefficients of the characteristic equation should be independently adjustable (as is the case with full state feedback). The result of minimizing the integral, then, specifies the "best" value for every coefficient of the characteristic equation, which in turn defines a standard form for the characteristic equation, as well as a pattern for the closed-loop roots.

Standard forms have also been formulated in terms of the frequency domain. For example, forms exist that optimize linear phase transition, maximum sharpness in the transition, and maximum flatness in the passband. The two most common such forms are the Binomial and the Butterworth. With respect to their step-response characteristics, the Binomial exhibits no overshoot, because the associated poles are real (critically damped) and thus there is no imaginary component. The Butterworth in contrast, has a slight overshoot and a shorter rise time due to the associated complex pair.

Most of the proposed performance indexes did not withstand the test of time. Use is still found for the integral of absolute error squared (IAES) and the integral of time times the absolute error (ITAE) and an occasional problem suggests the use of one of the others. Standard forms can also be used with pole placement.

The following forms can be used as templates for the coefficients of the desired characteristic equation.

Time Domain Specifications—Step Input

The IAES Standard Forms

$$s + \omega_0 \qquad \text{Eq. 10.6}$$

$$s^2 + \omega_0 s + \omega_0^2 \qquad \text{Eq. 10.7}$$

$$s^3 + \omega_0 s^2 + 2\omega_0^2 s + \omega_0^3 \qquad \text{Eq. 10.8}$$

$$s^4 + \omega_0 s^3 + 3\omega_0^2 s^2 + 2\omega_0^3 s + \omega_0^4 \qquad \text{Eq. 10.9}$$

$$s^5 + \omega_0 s^4 + 4\omega_0^2 s^3 + 3\omega_0^3 s^2 + 3\omega_0^4 s + \omega_0^5 \qquad \text{Eq. 10.10}$$

The ITAE Standard Forms

$$s + \omega_0 \qquad \text{Eq. 10.11}$$

$$s^2 + 1.4\omega_0 s + \omega_0^2 \qquad \text{Eq. 10.12}$$

$$s^3 + 1.75\omega_0 s^2 + 2.15\omega_0^2 s + \omega_0^3 \qquad \text{Eq. 10.13}$$

$$s^4 + 2.1\omega_0 s^3 + 3.4\omega_0^2 s^2 + 2.7\omega_0^3 s + \omega_0^4 \qquad \text{Eq. 10.14}$$

$$s^5 + 2.8\omega_0 s^4 + 5\omega_0^2 s^3 + 5.5\omega_0^3 s^2 + 3.4\omega_0^4 s + \omega_0^5 \qquad \text{Eq. 10.15}$$

Time Domain Specifications—Ramp Input

The ITAE Standard Forms

$$s + \omega_0 \qquad \text{Eq. 10.16}$$

$$s^2 + 3.2\omega_0 s + \omega_0^2 \qquad \text{Eq. 10.17}$$

$$s^3 + 1.75\omega_0 s^2 + 3.25\omega_0^2 s + \omega_0^3 \qquad \text{Eq. 10.18}$$

$$s^4 + 2.41\omega_0 s^3 + 4.93\omega_0^2 s^2 + 5.14\omega_0^3 s + \omega_0^4 \qquad \text{Eq. 10.19}$$

$$s^5 + 2.19\omega_0 s^4 + 6.5\omega_0^2 s^3 + 6.3\omega_0^3 s^2 + 5.24\omega_0^4 s + \omega_0^5 \qquad \text{Eq. 10.20}$$

Frequency Domain Specifications

The Binomial Standard Forms

$$s + \omega_0 \qquad \text{Eq. 10.21}$$

$$s^2 + 2\omega_0 s + \omega_0^2 \qquad \text{Eq. 10.22}$$

$$s^3 + 3\omega_0 s^2 + 3\omega_0^2 s + \omega_0^3 \qquad \text{Eq. 10.23}$$

$$s^4 + 4\omega_0 s^3 + 6\omega_0^2 s^2 + 4\omega_0^3 + \omega_0^4 \qquad \text{Eq. 10.24}$$

$$s^5 + 5\omega_0 s^4 + 10\omega_0^2 s^3 + 10\omega_0^3 s^2 + 5\omega_0^4 s + \omega_0^5 \qquad \text{Eq. 10.25}$$

The Butterworth Standard Forms

$$s + \omega_0 \quad \text{Eq. 10.26}$$

$$s^2 + 1.4\omega_0 s + \omega_0^2 \quad \text{Eq. 10.27}$$

$$s^3 + 2\omega_0 s^2 + 2\omega_0^2 s + \omega_0^3 \quad \text{Eq. 10.28}$$

$$s^4 + 2.6\omega_0 s^3 + 3.4\omega_0^2 s^2 + 2.6\omega_0^3 s + \omega_0^4 \quad \text{Eq. 10.29}$$

$$s^5 + 3.2\omega_0 s^4 + 5.24\omega_0^{22} s^3 + 5.24\omega_0^3 s^2 + 3.24\omega_0^4 s + \omega_0^5 \quad \text{Eq. 10.30}$$

Design Procedure

The design procedure for making a closed-loop system optimal according to one of the standard forms is presented next. It consists of selecting the closed-loop denominator transfer function coefficients of the system being designed, proportional to those of the selected standard form (e.g., IAES, ITAE). Following is the procedure:

1. Obtain the closed-loop transfer functions for the system being designed in terms of symbolic coefficients. At the end of this step, the designer will know the order of the system, which determines the closed-loop, denominator polynomial.
2. Select the designed standard form (e.g., IAES, ITAE).
3. Examine a normalized step response for the chosen standard form, to determine the normalized setting time.
4. Relate this information to the desired setting time of the designed system. This will determine the lowest-order term in the transfer function.
5. Set up a series of equations that equate the CLTF coefficients in the standard form and the system being designed.
6. Solve for the coefficients in the CLTF of the system being designed. Confirm that this is physically realizable.
7. Simulate the system to confirm that the step response matches, in a normalized way, the response of the standard form.

Example 10.1

A closed-loop, unity-feedback, position-control system is to drive a mechanical system, (Gp), whose transfer function follows. The system uses a hydraulic actuator with very fast dynamics, represented simply as K_a. This system will experience a sawtooth wave form as its command signal. Using the ITAE criterion for a ramp input, set up a system of equations that would yield the optimal controller gain, K_c.

Solution The plant of the system is:

$$G_P = \frac{1}{ms^2 + bs + k} \quad \text{Eq. 10.31}$$

and the OLTF is:

$$\text{OLTF} = \frac{K_c K_a}{ms^2 + bs + k} \quad \text{Eq. 10.32}$$

then the CLTF becomes:

$$\text{CLTF} = \frac{K_c K_a}{ms^2 + bs + k + K_c K_a} \quad \text{Eq. 10.33}$$

$$= \frac{\dfrac{K_c K_a}{m}}{s^2 + \dfrac{b}{m}s + \dfrac{k + K_c K_a}{m}}$$

Matching coefficients of the CLTF to the ITAE (ramp-input) form (see equation 10–17):

The constant term:

$$\omega_n^2 = \frac{k + K_c K_a}{m} \quad \text{Eq. 10.34}$$

and the s-term:

$$3.2\omega_n = \frac{b}{m} \quad \text{Eq. 10.35}$$

Two equations, and two unknowns (ω_n and K_c), are obtained. Thus, the system is solvable.

10.6 Location of the Compensator in the Loop

Before starting the design, a preliminary analysis is commonly undertaken. This gives some understanding of what the design must accomplish and often influences the choice of design methods. Such analysis usually results in a graphical portrayal of the dynamic response (e.g., step response, Bode diagram, root locus, and others).

Another decision that must be made is where to locate the compensator, because the compensator affects how the system behaves in response to changes in the input and disturbances. There are two principal locations where the compensator, G_C, can be placed: (1) next to the plant (cascade compensator) or (2) in the feedback loop (feedback compensator). Sometimes, a compensator can be placed in the input path, but this approach has no impact on the stability of the closed loops. Figure 10–1 shows the location of the compensator in the closed loops.

10.7 Design Methods for Cascade Compensation

Cascade compensation is the term used to describe the insertion of a filter in series (cascade) with the plant. The filter has been designed so that the cascaded combination, plus the feedback link, forms a closed-loop system that meets system specifications. We are interested in three design approaches:

1. Design using algebra
2. Graphical design using the Bode diagram
3. Graphical design using root loci

A requirement of all of these approaches is that the designer must know, quantitatively, what is wanted. For the algebraic design, one should know numerical values for the open-loop poles and zeros of the equivalent transfer function. For the Bode approach, one needs the Bode diagram for the plant, plus the desired PM, bandwidth, and error coefficient (gain). For the root locus approach, the designer must know the poles and zeros of the plant and the desired locations of a pair of dominant, closed-loop, complex roots.

10.8 Design Using Algebra

Imagine a system whose plant has two poles, one too close to the origin and, consequently, producing a very slow system. Is it possible to come up with a compensator that cancels out the slow pole and replaces it with one that is much faster? The

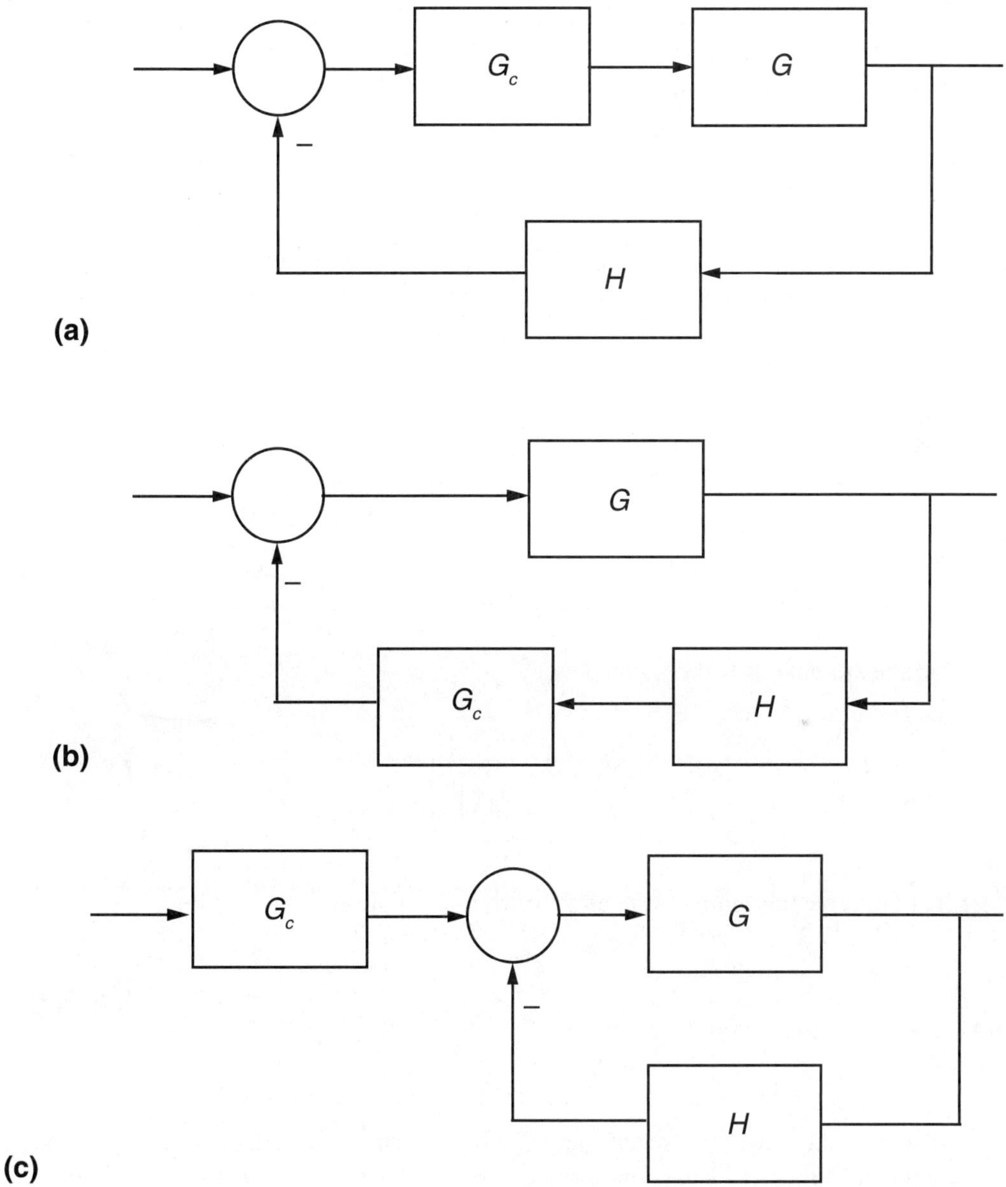

Figure 10–1 Possible locations of compensator, $G_c(s)$: (a) cascade compensation, (b) feedback compensation, and (c) input or prefilter compensation

algebraic design is based on this approach. One note of caution: This approach is unsuitable when the plant has unstable poles, because it is impossible to come up with an exactly numerically equal zero.

The compensated system will have an OLTF given by:

$$G_{eq} = G_c G_p \qquad \text{Eq. 10.36}$$

where G_c is the compensator and G_p is the plant's transfer function.

Because G_{eq} and G_p are known, we can solve for G_c:

$$G_c = \frac{G_{eq}}{G_p} \qquad \text{Eq. 10.37}$$

resulting in pole-zero cancellation, plus appropriate poles and zeros for G_c

Example 10.2

Let the system's plant be defined as:

$$G_p = \frac{1}{\left(\frac{s}{2}+1\right)\left(\frac{s}{10}+1\right)} \qquad \text{Eq. 10.38}$$

It is desirable to eliminate the effect of the slow pole at $s = -2$. Thus:

$$G_{eq} = \frac{1}{\left(\frac{s}{10}+1\right)\left(\frac{s}{20}+1\right)} \qquad \text{Eq. 10.39}$$

Solution Notice that we are using the Bode form of the OLTF, so that the open-loop gain is preserved. To obtain the compensator's transfer function, we divide:

$$G_c = \frac{G_{eq}}{G_p} = \frac{\left(\frac{s}{2}+1\right)}{\left(\frac{s}{20}+1\right)} \qquad \text{Eq. 10.40}$$

Notice that G_{eq} is determined from specifications and that its order depends on the number of specifications that can be described, numerically, as a transfer function. In practice, specifications are often given as limits, and this is not always convenient. Often, performance is also specified by a pair of dominant roots that may work well even though the system is of fourth or fifth order.

Example 10.3

Let the plant's transfer function be:

$$G_p = \frac{1}{s\left(\frac{s}{2}+1\right)\left(\frac{s}{10}+1\right)\left(\frac{s}{20}+1\right)} \quad \text{Eq. 10.41}$$

Solution The desired resulting (equivalent) transfer function is:

$$G_{eq} = \frac{1}{s\left(\frac{s}{5}+1\right)} \quad \text{Eq. 10.42}$$

Then:

$$G_c = \frac{G_{eq}}{G_p} = \frac{\left(\frac{s}{2}+1\right)\left(\frac{s}{10}+1\right)\left(\frac{s}{20}+1\right)}{\frac{s}{5}+1} \quad \text{Eq. 10.43}$$

This is not realizable, because there are excess zeros, but it can be approximated by introducing remote poles in the compensator. This solution will make the order of the compensator denominator greater or equal to its numerator.

Example 10.4

Consider a second order plant:

$$G_p = \frac{K_v}{s^2} \qquad \text{Eq. 10.44}$$

If we want a pair of dominant, closed-loop roots in a type 1 system, let the desired (compensated) open-loop transfer function be:

$$G_{eq} = \frac{K_v}{s\left(\frac{s}{\tau}+1\right)} = G_c G_p \qquad \text{Eq. 10.45}$$

Solution Solving for the compensator's transfer function, G_c, yields:

$$G_c = \frac{\frac{K_v}{s\left(\frac{s}{\tau}+1\right)}}{\frac{K_v}{s^2}} = \frac{s^2}{s\left(\frac{s}{\tau}+1\right)} = \frac{s}{\frac{s}{\tau}+1} \qquad \text{Eq. 10.46}$$

We can check our results by:

$$G_{eq} = G_c G_p = \frac{s}{\frac{s}{\tau}+1}\frac{K_v}{s^2} = \frac{K_v}{s\left(\frac{s}{\tau}+1\right)} \qquad \text{Eq. 10.47}$$

If G_p is of third order, G_c is not realizable, and a remote pole would be needed in G_c.

10.9 Bode Diagram Design for Cascade Compensation

As previously mentioned, there are typically two places to put the compensator: (1) in the feedforward loop (adjacent to *Gp*) and (2) in the feedback path. The first method is called cascade compensation; the second, feedback compensation. Feedback compensation might only involve a minor feedback loop. We will first use the Bode diagram as the design tool.

Bode Diagram Cascade Design—Lead Compensation

The first step is to get the Bode diagram of the uncompensated system, *Gp*. Consider a small, high-speed, positioning servo, like the one used in magnetic disk memory systems. The plant consists of the motor, which provides a torque, and has inertia and some friction. The transfer function model (neglecting friction) might be:

$$G_p = \frac{100}{s^2} \qquad \text{Eq. 10.48}$$

The Bode diagram for the plant model appears in Figure 10–2. The gain crossover is at $\omega = 10$, and there is no phase crossover. The PM is 0, as is the GM, and the closed-loop system is at the stability limit.

The first problem is to stabilize the system. Because the only available adjustment is the gain, which cannot stabilize this system, the gain and phase curves must be reshaped. The system must be "compensated," (i.e., some dynamic components must be introduced to change the transfer function). These components may be added

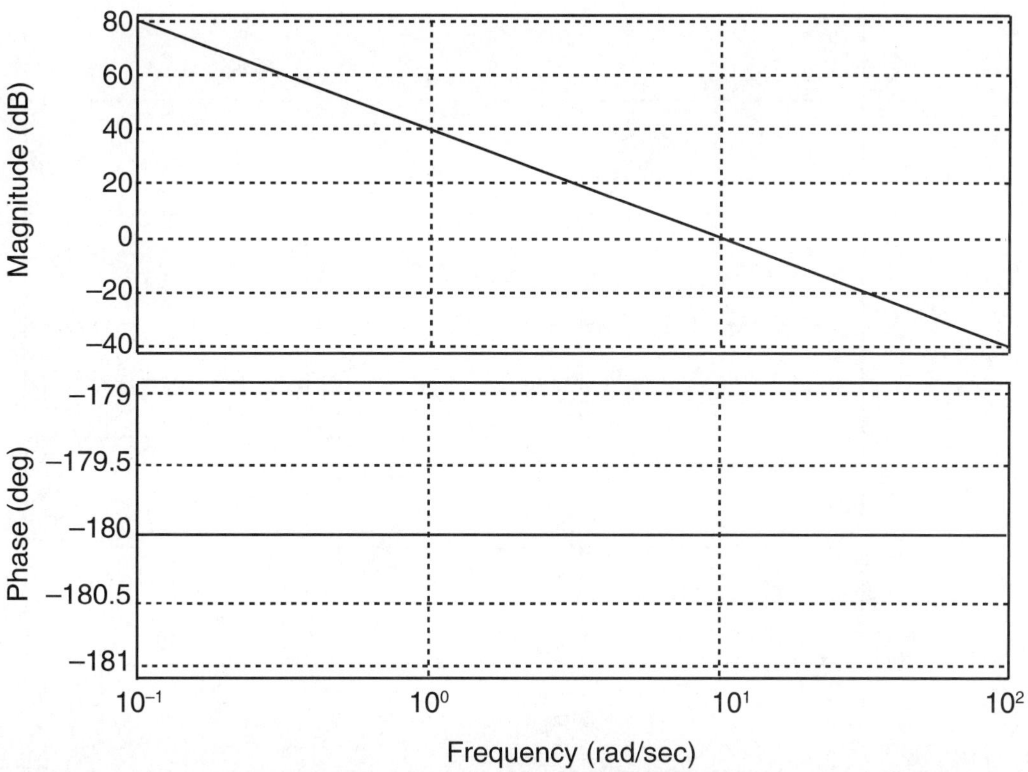

Figure 10–2 Bode plot of uncompensated system, $GH(s) = 100/s^2$

in cascade, or they may be put in a feedback loop. Assume that the designer chooses to use cascade compensation. Then a filter must be designed for the forward path.

When an uncompensated system is close to the stability limit, the use of one pole and one zero is usually enough to provide both stability and acceptable damping. In this case, the phase curve must be reshaped to provide a positive PM. Therefore, the zero is placed at a frequency lower than the original gain crossover frequency. For example, the zero might be placed at -3.5. A pole is needed for realizability and is placed at a frequency higher than the gain crossover frequency. In this example, the pole was placed at -35. The gain is adjusted to maintain the original error coefficient. The resulting open-loop transfer function for the compensated system is:

$$G_c G_p = \frac{100\left(\dfrac{s}{3.5} + 1\right)}{s^2\left(\dfrac{s}{35} + 1\right)} \qquad \textbf{Eq. 10.49}$$

The asymptotic Bode diagram of the compensated system appears in Figure 10–3, and the complete Bode diagram appears in Figure 10–4. The step response of the

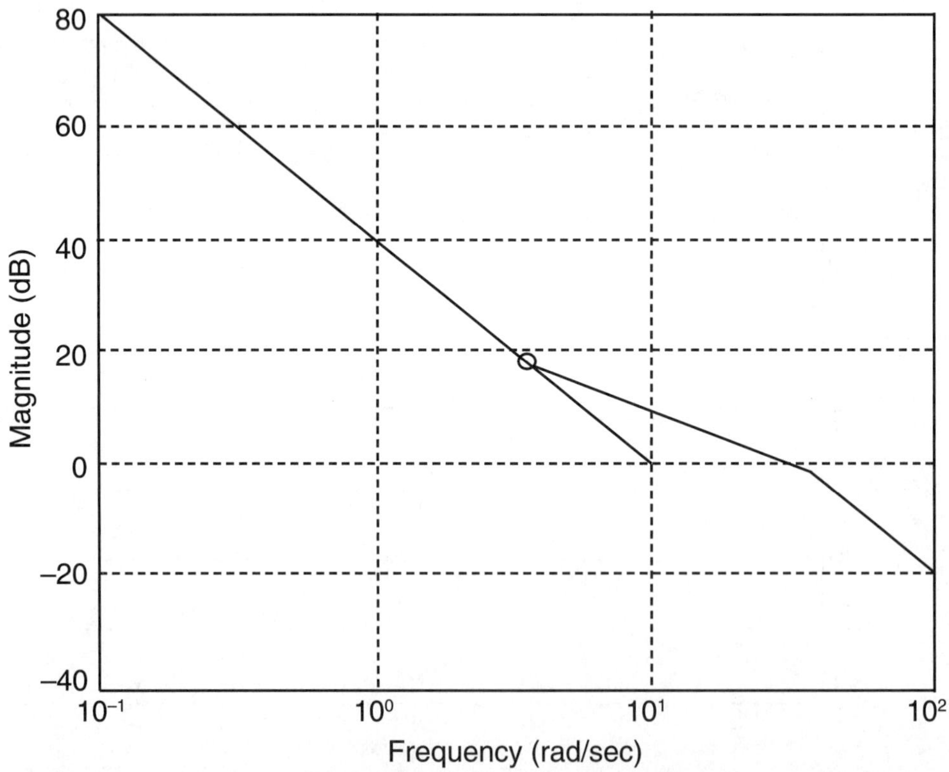

Figure 10–3 Asymptotic magnitude Bode plot for compensated system, $GH(s) = 100(s/3.5 + 1)/s^2(s/35 + 1)$

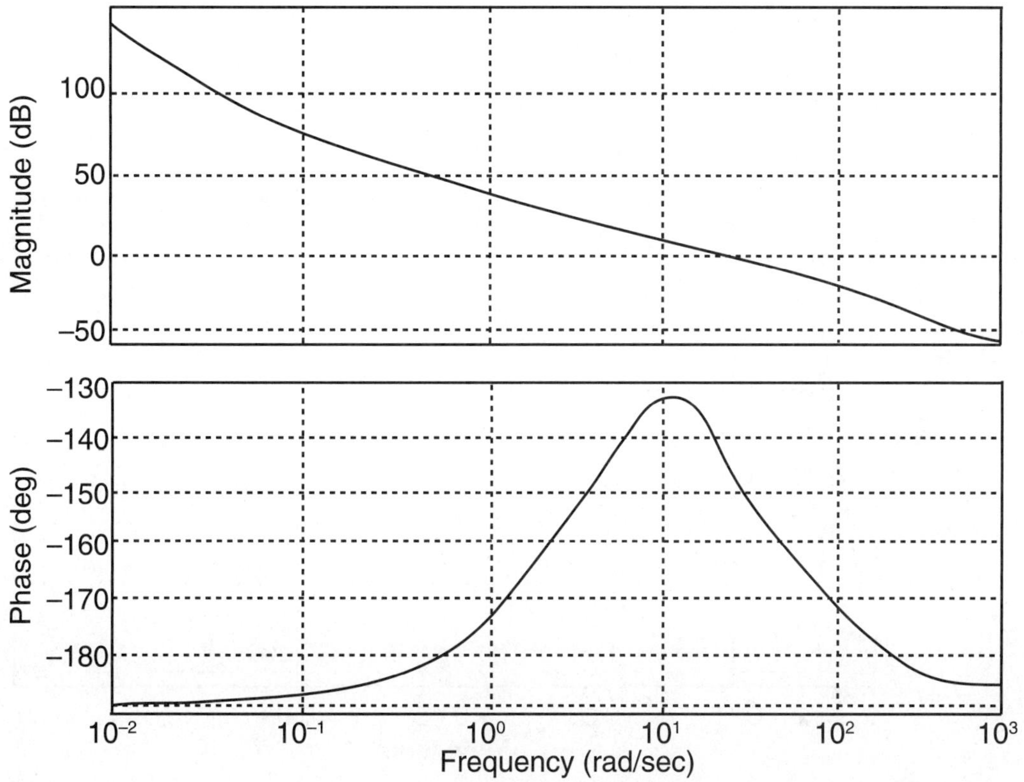

Figure 10–4 Complete Bode plot for compensated system, $GH(s) = 1{,}000(s + 3.5)/(s^3 + 35s^2)$

closed loop appears in Figure 10–5. The compensated system is stable and well damped. This result might be improved by additional trials. Analysis of the resulting Bode diagram shows:

Gain crossover frequency	~30
PM	~46 deg
Phase crossover frequency	~ none
GM	infinite
Damping ratio	.35
Natural frequency	~ 30
Settling time	~ 0.4 sec
Maximum overshoot	~ 1.25

If specifications are not completely satisfied, the zero and/or pole of the filter may be moved, or a second filter may be introduced.

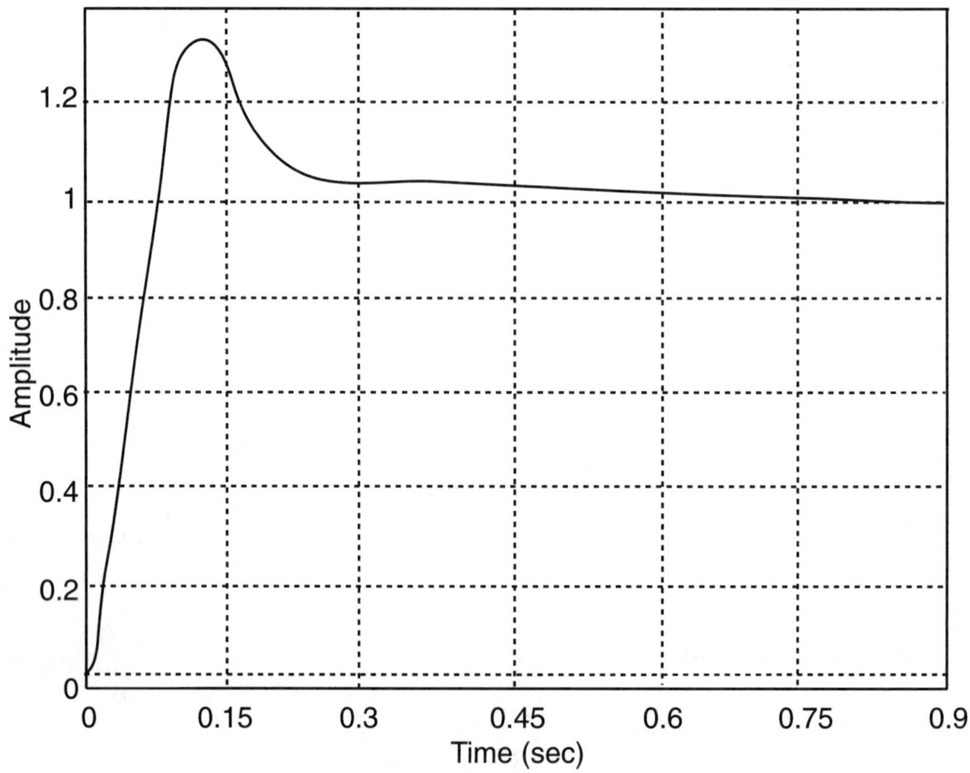

Figure 10–5 Closed-loop step response of compensated system

For many higher-order systems, the analysis assumes second-order dominance and uses the criteria for a second-order system. Consider a third-order system for which the open-loop transfer function is:

$$G(s) = \frac{15}{\left(\frac{s}{5}+1\right)\left(\frac{s}{25}+1\right)\left(\frac{s}{100}+1\right)} = \frac{187{,}500}{s^3 + 130s^2 + 3{,}125s + 12{,}500} \qquad \textbf{Eq. 10.50}$$

The transfer function is type 0, and the error coefficient is 15. The closed-loop transfer function is:

$$\frac{C}{R}(s) = \frac{187{,}500}{s^3 + 130s^2 + 3{,}125s + 200{,}000} \qquad \textbf{Eq. 10.51}$$

The roots of the characteristic polynomial are -117.88 and $-6 \pm j40.74$.

Thus, the system is stable but lightly damped, and probably quite oscillatory. The step response of the closed-loop appears in Figure 10–6, from which $M_{pt} = 1.5$ (this corresponds to $\varsigma = 0.2$). This would not be acceptable for most applications. The steady-state error for this type 0 system is:

$$E_{ss} = 1 - \frac{187,500}{200,000} = 0.0625 \qquad \textbf{Eq. 10.52}$$

The stability limit can be found using the closed-loop characteristic polynomial and the Routh criterion. The root locus for this system appears in Figure 10–7. It has three segments, two of which extend into the right half of the *s*-plane. Thus, if the gain is high enough, two roots would cross into the right-half plane, and the system would be unstable. The third segment shows that the real root is remote with respect to the complex roots, so second-order dominance is expected. In addition,

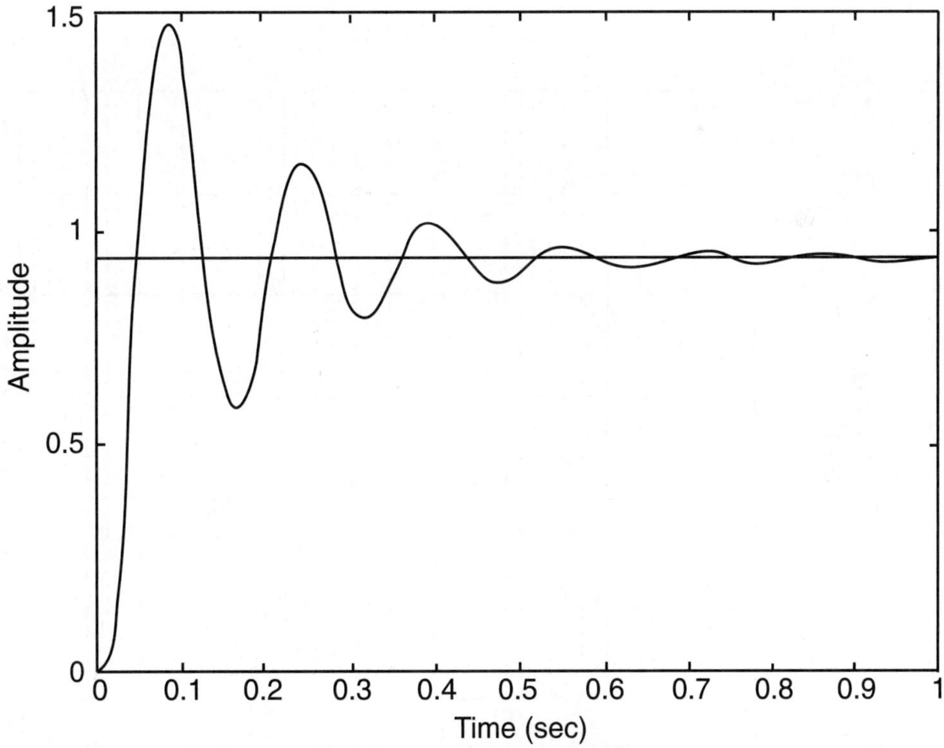

Figure 10–6 Closed-loop step response of uncompensated system

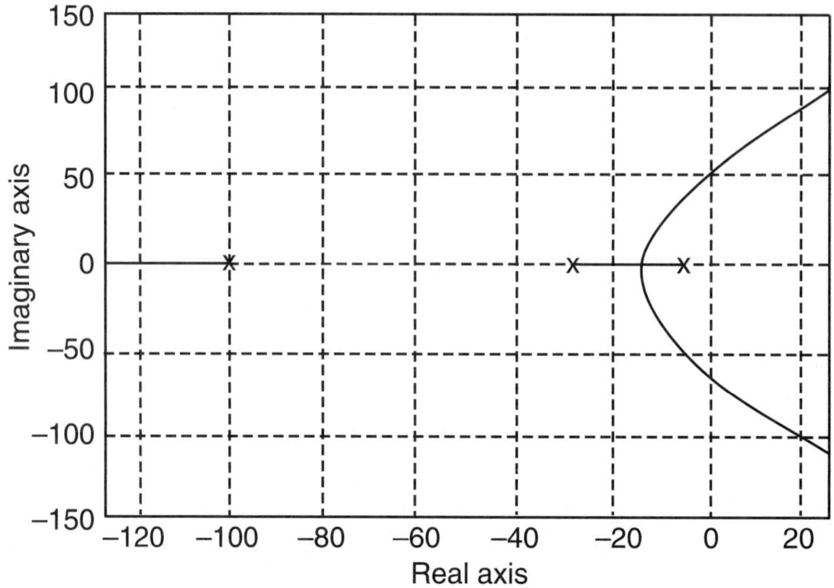

Figure 10–7 Root locus plot of uncompensated system

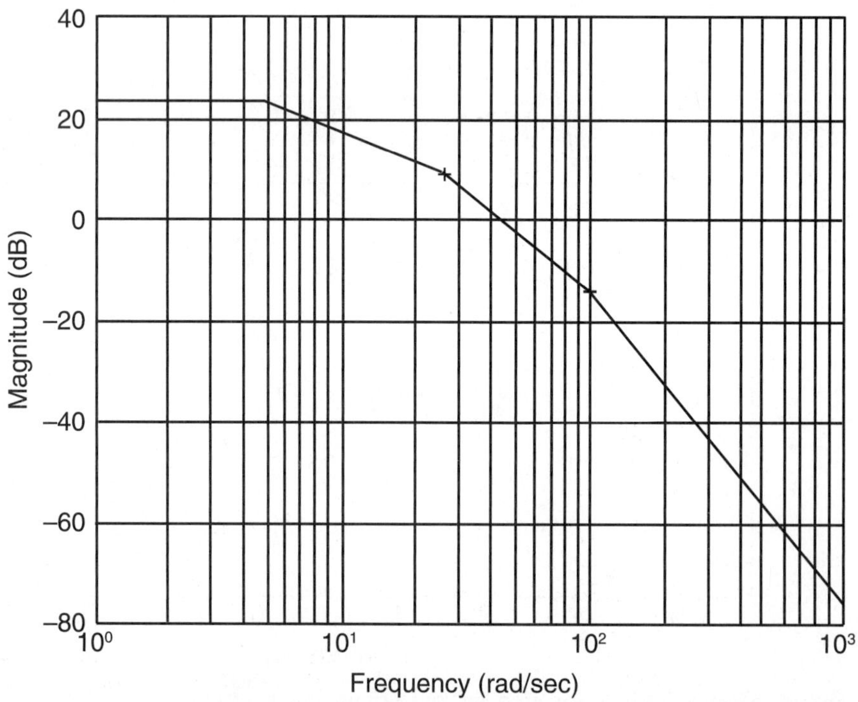

Figure 10–8 Magnitude asymptotic Bode diagram for $GH(s) = 187{,}500/(s^3 + 130s^2 + 3{,}125s + 12{,}500)$

analysis using second-order criteria should be valid. For example, the settling time is $T_s = \dfrac{4}{\sigma}$, where σ is the real part of the complex roots. Thus:

$$T_s = \frac{4}{6} = 0.667 \qquad \textbf{Eq. 10.53}$$

The asymptotic Bode diagram appears in Figure 10–8. It intersects the 0 db axis at $\omega = 43$ and in an asymptote with slope of -40 db/decade. This predicts a stable system, but the phase curve is needed for detailed analysis.

The complete Bode diagram appears in Figure 10–9, from which:

Gain crossover = 38 rad/sec

GM = 7 db

Phase crossover = 54 rad/sec

PM = 20 degrees

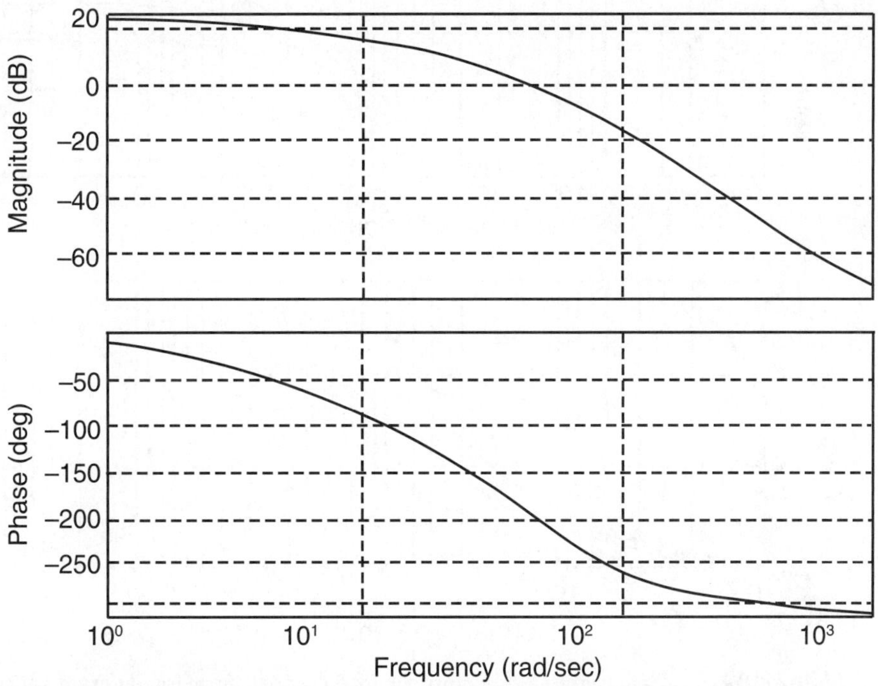

Figure 10–9 Bode diagram for uncompensated system, $GH(s) = 187{,}500/(s^3 + 130s^2 + 3{,}125s + 12{,}500)$

The design problem is to introduce damping (i.e., to reduce the overshoot and oscillation of the step response while retaining the error coefficient at $Kp = 15$. The asymptotic Bode diagram for the uncompensated system shows a gain crossover at 38 radians/second, in an asymptote with slope of -40 db/decade. The simplest compensation appears to be a cascaded compensator, using a high-pass filter to introduce some positive phase shift. Because the requirement is not severe, a single-0/single-pole filter of the form:

$$G = \frac{s/z + 1}{s/p + 1} \qquad p >> z \qquad \textbf{Eq. 10.54}$$

should be adequate.

This type of filter is called a lead compensator, and its generalized characteristics for different ratios of p/z, appear in Figure 10–10.

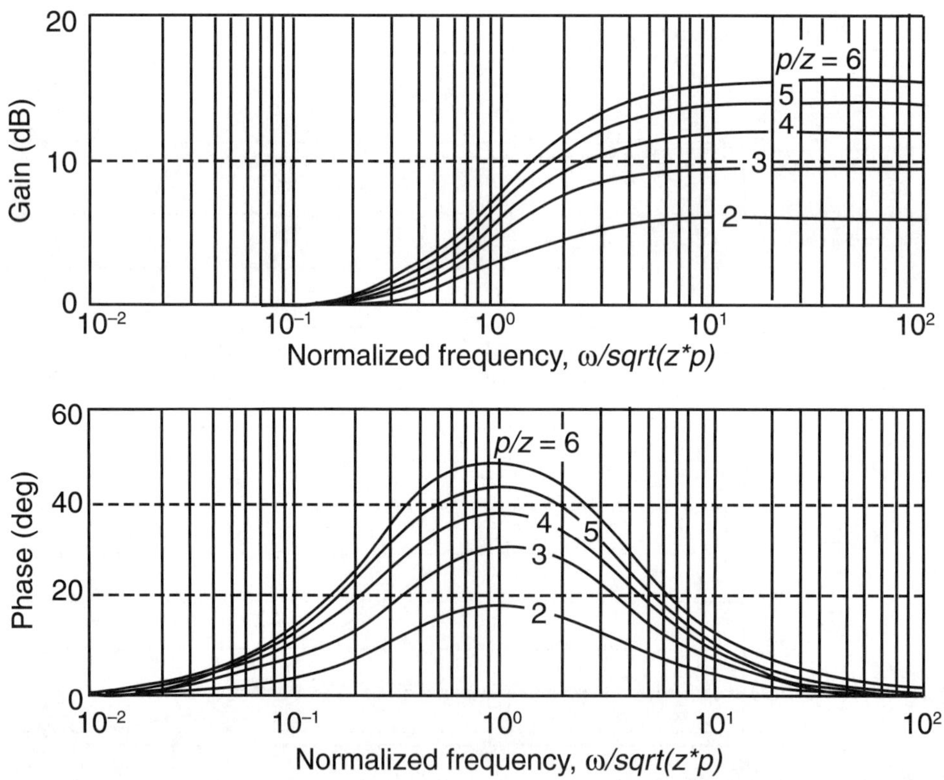

Figure 10–10 Generalized Bode plots for lead compensator, for different ratios of p/z

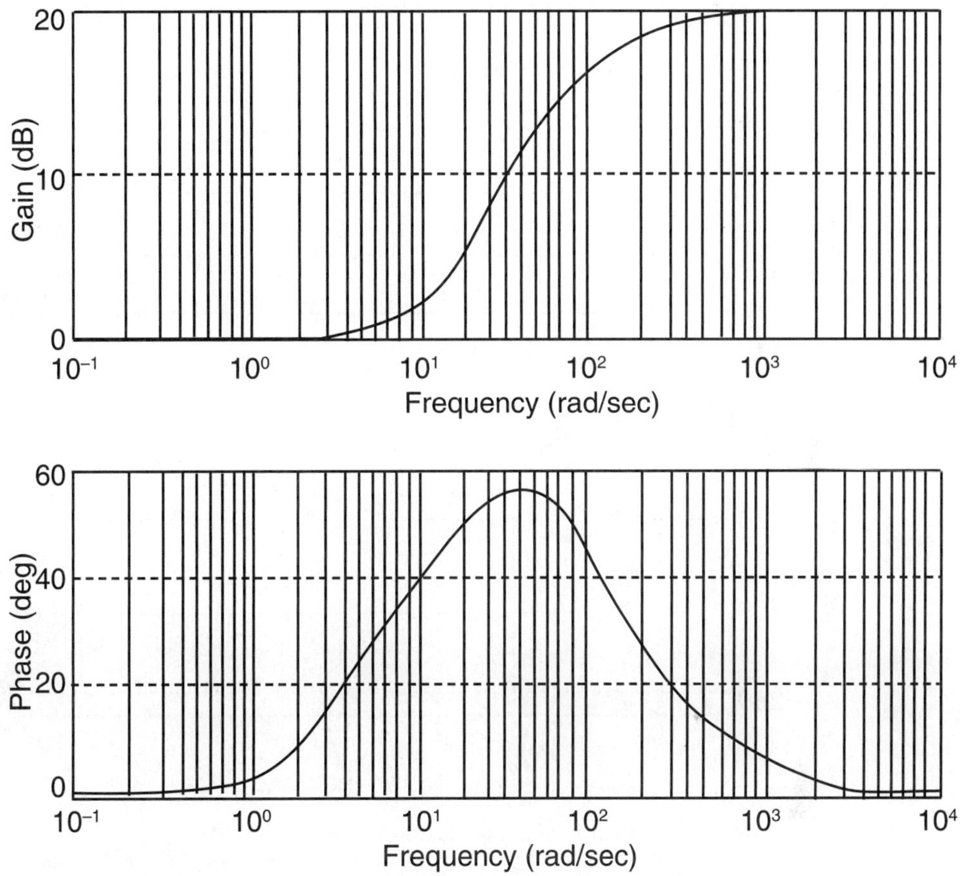

Figure 10–11 Lead compensator, $Gc(s) = (s/25 + 1)/(s/250 + 1)$

Inspection of the asymptotic Bode diagram for the uncompensated system suggests use of a 0 to cancel the pole at −25 and a pole at −250. The Bode diagram for this lead filter by itself appears in Figure 10–11. The Bode diagram for the compensated system appears in Figure 10–12, and the step response appears in Figure 10–13, which shows the improvement in the damping characteristics.

Bode Diagram Cascade Design—Lag Compensation

The steady-state accuracy of a feedback control is determined by the Bode gain. It is customary to name the Bode gain error coefficient, with a further definition depending on the type of input that will be applied to the system, in particular,

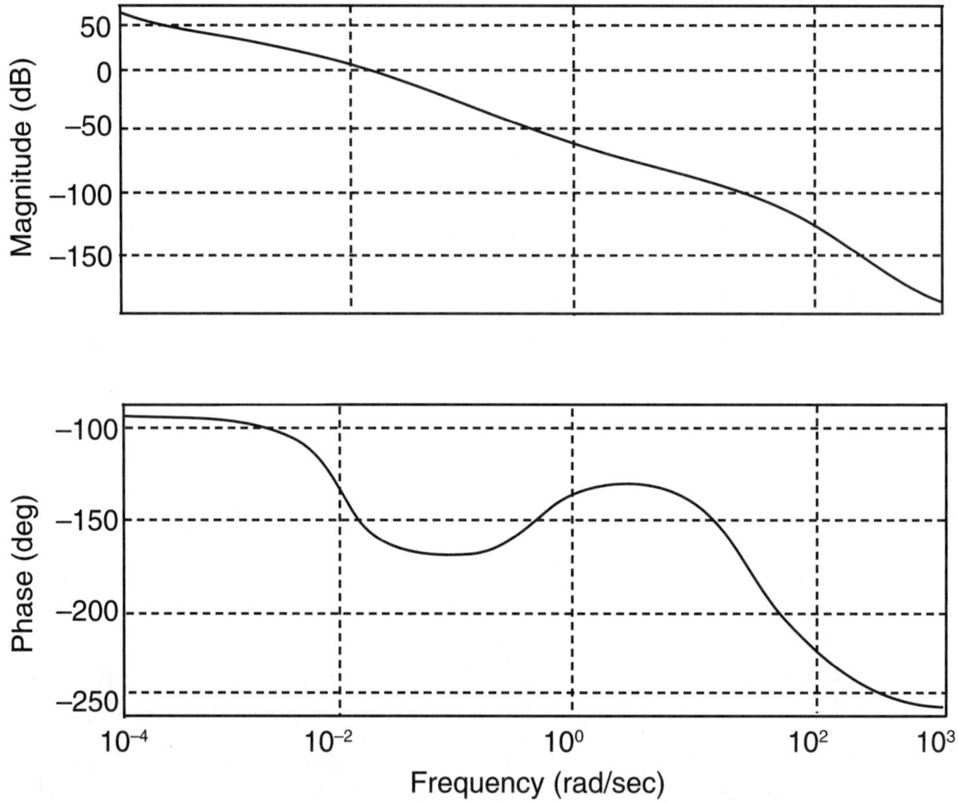

Figure 10–12 Bode plot of compensated system

the type of input that produces a finite, steady-state error (rather than 0 or ∞). For instance, with a type 0 system, subject to a step input, the Bode gain is referred to as the position error coefficient. Similarly, in a type 2 system, subject to a parabolic input, the Bode gain is referred to as the acceleration error coefficient. When high gain is required, the system is usually unstable and may be badly unstable. On the Bode diagram, this appears as a large, negative PM. The design problem is to insert a compensator (filter) that retains the high error coefficient but reduces the gain at high frequencies, thus moving the gain crossover to a new location, at which the phase provides an acceptable PM. The function of a LEAD compensator was to reshape the phase curve to obtain a suitable PM, but the basic function of the LAG compensator is to reduce the high-frequency gain. This

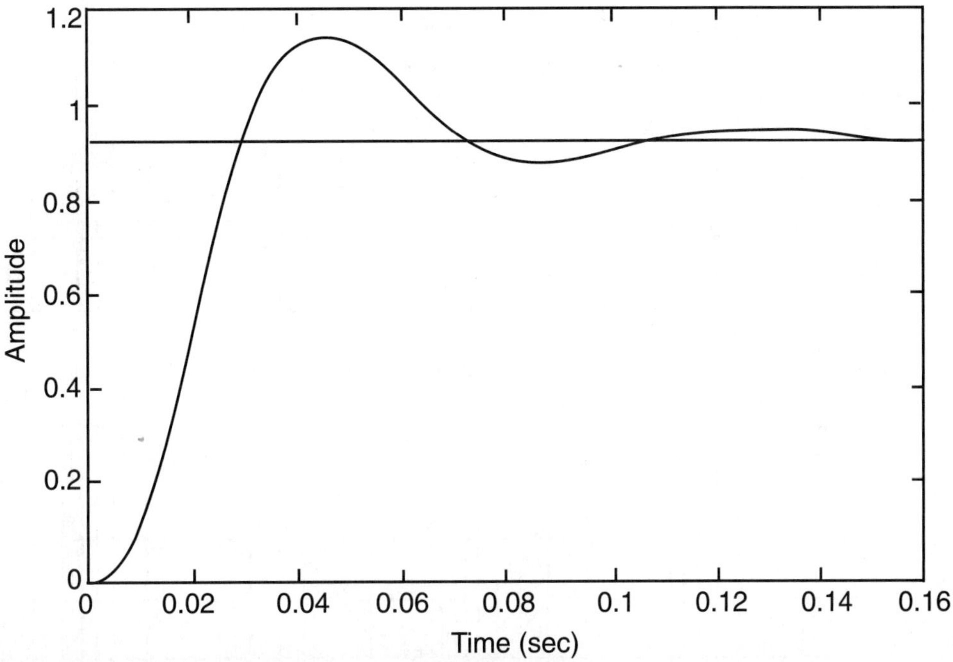

Figure 10–13 Step response of compensated system

moves the gain crossover to the left and reduces the bandwidth but provides an acceptable PM.

The lag compensator is simply a low-pass filter placed in series with the forward path of the feedback loop. The designer normally places it in the controller, where it accepts the error signal and processes it. A typical transfer function, called a lag compensator, has the general form of:

$$G = \frac{s/z + 1}{s/p + 1} \qquad p << z \qquad \textbf{Eq. 10.55}$$

A generalized Bode plot appears in Figure 10–14. Notice that the lag compensator passes low frequencies without attenuation but attenuates high frequencies. It also introduces phase shift at frequencies between p and z, plus some phase shift beyond this range.

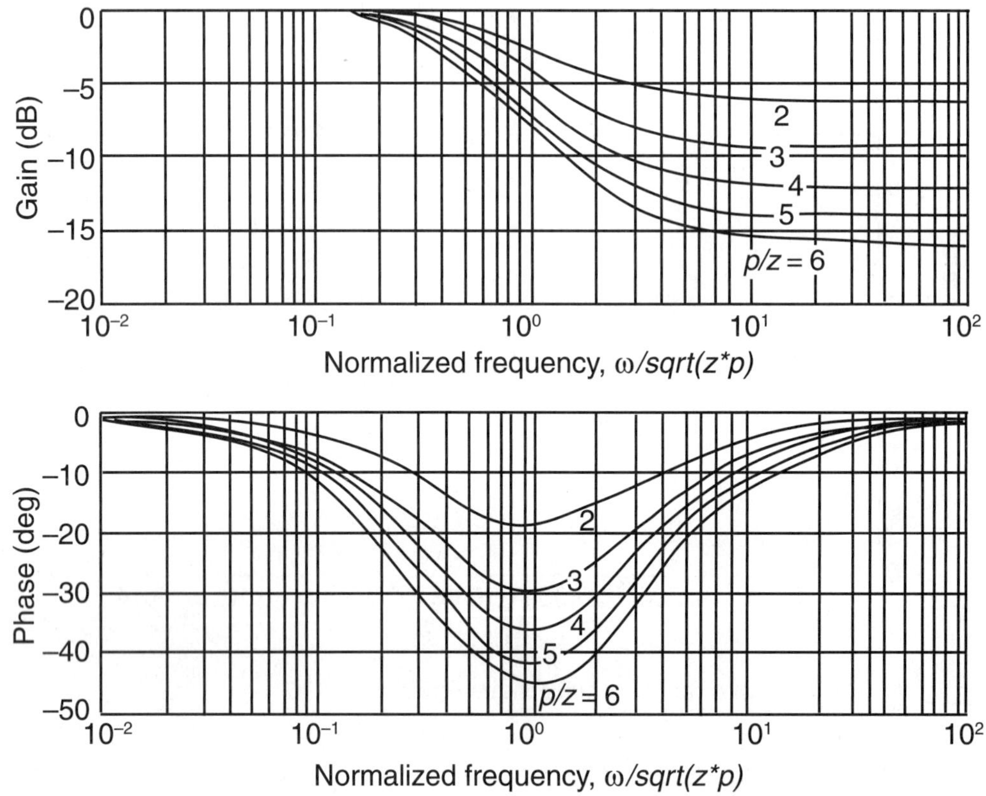

Figure 10–14 Generalized Bode plots for lag compensator, for different ratios of z/p

Example 10.5

Use MATLAB to obtain a Bode plot for the lag compensator:

$$G_C(s) = \frac{s/50 + 1}{s/5 + 1} \qquad \textbf{Eq. 10.56}$$

and confirm that it provides 20 dB of attenuation and that the maximum phase contrivution occurs at the geometric mean of the pole and the zero (i.e., z/p).

Solution The Bode diagram for the filter appears in Figure 10–15. Clearly, this filter attenuates at all frequencies greater than $\omega = p$. This filter also modifies the phase at frequencies from $\omega = p$ to $\omega = z$.

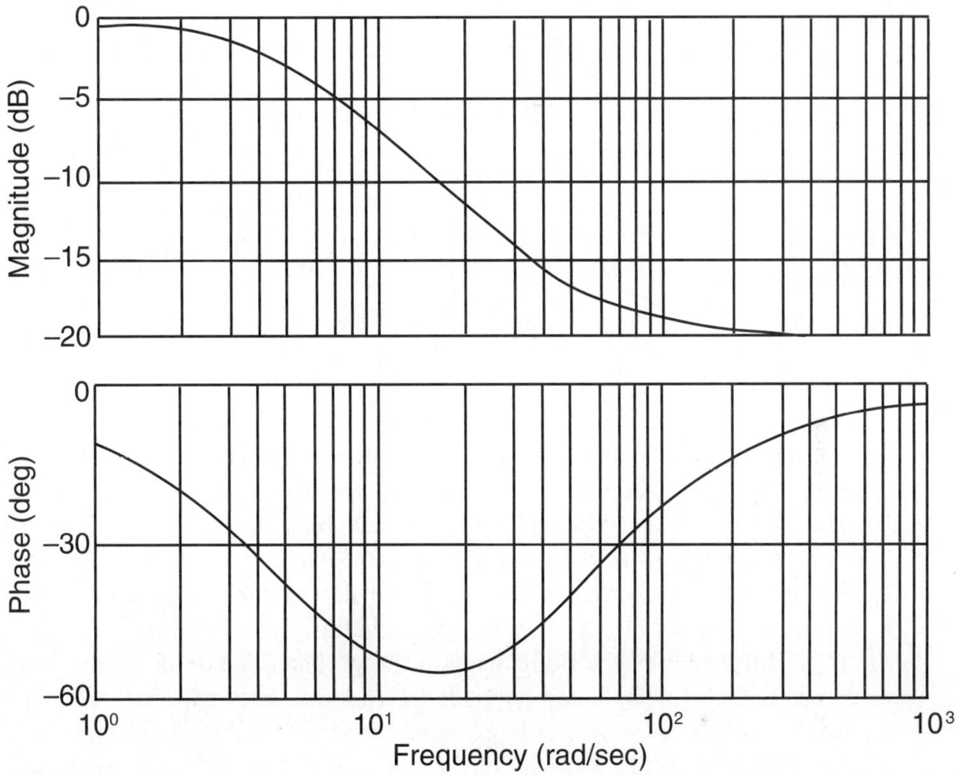

Figure 10–15 Bode plot for lag compensator, $Gc(s) = (s/50 + 1)/(s/5 + 1)$

Thus, with proper design, it is possible to use the attenuation characteristics while avoiding the effects of the phase shift. The lag filter is designed to attenuate its input signal by an amount to be calculated or estimated. To ignore the phase shift characteristics, the filter is designed so that both z and p are at lower frequencies than the desired gain crossover. Following are the design steps:

1. Obtain the Bode diagram.
2. Inspect the gain and phase curves, and choose a new gain crossover frequency such that an acceptable PM is obtained.
3. Pick a value for z. This will normally be at least 1 decade lower than the new gain crossover frequency.
4. Again inspect the Bode diagram, and determine the amount of attenuation needed to move the loop gain to the new gain crossover. Use this value to design the lag pole.

Example 10.6

Consider a fourth-order system with transfer function:

$$G(s) = \frac{250}{s\left(\frac{s}{10}+1\right)\left(\frac{s}{100}+1\right)^2} \quad \text{Eq. 10.57}$$

$$= \frac{25,000,000}{s(s+10)(s+100)^2} \quad \text{Eq. 10.58}$$

Solution Note that the gain is high. The Bode diagram for the uncompensated system appears in Figure 10–16. The gain crossover is at $\omega = 50$, and the PM is approximately -22 degrees. The system is unstable.

Stabilizing the system with lead compensation is not desirable, because several poles and zeros would be required, and gain problems and noise problems could occur. Using a lag filter alleviates such problems.

Inspection of the Bode diagram shows that a PM of 45 degrees might be obtained if the gain curve can be made to cross the 0 dB axis at $\omega = 5$. This requires that the gain at $\omega = 5$ must be reduced by 34 dB. If this is accomplished, the new gain crossover would be at $\omega = 5$, so the new PM would be approximately 50 degrees, which would be acceptable.

This reduction in gain must be accomplished without changing the phase at frequencies at and near the new gain crossover, because such changes in phase also change the PM. To obtain the desired result, both the zero and the pole of the filter can be placed at frequencies well below the desired gain crossover. Usually, the zero is placed at least 1 decade below the gain crossover, and the pole is placed at a still lower frequency to provide the necessary gain reduction. Thus, we shall place the zero at 0.6 rad/s, which is more than 1 decade below the desired gain crossover frequency of 5 rad/s. The pole is placed relative to the zero to obtain the desired attenuation. Because the attenuation of a lag compensator is p/z, then:

$$-34 \text{ dB} = 20\log\left(\frac{p}{z}\right) = 20\log\left(\frac{p}{0.6}\right) \quad \text{Eq. 10.59}$$

and solving for p results in the location of the pole: $p = 0.012$

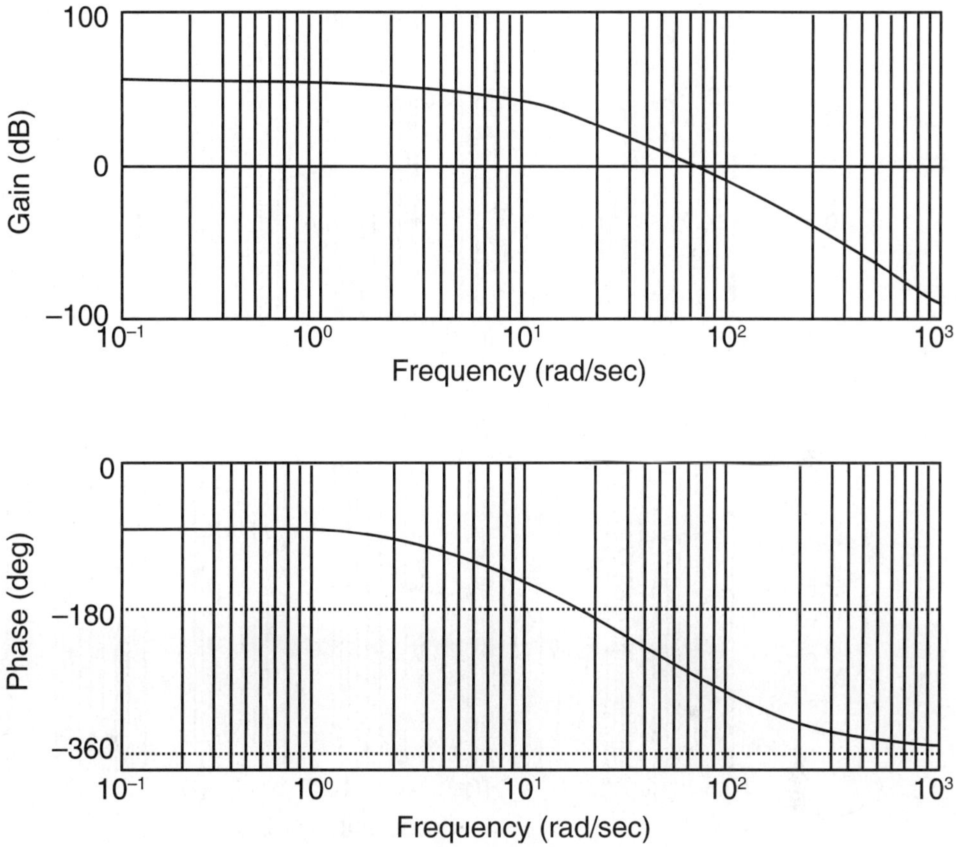

Figure 10–16 Bode diagram for uncompensated system, $G(s) = 2.5 \times 10^7/s(s + 10)(s + 100)^2$

The Bode diagram for the compensated system appears in Figure 10–17, using a compensator with $z = 0.6$ and $p = 0.012$. The gain crossover is at $\omega = 5$, and the phase margin is approximately 50 degrees. The characteristic polynomial is:

$$s^5 + 210s^4 + 12{,}002s^3 + 100{,}096s^2 + 334{,}130s + 200{,}000 \qquad \text{Eq. 10.60}$$

and its roots are:

1.0e + 002*

$-1.0038 + j0.0602$

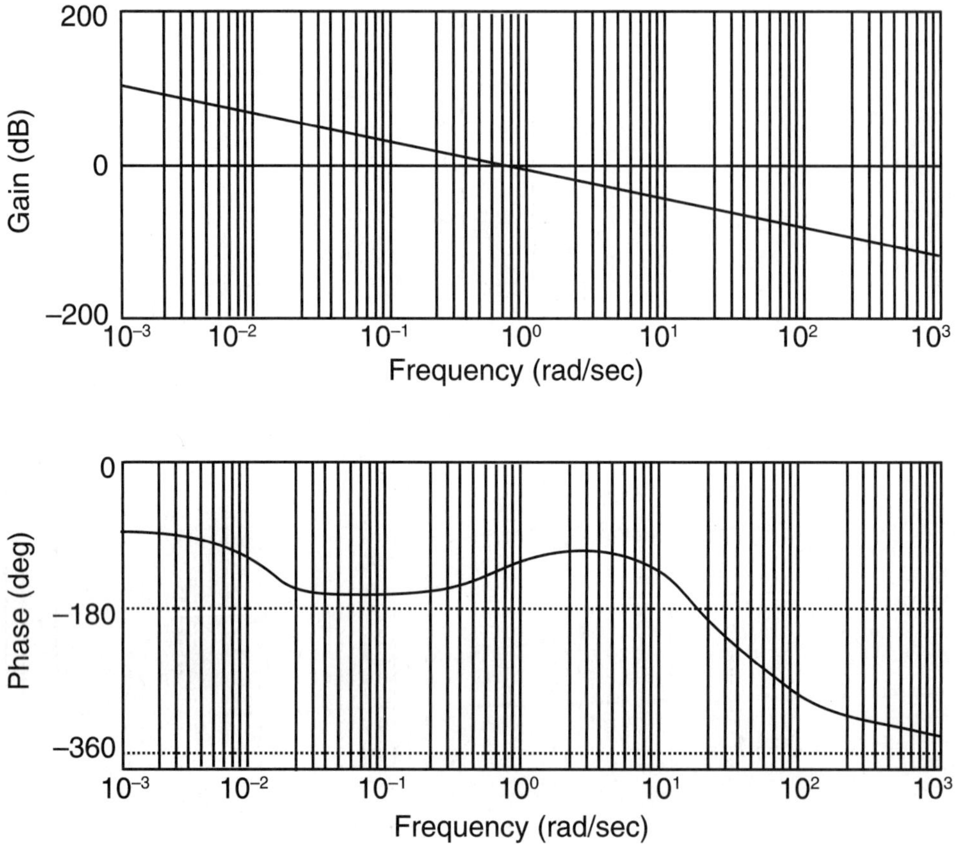

Figure 10–17 Bode diagram for compensated system, using compensator $Gc(s) = (s/0.6 + 1)/(s/0.012 + 1)$

$-1.0038 - j0.0602$

$-0.0425 + j0.0287$

$-0.0425 - j0.0287$

-0.0075

The step response appears in Figure 10–18. The root locus of the compensated system appears in Figure 10–19.

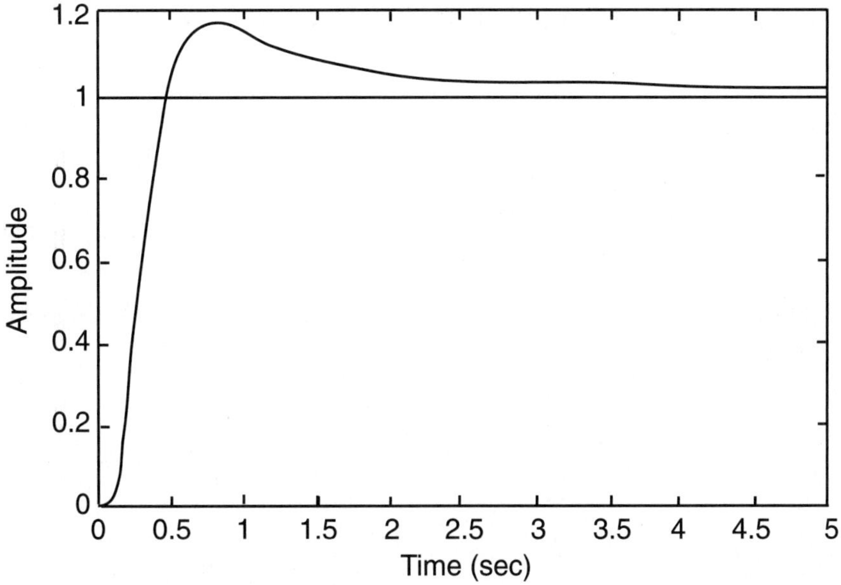

Figure 10–18 Step response for compensated system

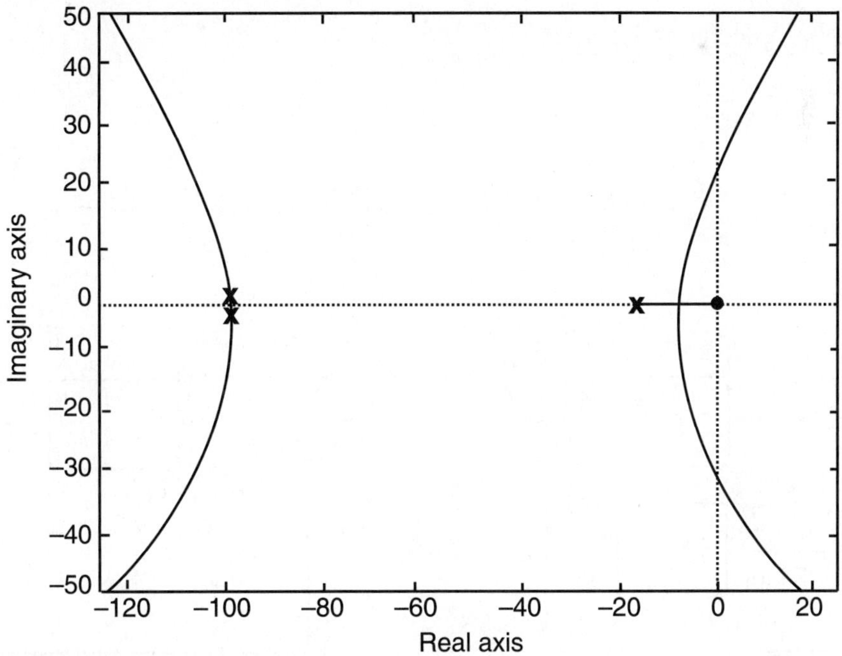

Figure 10–19 Root locus of compensated system

10.10 Bode Design of Feedback Compensation

Instead of placing a filter in cascade with the plant to achieve compensation, the designer may decide to measure an available signal, such as velocity, and feed it back to a summing junction. Normally, this signal must be amplified or attenuated and, if necessary, may be fed through a filter.

An algebraic approach to the design is possible, but it is not commonly used for two reasons:

1. Devices like accelerometers and tachometers are available. When one of these is to be used, its transfer function usually determines H.
2. Choice of G_{eq} is difficult.

There are cases in which this approach is viable. Figure 10–20 shows a block diagram used to derive the approximation. Equating the two blocks in the feedforward paths:

$$G_{eq} = \frac{G_P}{1 + G_P H} \qquad \text{Eq. 10.61}$$

Solving for H results in:

$$H = \frac{G_p - G_{eq}}{G_p G_{eq}} \qquad \text{Eq. 10.62}$$

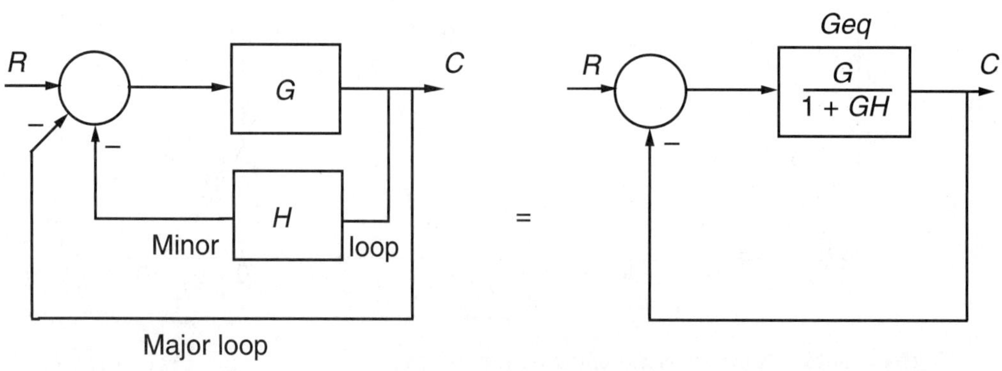

Figure 10–20 Block diagram used to define approximation

Example 10.7

A unity-feedback, second-order system with the transfer function:

$$G_p = \frac{100}{s(s+5)} \quad \text{Eq. 10.63}$$

produces a closed-loop step response that is very oscillatory, as shown in Figure 10–21. Use feedback compensation to obtain:

$$G_{eq} = \frac{32}{s(s+8)} \quad \text{Eq. 10.64}$$

and simulate using SIMULINK to confirm your results.

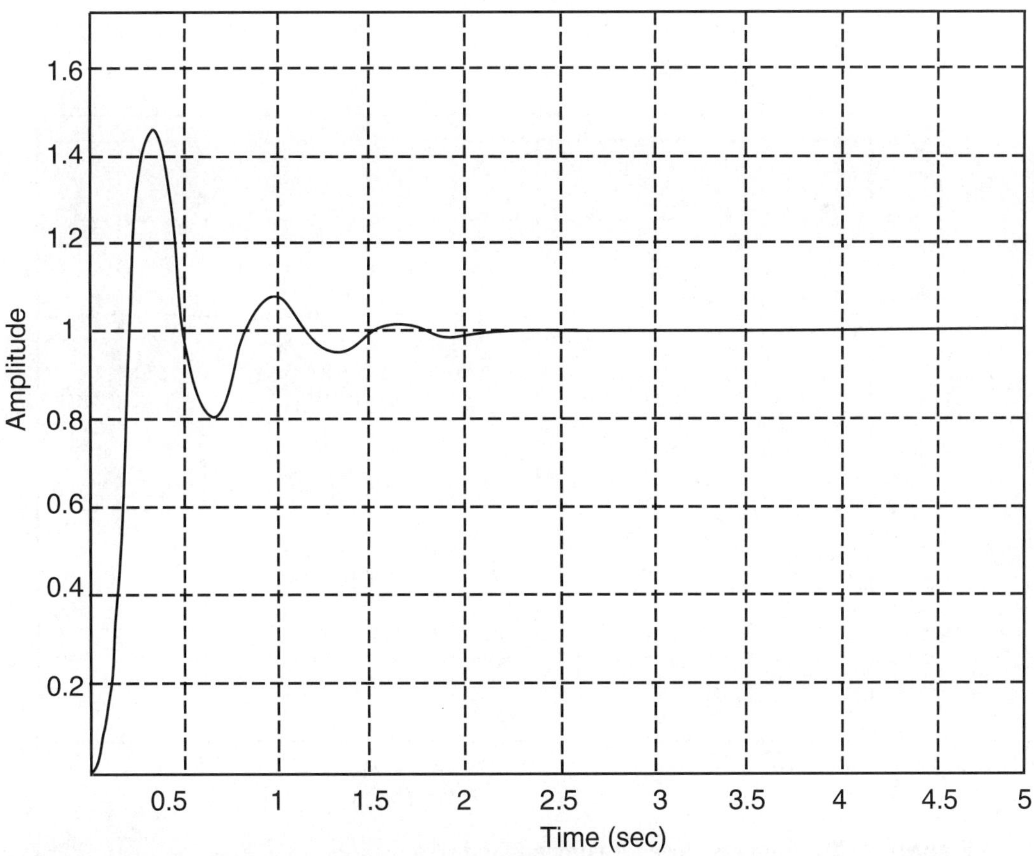

Figure 10–21 Step response for uncompensated system

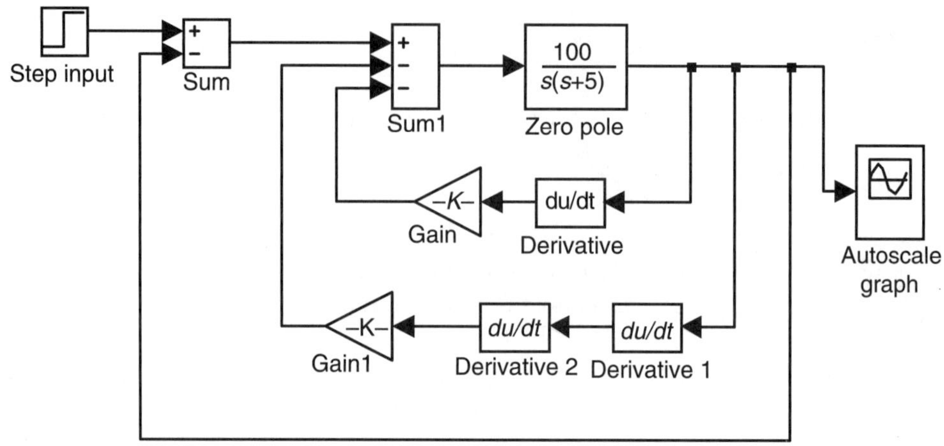

Figure 10–22 SIMULINK implementation of compensated system

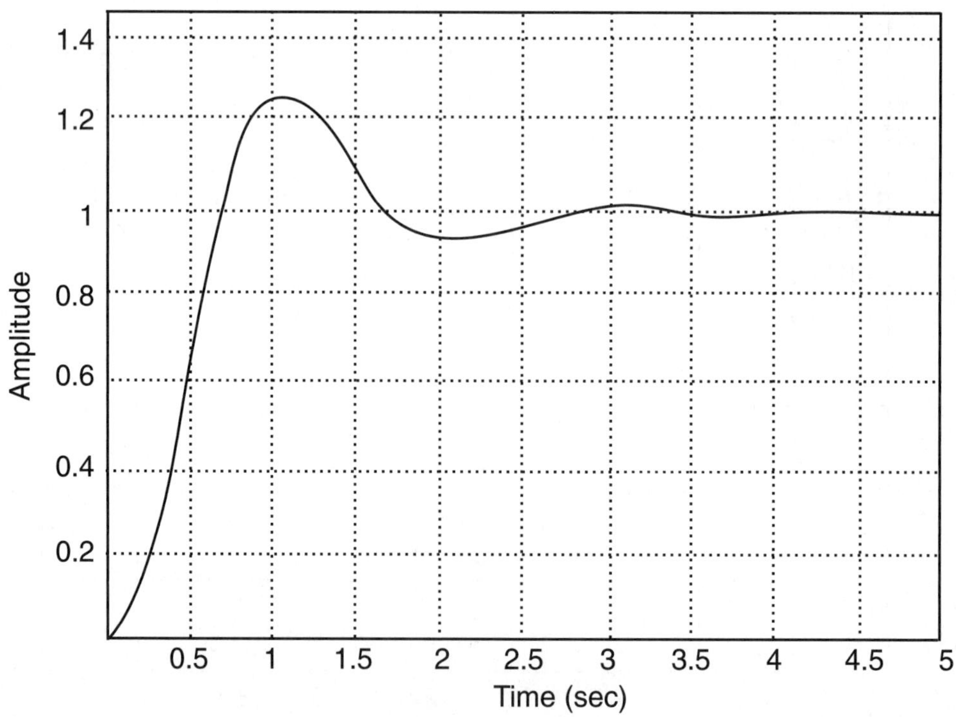

Figure 10–23 Step response for compensated system

Solution For the usual unity-feedback loop:

$$H = \frac{G_p - G_{eq}}{G_p G_{eq}} = \frac{68}{3,200}s^2 + \frac{640}{3,200}s \qquad \textbf{Eq. 10.65}$$

The closed-loop transfer function can not be evaluated using the Control Toolbox `feedback()` function, because the order of the H's numerator is higher than H's denominator. However, the closed-loop transfer function can be evaluated by applying the formula $G/(1+GH)$. To obtain a step response, the transfer function is implemented in SIMULINK (Figure 10–22), which produces the response in Figure 10–23, which is less oscillatory than the uncompensated system.

10.11 Graphical Approach for Bode Design of Feedback Compensation

If the Bode diagram is to be used, the algebraic relationships in a feedback loop give:

$$G_{eq} = \frac{G}{1 + GH} \qquad \textbf{Eq. 10.66}$$

To simplify graphical design manipulations, the following approximations are convenient:

when $GH << 1$, $G \sim G_{eq}$

when $GH >> 1$, $1 + GH \sim GH$ and $1/H \sim G_{eq}$

Further simplification provides:

If $GH < 1$, $G = G_{eq}$

If $GH > 1$, $1/H = G_{eq}$

Interpretation of these inequalities in terms of the asymptotic Bode diagram shows that whenever the G curve lies below the $1/H$ curve, $GH < 1$, so $G = G_{eq}$, and whenever the $1/H$ curve lies below the G curve, $GH > 1$, so $1/H = G_{eq}$. Thus, the asymptotic curve for $|G_{eq}|$ may be read inspecting the constructed Bode diagram. This is a very convenient technique, but care is required wherever the two curves cross.

Example 10.8

Design a compensator using the graphical techniques described previously for a system whose transfer function is:

$$G = \frac{10}{s(s+1)(0.1s+1)} = \frac{10}{s(s+1)(s/10+1)} \quad \textbf{Eq. 10.67}$$

Solution Figure 10–24 shows the asymptotic Bode magnitude diagram, and Figure 10–25 shows the true Bode magnitude diagram, including the phase curve. The PM and GM indicate that the system is almost unstable. Choice of the "desired" G_{eq} is expedited by considering possible locations for its gain crossover such that an adequate PM results. This implies that the gain crossover will be on an asymptote with −20 dB/decade slope. It is also desirable that the length of this asymptote be a

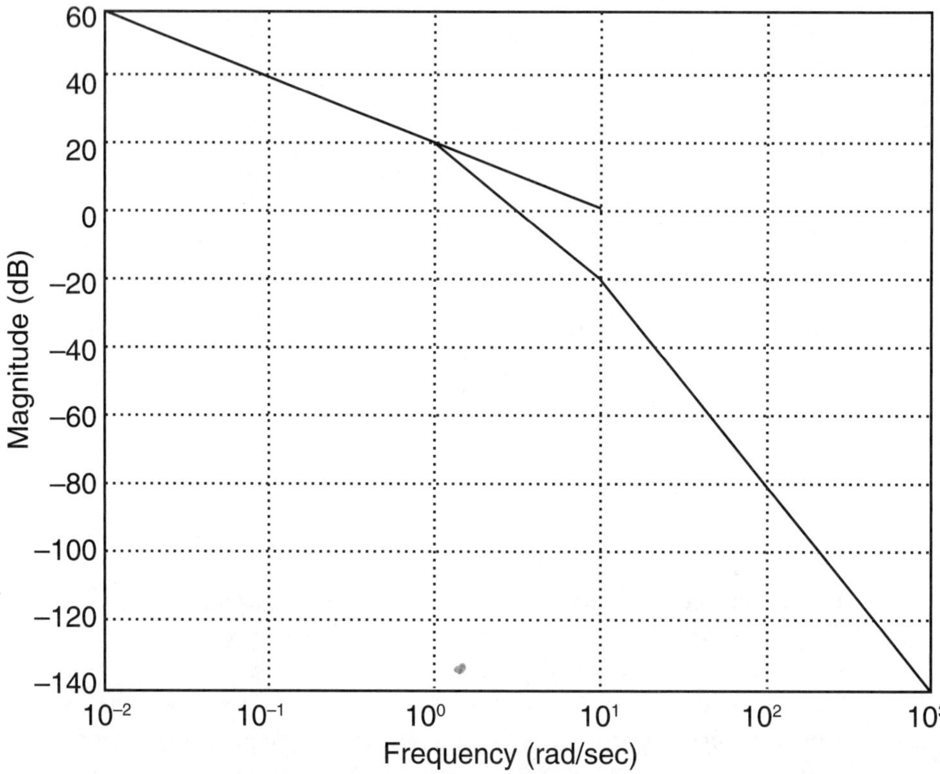

Figure 10–24 Asymptotic magnitude Bode diagram for $G(s) = 10/s(s+1)(0.1s+1)$

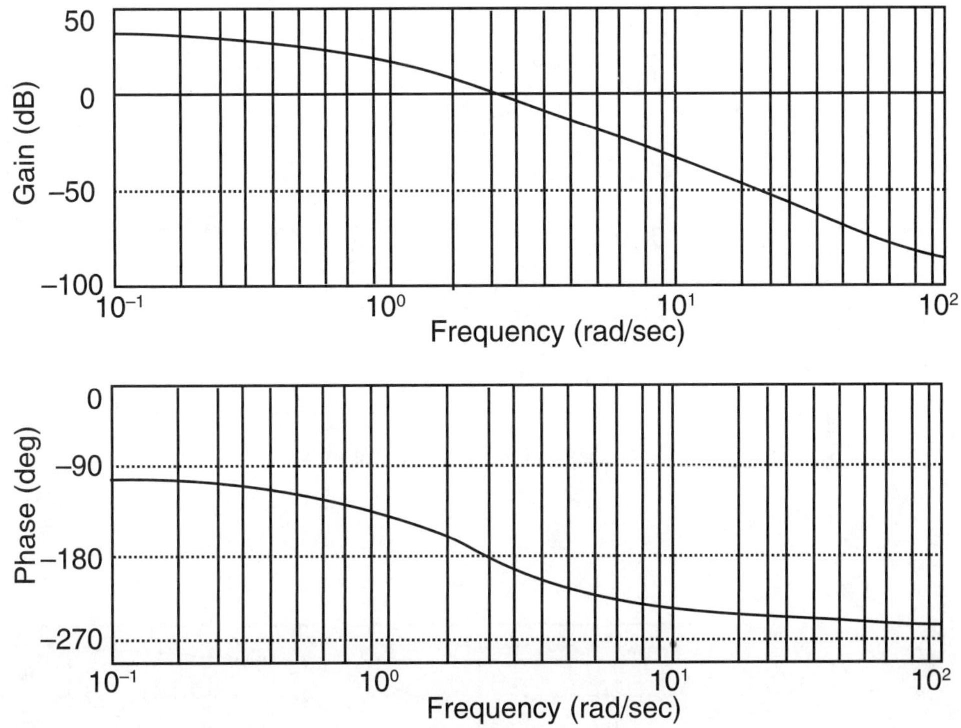

Figure 10–25 Bode diagram for $G(s) = 10/s(s + 1)(0.1s + 1)$

decade or more, if possible. For the case of velocity feedback, the desired G_{eq} is also $1/H$. The $1/H$ line is a single straight line with a slope of -20 db/decade. For this example, the $1/H$ line is chosen to cross the 0 db axis at $\omega = k_t = 0.5$, from which $H = 2$. If the phase curve is added and the PM is large, the bandwidth may be increased by raising the gain. This may also reduce the PM. Figure 10–26 shows the complete block diagram, and Figure 10–27 gives the step response. Note that the PM is large, and the step response is quite slow. A reduced feedback gain might improve the compensation.

For both positioning systems and tracking systems, velocity feedback can provide good damping. However, when used with tracking systems, velocity feedback also contributes to the steady-state error. To avoid this error, the designer may include a filter in the velocity feedback path. This is easily designed on the Bode diagram. After drawing the Bode diagram for the uncompensated system, the damping is improved by locating a suitable point for the new gain crossover and drawing the $1/H$ line through it, with slope of -20 db/decade. The $1/H$ line is extended to higher frequencies until it intersects the uncompensated G curve, as shown in Figure 10–28.

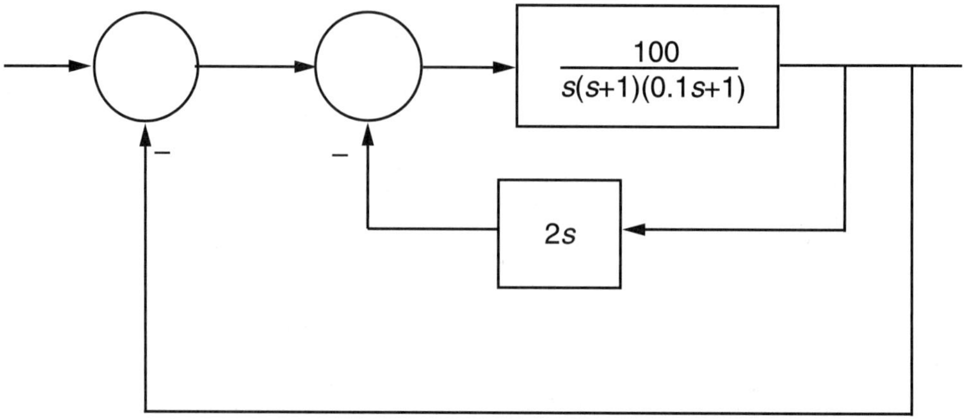

Figure 10–26 Block diagram for compensated system

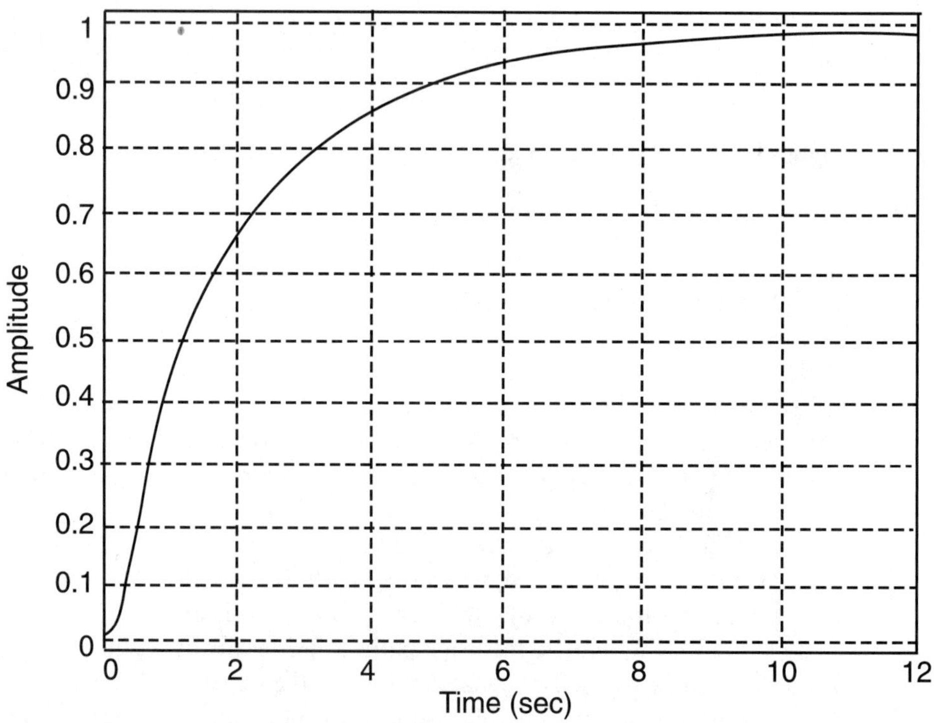

Figure 10–27 Step response of compensated system

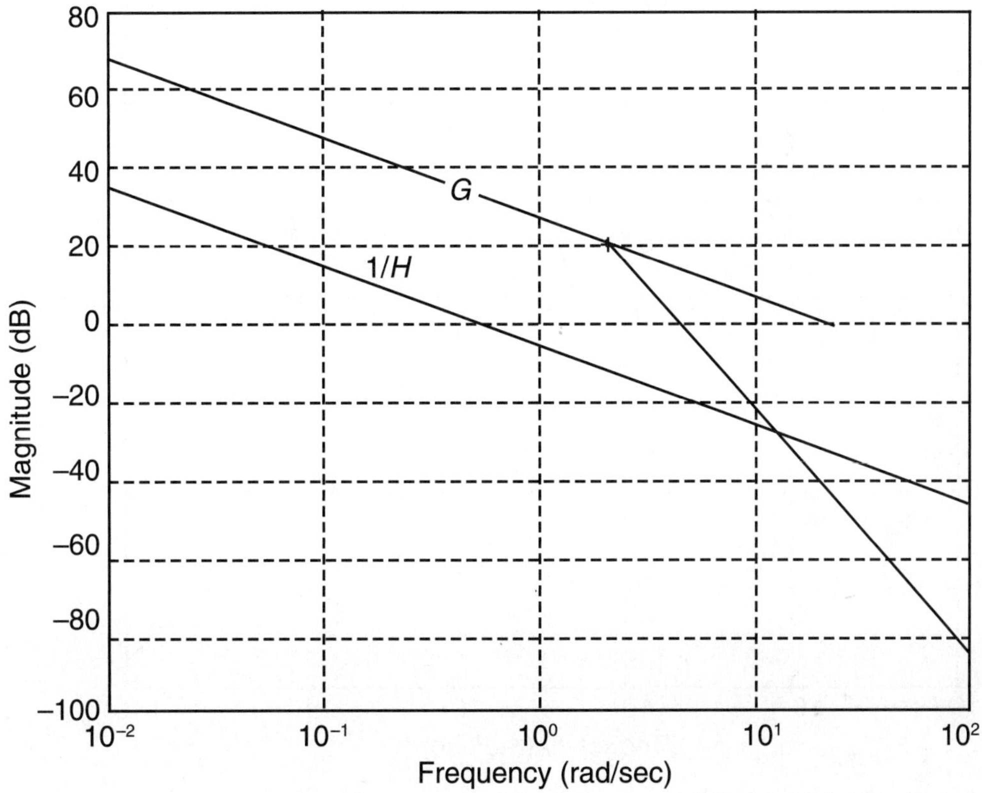

Figure 10–28 Plot of $1/H$ and G

The $1/H$ line is also extended in the low-frequency direction for a distance determined by the desired PM. Its slope is then changed to -40 dB/decade, or -60 dB/decade or more, which leads to a low-frequency intersection with the uncompensated G curve. This is shown in Figure 10–29, where the filter to be used in the feedback path is:

$$\frac{s+1}{s+0.1} \qquad \textbf{Eq. 10.68}$$

An alternate approach to the design is to draw the asymptotic Bode diagram for the uncompensated forward path, then superimpose the asymptotic Bode diagram for the desired system, including:

- An asymptote with slope of -20dB/dec, which lies below the uncompensated Bode diagram and intersects the 0 dB axis at a suitable frequency.

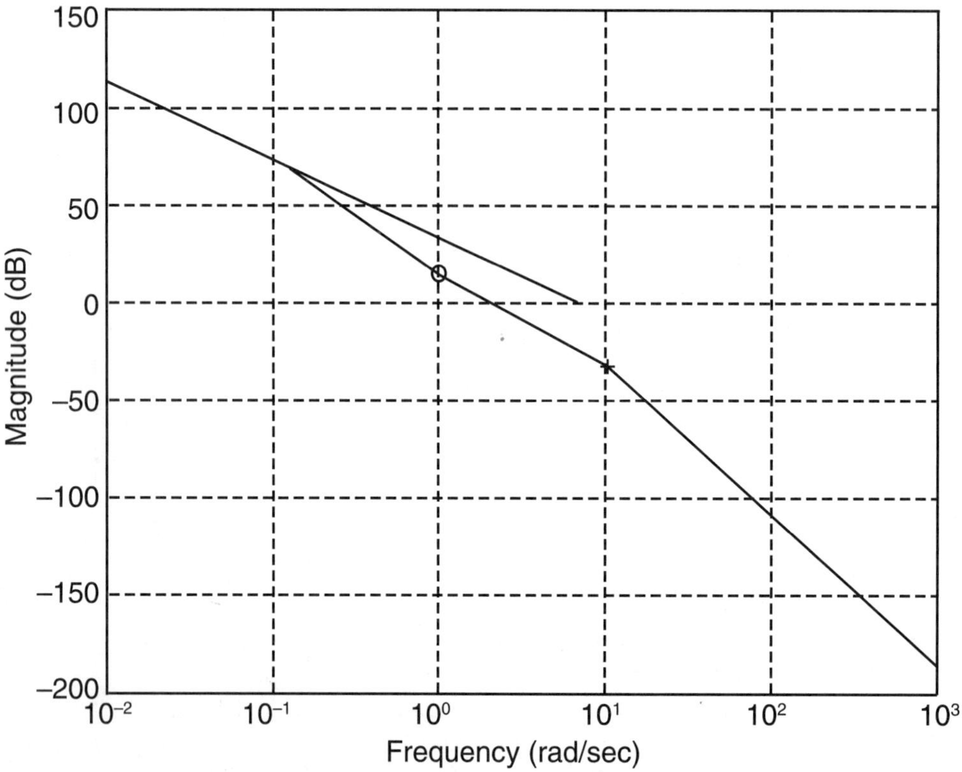

Figure 10–29 Asymptotic diagram

- This asymptote is extended toward higher frequencies until it crosses the original Bode diagram.
- The asymptote is also extended toward lower frequencies until its length provides suitable PM at the gain crossover.
- The slope of this low-frequency asymptote is changed to −40 (or −60 or −80) dB/decade and extended to cross the original diagram. The intersections may be interpreted as poles and zeros.

10.12 Comparison of Lead and Lag Compensation

After seeing the design techniques for both lead and lag compensation, it helps to reflect on the benefits and approach of each compensator. Lead compensation improves dynamic response by reshaping the phase plot in the Bode diagram. Lag compensation, in contrast, improves steady-state accuracy and does so by reshaping the magnitude plot in the Bode diagram.

10.13 Design Using Root Loci

Linear dynamic systems can be modeled using polynomials that can be factored. The singular points determined by factoring are called the roots of the polynomial. These roots may be plotted on the s-plane. If any parameter in the dynamic system changes value, all the roots also change value. Thus, the root points on the s-plane move and, in doing so, they trace curves called root loci. The design problem is to reshape these root locus curves, thus moving a chosen root to a desired location on the s-plane.

Early developments by Walter Evans and others used graphical methods to find construction rules by which root locus curves were defined on the s-plane. Today, the digital computer is used to factor the polynomials and plot the root loci. The graphical methods and the results of graphical construction permit easy sketching of the loci, which help in the use of the computer-gencrated curves, as described in Chapter 7.

Chapter 6 developed most of the important graphical procedures and results of system performance. If these results are to be used intelligently, the design engineer must know how the results are generated. In this treatment, only the use of the results is considered. The first step in designing with the root locus method is to factor the pertinent polynomials and plot the poles and zeros of the transfer function. Most transfer functions have only two polynomials: one for the numerator and one for the denominator. One may then obtain the root locus curves, either by graphical construction or with the computer. When graphical construction is used, the ordinate scale and abscissa scale must be identical, because angle measurements are needed.

We know that root loci start on poles and end on zeros or, if there are more poles than zeros, the excess loci start on poles but proceed to infinity. Asymptotes to the loci may be added to the root locus plot to aid in visualization and sketching.

Example 10.9

The plant of a small positioning system may be modeled as:

$$G_p(s) = \frac{100}{s^2} \qquad \textbf{Eq. 10.69}$$

Its root locus lies entirely on the imaginary axis. The closed-loop system has roots at $\pm j10$.

Solution The root locus of this system, drawn using MATLAB, appears in Figure 10–30. For many designs, the second step is to mark the point at which a root is to be placed. To stabilize the system and provide acceptable dynamic performance, we choose to place roots at $-0.5 + j0.5$, as shown in Figure 10–30. We cannot require a second root, except a conjugate.

In most cases, a root locus curve for the uncompensated system does not pass through the desired root point, as is the case in our sample system. Adding a filter to the system makes it possible to route a locus through the desired root point.

Each point in the root locus must satisfy the angle and magnitude condition. The angle of the uncompensated system at the chosen point is $-2(135)$ degrees or -270 degrees. However, the required angle is -180 degrees, so any compensator must contribute a net of $+90$ degrees at the chosen point. A possible solution is to use a simple cascaded lead filter ($z>p$ with z and p real). At the desired root point, the angle of $G_P(s)$ is:

$$\angle G_P(s) = \angle \frac{1}{s^2} = \theta_1 + \theta_2 = -270^\circ \qquad \text{Eq. 10.70}$$

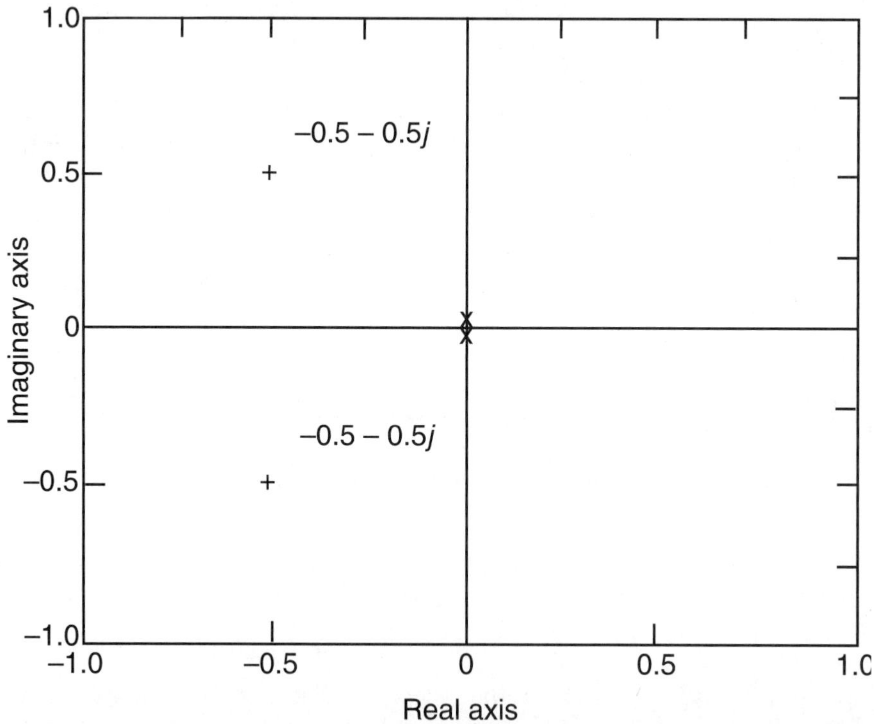

Figure 10–30 Root locus plot for $G(s) = 100/s^2$

Using a simple cascaded lead filter:

$$G_C(s) = \frac{s+z}{s+p} \qquad \textbf{Eq. 10.71}$$

and:

$$\angle G_c(s) = -\theta_z + \theta_p \qquad \textbf{Eq. 10.72}$$

And, by inspection, $\theta_p - \theta_z = 90°$. Then, the angle of the compensated system at the point is:

$$\angle G_P G_C = \theta_1 + \theta_2 + \theta_p - \theta_z = -180 \text{ degrees} \qquad \textbf{Eq. 10.73}$$

The designer is free to choose the location of the filter 0, as long as, as in this problem, it is greater than -2. Then the location of the pole, p, is determined by the requirement that $\theta_z - \theta_p = 90$ degrees. Choosing $z = -2$, the location of the pole should be $p \cong -22$. The root locus of the compensated system appears in Figure 10–31.

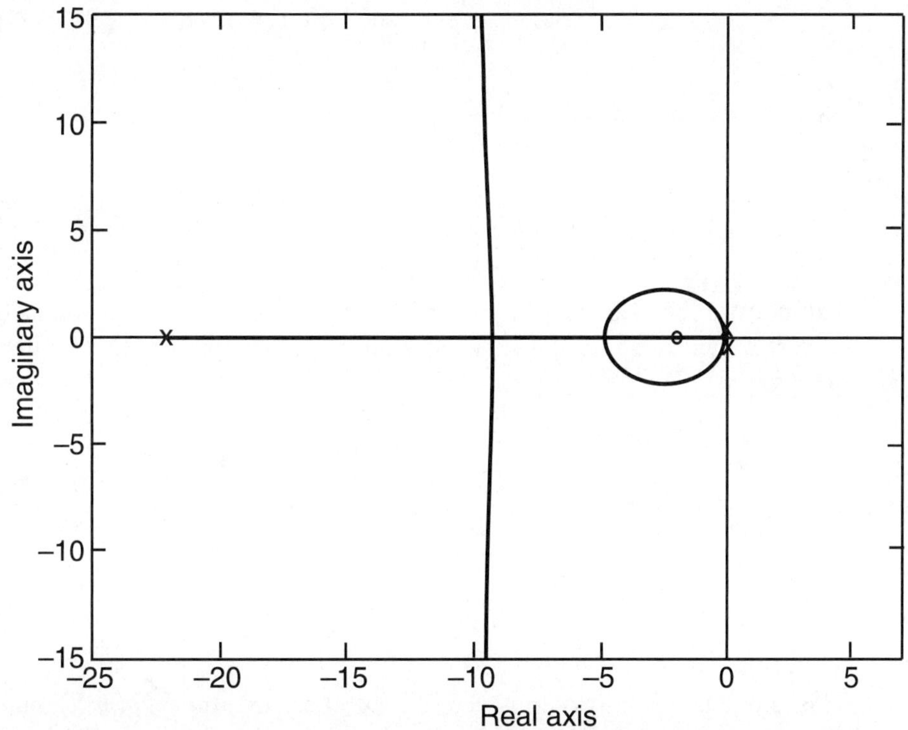

Figure 10–31 Root locus of compensated system, $G(s)Gc(s) = 100(s+2)/s^2(s+22)$

The next step is to adjust the gain so that the desired root point becomes a root. This is left as a homework exercise.

10.14 Partitioning

In many systems, the adjustable parameter is not the forward gain. It may be the gain in a feedback path or some other parameter, such as a change in load inertia. To treat such cases, the closed-loop characteristic polynomial is formed and partitioned (i.e., manipulated into transfer function form) and rearranged. The partitioning procedure is best demonstrated by example.

Example 10.10

Consider a second-order servo with tachometer feedback for damping. This means that:

$$G(s) = \frac{K}{s^2 + ps} \quad \text{and} \quad H(s) = 1 + ks \qquad \textbf{Eq. 10.74}$$

whose closed-loop transfer function is:

$$\frac{C}{R} = \frac{K}{s^2 + (p + Kk_t)s + K} = \frac{K}{s^2 + ps + Kk_t\, s + K} \qquad \textbf{Eq. 10.75}$$

Solution The characteristic equation is the denominator polynomial and is to be partitioned to apply the root locus method. Grouping the terms and setting them to 0:

$$(s^2 + ps + K) + Kk_t\, s = 0 \qquad \textbf{Eq. 10.76}$$

from which:

$$1 + \frac{Kk_t\, s}{s^2 + ps + K} = 0 \qquad \textbf{Eq. 10.77}$$

This is one partition (in terms of k_t). Two others are available (p, K), but they are not considered here. The root locus gain is Kk_t, but to adjust the damping, the

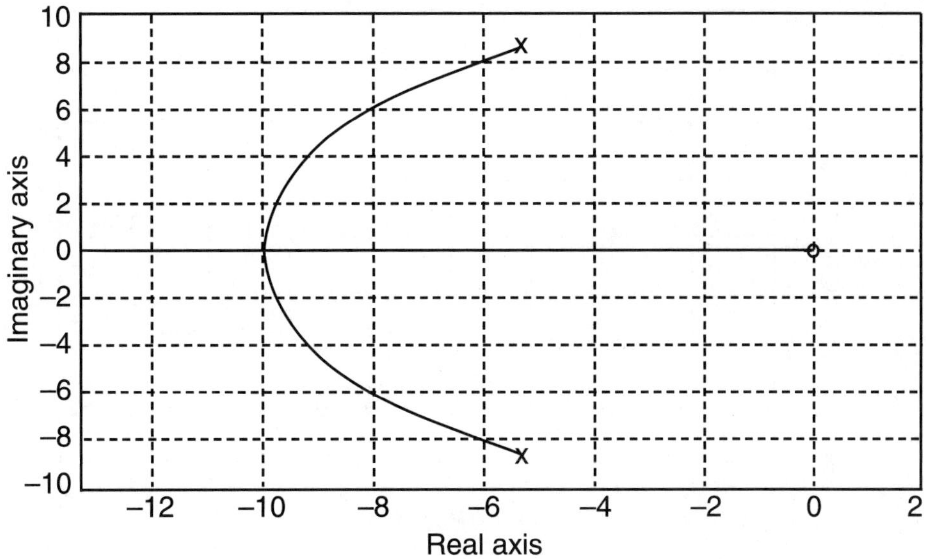

Figure 10–32 Root locus partitioning example, with $p = 10$ and $K = 100$

designer sets the forward gain, K, to a fixed number and adjusts the roots with the feedback gain, k_t. Consider the case in which $p = 10$ and $K = 100$:

$$\frac{100k_t\, s}{s^2 + 10s + 100} = -1 \tag{Eq. 10.78}$$

The root locus appears in Figure 10–32.

Example 10.11

A third-order system has a transfer function:

$$\frac{K}{s(s+1)(s+5)} \tag{Eq. 10.79}$$

Solution The root locus of the uncompensated system appears in Figure 10–33. As with any root locus, the loci begins at the open-loop poles when $K = 0$, and

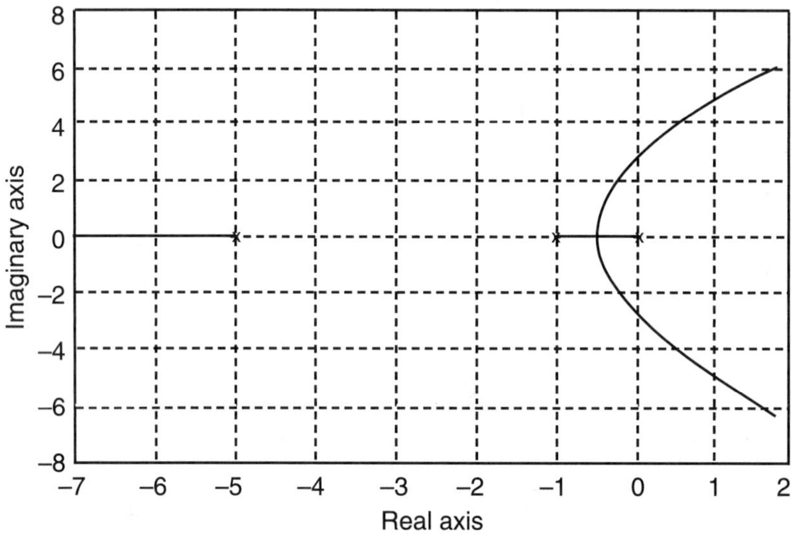

Figure 10–33 Root locus of uncompensated system (varying parameter is K)

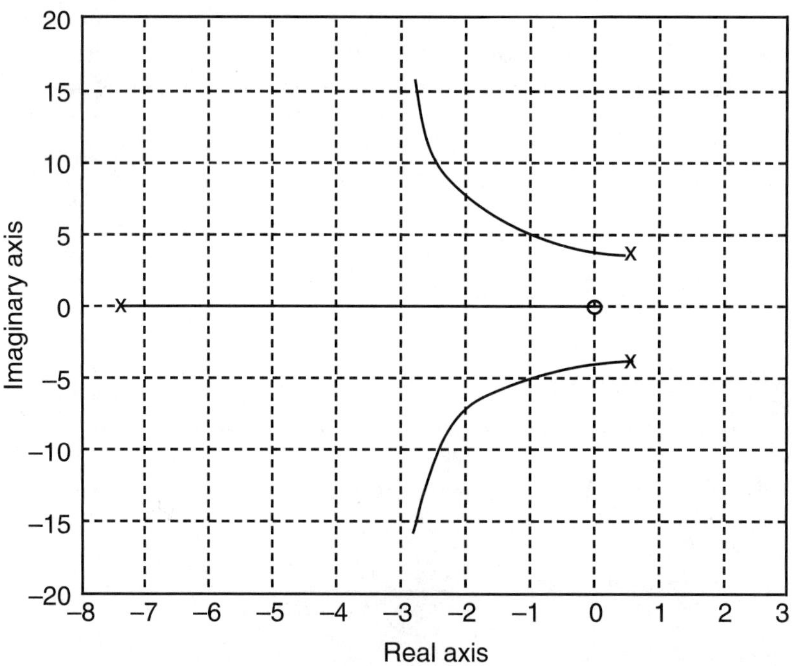

Figure 10–34 Root locus of compensated system, for $K = 100$ (varying parameter is k_t)

ends at the open-loop zeros when K = infinity. With velocity feedback added ($H(s) = 1 + k_t s$), the characteristic equation is:

$$s^3 + 6s^2 + (5 + 100k_t)s + 100 \quad \text{Eq. 10.80}$$

(for $K = 100$), from which:

$$\frac{100k_t s}{s^3 + 6s^2 + 5s + 100} = -1 \quad \text{Eq. 10.81}$$

The designer wants to keep $K = 100$ but must stabilize the system. The designer decides to use tachometer feedback. The resulting root locus appears in Figure 10–34.

10.15 Pole Placement

The dynamic response of a system is determined by the roots of its characteristic equation, also called the poles of the system function. Thus, if a model of the dynamic system is known, the designer needs a compensation scheme to place the system poles at desired locations. The common approach to pole placement is to use state variable feedback. In this treatment, we shall consider a third-order system. Then, using its characteristic equation directly, we will repeat the design using state variables to illustrate the procedures.

Example 10.12

Consider a third-order closed-loop system with the following transfer function:

$$G = \frac{100}{s(s+1)(s+10)} \quad \text{Eq. 10.82}$$

for which the characteristic equation is:

$$s^3 + 11s^2 + 10s + 100 \quad \text{Eq. 10.83}$$

Solution The roots of this polynomial are:

-10.9227

$-0.0387 + j3.0285$

$-0.0387 - j3.028$

These roots are not acceptable, because response is very oscillatory due to low damping. One solution would be to replace the roots with a third-order Butterworth polynomial. By choosing $w_n = 7.5$, the characteristic equation becomes:

$$s^3 + 15s^2 + 112.5s + 421.875 \quad \text{Eq. 10.84}$$

with roots at:

-7.5

$-3.75 + j6.4952$

$-3.75 - j6.4952$

The next step is to choose the compensation method. If we choose to place a compensator in cascade with the open-loop system, it can cancel the plant poles and replace them with poles that will satisfy the Butterworth requirement.

An alternate compensation method to allow pole placement is to feed back derivatives of the output signal, adjusting the gain in each feedback path. Such feedback is commonly used to improve damping in the system without attempting pole placement. The block diagram of the newly compensated system appears in Figure 10–35.

The characteristic equation for this compensated system is:

$$s^3 + (11 + 100k_2)s^2 + (10 + 100k_1)s + 100k_0 \quad \text{Eq. 10.85}$$

For this system, a binomial-type response is desired, such that the third-order binomial is:

$$s^3 + 15s^2 + 75s + 125 \quad \text{Eq. 10.86}$$

To find the values of gain that are needed, coefficients of like-powered terms are equated to the standard form, and the set of simultaneous equations is solved.

For a binomial type response:

$$11 + 100k_2 = 15 \quad k_2 = \frac{4}{100} = 0.04 \quad \text{Eq. 10.87}$$

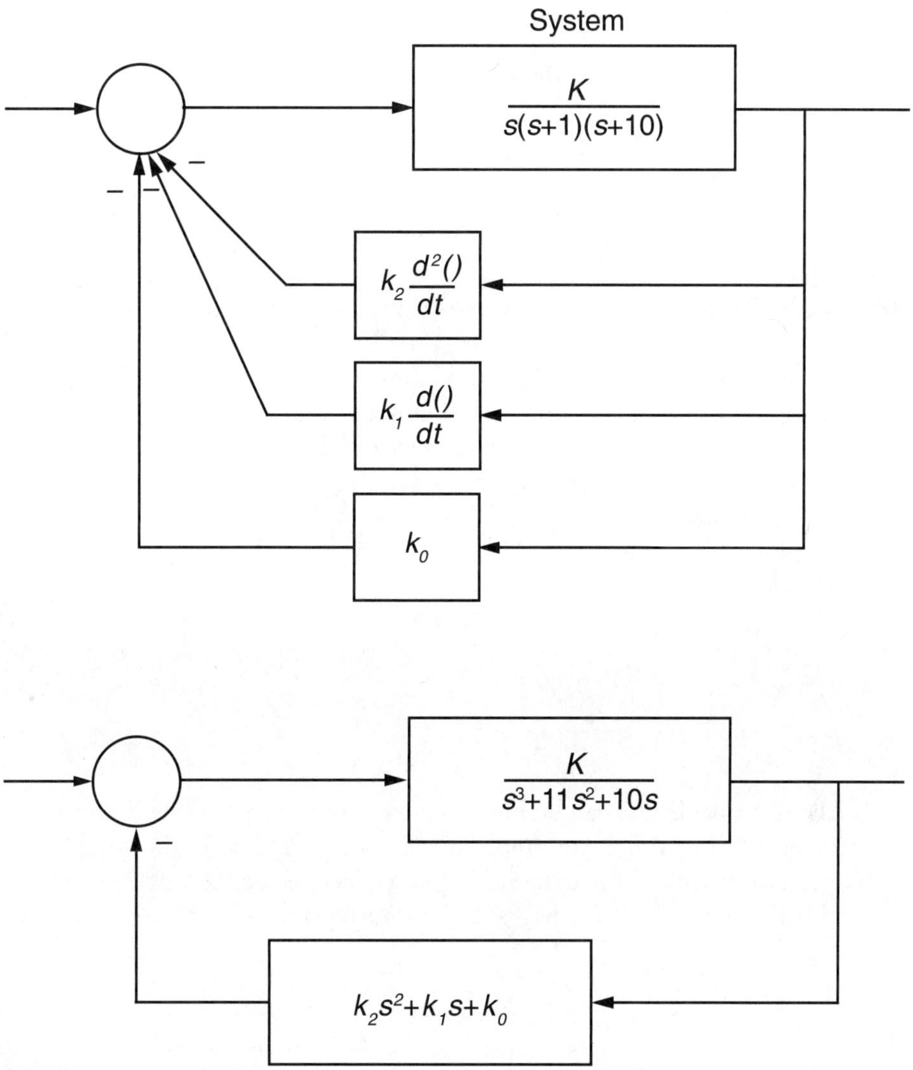

Figure 10–35 System compensation using pole placement

$$10 + 100k_1 = 112.5 \quad k_1 = \frac{102.5}{100} = 1.025 \qquad \textbf{Eq. 10.88}$$

$$100k_0 = 421.875 \quad k_0 = 4.21875 \qquad \textbf{Eq. 10.89}$$

For a Butterworth type response:

$$11 + 100k_2 = 15 \quad k_2 = 0.04$$

$$10 + 100k_1 = 75 \quad k_1 = 0.65$$

$$100k_0 = 125 \quad k_0 = 1.25$$

10.16 Pole Placement with State Variables

This section uses state variables to compensate the system. Consider the uncompensated system in Figure 10–36. The state equations for the uncompensated system may be obtained from the basic block diagram, shown in Figure 10–36. Notice that the constants, 10 and 11, are part of the plant to be controlled. The state variable assignment is:

$$C = x_1$$

$$\dot{x}_1 = x_2$$

$$\dot{x}_2 = x_3$$

$$\dot{x}_3 = -10x_2 - 11x_3 + 100U$$

All are available and can be measured and used as part of the dynamic system. The forward gain, $K100$, is the amplifier gain needed to obtain the required steady-state accuracy, as well as the dynamic response. When the feedback gains (k_1, k_2, k_3) are added to the basic block diagram, the state equations may be written:

$$\begin{bmatrix} \dot{x}_1 \\ \dot{x}_2 \\ \dot{x}_3 \end{bmatrix} = \begin{bmatrix} 0 & 1 & 0 \\ 0 & 0 & 1 \\ 0 & -10 & -11 \end{bmatrix} \begin{bmatrix} x_1 \\ x_2 \\ x_3 \end{bmatrix} + \begin{bmatrix} 0 \\ 0 \\ 100 \end{bmatrix} U \qquad \textbf{Eq. 10.90}$$

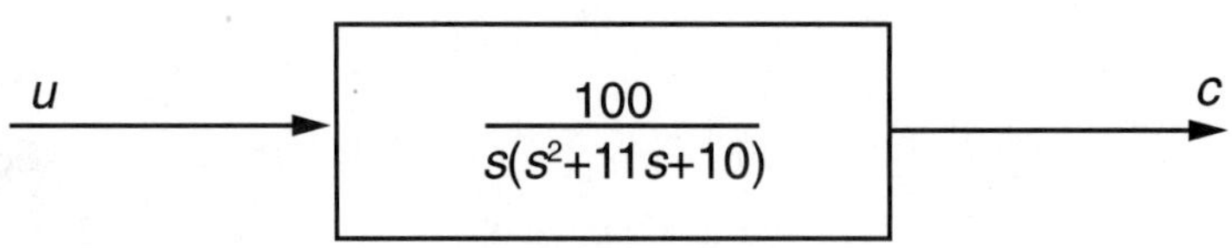

Figure 10–36 Uncompensated system, state-space formulation

where U is given by:

$$U = K\{R - k_t x\} = KR - K[k_1 \quad k_2 \quad k_3]\begin{bmatrix} x_1 \\ x_2 \\ x_3 \end{bmatrix} \qquad \text{Eq. 10.91}$$

Inserting into the state-space definition, taking the Laplace transform, and solving yields:

$$[sI - A] = \begin{bmatrix} s & -1 & 0 \\ 0 & s & -1 \\ 100k_1K & 10 + 100k_2K & s + 11 + 100k_3K \end{bmatrix} \qquad \text{Eq. 10.92}$$

Now, taking the determinant:

$$\det[sI - a] = s^3 + (11 + 100k_3K)s^2 + (10 + 100k_2K)s + 100k_1K \qquad \text{Eq. 10.93}$$

This is the characteristic polynomial, and roots can be placed at chosen points by assigning appropriate values to the feedback gains. This is accomplished by equating coefficients, as shown previously. While it is convenient to specify a Butterworth or another form, we may specify any set of roots. If, for example, we wanted roots at $-1+j1$, $-1-j1$, and -15, these specify the polynomial:

$$s^3 + 17s^2 + 32s + 30 \qquad \text{Eq. 10.94}$$

and the required gains are then:

$$Kk_1 = 0.3,\ Kk_2 = 0.16,\ Kk_3 = 0.06 \qquad \text{Eq. 10.95}$$

10.17 The Parameter Plane

The roots of the characteristic equation are determined by the parameters of the dynamic system. Most of the parameters (e.g., plant parameters) cannot be changed, but those of the "compensator" are adjustable. We may think of the design process as a procedure for producing roots at desired locations by adjusting the values of the parameters. The root locus method does this but allows only one adjustable parameter and requires graphical presentation of the data. The parameter plane method allows more than one adjustable parameter (usually two) and is completely algebraic. Graphical presentation of results is convenient for analysis. The procedures are most easily explained with an example.

Example 10.13

Consider a fourth-order polynomial with two adjustable parameters, α and β:

$$s^4 + (2 + 4\alpha)s^3 + (3 + 6\alpha + 5\beta)s^2 + (10 + 2\alpha + 11\beta)s + 3\beta = 0 \quad \textbf{Eq. 10.96}$$

Solution This can be partitioned:

$$s^4 + 2s^3 + 3s^2 + 10s + 50 + \alpha(4s^3 + 6s^2 + 2s) + \beta(5s^2 + 11s + 3) = 0 \quad \textbf{Eq. 10.97}$$

which is of the form:

$$D(s) + \alpha B(s) + \beta C(s) = 0 \quad \textbf{Eq. 10.98}$$

where:

$$D(s) = s^4 + 2s^3 + 3s^2 + 10s + 50 \quad \textbf{Eq. 10.99}$$

$$B(s) = 4s^3 + 6s^2 + 2s \quad \textbf{Eq. 10.100}$$

$$C(s) = 5s^2 + 11s + 3 \quad \textbf{Eq. 10.101}$$

Note that there are only two parameters, α and β, and each appears linearly in the coefficients. Design requires only simple algebra. We choose a desired root and substitute the value of s into the equations for B, C, and D, obtaining (for a desired real root) real numbers for B, C, and D. Then, the partitioned equation defines a straight line on the α, β parameter plane. Any point on this line supplies the desired root, but for a different α, β pair. Note that the remaining roots are defined by the chosen α–β pair but that their values are not known unless the polynomial is factored. The designer must choose a suitable value for s.

If the designer chooses a complex root (i.e., a complex value of s,) the original polynomial will have complex coefficients. By requiring that real and imaginary roots go to 0 independently, we obtain two linear equations:

$$B_R\alpha + C_R\beta + D_R = 0 \quad \textbf{Eq. 10.102}$$

$$B_I\alpha + C_I\,\beta + D_I = 0 \qquad \text{Eq. 10.103}$$

The Cramer's rule solution provides a unique α–β pair for the desired value of s. This procedure will be illustrated manually in the following example.

Example 10.14

Consider an unstable, third-order type 0 system. A cascade compensator is to be used, and the error coefficient is to remain unchanged. For this system:

$$G_p = \frac{72{,}000}{(s+10)^2(s+50)} \qquad \text{Eq. 10.104}$$

$$G_c = \frac{P(s+z)}{s(s+p)} = \frac{1}{\alpha}\frac{(s+\alpha p)}{(s+p)} \qquad \text{Eq. 10.105}$$

where:

$$\alpha = \frac{z}{p} \qquad \text{Eq. 10.106}$$

Specifications require that the settling time be about 2 seconds.

Solution Thus:

$$T_s = \frac{4}{\varsigma\omega_n} \quad \text{and} \quad \varsigma\omega_n = 2$$

If we choose $\varsigma = 0.4$, $\omega_n = 5$ or $s = -2.7 \pm j5.4$, the characteristic equation will be:

$$s^4 + 70s^3 + 1{,}100s^2 + 5{,}000s + p(s^3 + 70s^2 + 1{,}100s + 77{,}000) + A(72{,}000s) = 0 \qquad \text{Eq. 10.107}$$

where $A = 1/\alpha$. Substituting the chosen value for s, this becomes:

$$-22{,}547 - j6{,}357.12 + p(72{,}743.36 + j3{,}830.784) + A(194{,}400 + 385{,}776) = 0 \qquad \text{Eq. 10.108}$$

from which:

$$\sum \text{reals} = 72{,}743.36p - 194{,}400A - 22{,}547 = 0 \quad \textbf{Eq. 10.109}$$

$$\sum \text{Imag} = 3{,}830.784p + 38{,}5776A - 6{,}357.12 = 0 \quad \textbf{Eq. 10.110}$$

Solving simultaneously:

$$p = 0.3448 \quad A = 0.01305 \quad \alpha = 76.6 \quad \textbf{Eq. 10.111}$$

The transfer function of the compensator is:

$$G_c(s) = 0.01305\frac{s + 26.42}{s + 0.3448} \quad \textbf{Eq. 10.112}$$

and the closed-loop roots are at:

$$-14.67, -50.2, -2.694 \pm j5.354 \quad \textbf{Eq. 10.113}$$

Simple algebraic solution of simultaneous equations provides values for the two parameters (α and β), which are coordinates of a chosen complex root. If one considers design only, and is confident that the chosen value of s meets all the dynamic specifications, then the simple algebraic solution is a very desirable design step. Unfortunately, the contributions of all the roots must be considered, and the designer may need to revise the choice of the design value of s to compromise between conflicting dynamic contributions. The parameter plane gives a graphical view of the root/parameter relationship and thus provides a convenient basis for both analysis and design.

The coordinates of the parameter plane curves (usually α and β, though other symbols may be used) are the values of the adjustable parameters of the system. The chosen root point may be mapped onto the α, β plane using the characteristic polynomial as a mapping function, and a portion of the s-plane is mapped onto the same axes (i.e., a family of constant ς curves and a family of constant ω curves). The designer uses these curves to assess the impact of changes in α and β on the system dynamics. This requires that several lines be mapped from the s-plane onto the parameter plane (e.g., constant ς curves and constant ω curves).

Example 10.15

Consider a second-order system to be compensated with a cascaded filter, while keeping the error coefficient constant. Let:

$$G(s) = \frac{10}{s(s+1)} \quad \text{Eq. 10.114}$$

$$G_c(s) = \frac{p}{z}\left(\frac{s+z}{s+p}\right) = \alpha\left(\frac{s+p/\alpha}{s+p}\right) = \alpha\left(\frac{s+\beta/\alpha}{s+\beta}\right) \quad \text{Eq. 10.115}$$

where $\alpha = p/z$ and $\beta = p$

Solution The system has unity feedback, and the characteristic equation is:

$$s^3 + (\beta + 1)s^2 + (10\alpha + \beta)s + 10\beta = 0 \quad \text{Eq. 10.116}$$

Figure 10–37 shows the $\varsigma = 0$ curve on the α/β plane. Testing to find the stable area on this plane, we choose a test point at $\alpha = 1$, $\beta = 2$ and substitute these

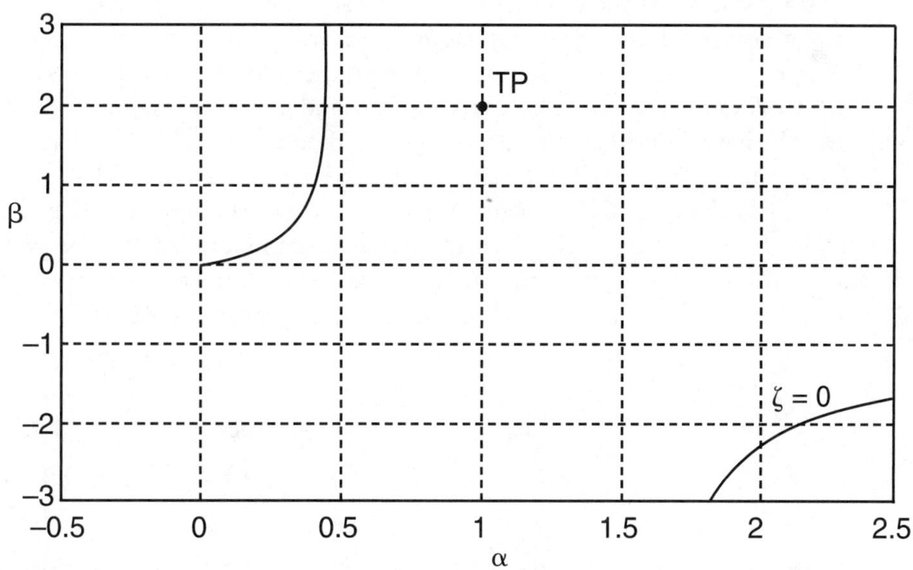

Figure 10–37 Parameter plane corresponding to $\varsigma = 0$

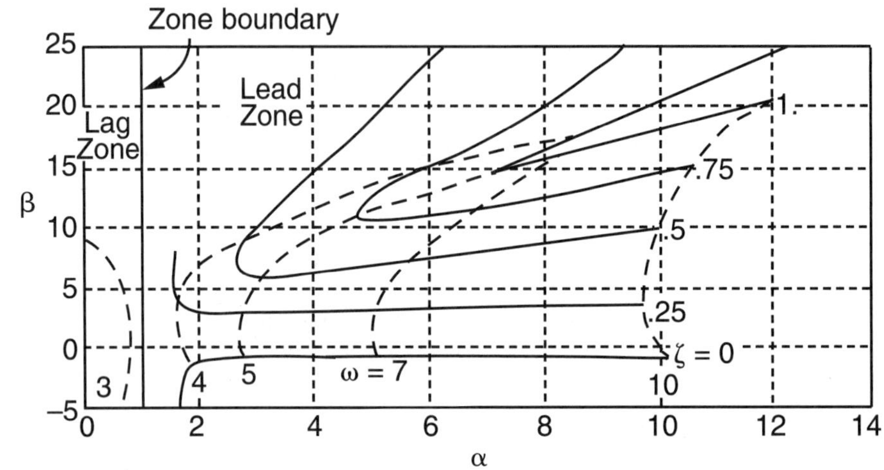

Figure 10–38 Expanded parameter plane

values into the characteristic equation. Use the Routh criterion, or factor the polynomial to find roots at:

$s = -2$

$s = 0.5 \pm j3.122$

This indicates that the test point is in the stable area.

Because the compensator has one pole and one 0, when $z = p$, then $\alpha = 1$, the compensator contributes nothing, although there will be an additional root at $-p$. For this example, the test point was at $\alpha = 1$ and $\beta = p = 2$, so there is a real root at -2. We have defined $\alpha = p/z$, and if $\alpha > 1$, then $z < p$ and the compensator introduces phase lead, while if $\alpha < 1$, then $z > p$ and the compensator is low pass (or phase lag). Thus, the vertical line for $\alpha = 1$ is a boundary line between root relocation zones for lead and lag. Figure 10–38 is an expanded parameter plane for the problem. For all points on the plane, the error coefficient is $K_v = 10$, and available operating points are determined by inspection. Four such points appear in Table 10–1.

α	β	ς	ω_n	**Root 1**	**Root 2**
1.7	2.8	0.25	4.0	−1.726	$-1.037 \pm j3.892$
2.5	2.8	0.25	6.0	−1.130	$-1.335 \pm j4.796$
2.8	8.0	0.5	4.0	−5.0	$-2.000 \pm j3.464$
6.5	8.05	0.5	8.0	−1.275	$-3.887 \pm j6.930$

Table 10–1 Root locations for four parameter values

10.18 The PID Controller

Many industrial controllers are designed for general, closed-loop control applications and are found in numerous applications in the control field. These controllers have a high degree of flexibility and are relatively easy to install and operate. Many of the controllers have what is normally called a proportional-integral-derivative (PID) controller. It operates on the error signal by producing an output that is proportional to the error, the integral of the error, and the derivative of the error. Two "parallel" PID implementations are available and shown in Figure 10–39a and Figure 10–39b. Implementation (a) uses three paths, each separately weighted by gains, P_P, P_I, and P_D. Implementation (b) uses gains that have physical significance to the dynamic of the process being controlled.

A mathematical representation of form (a) is:

$$o(t) = Kp(t) + K_I \int e(t)dt + K_D \frac{de}{dt} \qquad \textbf{Eq. 10.117}$$

and form (b) is:

$$o(t) = P\left[e(t) + \frac{1}{T_i}\int e(t)dt + T_d \frac{de}{dt}\right] \qquad \textbf{Eq. 10.118}$$

Converting from one form to the other can be easily performed:

$$\begin{aligned} Kp_P &= P \\ Kp_I &= \frac{P}{T_i} \\ Kp_D &= PT_d \end{aligned} \qquad \textbf{Eq. 10.119}$$

By taking the Laplace transform of this equation and grouping the terms into one transfer function, one can conclude that the general PID controller has a transfer function with a pole at the origin and two zeros. Thus, the transfer function for a PID controller (form [a]) is:

$$\frac{O(s)}{E(s)} = \frac{K_{ps} + K_I + K_D s^2}{s} \qquad \textbf{Eq. 10.120}$$

This form, which is equivalent to two zeros and one pole at the origin, is usually called the parallel implementation.

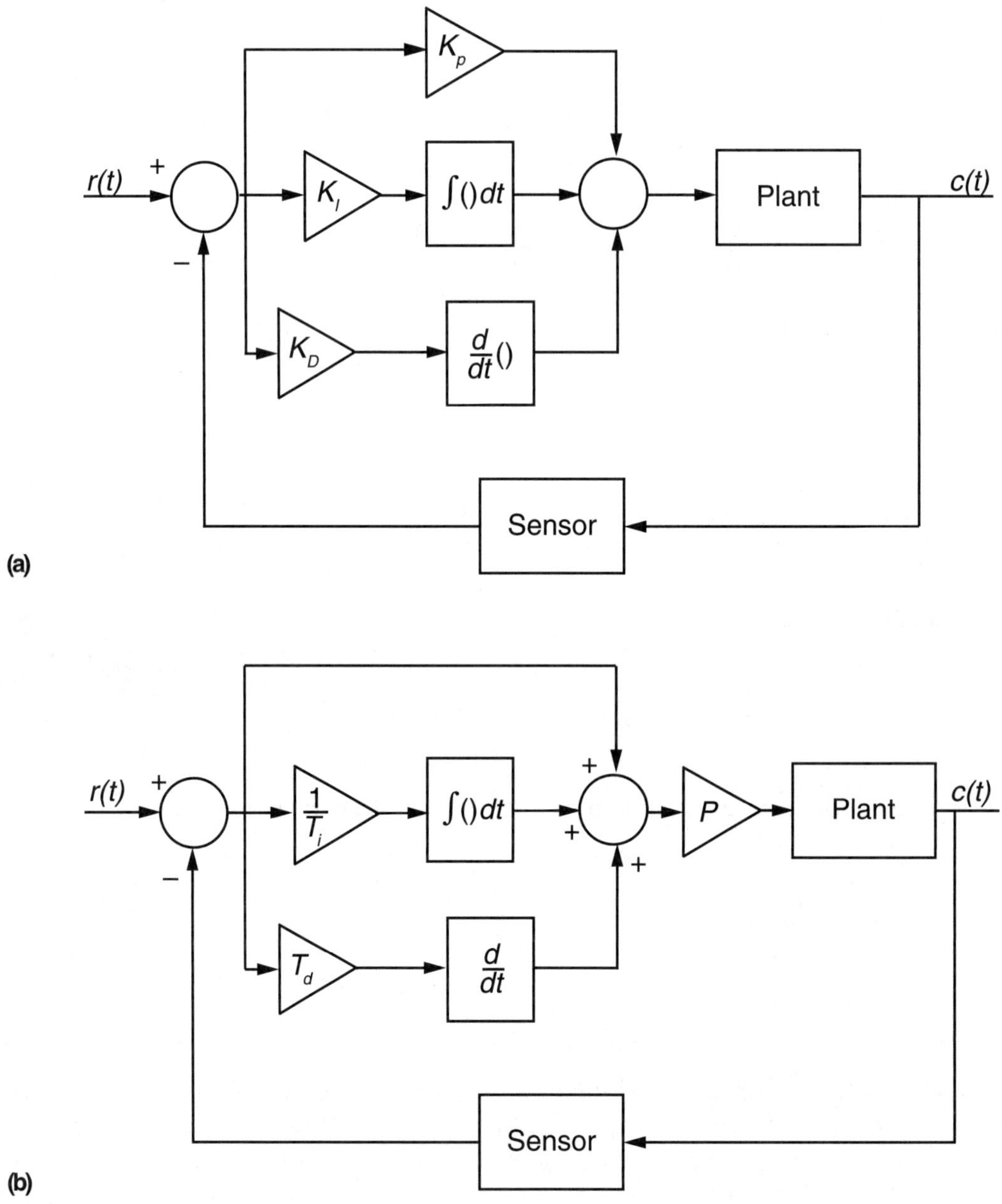

Figure 10–39 Parallel PID control implementations: (a) independent paths and (b) classical

Characteristic to improve	Recommended compensator type and corresponding transfer function	Compensator pole-zero arrangement and Bode Plots	Compensator-equivalent (or similar) PID controller type and corresponding transfer function	Graphical representation of PID Controller pole-zero arrangement
Dynamic response (decrease Rise time, or Settling time or P.O.)	**lead compensator** $\dfrac{\frac{s}{z}+1}{\frac{s}{p}+1}$ $p >> z$	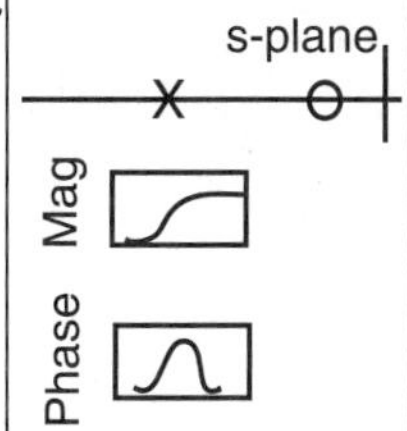	**PD** $\theta(t) = K_p e + K_D \dfrac{de}{dt}$	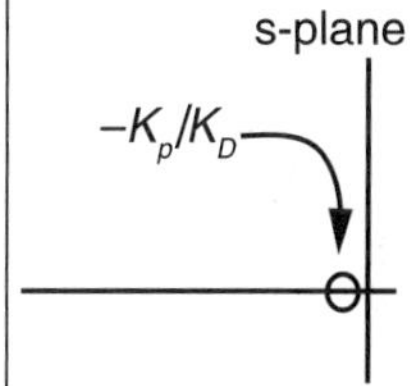
Steady-state accuracy (decrease steady-state error)	**lag compensator** $\dfrac{\frac{s}{z}+1}{\frac{s}{p}+1}$ $z >> p$	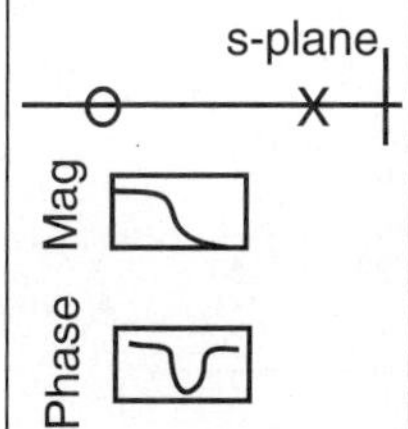	**PI** $\theta(t) = K_p e + K_I \int e dt$	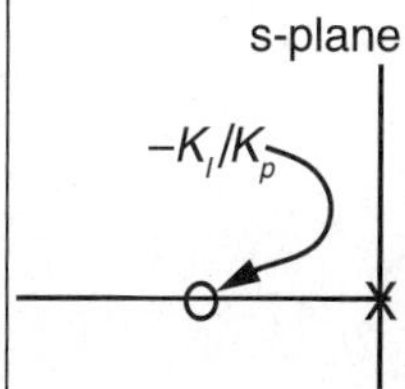

Table 10.2 Comparison of lead and lag compensators with PID controller

It also helps to compare two special cases of the PID controller: (1) the proportional + derivative (PD) and the proportional + integral (PI) to the compensators previously discussed, the lead and lag filters. Table 10–2 shows this comparison.

Another implementation, shown in Figure 10–40,[4] does not differentiate changes in the input signal. The system response can be significantly different between these two implementations, especially if the setpoint is changing. This will be demonstrated later.

Another implementation is called the "series" PID implementation, in which the input signal goes through a pseudodifferentiator (in reality, a lead compensator), after which it goes into the PI block. Figure 10–41 shows the equivalent transfer function representation. The parameter α part of the controller has a range of values of 0.05 to 0.1.[5]

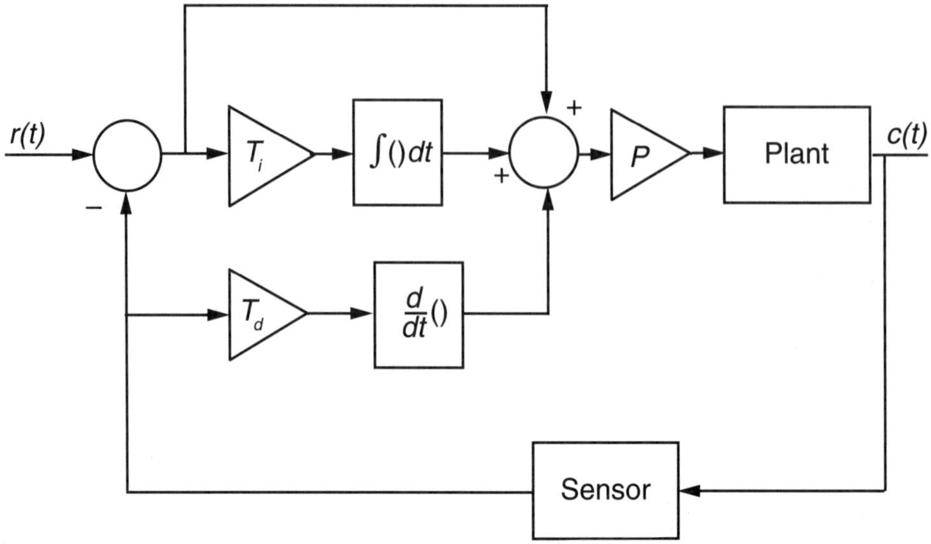

Figure 10–40 Alternate parallel PID implementation in which the derivative does not differentiate immediate changes in setpoint

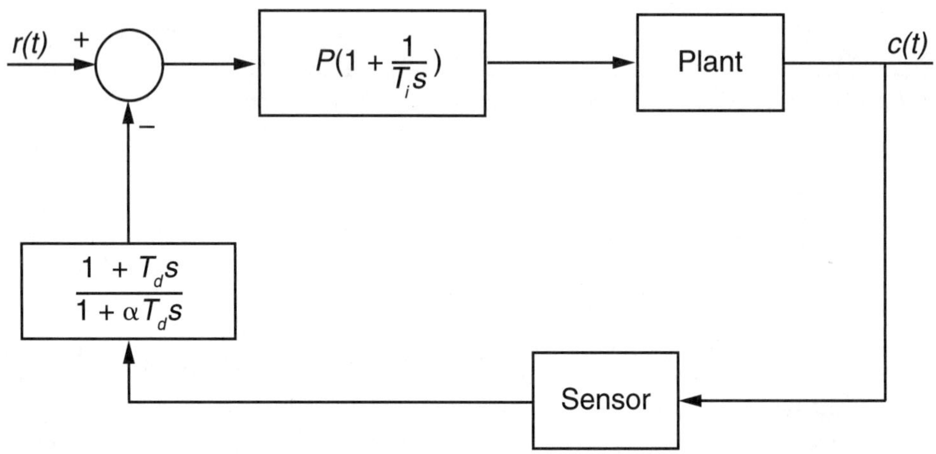

Figure 10–41 Series PID implementation form

10.19 The Ziegler-Nichols Method

This section presents an experimental method for selecting the coefficients of a controller that is particularly useful when little knowledge is available about the plant. It was developed by John Ziegler and Nathaniel Nichols and published in 1942 by ASME. As a result of their many experiences with control systems, analog computer simulations,[6] and their stroke of genius, Ziegler and Nichols were able to come up with some guidelines for determining the parameters for the PID equation.

They developed two methods. The first consists of increasing the system gain until the system is brought to marginal stability (a stable, but oscillating system). The second consists of applying a step change in the setpoint and extracting information from the response. We will discuss only the first method. The method is very effective at arriving at satisfactory controller settings with a relatively simple experimental procedure. It also works very well for other control strategies [P, PI]. Their method attempts to produce what is called a quarter decay response (QDR). This means that the amplitude of the output decreases one-quarter with each subsequent cycle, whether due to a disturbance step or a change in setpoint, as shown in Figure 10–42.

10.20 Ziegler-Nichols Tuning Procedure

The method consists of increasing the gain until the system goes into sustained oscillation (amplitude neither increases nor decreases). The specific gain that produces this behavior is defined as P_u (ultimate gain), and the period of oscillation (1/frequency of oscillation) is called T_u (ultimate period). The resulting parameters generally produce a system with acceptable behavior. This means that the system is stable and reasonably fast. Table 10–3 shows the recommended parameters selection guidelines proposed by Ziegler and Nichols, Corripio, and St. Clair.[7] One should consider these parameter values as starting points, a basis from which tweaking can begin. An overview of the assumptions made for developing these techniques is provided by Franklin et al.[8]

Notice that St. Clair's coefficients are gentler, that is, they produce a system with lower PO. Compare the two sets of coefficients with the last row in the table, which shows coefficient limits that produce no effect. In the following example, we will use the coefficients suggested by Ziegler-Nichols and St. Clair to arrive at the PID parameters.

Table 10–4 and Table 10–5 provide the controller parameters for a PI and simply a P implementation.

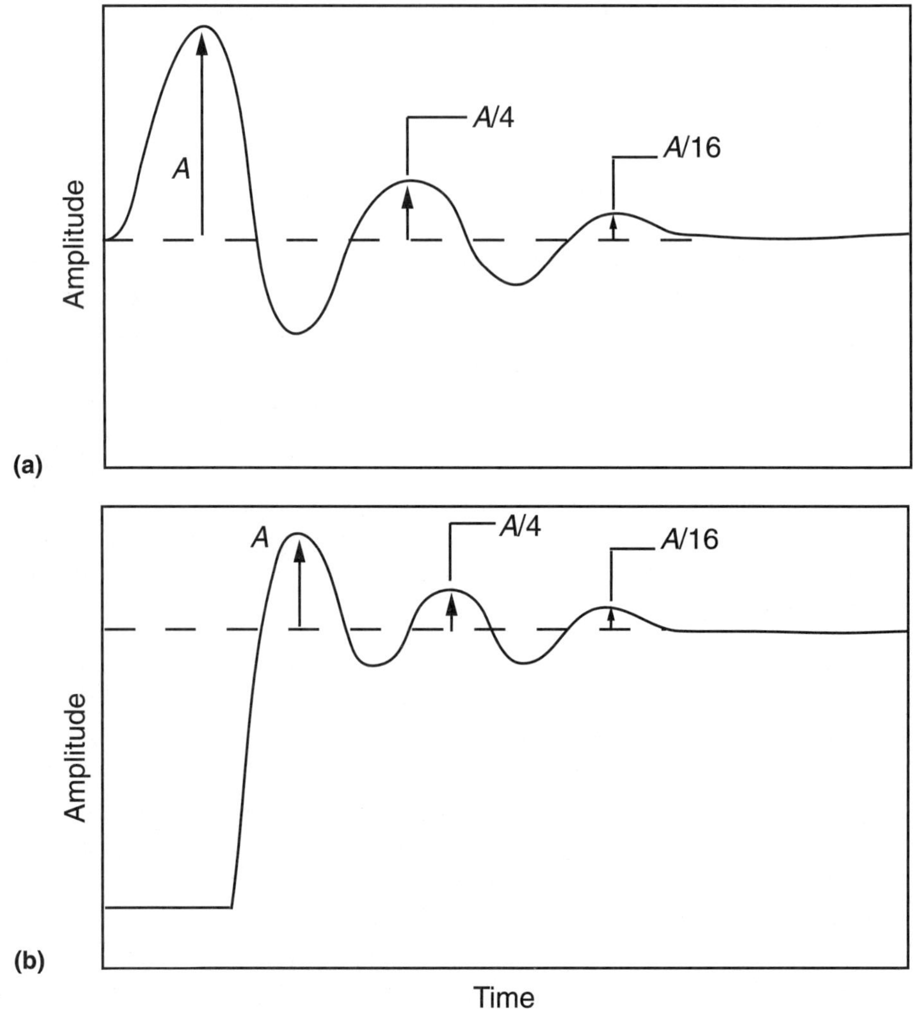

Figure 10–42 Quarter decay response from (a) a disturbance change and (b) a change in setpoint

Method	Gain, P	Integral Time, T_i	Derivative Time, T_d
Ziegler-Nichols	0.6 P_u	$T_u/2$	$T_u/8$
Corripio	0.75 P_u (parallel) 0.6 P_u (series)	$T_u/1.6$ (parallel) $T_u/2$ (series)	$T_u/10$ (parallel) $T_u/8$ (series)
St. Clair (conservative)	0.25 P_u	1.2 T_u	$T_u/8$
Parameter values producing least effect	0	∞	0

Table 10–3 Recommended PID coefficient values for a controller

Method	Gain, *P*	Integral Time, T_i
Corripio	0.45 P_u	T_u/1.2

Table 10–4 Recommended PI coefficient values for a controller

Method	Gain, *P*
Corripio	0.5 P_u

Table 10–5 Recommended *P* coefficient values for a controller

Example 10.16

Consider the control system in Figure 10–43. The step response to this system appears in Figure 10–44. Notice that the PO is over 40 percent and that it has a significant steady-state error.

Solution To apply the Ziegler-Nichols technique, the gain is increased until the system goes into sustained oscillation. This condition, corresponding to a $P_u = 2.13$, appears in Figure 10–45. Notice that $T_u = 1.18$s. These values can now be used to obtain the PID coefficients, and the result appears in Table 10–6.

By implementing the parameters derived from St. Clair's recommendations, shown in the block diagram of Figure 10–46, an acceptable step response (Figure 10–47) is obtained. If the Ziegler-Nichols coefficients are used, the response in Figure 10–48 is obtained. Notice that the later implementation produces a higher PO.

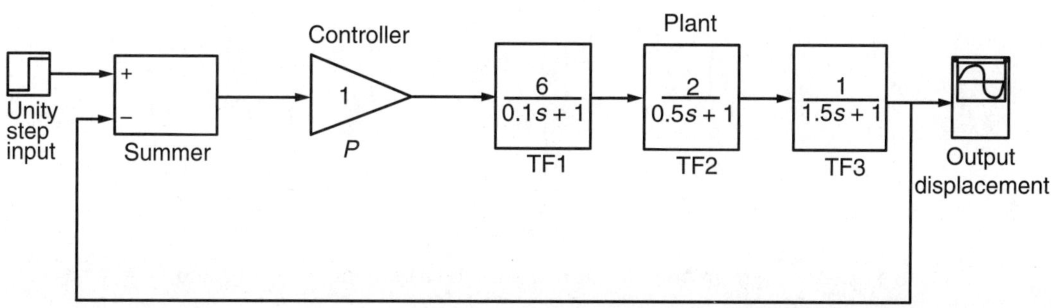

Figure 10–43 Sample system to be tuned using Ziegler-Nichols method

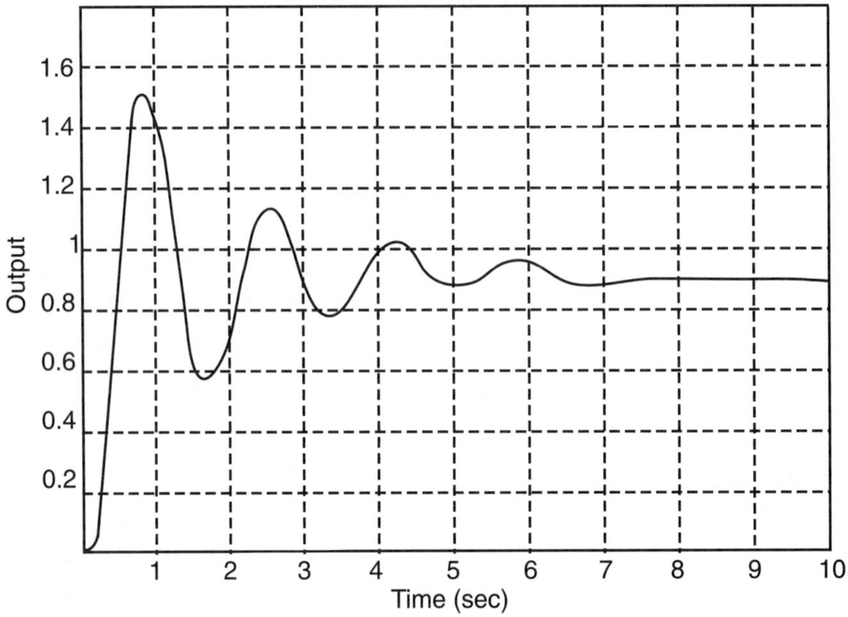

Figure 10–44 Original system response

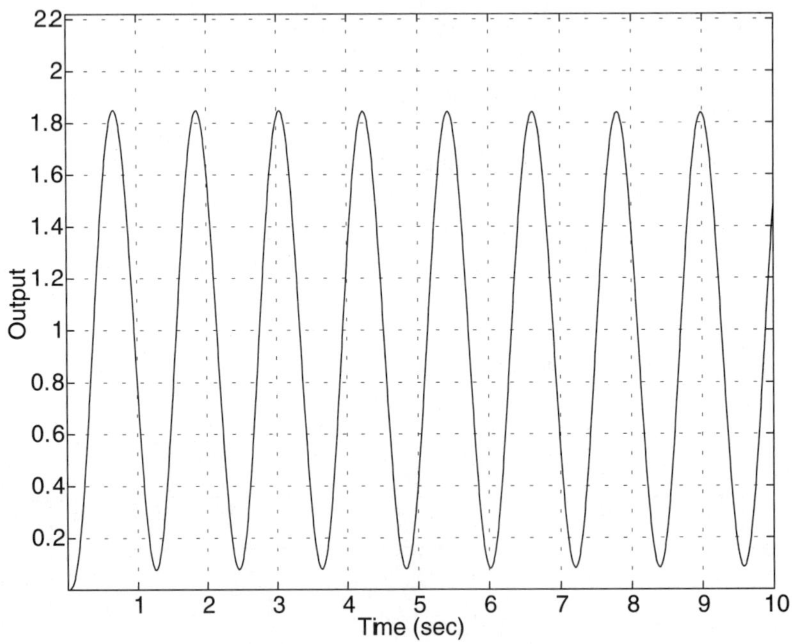

Figure 10–45 System behavior at marginal stability point, $P_u = 2.13$ and $Tu = 1.18$s

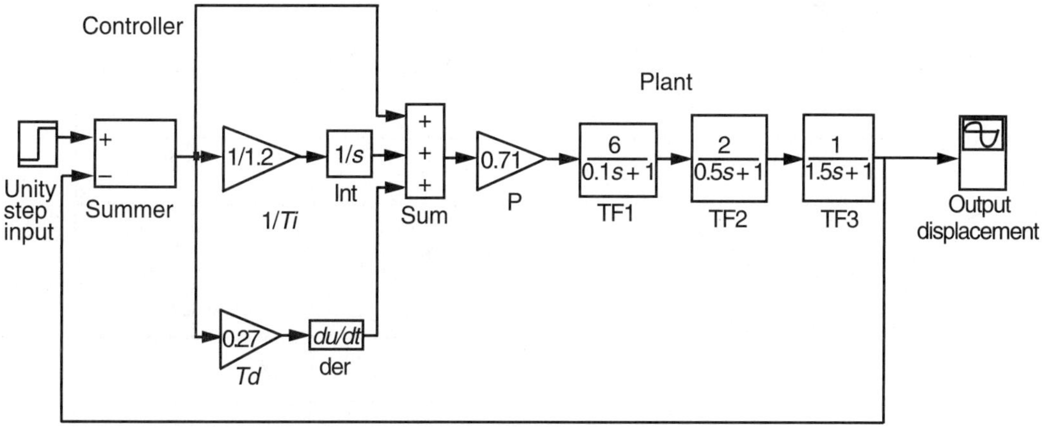

Figure 10–46 SIMULINK block diagram with PID implementation

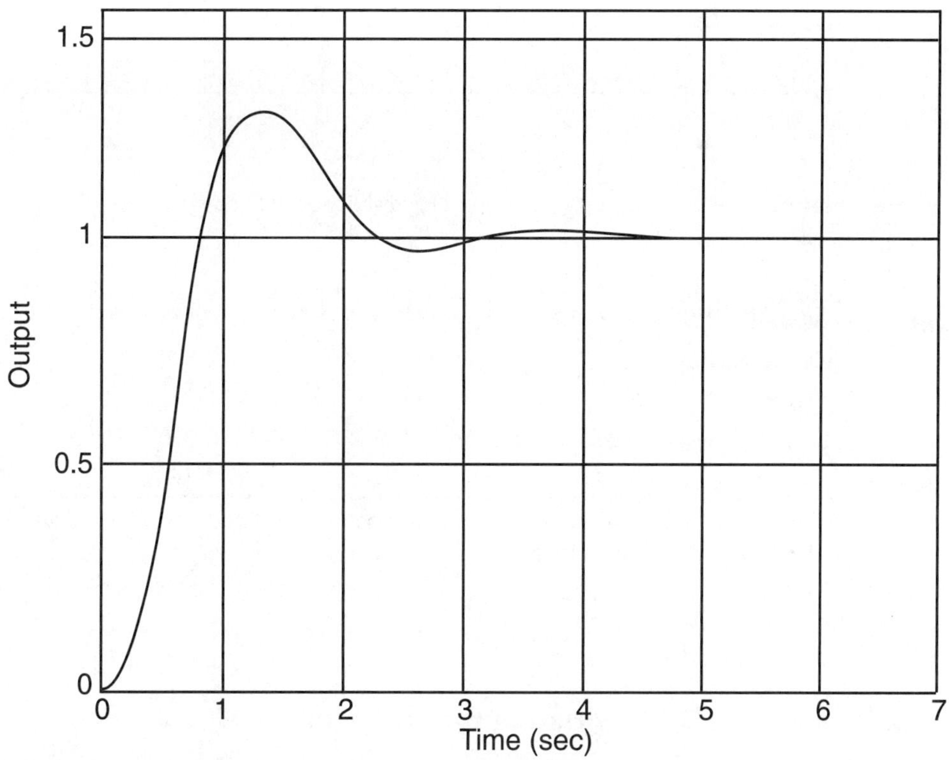

Figure 10–47 Step response using St. Clair PID coefficients

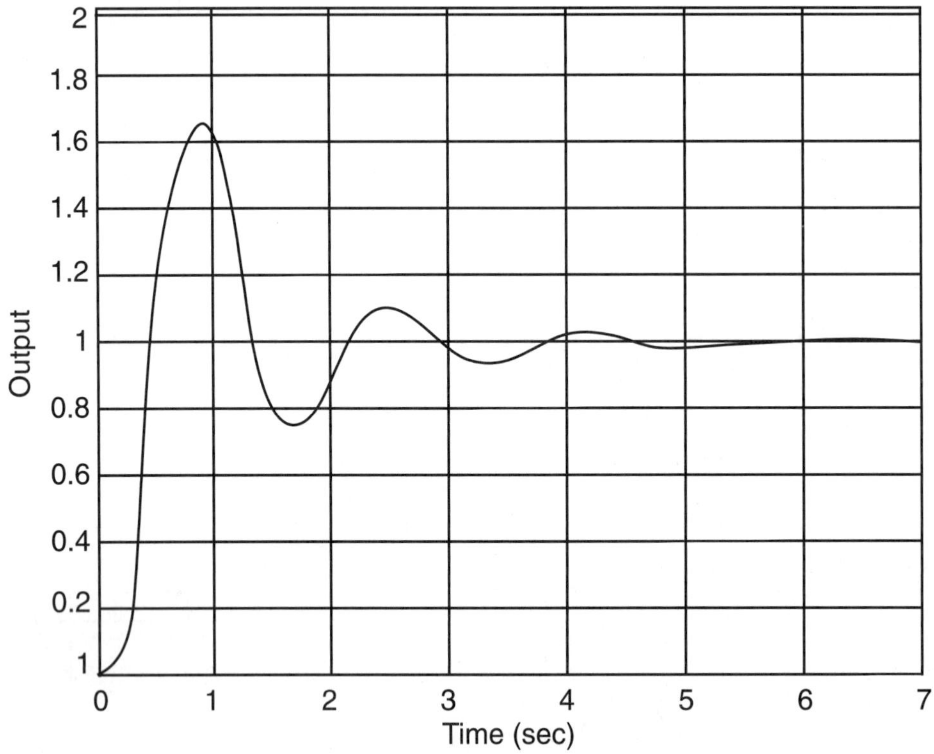

Figure 10–48 Step response using Ziegler-Nichols PID coefficients

Method	P	T_i	Td
Ziegler-Nichols	$0.6\ P_u$ 1.28	$0.5\ T_u$ 0.59	$0.125\ T_u$ 0.148
St. Clair (conservative)	$0.25\ P_u$ 0.533	$1.2\ T_u$ 1.42	$0.125\ T_u$ 0.148

Table 10–6 Ziegler-Nichols and St. Clair coefficients based on $P_u = 2.13$ and $T_u =$ 1.18s, obtained from simulation

10.21 Effect of Differentiating the Setpoint

Comparing Figure 10–39b and Figure 10–40, one can see that changes in the setpoint $r(t)$ are not immediately differentiated in the later implementation. These two implementations exhibit significant differences if the input is changing. We will use

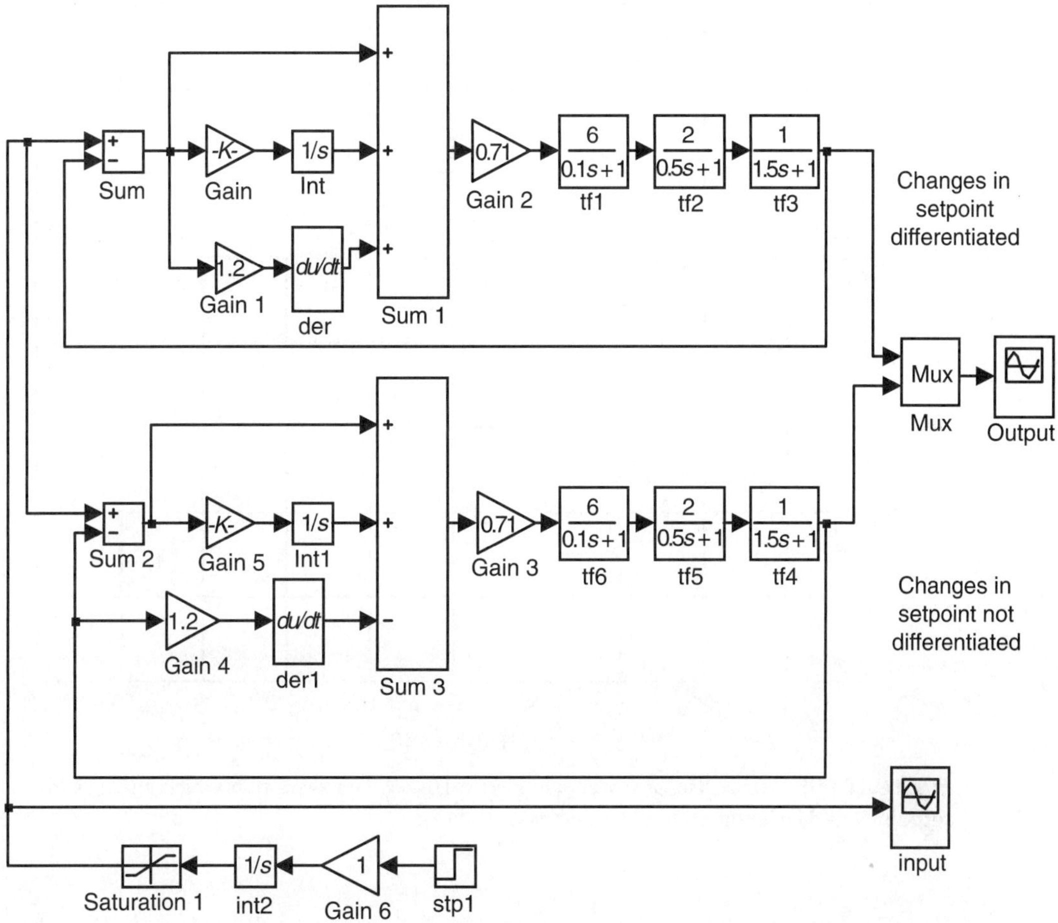

Figure 10–49 SIMULINK diagram showing two *PID* implementations on the same dynamic system: (a) setpoint signal (input) is differentiated and (b) input is not differentiated. The test input signal is a "slanted" step: It increases linearly from 0 to 1 in 1s.

SIMULINK simulations to convey the idea. Figure 10–49 shows the two systems, both excited with the same input. For the system on top, changes in setpoint are immediately differentiated, while for the bottom one, they are not. In the first simulation, the system is excited with a step, and the response of both systems are almost identical (simulations not included here). For the second simulation, the input is a slanted step; the input increases linearly from 0 to 1 in one second. Figure 10–50 shows the response of both systems. Note that the one that differentiates changes in setpoint exhibits much larger overshoot. There are many other techniques for tuning controllers. For a more in-depth study of this field, see the references for Corripio, Astrom, and Wittenmark, listed at the back of the book.[9,10]

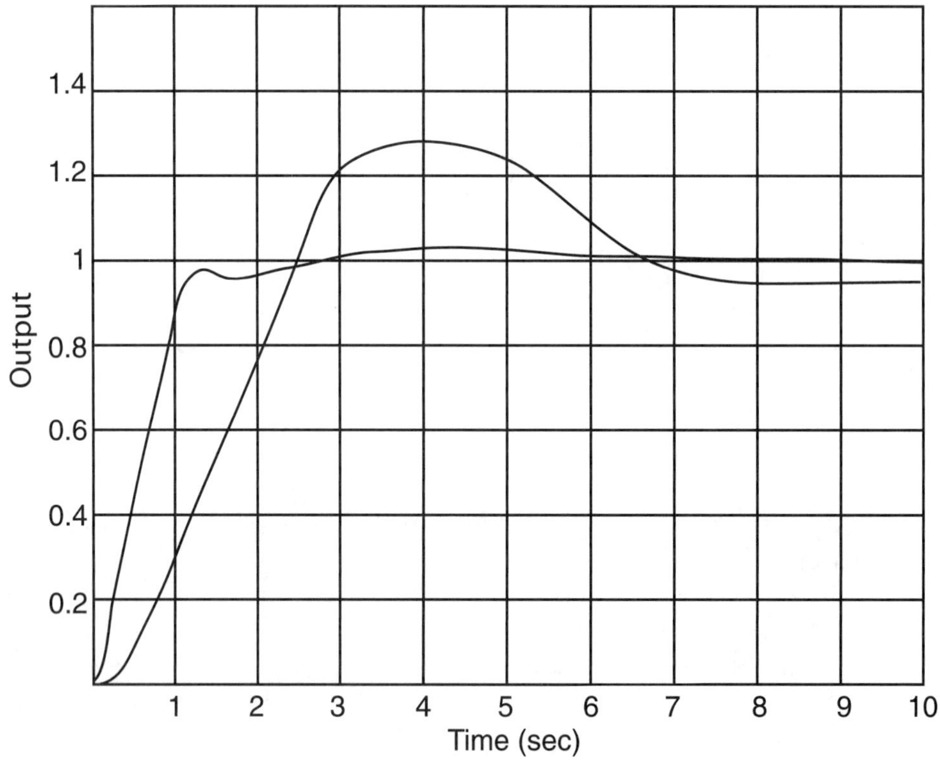

Figure 10–50 Response of systems in Figure 10–49. The system in which changes in setpoint are differentiated has the higher overshoot.

Kim also compares the performance of various techniques. He also presents other tuning techniques that can be performed in the open-loop mode.[11]

Summary

1. Designing a control system is more difficult than analyzing it.
2. There are many approaches for designing a control system. Performance indexes indicate the cumulative departure from some preestablished ideal behavior. They provide an objective means of comparing systems.
3. Lead and lag filters are compensators used to improve the behavior of a dynamic system. Lead filters typically improve dynamic response, while lag filters improve steady-state error.
4. The Bode and root locus plots can be used to design control systems.

Idea Portfolio

1. Propose a controller for this system. What type of controller would you use (PID, lead compensator, lag compensator)? What behavior is desired of the system? Determine what is a desire and what is a need. Translate these devices and needs into a set of specifications.
2. Prepare a formal request-for-bid you would submit to a controls firm so that it can design and build a controller.
3. Estimate how long it would take to complete this project. Create a PERT/GANTT chart that shows the steps and the duration to complete the project.

Problems

1. A closed-loop, position control system is to drive a mechanical system (plant), as shown in the following:

$$G_P = \frac{1}{ms^2 + bs + k}$$

   It uses a hydraulic actuator that has very fast dynamics, so it is simply represented as K_a. This system will experience a command signal resembling step input. Using the ITAE criterion for a step input, set up a system of equations that would yield the optimal controller gain, K_c.

   Use the following values for the other parameters: $m = 5$ kg, $b = 15$ N/(m/s), $k = 65\ N/m$, and $K_a = 0.002\ N/m$.
2. Set up a SIMULINK simulation, including the calculation of the ITAE performance index. Simulate for various values around the value obtained in Problem 1, and confirm your results. (Note: It is possible to set up a DO loop in MATLAB that repeatedly calls SIMULINK.)
3. A system has a plant with the following transfer function:

$$G_p = \frac{10}{\left(\frac{s}{2}+1\right)\left(\frac{s}{5}+1\right)\left(\frac{s}{10}+1\right)}$$

and it is desired that the resulting, compensated system have the following transfer function:

$$G_{eq} = \frac{10}{\left(\frac{s}{5}+1\right)\left(\frac{s}{10}+1\right)\left(\frac{s}{20}+1\right)}$$

Using the algebra design approach, determine the compensator's transfer function. Use SIMULINK to confirm that the compensated system is faster than the original (uncompensated) system.

4. A unity feedback system has the following forward transfer function:

$$G(s) = \frac{400}{s(0.2s+1)}$$

- Determine its PM and GM.
- Using the Bode diagram, design a lead compensator so that the phase margin is increased to close to 40 degrees. Retain the same error coefficient.

5. Design a lead compensator to improve the performance of a system whose forward transfer function is:

$$G(s) = \frac{3}{s^2(s+1)}$$

Your design should provide at least 20 degrees PM.

6. Design a lag compensator to improve the performance of a system whose forward transfer function is:

$$G(s) = \frac{90}{s(0.01s+1)^2}$$

Your design should provide at least 20 degrees PM.

7. Determine the appropriate value of k_s for the system shown in Figure 10–51 so that a minimum phase margin of 25 degrees is obtained. After completing the design, use block diagram reduction to obtain a G_{eq}. Then, perform Bode analysis on G_{eq} and confirm that the adequate phase margin was achieved.

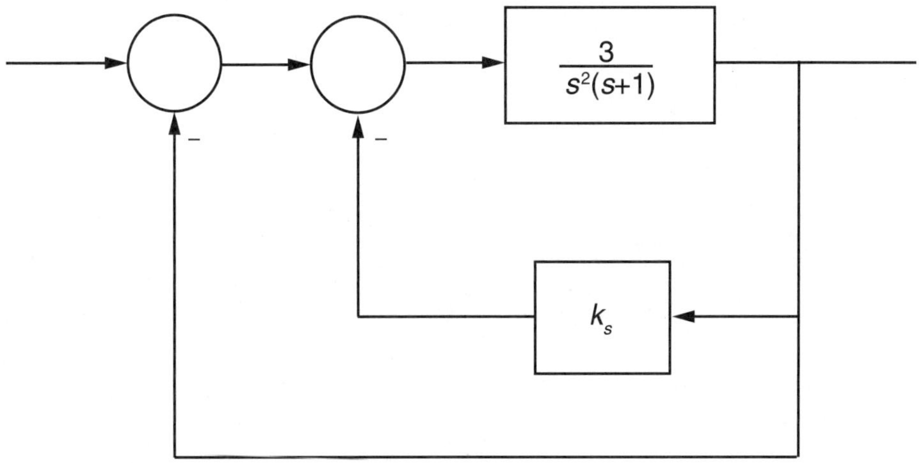

Figure 10–51

8. Repeat the previous design problem, but this time use the root locus method.
9. Design a lead compensator that will half the settling time while retaining the same damping ratio, for a system whose forward transfer function is:

$$G(s) = \frac{10}{(s+1)(s+5)}$$

10. Design a lag compensator to increase the error constant to 5 or the system whose forward transfer function follows:

$$G(s) = \frac{1}{s(s+1)(s+2)}$$

11. Notice that the original, uncompensated error constant (Bode gain) for the previous problem requests to increase the error constant to 5. By what factor will this change the steady-state error?
12. Simultaneously simulate the compensated and uncompensated systems of the previous problem to confirm your design.
13. An ideal instrument servo is to be damped by various feedback. The transfer function of this servo is $625/s^2$, and there are two feedback loops: (1) a unity feedback (for position) and (2) (K_s) for velocity feedback. Maximum overshoot to a step input is to be less than 25 percent. Design the value of K.

14. The transfer functions for a plant and compensator are:

$$G(s) = \frac{K}{s+3} \text{ and } G_c(s) = \frac{s+a}{s}$$

Find a and K so that the overshoot to a step is 4 percent and the settling time is 2 seconds.

15. A unity feedback system has a plant of:

$$Gc(s) = \frac{40}{s(s+2)}$$

and a compensator:

$$Gc(s) = \frac{1+\alpha\tau s}{1+\tau s}$$

For a crossover frequency of $\omega_c = 8$ radians/second and a PM of 45 degrees:
- Determine whether a lead or a lag compensator is needed.
- Find α and τ and the numerical expression for $Gc(s)$.

16. Assume that you are in the process of tuning a PID controller. Say that you increased the system gain until the system became marginally stable, as shown in the following. The gain in this situation was 24. Using the Ziegler-Nichols tuning rules, determine the appropriate PID gains (*P, Ti, Td*), for the Ziegler-Nichols controller implementation.
17. Consider a system whose forward transfer function is:

$$G(s) = \frac{7}{(3s+1)(6s+1)(12s+1)}$$

Using Ziegler-Nichols tuning rules, design a PID compensator. Using SIMULINK, compare the systems response to a step input, before and after compensation.

18. Repeat Problem 14, this time using St. Clair's coefficients.
19. Compare the implementations of Problem 17 and Problem 18 by simulating them side by side. Comment on the differences between the two.
20. Investigate the effects of differentiating changes in set point. Implement side by side, the design obtained in Problem 17 with changes in set point that are:
    - Differentiated (see Figure 10–39b)
    - Not differentiated (see Figure 10–40)
21. Repeat Problem 21, but this time using St. Clair's coefficients.

11

Implementing the Controls Scheme with Hardware: PLCs

Learning Objectives

In this chapter, you will learn how to:

- Identify the various controller technologies available to implement the control function.
- List the advantages and disadvantages of various controller technologies.
- Describe the characteristics and applications of a PLC.
- Implement a PID control task using a PLC.

11.1 Introduction

Now that you have learned about the schemes/control laws needed to achieve a specified control action, the next task is to implement that action by determining what hardware will be used for the control function. There are many hardware implementations, so the selection process is not always easy. The following factors affect the hardware that will be used to perform the control task.

11.2 Factors Affecting System Selection

- Engineer's experience with a system
- Support system
- Unit cost
- Difficulty modifying/expanding the system

- Tolerance to harsh environments (e.g., temperature, vibration)
- Existing software/hardware/development system
- Room for expansion
- Accuracy
- Communication/interface/reporting capabilities
- Customer's existing systems (similar systems already installed)
- Factory support

11.3 Sample Implementation Platforms

The following platforms can be used for implementing the central function, which has been referred to in previous diagrams as the "controller." Selecting the best platform is not always easy since there are many factors affecting the decision-making process. In addition, most engineers have not been exposed to all of the platforms during their education.

Digital Signal Processing Chips

Digital signal processing (DSP) chips are ICs that are useful in high-speed applications, including complex control, self-tuning, or adaptive control systems, video, and audio. They usually require other ICs (memory, D-to-A, etc.) to form a complete controller.

Electronic Analog Circuits

Electronic analog circuits use passive and active electronic components, including resistors, capacitors, and operational amplifiers (op-amps). One can build filters (e.g., low-pass, high-pass, notch) and compensators (e.g., lead compensators, lag compensators, lead-lag compensators). There are op-amps whose gain can be digitally selected.[1]

Fluidics

Fluidics, a technology invented in 1959 can perform analog and digital computation using gases and liquids, with no or few moving parts.[2] Components have been built ranging in size from 5 feet to 0.002 inches. Such components have been used in the aerospace, automotive, industrial, and medical fields, among others. Presently, there is an interest and effort for building microfluidic sensors and circuits using microelectronics manufacturing techniques.

Field Programmable Gate Arrays

A field programmable gate array (FPGA) is similar to a DSP chip but different in that the designer can design the architecture of the device in addition to the program. This allows an architecture that is optimal for the control task.

Mainframe Computer

A mainframe computer is particularly useful when the application involves a large and distributed dynamic system, such as a refinery.

Microcontrollers

Microcontrollers are computer-on-a-chip types of devices, usually using 8-bit words. They are very inexpensive and used extensively in the automobile industry. It is estimated that a high-end car might have as many as thirty microcontrollers that control various functions on the automobile. Some commonly used microcontrollers include:

- HC11, HC12, 6805 (Motorola)
- 8051 (originally Intel, now Dallas Semiconductor, and others.)
- PIC family (MicroChip)

PC Based

Controllers using the PC as a base can take advantage of all the features already in the PC environment. Many manufacturers allow the use of software tools in their operating environment. Examples are:

- LabView
- MATLAB Realtime Toolbox
- Data Translation
- LabTech Notebook, Control

PLCs

PLCs presently provide an excellent platform for performing control tasks. Not only are they relatively easy to program, they merge analog control and discrete control in the same piece of hardware. They are used in many industries, such as automotive, injection molding, manufacturing, and food processing. There is a growing range of capabilities and prices. Some are being used in such applications as controlling the handicap ramps in city buses.

Pneumatics and Hydraulics

For many years, pneumatics and hydraulics dominated the industrial control market. Today, they play an important role, even though a large portion of their previously dominated market share has been taken over by electronic devices. Pneumatics continues to be a leader in low-force actuation, while hydraulics dominates in the field of high-load applications.

11.4 Comparison of Various Approaches for Implementing the Control Task

Selecting an approach is not always easy. We have produced a comparison that hopefully helps you get a feel for the different approaches. Table 11–1 shows approximate unit cost, difficulty to modify the program, additional hardware required (e.g., enclosure, power supply), and ruggedness (tolerance to harsh environments, such as temperature, vibration, and radiation, for various types of controllers). These are the personal opinions of the author and many change quite rapidly due to the evolving nature of today's technology and the perspective of the evaluator.

11.5 Introduction to PLCs

PLCs: The Backbone of the Manufacturing Plant

The PLC is often called the backbone of any manufacturing operation. Visit any automotive plant and you will see hundreds of PLCs connected to one another and possibly to a supervisory computer or PLC. The PLC provides extensive

Controller Type	Unit Cost ($)	Development Time 1 = low 10 = high	Difficulty to Modify 1 = low 10 = high	Additional Hardware Required ($) 1 = low 10 = high	Ruggedness (Tolerance for Harsh Environments) 1 = low 10 = high
DSP	15	6	7	5	4
ANALOG ELEC.	5	6	9	5	5
FLUIDICS	100	8	10	6	10
FPGA	20	7	7	5	4
HUMAN OPERATOR	25,000	1.5	3	1	5
MAIN FRAME	35,000	8	5	1	1
MICRO CONTROLLER	10	7	9	5	5
PC BASED	1,500	2	5	1	3
PLC	550	2.5	1	3	5
PNEUMATICS HYDRA	700	8	10	6	7

Table 11–1 Various platforms for implementing a control task with hardware

communication and display capabilities. It provides a programming language technicians and electricians easily understand.

PLC History Arizona State University East Professor Edward McBrien experienced firsthand the development of the PLC and contributed the following background information on the PLC. The modern PLC has developed through the natural evolution and convergence of various disciplines. The desire of consumers to obtain lower-cost and higher-quality goods, coupled with the desire of workers to obtain higher wages, provides the incentive for manufacturers to find improved ways to produce useful goods and services.

The availability of relatively low-cost electrical energy enabled manufacturers to develop more complex equipment controlled by push buttons, limit switches, and relays. As production rates rose and product tolerances became tighter, keeping the moving parts of these items in good adjustment became increasingly more difficult. Even devices designed for millions of operations frequently caused production shutdowns.

In the middle of the last century, manufacturers were developing various types of magnetic and capacitative types of sensors to replace limit switches. Both Westinghouse Electric and General Electric developed magnetic amplifier systems that could implement the AND, OR, NOT, and LATCH functions performed by relays. As transistor costs decreased, the Square-D company developed a transistorized NOR module that could perform any of the logic functions by placing jumpers appropriately. The major weakness of these systems was that they required special power supplies and could not drive the associated loads directly. The development of silicon controlled rectifiers (SCR) and Triacs has eased the load-interfacing problem.

The development of printed circuits and integrated circuits enabled designers to further reduce costs and produce more complex control circuits. This set the stage for the arrival of the microprocessor and made the PLC possible. Early PLCs merely evaluated a series of Boolean statements and changed outputs according to input conditions.

The PLC revolutionized control system design procedures. Before the arrival of the PLC, the designer prepared a ladder, or logic diagram, and converted it to a wiring diagram that could be used in the shop or field. Any change in the system required a revision of both types of diagrams and detailed instructions to the shop. With PLC control, the procedure is reversed. The shop can connect inputs and outputs to the most convenient set of terminals, and the designer merely needs to enter the proper terminal designation into the program. Any system changes merely require changes in the program. Only the addition or deletion of equipment requires field changes.

By another account, PLCs began their use specifically in the automotive industry.[3]

In 1966, the Hydramatic Division of General Motors generated the first specification for a programmable controller. In 1969, Gould (then Modicon) delivered the first production PLCs to General Motors. The concept met with instant acceptance and, in 1971, the first PLC outside the automotive industry went into production.

These units were little more than very sophisticated switches, but as with most electronic devices, their capabilities and sophistication grew rapidly. In 1973, so-called "smart PLCs" became available. These devices could compare input and output (I/O) signals (feedback), and interface with printers, CRTs, and other devices. By 1975, analog PID PLCs could interface with analog input devices, such as thermocouples and pressure transducers. Previous devices could respond only to off-on signals, such as limit switches and optical sensors.

Soon, additional companies were involved in the PLC business, including Allen-Bradley, Modicon, and Reliance Electric (later bought by Allen-Bradley).

The PLC revolutionized the way in which industrial controls were implemented. Originally, the set of specifications dictated the wiring scheme. The thinking here included how the input switches would activate the coils in relays (or contactors) and then other switches or relays. This scheme was converted to a ladder logic wiring diagram (because it looked like the rungs of a ladder) and given to electricians to implement. A small change in the specification would force a complete repetition of this procedure.

As an example of this type of wiring, consider a two-switch press. Two switches are used for safety to make sure that the operator has both hands away from the workpiece. Consequently, hands will be unharmed. Figure 11–1 shows the wiring diagram.

As logic and programming devices became available, those in the PLC business decided to capitalize on them. A huge population of engineers, electricians, technicians, and mechanics understood wiring diagrams. Consequently, the developers proposed the ladder logic programming (graphical code that looks like a wiring diagram) as the "language" to program the PLC. This approach permitted the large, existing population of technical personnel to continue using their vast knowledge of wiring. However, some enhancements had to be made to the original ladder logic wiring diagram function/logic to make it amenable to the new programming language. The new logic operation can be described as the original wiring diagram put inside an infinite do-loop. This is shown in Figure 11–2. Note that there are now other more sophisticated languages used to program PLCs. In addition to ladder logic programming, considered a low-level programming language, there are other alternatives for the more powerful PLCs.

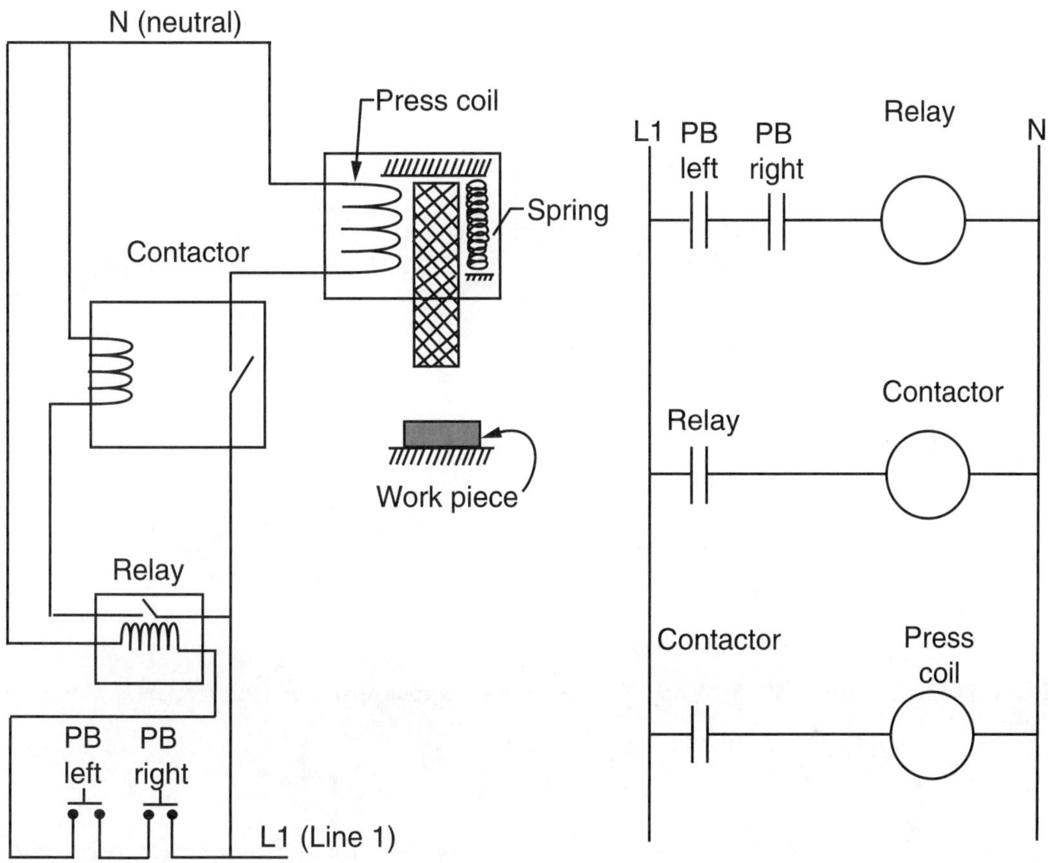

Figure 11–1 Two-switch (for operator safety) press wiring schematic and diagram

For the Rockwell Automation family (the parent company of both Allen-Bradley and Rockwell Software), these high-level programming environments include:

- Sequential flow charts (SFCs), lending themselves to batch-type processes
- Function blocks, lending themselves to control-task implementation for proven control applications by means of a graphical interface that allows hierarchical conditions and custom control rules to be specified
- Structure text, where the programming takes the form of native languages (e.g., Visual Basic, C, and C++)

Perhaps the most attractive feature of the new programming environments is the ability to mix the preceding programming modes with ladder logic in the same application.

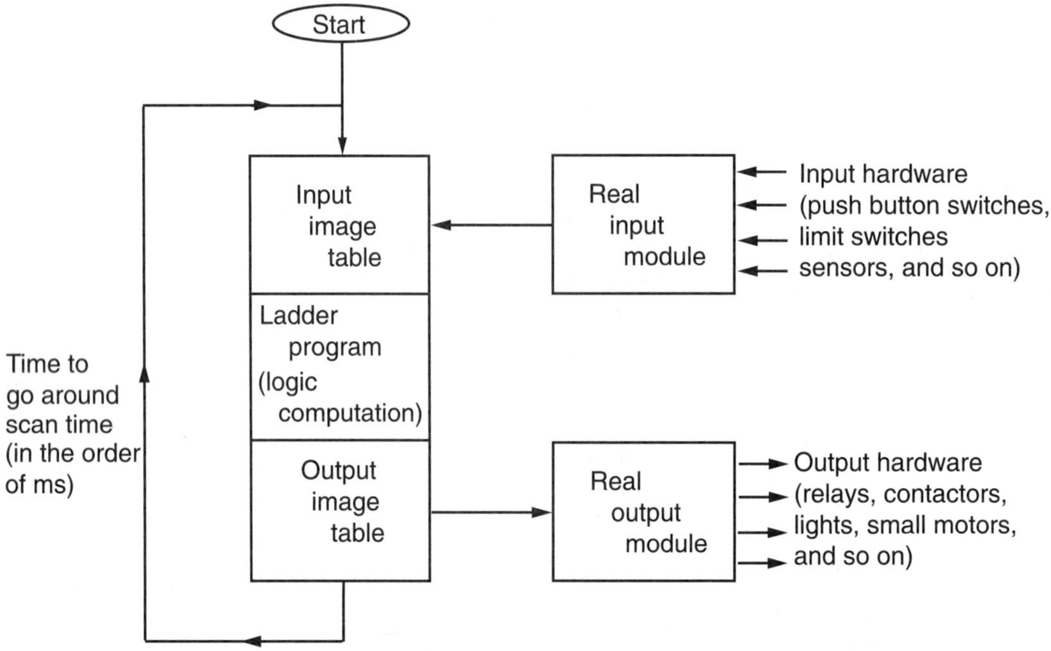

Figure 11–2 Schematic of a PLC input/logic processing/output operation (scan cycle)

PLC Wiring To do wiring of the control task for the two-switch press, the process becomes much simpler. Figure 11–3 shows the ladder logic diagram (program) and the wiring diagram.

PLC Type in This Book There are many available types of PLCs. This book focuses on particular types and manufacturers: the SLC500 (Small Logic Controller) family and the MicroLogix family,[4] made by Allen-Bradley (Madison, Wisconsin). Both use the same programming software, the RSLogix 500 (Rockwell Software, Milwaukee, Wisconsin). Figure 11–4 shows them and some of the optional add-on modules.

So far, the PLC's ability to perform on-off control has been presented. The next section addresses the use of the PLC for performing closed-loop PID control. The SLC500 and MicroLogix PLCs in the RSLogix 500 environment use the PID instruction as one of the "steps" in carrying out the control task. It is not necessary to know all the instructions of ladder logic programming to be able to perform closed-loop control.

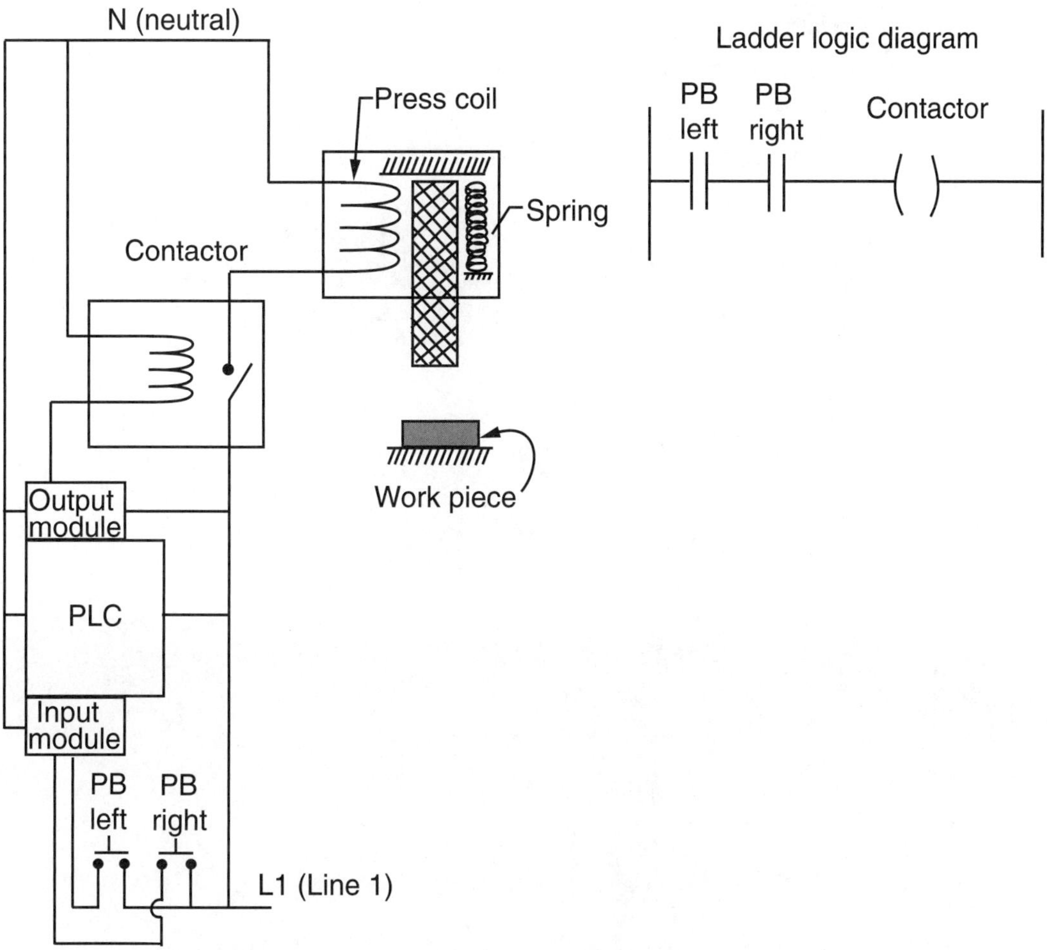

Figure 11–3 Two-switch (for operator safety) press circuit, using a PLC

The PID function can be implemented in all the pictures in this family: the SLC500s and the MicroLogix 1000, 1200, and 1500. For those interested in evaluating the results of one PID implementation over another, the MicroLogix 1500 LRP is particularly well suited. It has built-in memory for data acquisition, which allows retrieval of process variables.

This company produces controllers that tremendously expand the capabilities for controlling feedback continuous process. Some of these controllers (e.g., Control Logix, Process Logix) can perform much more sophisticated filtering and compensation tasks. In addition, they can execute autotuning tasks that produce optional PID coefficients.

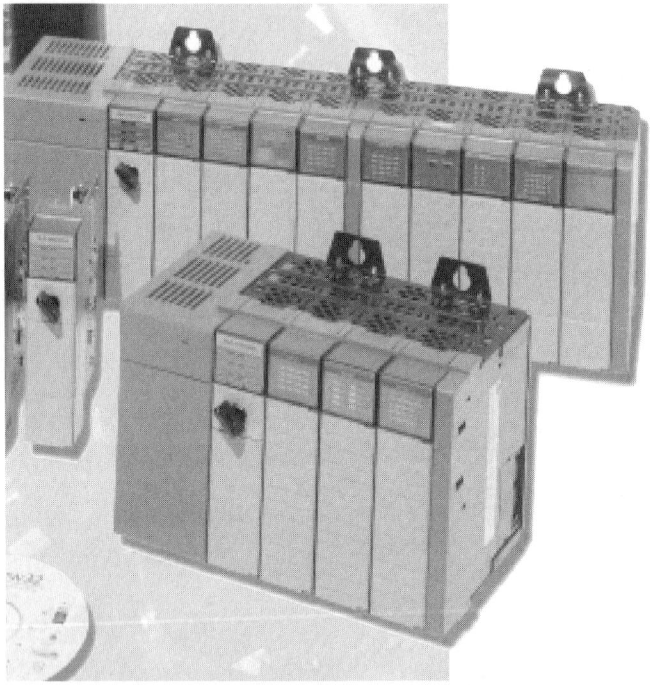

(a)

(b)

Figure 11–4 Two examples of Allen-Bradley's PLCs: (a) SLC500 family and (b) MicroLogix 1500 (Courtesy of Allen-Bradley)

11.6 Implement PID Closed-Loop Control with the PLC

As indicated in the previous section, there are many platforms in which the control function can be implemented. We have chosen to use the PLC for several reasons:

- It is one of the most common controllers in industry. Its tolerance to industrial environments, and its programming and communication versatility, make it an ideal choice. It is possible to program and monitor a PLC through the Internet.
- It gives the user an accessible, hands-on environment for setting up, programming, and tuning a feedback control system.

- In addition to continuous control, it can perform on-off control actions, which constitute most situations in industry, extremely well.
- It provides a programming environment that is easy and understandable to most engineers, particularly for those whose educational background is not computing or programming. The necessary programming language is minimal for a PID loop and easy to learn.

The main purpose for introducing the PLC in this book is to implement PID control, discussed in Chapter 10, Equations 10.117 and 10.118. One way of thinking about a PID controller in a digital machine is to envision a computing block embedded in an infinite do-loop, which performs the digital-equivalent calculation (see Chapter 12) on the error:

$$\text{output of PID} = K_P e + K_I \int e \, dt + K_D \frac{de}{dt} \qquad \textbf{Eq. 11.1}$$

where K_P is the proportional gain, K_I is the integral gain, and K_D is the derivative gain. The preceding equation is most often realized as:

$$\text{output of PID} = P_P \left[e + (1/T_i) \int e \, dt + T_d \frac{de}{dt} \right] \qquad \textbf{Eq. 11.2}$$

where P_P is the proportional gain, T_i is the reset term, and T_d is the rate term. These last sets of parameters are used in the Allen-Bradley PID instruction.

Basic Control Nomenclature

Industry uses PLCs heavily, especially for on-off control. A subset of this group uses the PLC for batch processing, which includes on-off and continuous (process) control. Common applications include:

- A dairy, to pasteurize milk and convert it to its various by-products (e.g., cheese, eggnog)
- A pasta manufacturing plant
- A distillation column
- The control of an oven used in the microelectronics industry
- The ramp of a bus that lifts a wheelchair-bound passenger

Historically, PLCs have been used more to perform process control tasks than the type of control activities presented to date (i.e., position control). As a result, the nomenclature used by PLC manufacturers reflects the terminology used in the process control industry. Familiarize yourself with this nomenclature, because it will be used in the programming environment.

The following application introduces this terminology. Consider the level control system in Figure 11–5. In this system, the valve is open or closed to maintain the level constant, despite a varying input flow (disturbance). Notice that the signs of the signals arriving to the summing junction are deliberately left out. These signs can be set via software to accommodate a valve that closes with increasing input or opens with increasing input. These actuators are referred to as forward acting or reverse acting.

Now we will relate the variables in Figure 11–5 to the nomenclature used earlier throughout this book:

Set point—The set point is the line desired value of the output. It has been called earlier $r(t)$ (or $R(s)$).

Process variable—The process variable is the output of the system, the variable we wish to control. We can talk about the physical quantity (level) or the signal produced by the level sensor (level detector), typically a voltage. Later, we will establish a nomenclature that will allow us to differentiate between them.

Error—Error is the difference between the set point and the process variable. If the system produced an output identical to our command, the error would be 0. In our

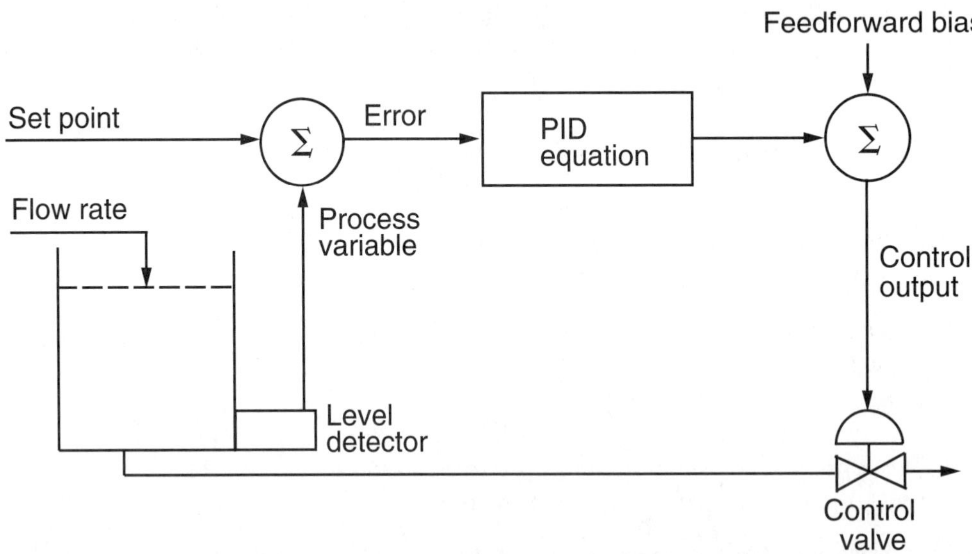

Figure 11–5 Closed-loop system using process control terminology (adapted from the reference manual for MicroLogix 1200 and MicroLogix 1500 programmable controllers.)

earlier treatment, we have called the error $e(t)$ (or $E(s)$). We learned in Chapter 4 that a type 0 system exhibits a small steady-state error for a constant set point. In contrast, a type 1 system exhibits 0 error for a constant set point. When the PID equation contains an I term, it will also have no steady-state error.

Feed-forward bias—The feed-forward bias is a constant "fudge" factor that improves steady-state error for a particular type of input, for a type 0 system. Even if an integral term is part of the control equation, this feed-forward bias decreases the necessary contribution of the integral term. This, in turn, minimizes problems with windup (a case in which it is difficult to reduce the contribution of the integrator quickly). Therefore, feed-forward bias is useful in systems that are naturally skewed in one direction. An example is a position control system that is holding a load (weight) horizontally. It might help to introduce a feed-forward bias that overcomes the weight.

Control output—Control output is the output of the controller, typically a voltage, usually to some sort of driver that interacts with the physical system. In our previous treatment, we have not issued a name for this variable. In the PID presentation in this chapter, this is $o(t)$ (or $O(s)$).

Control valve—The term *control valve* refers to the physical component in the system that can change the output. The specific control element used, of course, depends on each system.

Input Signal Types and Ranges

The *A*/*D* conversion, which occurs in the Analog Input modules of the MicroLogix family, allows the user to select from various ranges in current or voltage. These settings are specified through software, as well as in the specific wiring termination at the PLC. Available ranges are:

Current

4 to 20 mA

0 to 20 mA

Voltage

−10V to 10V

0 to 5V

0 to 10V

1 to 5V

Filter Frequency (Hz)	Cut-Off Frequency (Hz)	Step Response (mS)
50	13.1	60
60	15.7	50
250	65.5	12
500	131	6

Cut-off frequency corresponds to the -3 dB point. Step response is the time required for the analog input to reach 100 percent of its expected final value.

Table 11–2 Characteristics of digital filters

Low-Pass Filtering

Because most real-world signals have noise, especially in an industrial environment, the MicroLogix family provides a digital low-pass filter. The user can select the sampling and the cut-off frequencies (choices appear in Table 11–2). Bode plots for these filters appear in Figure 11–6.

Internal Representation of Analog Signals

To facilitate ease of use, Allen-Bradley has the following forms for representing analog numbers internally:

- Raw/proportional data
- Engineering unit
- Scaled-for-PID
- Percentage full range

Each of these has its own numerical representation inside the PLC. For integer arithmetic, the SLC 500 and MicroLogix families use 16 bit, signed integer architecture. The most significant bit (MSB) carries the sign bit, and the remaining 15 bits carry the magnitude of the number. This allows a range of −32,768 to 32,767 (approximately $+/-2^{15}$). Table 11–3 shows the many choices. For our example, we will focus on the specific internal representation that is most closely geared toward PID, specifically:

Type of signal	Single-sided voltages
Range	0–10V
Internal representation	Scaled-for-PID

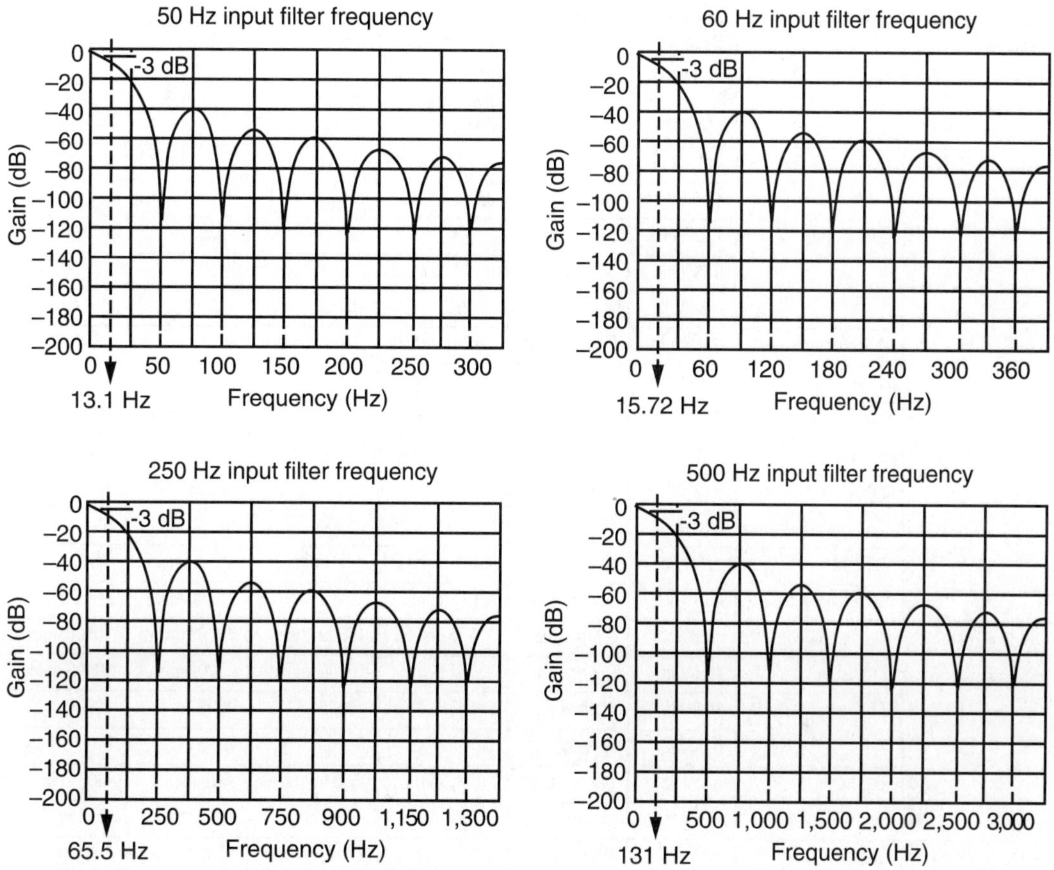

Figure 11–6 Available digital filters in the MicroLogix 1500

The internal representation of the scaled-for-PID selection is:

Input	internal integer representation inside PLC
0V	0
10V	16,383

Note that the actual internal numerical range is much wider (−32,768 to 32,767). However, the purpose of operating in a subset of the range is to provide ample computational room that would prevent tripping an overflow or underflow condition. The occurrence of an overflow or underflow condition could halt the processor and stop PLC operation. In PLC nomenclature, an overflow or underflow event

1769-IF4 Input Range	Input Value	Example Data	Input Range Condition	Raw/ Proportional Data	Engineering Unit	Scaled-for-PID	Percent Full Range
				Decimal Range	Decimal Range	Decimal Range	Decimal Range
-10V to +10V dc	Over 10.5V dc	+11.0V dc	Over-range	32767 (max.)	10500 (max.)	16793 (max.)	N/A
	+10.5V dc	+ 10.5V dc	Over-range	32767 (max.)	10500 (max.)	16793 (max.)	N/A
	-10V to +10V dc	+10.0V dc	Normal	31206	10000	16383	N/A
		0.0V dc	Normal	0	0	8192	N/A
		-10.0V dc	Normal	-31206	-10000	0	N/A
	-10.5Vdc	-10.5V dc	Under-range	-32767 (min.)	-10500 (min.)	-410 (min.)	N/A
	Under -10.5V dc	-11.0V dc	Under-range	-32767 (min.)	-10500 (min.)	-410 (min.)	N/A
0V to 5V dc	Over 5.25V dc	5.5V dc	Over-range	32767 (max.)	5250 (max.)	17202 (max.)	10500 (max.)
	5.25V dc	5.25V dc	Over-range	32767 (max.)	5250 (max.)	17202 (max.)	10500 (max.)
	0.0V dc to 5.0V dc	5.0V dc	Normal	31206	5000	16383	10000
		0.0V dc	Normal	0	0	0	0
	-0.5V dc	-0.5V dc	Under-range	-3121 (min.)	-500 (min.)	-1638 (min.)	-1000 (min.)
	Under -0.5V dc	-1.0V dc	Under-range	-3121 (min.)	-500 (min.)	-1638 (min.)	-1000 (min.)
0V to 10V dc	Over 10.5V dc	11.0V dc	Over-range	32767 (max.)	10500 (max.)	17202 (max.)	10500 (max.)
	+10.5V dc	10.5V dc	Over-range	32767 (max.)	10500 (max.)	17202 (max.)	10500 (max.)
	0.0V dc to 10.0V dc	10.0V dc	Normal	31206	10000	16383	10000
		0.0V dc	Normal	0	0	0	0
	-0.5V dc	-0.5V dc	Under-range	-1560 (min.)	-500 (min.)	-819 (min.)	-500 (min.)
	Under -5.0V dc	-1.0V dc	Under-range	-1560 (min.)	-500 (min.)	-819 (min.)	-500 (min.)
4 mA to 20 mA	Over 21.0 mA	22.0 mA	Over-range	32767 (max.)	21000 (max.)	17407 (max.)	10625 (max.)
	21.0 mA	21.0 mA	Over-range	32767 (max.)	21000 (max.)	17407 (max.)	10625 (max.)
	4.0 mA to 20.0 mA	20.0 mA	Normal	31206	20000	16383	10000
		4.0 mA	Normal	6241	4000	0	0
	3.2 mA	3.2 mA	Under-range	4993 (min.)	3200 (min.)	-819 (min.)	-500 (min.)
	Under 3.2 mA	0.0 mA	Under-range	4993 (min.)	3200 (min.)	-819 (min.)	-500 (min.)
1.0V to 5V dc	Over 5.25V dc	5.5V dc	Over-range	32767 (max.)	5250	17407	10625
	+5.25V dc	5.25V dc	Over-range	32767 (max.)	5250	17407	10625
	1.0V to 5.0V dc	5.0V dc	Normal	31206	5000	16383	10000
		1.0V dc	Normal	6243	1000	1	1
	0.5V dc	0.5V dc	Under-range	3121 (min.)	500	-2048	-1250
	Under 0.5V dc	0.0V dc	Under-range	3121 (min.)	500	-2048	-1250

Table 11–3 Valid Input Data

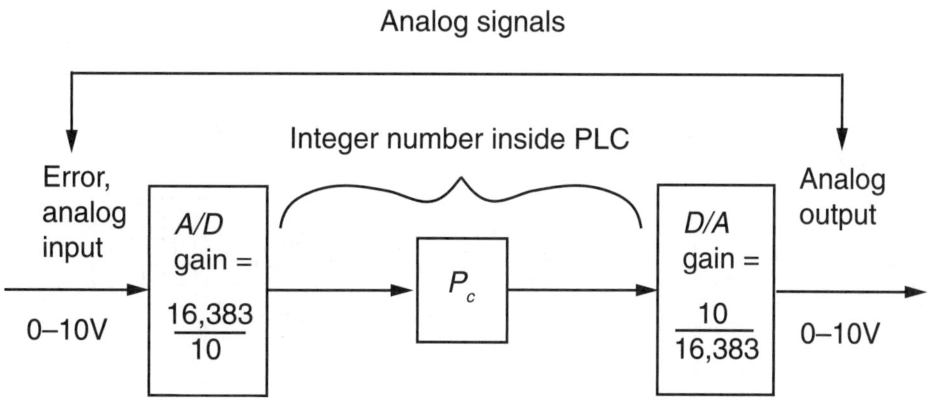

Figure 11–7 Analog and integer representation inside PLC

"faults" the PLC. On the other hand, it is possible to program an "error trap" that would allow the PLC to continue operating despite an overflow or underflow condition.

Does the numerical value of the gain implemented inside the PLC differ from the gain used in the simulation? The answer is no. Consider the process in Figure 11–7. It shows the *A/D* process, the amplification of the error by the gain *P* (see PID Equation 11.2), and, finally, the *D/A* conversion. Notice that the gains associated with the *A/D* and *D/A* cancel each other out, producing a product equal to 1. Thus, *Kp* does not need to be altered as a result of the integer environment in which it exists. Allen-Bradley documentation refers to the gain *P* as *Kc*.

Overall Programming and Monitoring Environment

It might help to glimpse how the PLC communicates and interacts with other equipment. Interaction depends strongly on the communication protocol used, because Allen-Bradley offers numerous communication protocols, allowing one PC to communicate with many networked PLCs. For the sake of simplicity, we will focus on a MicroLogix 1500 LRP, which has two serial RS232 communication ports (DF1 communication protocol, in Allen-Bradley communication language). One of the ports (Port 0) is connected to the PC, the other (Port 1) to a human-machine interface (HMI), Panel View 330. HMIs are also called man-machine interfaces (MMIs). Notice that the RS232 protocol is a port-to-port connection. The interrelation of these components appears in Figure 11–8.

The PC provides the environment in which the PLC can be programmed using a high-level, graphical programming environment (RSLogix 500). RSLogix

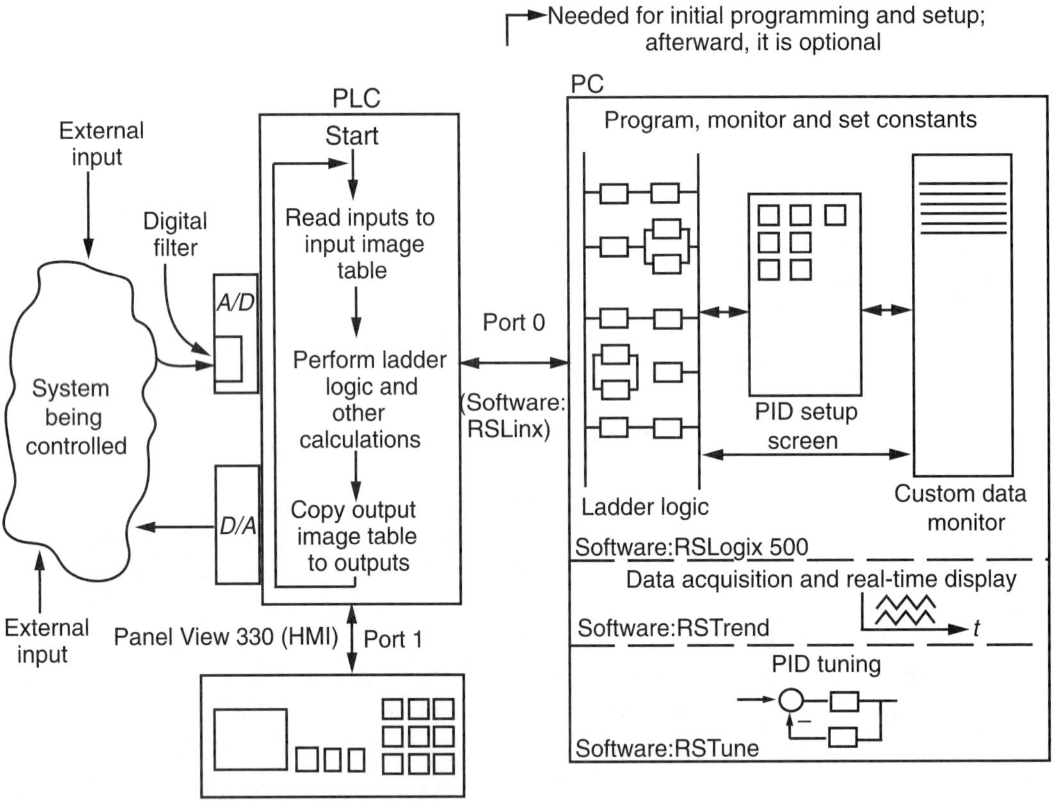

Figure 11–8 Relationship between system, PLC, and PC

can also be used to monitor what is happening inside the PLC and to alter constant values in the PLC memory while the PLC is "running" ("RUN" mode, in Allen-Bradley nomenclature). Additional software packages allow data acquisition and real-time display and trending (RS Trend). Also, RSTune facilitates the tuning of a PID control loop. In addition, the MicroLogix 1500 LRP has some internal memory designated for data acquisition and storage. This feature allows the PLC to gather and store a significant quantity of data, which can be retrieved later.

When a program is created in the PC, the process of sending it to the PLC is called "downloading" the program. When a program that resides in the PLC is brought into the PC, the process is called "uploading." (Note: This nomenclature is not the same in other manufacturer's PLCs.) When this operation is done, any comments or labels in the original programming will not be recovered, because that support information does not reside in the PLC memory, only in the programming environment.

In an effort to facilitate the implementation of a PID loop, Allen-Bradley has made significant efforts to make the development as user-friendly as possible, especially to the operator who is overseeing the PLC operation. With this in mind, the company has implemented various ways of representing the process variables. To illustrate this, consider a position control system whose job is to move a part within a range of 0 to 500 mils (0 to 0.5 in). Assume that a linear potentiometer, having the range of one inch, is used to measure the position of the part. Because the PID instructions are to be used in this ladder logic program, the "scaled-for-PID" internal representation will be used in the *A/D* and *D/A*. This position signal, which in our case is the process variable, can be represented in the following ways:

Actual mils (*PV*)	0–500
Percent of range (*S_PV*)	0–100
Output voltage produced by the position sensor (*E_PV*)	2.29V–9.77V
Internal, integer number produced by the *A/D* in the PLC (*I_PV*)	3,750–16,000

An analog amplification circuit could be included, but for this application it is unneeded. It is obvious that the programmer has the option to decide what would be the representation most understandable to the operator. Say that in our position control example the "percentage of range format" is most reasonable. Another choice is to use the units that reflect the numerical value of the *PV*. This type of representation is called "scaled engineering units."

In doing this, the programmer must perform some calculations for communicating to the PLC the relationship between the numbers displayed to the operator and those used internally by the PLC in performing the computations.

We propose the following notation to distinguish among the various representations of a given variable. The naming convention for the variables includes up to three components:

1. A qualifier indicative of the form in which the variable appears:
   - No prefix, indicating the actual physical variable.
   - *E*, for the electronic (analog) representation, which is usually the signal coming out of the sensor.
   - *I*, the integer number produced by the *A/D*
   - *S*, for scaled, which could be in units that facilitate the understanding of the operator. A choice of either "percentage of range format" or "scaled engineering units" is available.
2. A qualifier indicative of the location in the control loop where it appears:
   - *SP*, for set point.
   - *E*, for error.
   - *CV*, for control output variable.
   - *PV*, for process variable.

3. A qualifier that captures the name or nature of the variable (optional). This might be very useful when the system includes multiple inputs and outputs.

Following are some examples that do not include name/nature qualifiers:

Input

PV	Process variable (actual) (e.g., mils if we are working with displacement).
E_PV	Electronic signal produced by the sensor measuring the process variable.
I_PV	Internal process variable, as it appears inside the PLC. It occurs in the form of an integer that can range in value from 0 to 16,383 (scaled-for-PID representation).
S_PV	Scaled process variable. The units of this variable can be anything that would facilitate understanding by the operator.

For integer arithmetic the SLC 500 and MicroLogix families use a 16 bit, signed integer architecture. The most significant bit (MSB) carries the sign bit and the remaining 15 bits carry the magnitude of the number. This allows a range of $-32{,}768$ to 32,767 (approximately $+/-2^{15}$).

The variable *S_PV* warrants some additional clarification; several alternatives are available for its definition. If the numerical value of the PC is adequate to internal integer representation of the PLC, then *S_PV* can assume the same numerical values. For example, if the *PV* ranges from 10 to 15 psi, a 1-to-1 correspondence would be inappropriate, because a 0-to-15 would provide insufficient resolution. Remember that we are using an internal integer representation. The programmer could overcome this problem by choosing an *S_PV* that is 100 times the *PV*. For instance, a *PV* of 3 psi would be represented as 300 and a *PV* of 15 psi as 1,500. In contrast, the PLC programmer might choose to represent the process variable using a percentage of full range (*P_PV* notation). Using this approach, *PV* of 3 psi would be represented as 20 percent [from $3/15 \times 100 = 20$ percent]. Clearly, each approach has its strength and weakness. The first approach provides insight into the working variables of the physical process but does not communicate the relative position between the normal and abnormal operating ranges. The second approach indicates the value with respect to the operating range but fails to provide insight into the actual value of the variable. The scenario is similar to the temperature gauge in your car. Do you prefer a dial indicator (gauge) that indicates actual temperature or one that positions a needle in a range composed of a normal (black) and overtemperature (red) region? The operator who understands the specifics of the thermodynamic process may better appreciate the first approach, while another might prefer the second method.

Output

P_CV This is the control variable output of the PLC's *D/A*, expressed as a percentage of maximum. Because we are using 0–10V output, *P_CV* = 100 corresponds to an output of 10V and *P_CV* = 25 corresponds to an output of 2.5 V.

I_CV This is the control variable as it appears inside the PLC. It occurs in the form of an integer that can range in value depending on the internal representation selected. For example, for "scaled-for-PID," the range is from 0 to 16,383.

With PID instruction, one can work with *S_PV* or the *I_PV*. With the first approach, all the dialog with the PID instruction occurs using easy-to-relate, scaled engineering units (either same units as the *PV*, a 100X multiple of the *PV*, or a percentage of full range). With the second approach, one must deal with the number produced by the *A/D* in the range of interest. In the following, the first approach is used, namely *S_PV* expressed as a percentage of range. Therefore, *S_PV* of 50 percent corresponds to a *PV* of 250 mils. Notice that we could have chosen the *S_PV* to have the same units as the *PV*, but we chose not to.

For the position control example, Figure 11–9 shows the relationship between the scaled process variable, *S_PV*, and the internal process variable, *I_PV*. Remember that the *S_PV* is the number that is meaningful to the operator; *I_PV* is the interval integer number used by the PLC in its computations (produced by the *A/D* of the PLC). Allen-Bradley provides an instruction SCP (scale with parameters) that facilitates this transformation. As suggested previously, it might be desirable, especially when there are several process variables, to add another qualifier to designate the name of the process variable. For instance, *S_PV_Wax_Temp* would denote "Scaled Process Variable Wax Temperature."

Next, it is necessary to determine the *A/D*'s output for the entire range in the process variable. The *PV* is measured by an independent device (e.g., a micrometer or a gauge block, for the displacement example). (If this is not possible, go to the sensor data sheet and calculate what will be the resulting voltage for a given input.) The *I_PV* is recorded by looking at the *A/D* output in the input data table (or a custom data monitor). For our example of displacement measurement, the endpoints of the *PV* are:

$S\_PV_1 = 0$ percent $\quad (PV_1 = 0 \text{ mils}) \quad I\_PV_1 = 3{,}750$

$S\_PV_2 = 100$ percent $\quad (PV_2 = 500 \text{ mils}) \quad I\_PV_2 = 16{,}000$

Because we are sure that the sensor is producing a voltage the *A/D* can read, we have ignored *E_PV*. Once (S_PV_1, I_PV_1) and (S_PV_2, I_PV_2) are known, S_{min} and S_{max}, the numbers required by the PID screen can be calculated. Figure 11–10

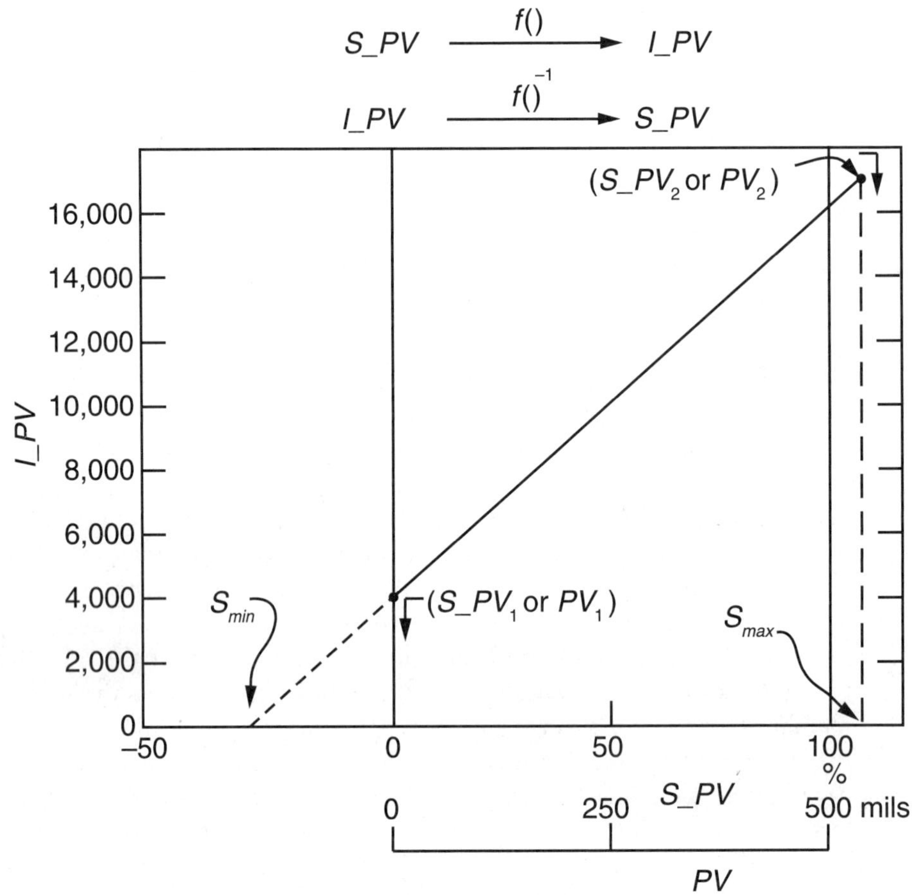

Figure 11–9 Relationship between *PV*, *S_PV*, and *I_PV*

shows the setup screen for the PID block in the MicroLogix 1500. S_{min} and S_{max} correspond to the *S_PV* that would cause the *A/D* to produce outputs of 0 and 16,383. The *PV* might not be able to possess this entire range due to physical constraints, but the PLC uses these numbers as a reference.

The relationship between *I_PV* and *S_PV* (see Figure 11–9) can now be represented as:

$$I\_PV = m^*S\_PV + b \qquad \textbf{Eq. 11.3}$$

Thus, m and b can be calculated from:

$$m = (I\_PV_2 - I\_PV_1)/(S\_PV_2 - S\_PV_1) \qquad \textbf{Eq. 11.4}$$

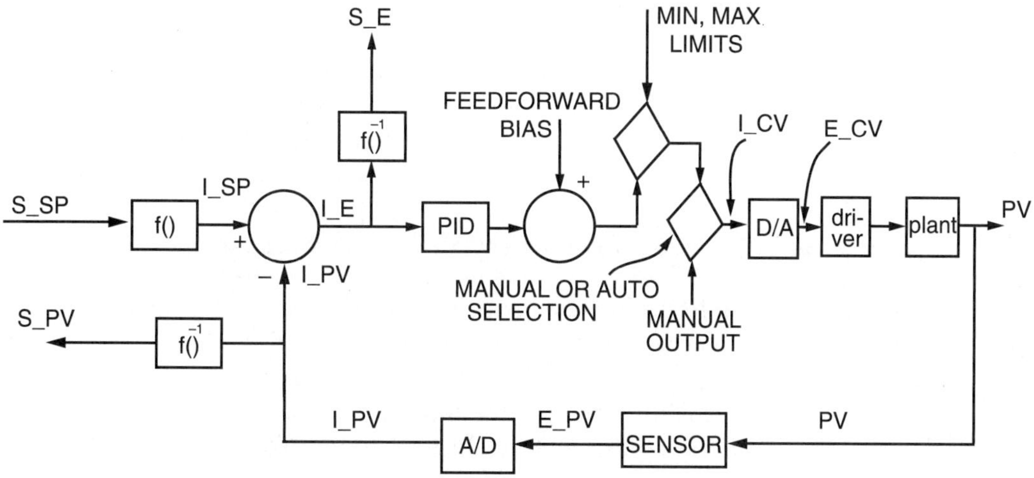

Figure 11–10 Relationship between actual variables, scaled variables, and internal variables

and:

$$b = I\_PV_1 \qquad \text{Eq. 11.5}$$

Then, $S_{\min}$ can be calculated from:

$$0 = m^* S_{\min} + b \qquad \text{Eq. 11.6}$$

and $S_{\max}$:

$$16{,}383 = \mathrm{m}^* S_{\max} + b \qquad \text{Eq. 11.7}$$

To select the integer representation as the input mode to the PID instruction, enter zeros for both $S_{\min}$ and $S_{\max}$. Let us establish a notation. If *S_PV* is given and *I_PV* is desired, the mathematical unit-conversion operation embedded in Figure 11–9 is called $f(\ )$. If *I_PV* is given and *S_PV* is desired, the unit-conversion operation is called $f(\ )^{-1}$. Remember, that to go between *I_PV* and *S_PV*, use the instruction SCP (scale with parameters). Now we can represent an overall diagram, illustrating the complete process, as is shown in Figure 11–10. Notice that all the output variables are "scaled" as a percentage of the maximum.

When referring to PID instruction, two items must be discussed. The first is the PID instruction appearing in the ladder logic. The second is the setup screen activated by double-clicking on the "Setup Screen" label of the PID instruction. Figure 11–11 shows a typical instruction. The PID file is the data file where all the variables associated with this instruction appears. Notice that as soon as the first PID instruction is used, the RSLogix 500 software creates a new set of data tables for PID instructions.

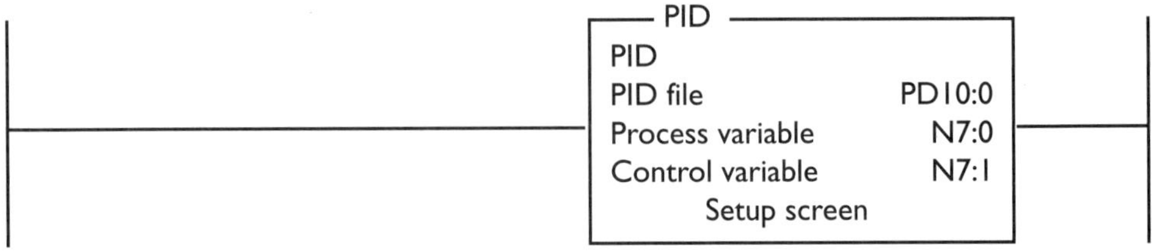

Figure 11–11 PID instruction as it appears in a ladder logic program.

The next item in the instruction is the process variable. This should point to the *A/D* address or a location in the integer file (N7:0 in this example). This allows the programmer to perform some sealing (biasing before using the value in the PID instructions). The next item is the control variable, the result of the PID operation. This could point to the *D/A* or to a memory location (N7:1 in this example) for some postprocessing.

The following section briefly describes the parameters in the PID setup screen (Figure 11–12). For a more specific description, see Appendix D, which is taken from Allen-Bradley's instruction set ("Process Control" section).

When referring to the PID setup screen, in Figure 11–12, one should note that these values can be entered manually (while the PLC is in "Run" or "Program" mode) or individually. In other words, the PLC can modify its parameters while running. Each of the input parameters is simply a value in a memory location (accessible through the PID data file or setup screen).

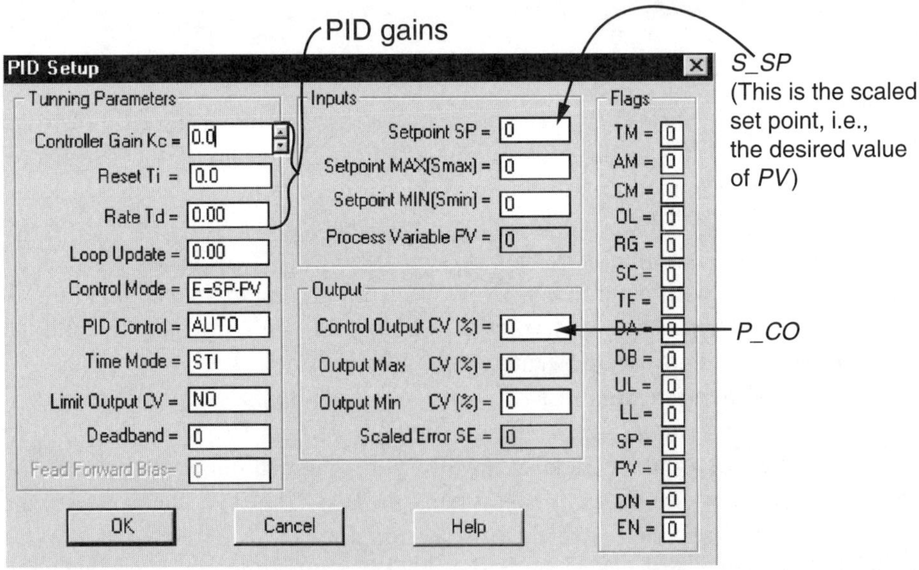

Figure 11–12 MicroLogix 1500 PID setup screen

Tuning Parameters

Controller gain Kc—The controller gain K_c is the proportional gain for the PID equation (*P* in Equation 11.2). When the gain range (RG) bit is set, the actual gain used in the PID operation is 1/10 the entered value. That is, if the operator enters 58, a gain of 5.8 will be used (see Appendix D).

Reset Ti and Rate Td—PID coefficients in Equation 11–2 refer to Appendix D for appropriate values.

Loop update—The loop update variable defines how often the PID calculation is performed. The smallest number that can be entered, 1, corresponds to 10 ms.

Control mode—Control mode has two possible values and determines the sign in the feedback summing junction. This can be set according to how the control component behaves. For instance, a value opens or closes when the control signal increases. The two choices are often called forward or reverse acting.

PID control—The PID control is where the PID instruction is activated. When set to "Auto," the PID instruction calculates the output using Equation 11–2. When set to "Manual," the output is determined by the "Control Output CV(%)" of the PID setup screen. This feature is very useful in the initial setup of the controller, because the PLC output can be set by this value.

Timed mode—Timed mode sets the interrupt scheme for triggering the execution of the PID instruction. It has two values: (1) Timed and (2) STI (see Appendix D).

Limit output CV—It is possible to establish limits on the control variable. Here, that option is activated. "Yes" imposes the limits specified in "Output Max CV(%)" and "Output Min CV(%)."

Deadband—Deadband raises and lowers (relative to the set point) when the control action occurs.

Feed forward—Feed forward introduces a constant bias in the output.

Inputs

Set point PS—is where the set point is entered. There are three choices: (1) enter a percentage (e.g., if we want the position to be 250 mils, we would enter 50), (2) enter scaled engineering units (in this example it is possible to use a 1-to-1 ratio so 250 would be entered for 250 mils), or (3) enter an integer number indicative of the number produced by the *A*/*D* when the *PV* is at the set point of 250 mils. To select choices (1) or (2), Setpoint Max (Smax) and Setpoint Min (Smin) must be defined according to the procedure in Figure 11–9 and Equations 11–3 to 11–7. To select choice (3), enter 0 for both Setpoint Max (Smax) and Setpoint Min (Smin).

Setpoint Max (Smax) and Setpoint Min (Smin)—See Equations 11–3 to 11–7.

Process variable—shows the value of the *PV* is expressed as *I_PV*). This is an output and not editable.

Outputs

Control output CV(%)—defines the output of the PLC. A 100 will make the *D/A* produce the maximum output, while a 0 will make the *D/A* produce the minimum value. This is only active when the **PID control** is set to "Manual."

Output Max CV(%) and Output Min CV (%)—These settings establish the limits on the output of the PLC. This option is only active when **Limit Output CV** is set to "Yes."

Scaled Error SE—shows the error relative to the range of the *PV*.

Selection of the PID Tuning Parameters

Two paths are available for selecting the coefficients of the PID equation. One is to use the results obtained through the design process and confirmed by simulation of the model (following the process presented in the previous chapters of this book). Another approach, an empirical one, often called "tuning" the PLC, consists of the Ziegler-Nichols method and its variations in which the controller is tuned after it is connected to the plant.

Summary

1. There are many platforms for implementing a control function. The choice depends on the complexity of the control task, the final cost, development time, and communication capabilities, among other factors.
2. To program a PLC, a graphical programming language/interface called ladder logic programming is used. It has its origins in ladder-looking wiring diagrams, used to describe the original relay logic used to perform control actions.
3. The communications capability offered by the PLC and the Internet allows an engineer on one continent to program, monitor, and interrogate a PLC on another continent.
4. Today's PLCs come in racks with extensive add-on and upgrade capabilities.

Idea Portfolio

1. Is PLC an appropriate controller for this application? What other controller would be more suitable?
2. How many of these units do you expect to build?
3. How many analog inputs and outputs are needed?
4. How many discrete inputs and outputs are needed?

Problems

1. List the industries in your area that, in your opinion, use PLCs.
2. Define the following terms:
    - Set point
    - Process variable
    - Error
    - Control output
3. List the various input ranges the MicroLogix 1500 PLCs can handle.
4. For the 0 to 10V DC range of an analog input module, using the scaled-for-PID internal representation, define the internal number that corresponds to:
    - 10.5V
    - 10V
    - 5.23V
    - 0V
    - −0.5V
5. Assume that you have a kiln being controlled by a PLC. Draw a schematic that shows all the components in the system.
6. Consider the following applications:
    - Handicapped ramp in a city bus
    - Oven used in your local bakery
    - Ignition controller in a small tractor

    Perform an economic use analysis to determine which controller you will use in each application.
7. The filter in the analog input module of the MicroLogix 1500 family is a(n) __________ filter.
    - Analog
    - Digital

8. How would you further classify the filter in the analog input module of the MicroLogix 1500 family?
    - Low pass
    - High pass
    - Band pass
    - Band stop
9. Define what "uploading" and "downloading" mean in a PLC according to Allen-Bradley's nomenclature.
10. Explain the purpose of HMIs.
11. Assume that a PLC is controlling the level in a tank of water that is 20 feet tall. The desired level is 12 feet. The operators like to view the measurement in terms of the percentage of a full tank. The pressure transducer is a 0–10psi transducer producing 0–5V. The *A/D* is set to 0–5VDC in Raw/Proportional mode. Indicate the corresponding number for the variables below when the level is at exactly 12 feet. Make sure to indicate units.
    - *PV* = __________
    - *I_PV* = __________
    - *S_PV* = __________
12. Assume that a sinusoidal signal is applied to a MicroLogix 1500 with an analog to digital converter set in the 0–10VDC range. The sinusoidal signal has the form:

$$V_C(t) = 4 + 2\sin(2\pi \cdot 5t)$$

The antialliasing filter is set to 50 Hz sampling frequency setting (cut-off frequency is 13.1 Hz).

- Determine the equivalent analog signal that is received by the PLC (i.e., the output of the filter if the filter were an analog filter).
- Realizing that the filter is processing numbers, because it is a digital filter, express the equivalent analog numbers inside the PLC for the:
    - Average value of the signal.
    - Minimum value of the signal.
    - Maximum value of the signal.

13. Define "scan time" in a PLC.
14. Determine which of the following applications would be candidates to be controlled by a PLC. Justify your answers.
    - Industrial oven
    - Communications network
    - Press

- Distilling column
- Computer router
- Hardisk drive controller
- Micropet
- Industrial coffeemaker
- Household coffeemaker

15. Under which conditions is a current analog input favored over a voltage analog input?
16. Which form (format of the defining equation) is used in a PLC?
17. Assuming that a PID controller used in a mathematical model has the following settings:
    - $K_P = 5.5$
    - $K_I = 2.1$
    - $K_D = 0.3$

    What are the equivalent PID parameters in the representation used by Allen-Bradley to be entered in the PLCs PID setup screen (K_c, T_i, T_d).
18. Search the Web for current applications of fluidics technology.
19. Think of an application for which fluidics would be the most feasible technology.
20. At a particular frequency, a filter provides -15 dB of attenuation. Express this number as a gain.
21. Why do you think a filter has a Bode plot with the shape shown in Figure 11–6 instead of a monotonically decreasing curve?
22. Where does the term "ladder logic programming" come from?
23. How many bits are used in the internal representation of analog signals in the MicroLogix 1500 family of PLCs?
24. Define all the variables in the snapshot of the PID drop-down menu in Figure 11–10. Visit and download any necessary documents from the Allen-Bradley Web site at http://www.ab.com.
25. Assume that the *PV* in an environmental chamber ranges from 75°F to 125°F. Propose four ways of representing this process variable using "scaled engineering units." Indicate the resolution associated with the set point for each case. Resolution is the smallest quantity by which the set point can be incremented.

12

Introduction to Digital Control Systems

Learning Objectives

In this chapter, you will learn how to:

- Mathematically define the z-transform.
- Take the z-transform of a time function.
- Describe the dynamic characteristics of a sampler and a zero-order-hold.
- Use the z-transform to capture a difference equation.
- Use SIMULINK to simulate dynamic systems containing difference equations.
- Tell if a microcontroller is sufficiently fast enough to replace an analog controller.

12.1 Historical Perspective

In the past 30 years, many of the analog control systems (electronic, pneumatic, and mechanical) have been replaced by digital computers with analog-to-digital and digital-to-analog converters. There are several driving factors:

- Changes in the control strategy or set point are easily accomplished via software.
- Due to the digital manner in which information is stored, it is not prone to drift or change due to environmental conditions (e.g., temperature, component aging).
- Increase in the processing speed of the microcontroller.
- Information is in a format that is easily displayed and communicated, regardless of the distance involved.

- Due to the proliferation of electronics in every facet of life, and the associated increase in production numbers (microelectronic manufacturing), component costs have decreased. Unfortunately, this has generally not been the case with other technologies (e.g., hydraulic systems) and is another reason other technologies attempt to use electronics as much as possible.

In the middle of the twentieth century, few applications could justify the expense of a computer; only refineries and power-generating plants could afford such an expense. However, in the past three decades, with the advent of the microprocessor and microcontroller, the scenario has changed. A modern, high-end car may have as many as thirty microcontrollers doing various tasks. Three examples of systems in an automobile that use microcontrollers are the antilock function, the "electronic carburetor" (the fuel-injection system), and the climate control.

12.2 Introduction

The computer handles information differently than its analog counterpart. Whereas the analog controller possesses and acts on information continuously, the digital computer does not. Information is accessed and updated at discrete points in time. This presents some challenges to control engineers because the tools they used in the past might not always apply. Two possible design scenarios face the control engineer. These depend on the computer's sampling rate, the available computing power, and the dynamics associated with the system being controlled. Sampling rate (samples/second) is the frequency at which the computer reads and processes the analog input signal(s) and produces an analog output signal. Which of the two scenarios the control engineer will encounter is determined by the following conditions:

- Sampling rate is relatively slow compared with the dynamics of the plant/system being controlled
- Sampling rate is extremely fast relative to the dynamics of the plant or system being controlled

When control engineers face the first scenario, they will have to analyze and design the system from scratch using a new set of design tools. These tools differ from what they had used for continuous systems. This approach will require an extensive use of mathematical tools that are outside the scope of this book. Specifically, most of the work is done with the aid of the z-transform, a mathematical tool that is to discrete systems what the Laplace transform is to continuous systems.

When engineers face the second scenario, the analysis and design can still be done in the time domain and then converted to the discrete domain with the aid of some of the basic principles of the z-transform presented in this chapter. Those interested in handling the first scenario should take a semester-long course in digital control.

The goal of this chapter is to introduce digital systems and the z-transform, which provides a path for converting a continuous design to a discrete one. A final component of this chapter is a process for evaluating the resulting digital controller to confirm that it meets design objectives.

12.3 The Z-Transform

Consider a system whose block diagram appears in Figure 12–1. The signal $r(t)$, a continuous signal, is sampled every T seconds by a switch that is only closed for an infinitesimal period. Thus, the signal $r^*(t)$, on the other side of the switch, exists only when the switch is momentarily closed. Mathematically, $r^*(t)$ can be expressed as:

$$r^*(t) = \sum_{k=0}^{\infty} r(kT)\delta(t - KT) \qquad \text{Eq. 12.1}$$

The time-delayed, Dirac delta function was introduced in Chapter 2.

If we take the Laplace transform, considering that a continuous integration can be changed to a summation, then:

$$R^*(s) = \mathcal{L}\{r^*(t)\} = \sum_{k=0}^{\infty} r(kT)e^{-ksT} \qquad \text{Eq. 12.2}$$

and we define the variable z as:

$$z = e^{sT} \qquad \text{Eq. 12.3}$$

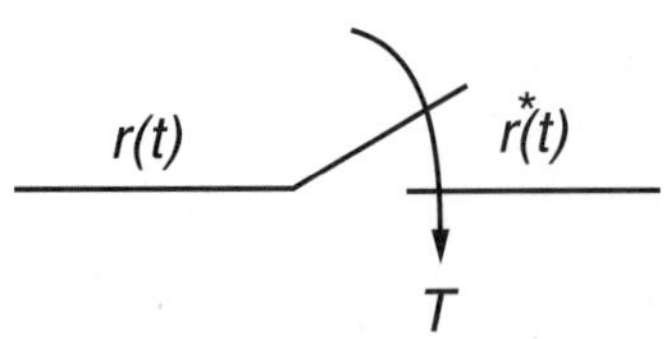

Figure 12.1 Sampling Process

that provides a mapping from the *s*-plane to the *z*-plane. The new transform, called the z-transform, is defined as:

$$R(z) = Z\{r^*(t)\} = R^*(s)\Big|_{z=e^{sT}} \qquad \text{Eq. 12.4}$$

$$R(z) = \mathcal{L}\{r^*(t)\}\Big|_{z=e^{sT}} \qquad \text{Eq. 12.5}$$

Example 12.1

A sampler samples every 2 seconds the signal $r(t)$, as shown in Figure 12–2. Obtain the z-transform of $r^*(t)$, $R(z)$.

Solution First, obtain $r^*(t)$:

$$r^*(t) = r^*(kT) = 0\delta(t) + 2\delta(t - 1 \cdot 2) + 2.5\delta(t - 2 \cdot 2) \qquad \text{Eq. 12.6}$$
$$+ 5\delta(t - 3 \cdot 2) + 7\delta(t - 4 \cdot 2) + 7\delta(t - 5 \cdot 2) + 5.5\delta(t - 6 \cdot 2)$$
$$+ 4.5\delta(t - 7 \cdot 2) + 2\delta(t - 8 \cdot 2) + .5\delta(t - 9 \cdot 2)$$

Then:

$$R(z) = 0 \cdot z^0 + 2z^{-1} + 2.5z^{-2} + 5z^{-3}$$
$$+ 7z^{-4} + 5.5z^{-6} + 4.5z^{-7} + 2z^{-8} + 0.5z^{-9} \qquad \text{Eq. 12.7}$$

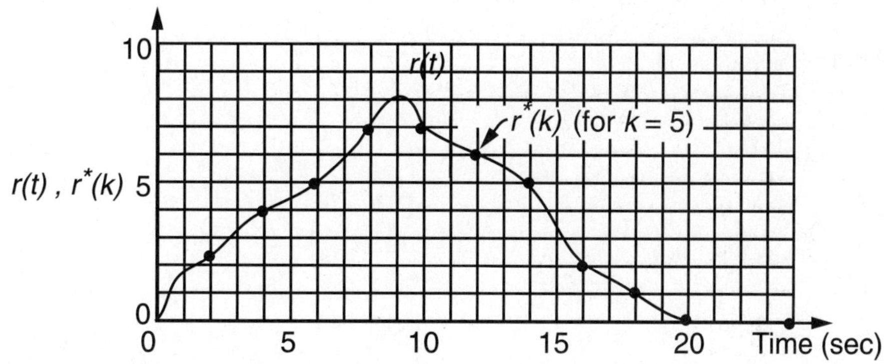

Figure 12.2 Time series

Example 12.2

Determine the z-transform of the sampled unit step function:

$$Z\{u^*(t)\} = \sum_{K=0}^{\infty} u(KT)z^{-k} \qquad \text{Eq. 12.8}$$

Solution Note that $u(kT) = 1$ for $K \geq 0$. Thus:

$$U(z) = 1 + z^{-1} + z^{-2} + z^{-3} + z^{-4} \ldots. \qquad \text{Eq. 12.9}$$

Remembering the geometric series:

$$\frac{1}{(1 - ax)} = 1 + ax + (ax)^2 + (ax)^3 + \ldots. \qquad \text{Eq. 12.10}$$

We can express $U(z)$:

$$U(z) = \frac{1}{1 - (1)(z^{-1})} = \frac{z}{z - 1} \qquad \text{Eq. 12.11}$$

This procedure can be applied to any time function as was done with the Laplace transform. Table 12–1 shows some of the most common functions.

Time Function *f*(*t*)	Laplace Transform *F*(*s*)	Z Transform *F*(z)
Impulse, $A\delta(t)$	A	A
Delayed unit impulse, $A\delta(t - KT)$	Ae^{-KTs}	Az^{-K}
Step, $A\,u(t)$	$\frac{A}{s}$	$\frac{Az}{z - 1}$
Ramp, $A\,t\,u(t)$	$\frac{A}{s^2}$	$\frac{ATz}{(z - 1)^2}$
Exponential decay, $Ae^{-bt}\,u(t)$	$\frac{A}{s + b}$	$\frac{Az}{z - e^{-bT}}$
Sine wave, $A \sin \omega t\, u(t)$	$\frac{A\omega}{s^2 + w^2}$	$\frac{Az \sin \omega T}{z^2 - 2z \cos \omega T + 1}$

Table 12–1 Common functions with their corresponding Laplace and z-transformer

12.4 Capture a Difference Equation Using the Z-Transform

Difference equations are algebraic expressions that capture the dynamic characteristics of systems. They are to the discrete world what differential equations are to the continuous world.

Consider a difference equation of the form:

$$e[KT] = ae[(K-1)T] + bu[KT] + cu[(K-1)T) \qquad \textbf{Eq. 12.12}$$

A form that can be simplified by dropping the sampling interval form from the expression:

$$e[K] = ae[(K-1)] + bu[K] + cu[(K-1)] \qquad \textbf{Eq. 12.13}$$

Notice that this is only a symbolic simplification, because $e[\ ]$ and $u[\ ]$ exist only at multiple intervals of the sampling period, T. Taking the z-transform of the previous expression results in:

$$E(z) = aE(z)z^{-1} + bU(z) + cU(z)z^{-1} \qquad \textbf{Eq. 12.14}$$

or, in transfer function format:

$$\frac{E(z)}{U(z)} = \frac{1 - az^{-1}}{b + cz^{-1}} \qquad \textbf{Eq. 12.15}$$

or:

$$\frac{E(z)}{U(z)} = \frac{z - a}{bz - c}$$

$$= \frac{1}{b}\left(\frac{z - a}{z + \frac{c}{b}}\right) \qquad \textbf{Eq. 12.16}$$

The preceding z-transfer function captures in the z-domain what the difference equation captures in the time domain. To determine the stability of a z-transfer function, we also look at the roots of the transfer function's denominator polynomial. However, unlike the s-transfer function, stability is ensured when the roots of the z-transfer function denominator polynomial lies inside the unit circle.

To simulate such a system in SIMULINK, create a discrete zero-pole transfer z-transfer function block, and use a step input to excite it. Place a scope at the output. For the simulation, use the following values: $a = 0.5$, $b = 2$, and $c = 1$. Enter the value of T in the z-transform function block. Use $T = 0.1$ and simulate the system for 25. There is no need for other components, because a discrete transfer function has zero-order-holds at the output. When you simulate the system, you should see a discrete response that decays. This is in accord with our previous statement, because the discrete transfer function has a pole at:

$$z = -\frac{c}{b} = -\frac{1}{2} \qquad \text{Eq. 12.17}$$

which is inside the unit circle.

Now if we change the value of c to 4, the new pole would be at $z = -2$, which is outside the unit circle. Resimulate the system with the new coefficients, and see what happens.

12.5 Zero-Order-Hold

A zero-order-hold (ZOH) is a device that "holds" the value it receives from the sampler. It captures and remembers that value for the following T seconds, where T is the sampling period (=1/sampling rate). When you apply an impulse to a system, the Laplace transform of the resulting output is the system's transfer function. We will use this concept to derive the transfer function for a ZOH. The sampler and the resulting output appear in Figure 12–3. The output can be expressed as:

$$c(t) = u(t) - u(t - T) \qquad \text{Eq. 12.18}$$

$$C(s) = \frac{1}{s} - \frac{1}{s}e^{-sT} = \frac{(1 - e^{-sT})}{s}$$

Thus, $G_o(s)$, the transfer function of a ZOH is:

$$G_o(s) = \frac{1 - e^{-sT}}{s} \qquad \text{Eq. 12.19}$$

12.6 Continuous-to-Discrete Transfer Function Transformation

Now, using the methods of Chapter 10, we are ready to tackle the transformation of a transfer function that has been designed in the continuous domain, into a discrete transfer function. Basically, we will transform a transfer function (in terms

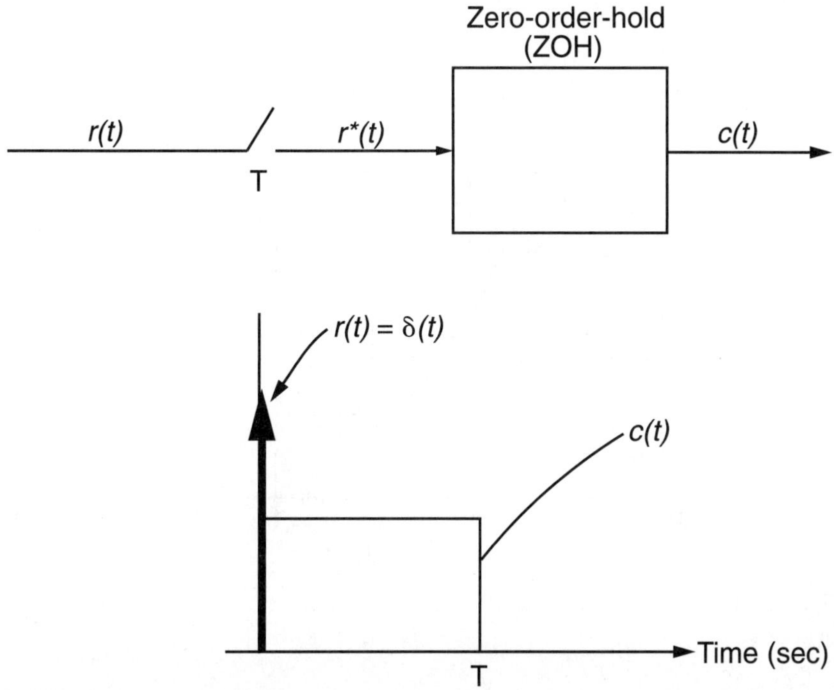

Figure 12.3 Zero-order-hold and its response to an impulse

of s) into a transfer function in terms of z. We will present several approaches for doing this, some based on time-domain characteristics (using the numerical approximation of the derivative and integral). We will also use the Laplace transform of the output to a ZOH that is excited with a step. As a vehicle for evaluating the accuracy of this procedure, we will compare the outputs of the discrete transfer functions (in terms of z) with the outputs of the continuous transfer functions (in terms of s). To perform this comparison, we will excite the continuous and discrete transfer functions with the same excitation. We have chosen the function:

$$c(t) = 5.22\,t\,e^{-5t^2} \qquad \textbf{Eq. 12.20}$$

This function appears in Figure 12–4. The specific, continuous transfer function we will try to approximate with a discrete one is the lead (or lag) filter, used in Chapter 10.

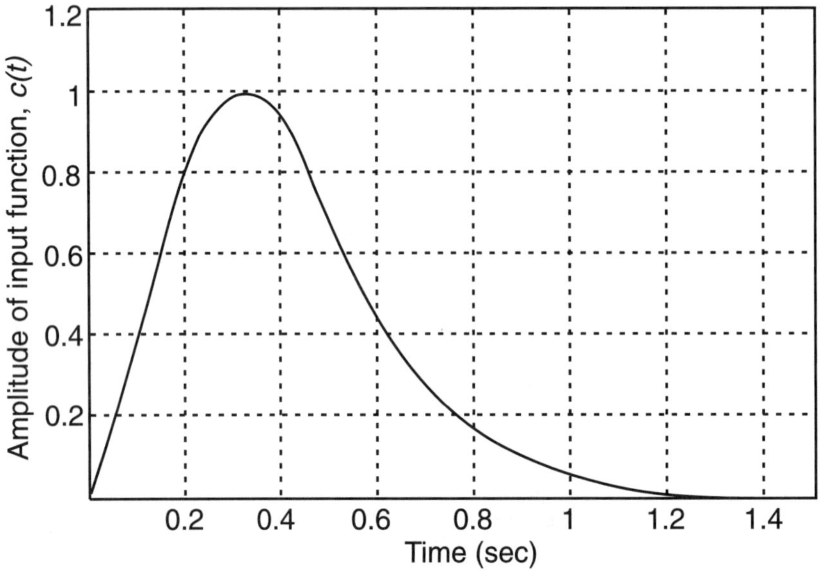

Figure 12.4 Input function used to excite the continuous and discrete transfer functions

12.7 Backward Rule Approach for Obtaining the Discrete Transfer Function (Direct Substitution)

Consider the transfer function corresponding to a lead- or lag-filter:

$$D(s) = \frac{U(s)}{E(s)} = \frac{\frac{s}{a}+1}{\frac{s}{b}+1} = \frac{b}{a}\frac{s+a}{s+b} \qquad \text{Eq. 12.21}$$

or

$$U(s)(s+b) = \frac{b}{a}E(s)(s+a)$$

$$sU(s) + bU(s) = \frac{b}{a}\{sE(s) + E(s)a\}$$

We can return to the time domain:

$$\dot{u}(t) + bu(t) = \frac{b}{a}\{\dot{e}(t) + ae(t)\} \qquad \text{Eq. 12.22}$$

The derivative can be approximated with what is normally called the backward difference, or:

$$\dot{e}(t) = \frac{de(t)}{dt} \approx \frac{e[KT] - u[(K-1)T]}{T} \qquad \text{Eq. 12.23}$$

and:

$$\dot{u}\,(t) = \frac{du}{dt} \approx \frac{U[KT] - U[(K-1)T]}{T} \qquad \text{Eq. 12.24}$$

where T is the sampling interval.

Substituting $e(KT)$ and $u(KT)$ for $e(t)$ and $u(t)$, together with the two previous expressions for their corresponding derivatives, results in

$$\frac{u[KT] - u[(K-1)T]}{T} + bu[KT] = \frac{b}{a}\left\{\frac{e[KT] - e[(K-1)T]}{T} + ae[KT]\right\} \qquad \text{Eq. 12.25}$$

To simplify our math, we will drop the T term inside the brackets:

$$\frac{u[K] - u[(K-1)]}{T} + bu[K] = \frac{b}{a}\left\{\frac{e[K] - e[(K-1)]}{T} + ae[K]\right\} \qquad \text{Eq. 12.26}$$

Multiply by Ta:

$$au[K] - au[(K-1)] + abTu[K] = be[K] - be[(K-1)] + abTe[K] \qquad \text{Eq. 12.27}$$

When we simplify and collect terms, we end up with a difference equation:

$$u[K] = \frac{1}{1+bT}u[K-1] + \frac{b+abT}{a+abT}e[K] - \frac{b}{a+abt}e[K-1] \qquad \text{Eq. 12.28}$$

Take the z-transform:

$$U(z) = \frac{1}{1+bT}U(z)z^{-1} + \frac{b+abT}{a+abT}E(z) - \frac{b}{a+abT}E(z)z^{-1} \qquad \text{Eq. 12.29}$$

By expressing the result in terms of a transfer function in terms of z, we get:

$$\frac{U(z)}{E(z)} = \frac{\left(\frac{b + abT}{a + abT}\right) - \left(\frac{b}{a + abT}\right)z^{-1}}{1 - \left(\frac{1}{1 + bT}\right)z^{-1}} \qquad \text{Eq. 12.30}$$

This is a z-transfer function that can be easily implemented in SIMULINK. For the simulation, we use the values $a = 1$, $b = 20$, and $T = 0.01$s.

For these values, the discrete transfer function becomes:

$$\frac{U(z)}{E(z)} = \frac{16.8333 - 16.667z^{-1}}{1 - 0.8333z^{-1}} \qquad \text{Eq. 12.31}$$

The preceding transfer function and the continuous transfer function are implemented in SIMULINK and shown in Figure 12–5. An integration step of 0.005s is used in the simulation. The resulting response appears in Figure 12–6. Notice the excellent agreement between the outputs of the z-transfer function implementation and of the continuous one. Notice that in SIMULINK "each discrete block is assumed to have a sampler at its input and a ZOH at its output".[1]

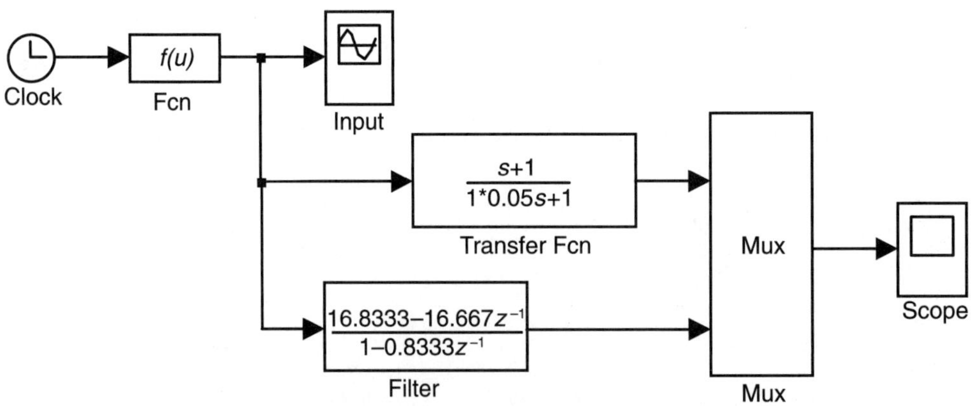

Figure 12.5 SIMULINK diagram of continuous and discrete blocks (time-domain implementation)

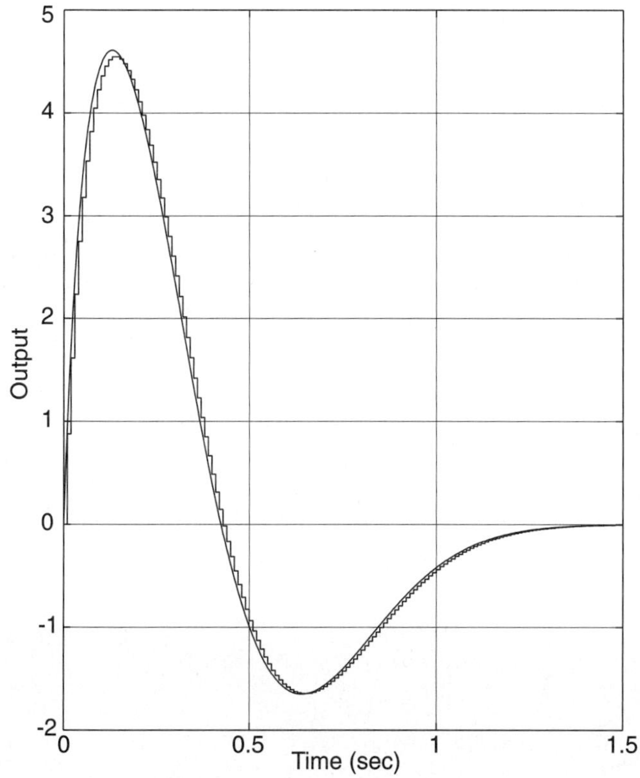

Figure 12.6 Response of continuous and discrete block (time-domain implementation)

12.8 Backward Rule Approach for Obtaining the Discrete Transfer Function (Using a Transformation)

One can reflect on the procedure performed in the previous section:

1. Start with a continuous transfer function (s-domain).
2. Convert the transfer function to a differential equation (time domain).
3. Approximate derivatives with the backward-difference approximations.
4. Convert the result to a difference equation.
5. Take the z-transform of difference equation to convert into a z-based transfer function.

It is possible to arrive at the same result by obtaining a relationship between s and z from the definition of the derivative. The backward difference form of the derivative is:

$$\dot{e}(t) \cong \frac{e(t) - e(t - T)}{T} \qquad \textbf{Eq. 12.32}$$

Taking the Laplace transform, assuming that all the initial condition is 0:

$$sE(s) = \frac{E(s) - E(s)e^{-sT}}{T} \qquad \text{Eq. 12.33}$$

or:

$$E(s)s = E(s)\left\{\frac{1 - e^{-sT}}{T}\right\}$$

By equating the terms following $E(s)$ in the previous equation:

$$s = \left\{\frac{1 - e^{-sT}}{T}\right\} \qquad \text{Eq. 12.34}$$

and applying the definition:

$$z = e^{sT} \qquad \text{Eq. 12.35}$$

results in:

$$s = \frac{1 - z^{-1}}{T} = \frac{z - 1}{Tz} \qquad \text{Eq. 12.36}$$

This means we can establish a mapping that allows us to convert an s-transfer function to a z-transfer function by making the simple substitution:

$$s \leftarrow \frac{z - 1}{Tz} \qquad \text{Eq. 12.37}$$

Example 12.3

Use the previously derived backward difference approximation to the derivative to convert the following transfer function to a z-transfer function:

$$D(s) = \frac{bs + a}{as + b} \qquad \text{Eq. 12.38}$$

Solution

$$D(z) = \frac{b\left(\dfrac{z-1}{Tz} + a\right)}{a\left(\dfrac{z-1}{Tz} + b\right)}$$

Eq. 12.39

Multiply by Tz:

$$D(z) = \frac{\dfrac{b}{a}\{z - 1 + aTz\}}{\{z - 1 + bTz\}}$$

$$= \frac{b}{a}\frac{\{z(1 + aT) - 1\}}{\{z(1 + bT) - 1\}}$$

Eq. 12.40

Multiply by z^{-1}:

$$D(z) = \frac{\dfrac{b}{a}\{(1 + aT) - z^{-1}\}}{\{(1 + bT)\ 2\ z^{-1}\}}$$

Eq. 12.41

Divide by $(1 + bT)$:

$$D(z) = \frac{\dfrac{b}{a}\left\{\dfrac{(1 + aT)}{(1 + bT)} - \dfrac{1}{(1 + bT)}z^{-1}\right\}}{\left\{1 - \dfrac{1}{(1 + bT)}z^{-1}\right\}}$$

Eq. 12.42

Notice that this can be substituted into the previous transfer function:

$$D(z) = \frac{\left(\dfrac{b + abT}{a + abT}\right) - \left(\dfrac{b}{a + abT}\right)z^{-1}}{1 - \left(\dfrac{1}{1 + bT}\right)z^{-1}}$$

Eq. 12.43

which is the result obtained when the derivative definition was used (see Section 12.7).

12.9 Laplace Transform Approach for Obtaining the Discrete Transfer Function

The following approach produces a discrete transfer function approximation that is slightly different than the one previously derived. Some of the intermediate steps can be facilitated with MATLAB's Symbolic Math toolbox. However, this development does not require it. Consider the block diagram in Figure 12–7. This shows the sampler connected to a ZOH, which "holds" the input signal and presents it to the continuous transfer function. The transfer function relating the output and the input is:

$$\frac{C(s)}{R^*(s)} = G(s) = G_o G_p = \frac{1-e^{-sT}}{s}\cdot\frac{b}{a}\cdot\frac{s+a}{s+b}$$
$$= \frac{b}{a}(1-e^{-sT})\frac{s+a}{s(s+b)}$$

Eq. 12.44

Expanding into partial fraction expansions:

$$G(s) = \frac{C(s)}{R^*(s)} = \frac{b}{a}(1-e^{-sT})\left\{\frac{\left(\frac{a}{b}\right)}{s} + \frac{\left(\frac{b-a}{b}\right)}{s+b}\right\}$$

Eq. 12.45

Now we take the Z transform:

Eq. 12.46

$$\frac{C(s)}{R^*(s)} = \frac{b}{a}(1-e^{-sT})\left\{\frac{\left(\frac{a}{b}\right)}{s} + \frac{\left(\frac{b-a}{b}\right)}{s+b}\right\}$$

$$G(z) = Z\left[G(s)\right] = \frac{b}{a}(1-z^{-1})Z\left\{\frac{\left(\frac{a}{b}\right)}{s} + \frac{\left(\frac{b-a}{b}\right)}{s+b}\right\}$$

$$= \frac{b}{a}(1-z^{-1})\left\{\frac{a}{b}\cdot\frac{z}{z-1} + \frac{b-a}{b}\cdot\frac{z}{z-e^{-bt}}\right\}$$

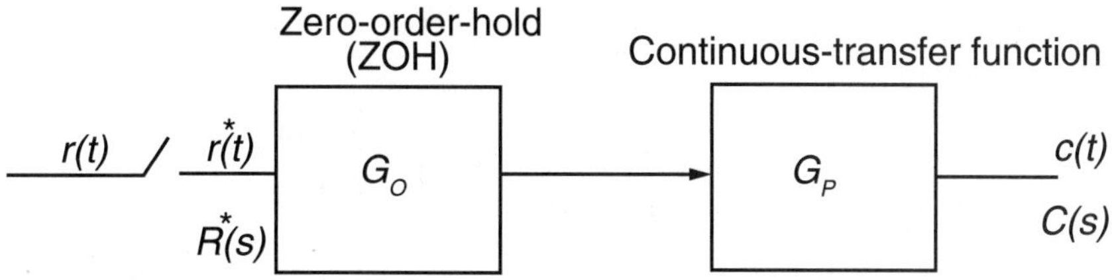

Figure 12.7 Sampler, zero-order-hold, and continuous-transfer function

This simplifies to:

$$G(z) = 1 + \frac{b-a}{a} \cdot \frac{z-1}{z-e^{-bT}} \qquad \textbf{Eq. 12.47}$$

Multiply the last term by $\frac{z^{-1}}{z^{-1}}$:

$$G(z) = 1 + \frac{b-a}{a} \cdot \frac{1-z^{-1}}{1-e^{-bT}z^{-1}} \qquad \textbf{Eq. 12.48}$$

This transfer function can be further simplified to:

$$G(z) = \frac{\left(\frac{b}{a}\right) - \left(\frac{b-a}{a} + e^{-bT}\right)z^{-1}}{1-e^{-bT}z^{-1}} \qquad \textbf{Eq. 12.49}$$

And, for the chosen values $a = 1$, $b = 20$, and $T = 0.01s$, the result is:

$$G(z) = \frac{20 - 19.8187z^{-1}}{1 - 0.8187z^{-1}} \qquad \textbf{Eq. 12.50}$$

This implementation appears in Figure 12–8. The output of both blocks (continuous and discrete) appear in Figure 12–9. Notice that the discrete transfer function approximates fairly well the continuous one, though not as well as the one derived using the time-domain approach.

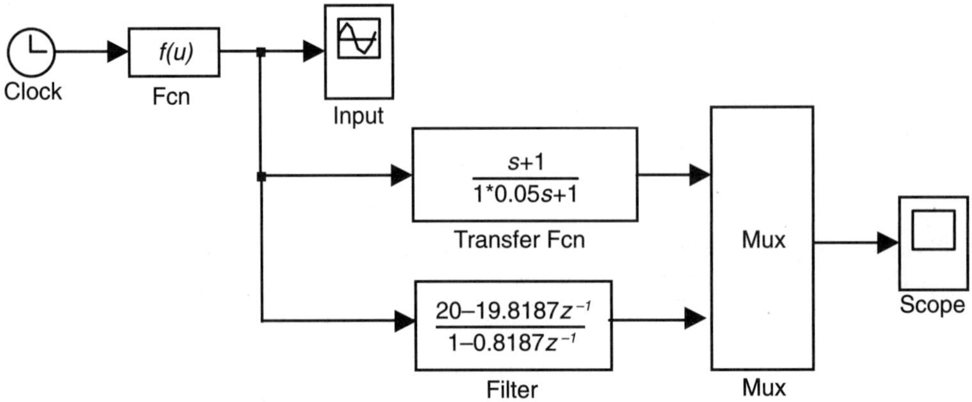

Figure 12.8 SIMULINK diagram showing forcing function exciting continuous and discrete blocks (Laplace transformation method)

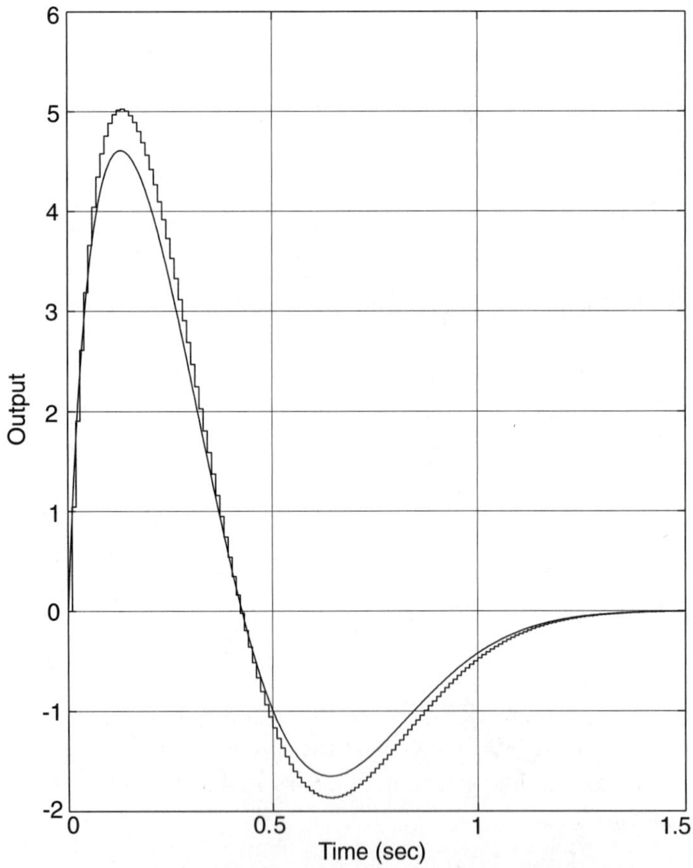

Figure 12.9 Response of continuous and discrete transfer function (Laplace domain implementation)

12.10 Other Approximations

There are other ways of obtaining similar transformations, based on different approximations of the derivative or the integral. Using the Forward Difference approximation of the derivative:

$$\dot{e}(t) = \frac{e[(K+1)T] - e[KT]}{T} \qquad \textbf{Eq. 12.51}$$

Taking the Laplace transform and performing the coefficient equivalence procedure performed previously results in the mapping:

$$s \leftarrow \frac{z-1}{T} \qquad \textbf{Eq. 12.52}$$

In a similar fashion, we can use the trapezoidal approximation to integration to obtain the following mapping:

$$s \leftarrow \frac{2}{T}\frac{z-1}{z+1} \qquad \textbf{Eq. 12.53}$$

There are other approximation methods as noted in Table 12–2. For further information, see the reference list provided at the back of this book.

Transformation Name	Source	Resulting Transformation
Backward difference	Backward difference approximation to the derivative	$s \leftarrow \frac{z-1}{Tz}$
Forward difference (also called Euler's method)	Forward difference approximation to the derivative	$s \leftarrow \frac{z-1}{T}$
Trapezoidal (also called Tustin's method or bilinear transformation)	Trapezoidal approximation to integration	$s \leftarrow \frac{2}{T}\frac{z-1}{z+1}$
Zero-order-hold	Laplace transform analysis of the zero-order-hold	Does not exist in general form

Table 12–2 Continuous to Discrete Conversion Methods

12.11 Using MATLAB to Perform the Continuous-to-Discrete Conversion

MATLAB has built-in functions that can perform the preceding operations, specifically, the function `c2dm` (continuous-to-discrete model). To illustrate this process, we have created the script file c2cm.m:

```
% script file c2d_exam.m
% transform a continuous-to-discrete transformation
% using the zero-order-hold approximation
% define coefficients of transfer function
a=1;
b=20;
T=0.01;
% define numerator and denominator of the transfer function
num=(b/a)*[1 a];
den=[1 b];
% perform the continuous-to-discrete model conversion
[num_d,den_d] = c2dm(num,den,T,'zoh'):
```

which when executed produces the following output:

```
>>c2dm_exa
num_d =
   20.0000 -19.8187
den_d =
   1.0000 -0.8187
```

which is the same result obtained manually in Section 12.9.

MATLAB offers various conversion schemes besides the ZOH approximation. The choices are:

'zoh'	Convert to discrete time assuming a ZOH on the inputs.
'foh'	Convert to discrete time assuming a first-order-hold on the inputs.

'tustin' Convert to discrete time using the bilinear (Tustin) approximation to the derivative.

'prewarp' Convert to discrete time using the bilinear (Tustin) approximation with frequency prewarping. Specify the critical frequency with an additional argument (i.e., C2DM(A,B,C,D,Ts,'prewarp',Wc).

'matched' Convert the SISO system to discrete time using the matched pole-zero method.

The preceding definitions are obtained from MATLAB's help utility by typing:

```
>>help c2dm
```

12.12 PID Control

One of the most common control schemes is the PID. We have seen its application in Chapters 10 and 11. In your system simulation, you might want to represent the controller as accurately as possible. SIMULINK allows you to simulate the PID controller in both the continuous and discrete formats. Figure 12–10 shows both implementations.[2]

12.13 Procedure for Transforming a Continuous Design to a Discrete Implementation

This section presents an overall method for transforming a continuous design to a discrete one. It assumes that the existing continuous design performs adequately. The method can also be used for a new system. In this case, the design effort would proceed as if it were to be implemented with a continuous controller. Once the design has been completed and verified, it is converted to a discrete one. The underlying assumption is that the digital controller that will be used is much faster than the dynamic system that is to be controlled. A block diagram outlining the process appears in Figure 12–11. Franklin, Powell, and Workman recommend that, for the discrete system to provide adequate performance, the sampling rate (*1/T,* where *T* is the sampling period) must be at least thirty times faster than the closed-loop bandwidth, *BW,* of the system.[3]

If the continuous transfer function is relatively simple, for instance a PID loop, an alternate rule of thumb is that the sampling period must be 1/5 to 1/10 of

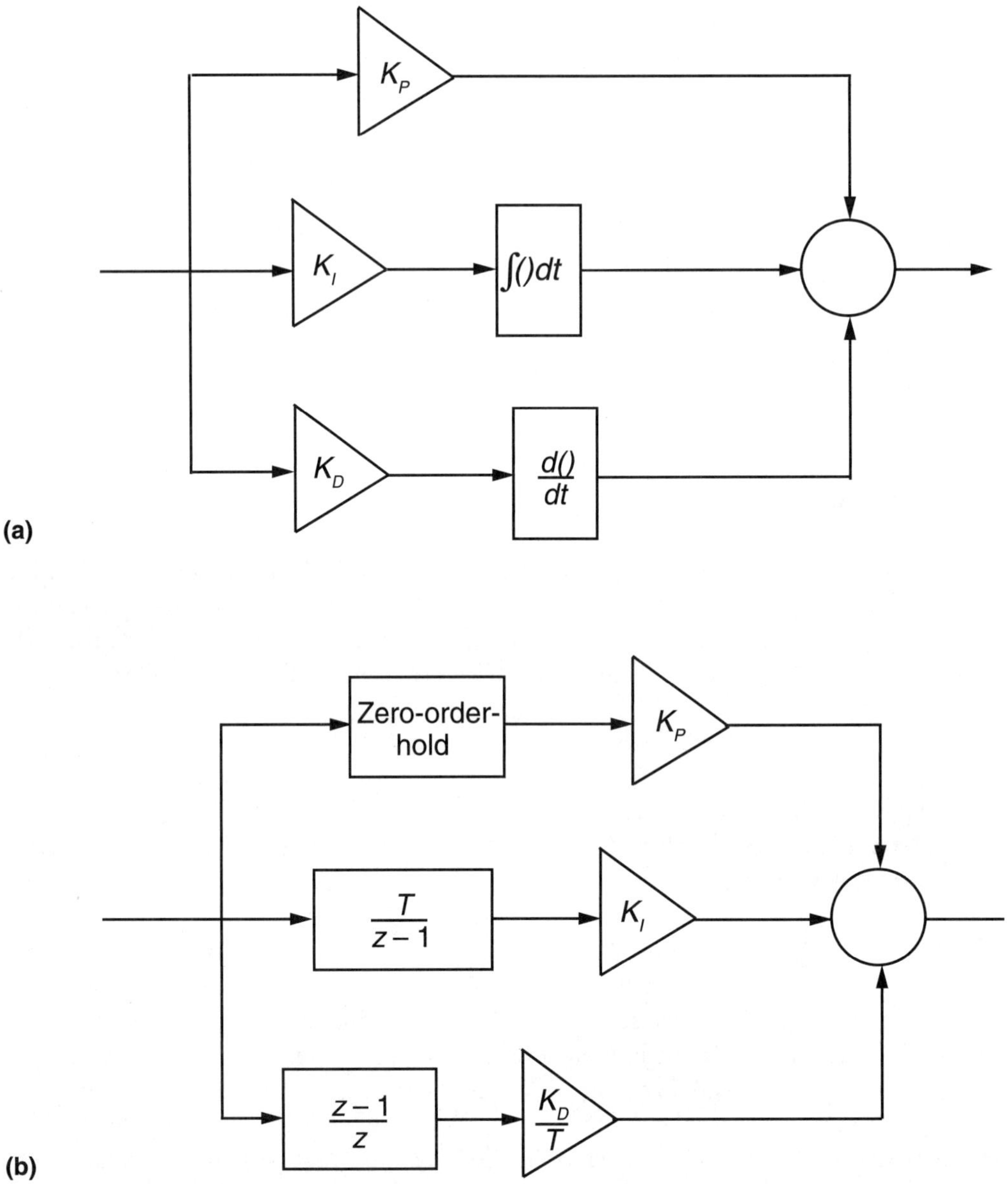

Figure 12.10 SIMULINK PID implementation in (a) continuous and (b) discrete blocks

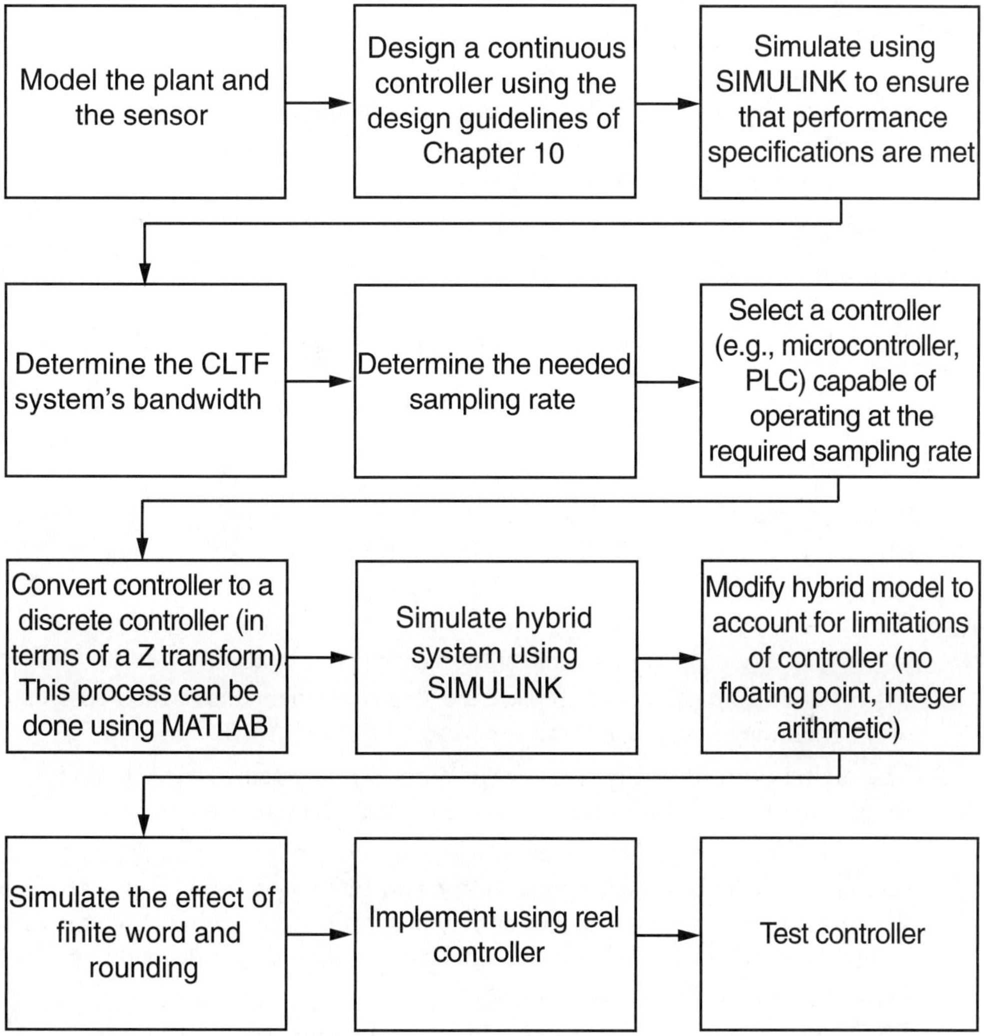

Figure 12.11 Discrete flowchart implementation

the ultimate period, *Tu*. As you might remember from Chapter 10, the ultimate period, *Tu,* is the period that results when the gain of the system is increased until the system experiences sustained oscillations. This guideline generally produces a sampling frequency that is slower than the guidelines presented in the previous paragraph.

Example 12.4

A system has a bandwidth of 20 Hz. Determine the minimum sampling frequency (or the maximum sampling period) of the microcontroller if some of the control actions are to be implemented with a digital controller.

Solution

$$f_s = 30 \cdot BW = 30(20\text{ Hz}) = 600\text{ Hz}$$

$$T = \frac{1}{f_s} = \frac{1}{600\text{ Hz}} = 1.66\text{ ms} \qquad \textbf{Eq. 12.54}$$

Thus, the sampling period must be at least 1.66 ms.

Summary

1. Many digital controllers have replaced analog controllers due to their cost, ease of programming or reprogramming, and stability.
2. The digital controller only accesses the output of a system at discrete points in time, rather than continuously, as its analog controller counterpart.
3. If the sampling rate (the frequency at which the controller samples the output) is very fast compared to the dynamics of the system being controlled, it is possible to convert a continuous compensator or transfer function and implement it with a digital controller. To do this, the sampling rate should be at least thirty times the bandwidth of the continuous system.
4. The z-transform is the mathematical tool used to analyze discrete systems and to capture a difference equation.
5. The z-transform is to discrete systems what the Laplace transform is to continuous systems.

Idea Portfolio

1. What would be the implication of converting the continuous system to a discrete system?

2. What is the bandwidth of the continuous system?
   - How would you determine it analytically (closed form)?
   - How would you determine it through simulation?
   - How would you determine it experimentally?
3. Having determined the bandwidth, what would be the minimum sampling rate of the system?
4. Create a complete block diagram of the system, considering that the controller to be used is discrete machine.
5. Where is the first-order-hold on the diagram?
6. What information is needed to simulate the system?

Problems

1. Obtain the z-transform of $f(t)$, which is shown in Figure 12–12.
2. Obtain the z-transform of $f(t) = tu(t)$. Do not use the answer from Table 12–1.

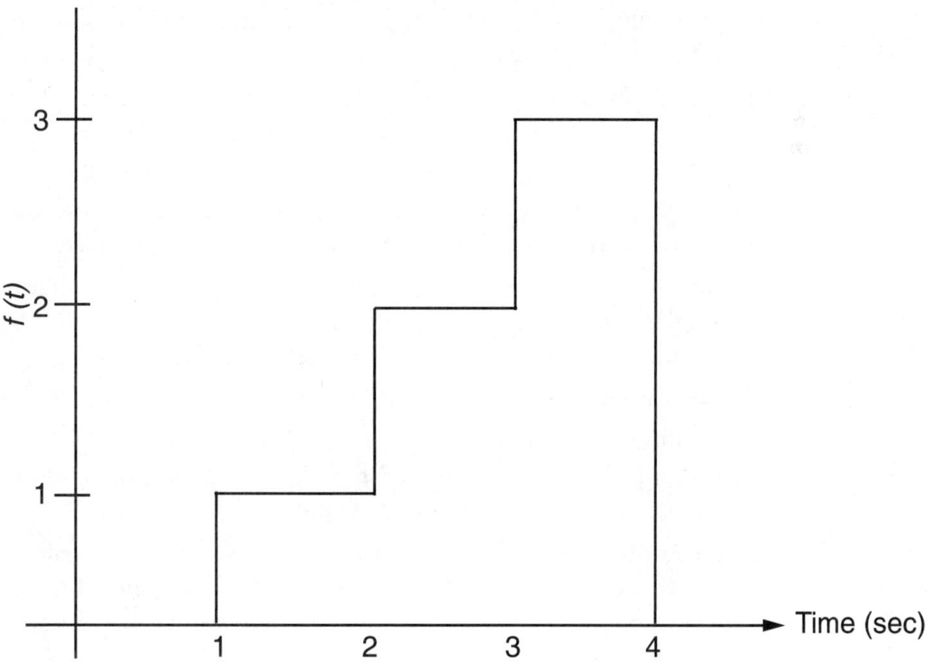

Figure 12.12 Time Series

3. Obtain $m[K{\bullet}T]$ for the following values of K, given $M(z)$:

$$M(z) = 22z^{-1} + 13.5z^{-2} + 7.5z^{-3}$$

- $m[0{\bullet}T] =$
- $m[1{\bullet}T] =$
- $m[3{\bullet}T] =$
- $m[4{\bullet}T] =$

4. Consider a low-pass filter with the transfer function:

$$G(s) = \frac{1}{\tau s + 1}$$

If the input is $e(t)$ and the output is $x(t)$, obtain a difference equation (in terms of τ and T) of the form:

$$x[K] = a\ x[K-1] + b\ e[K] + c\ e[K-1]$$

using the discrete definition of the derivative (time-domain approach).

5. Using the values $\tau = 2$ and $T = 0.05$s, propagate the difference equation of Problem 4 for $e(t)$ = unit step.
6. What are the general guidelines for selecting a sampling rate; such that a one-for-one, discrete-for-continuous, transfer function transformation is feasible?
7. For Problem 11 of Chapter 10, implement the PID controller using the discrete implementation (time-domain approach). Simulate both implementations (continuous and discrete) side by side, and compare results.
8. For Problem 11 of Chapter 10, implement the Laplace approach using the discrete implementation (time-domain approach). Simulate both implementations (continuous and discrete) side by side, and compare results.
9. Calculate the closed-loop bandwidth of Problem 17, Chapter 10.
10. Implement the continuous compensator design obtained in Problem 9, Chapter 10, with a discrete compensator. Simulate both implementations (continuous and discrete) side by side, and compare results.
11. Implement the continuous compensator design obtained in Problem 10, Chapter 10, with a discrete compensator. Simulate both implementations (continuous and discrete) side by side, and compare results.

13

Case Study: A Position Control System Using a DC Solenoid

Learning Objectives

In this chapter, you will learn how to:

- Identify the process a control engineer goes through when designing a control system.
- Describe the dynamic characteristics of a DC solenoid.
- Discuss the testing process that occurs for every new control system.
- Formulate a set of control system specifications.

13.1 Introduction

Many control engineers decided to pursue control engineering when they were exposed to the control aspects of a problem. The goal of this chapter is to propose a specific problem that can provide such an in-depth opportunity for the student. The instructor can assign this problem as a special one or use it as a lecture example. When the necessary hardware is available, this problem is an excellent hands-on, laboratory exercise.

13.2 Process Overview

This problem is presented in phases. Try not to look at phases that have not yet been covered.

13.3 Problem Definition

Assume that you are a control engineer and that an animation/special effects company wants a drive system that will move the jaw of a plastic figure proportionally to an electrical command signal. The head is connected to a plastic-tube-rod mechanism that goes to the system that you will design (see Figure 13–1). Your first task is to come up with different ways of accomplishing the job by sketching possible implementations. There are no technology or method restrictions.

After this phase is completed, feel free to discuss with your classmates and friends your best implementations, your reasons for selecting those approaches, and the pros and cons of each. Also, perform some quick engineering economics and project planning to estimate the following resources:

- Staff hours required (engineering, manufacturing)
- Component list and cost

Identify a reasonable set of specifications. Is a finite error permitted? Do you need integral control? What is a realistic time constant?

13.4 Driver Selection and Implementation

Assume that the instructor, taking the role of the chief engineer, suggests that a DC solenoid, with some enhancements, will accomplish the job. Keep in mind that a

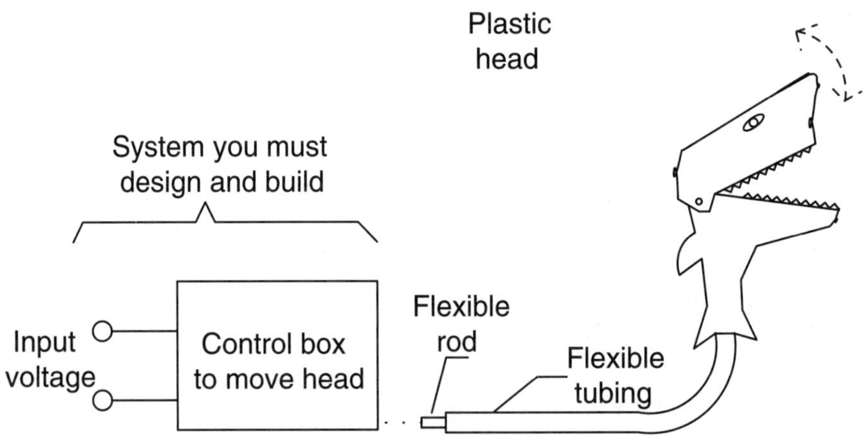

Figure 13–1 System that must be designed

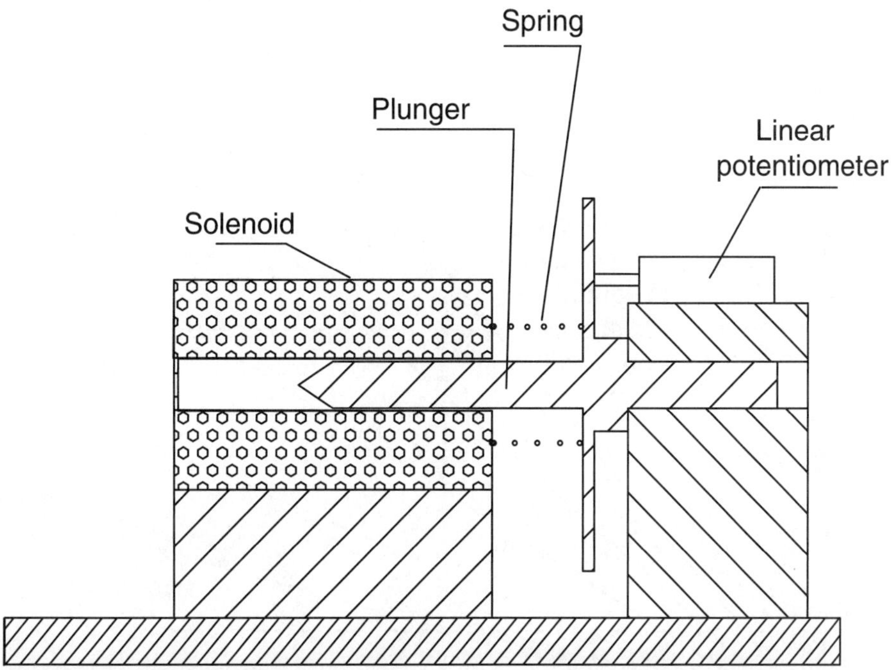

Figure 13–2 Assembled DC solenoid

DC solenoid is normally a two-position device. Further, the instructor has a solenoid/spring arrangement available and assembled. The instructor wants you to evaluate Figure 13–2. Your task is to do some research and determine the characteristics of the DC solenoid.[1,2] Be ready to share your findings with the rest of the class. Note that when the solenoid is activated, it pulls the plunger inward.

13.5 Testing

To perform this task, ideally you have access to hardware. It is possible to complete this task without it, however. Assume that you want to explore how the hardware in Figure 13–2 will respond to a voltage applied to the solenoid using a power supply. Notice that the solenoid compresses the spring when activated. The goal is to apply a voltage to the solenoid that makes the plunger move into the halfway position.

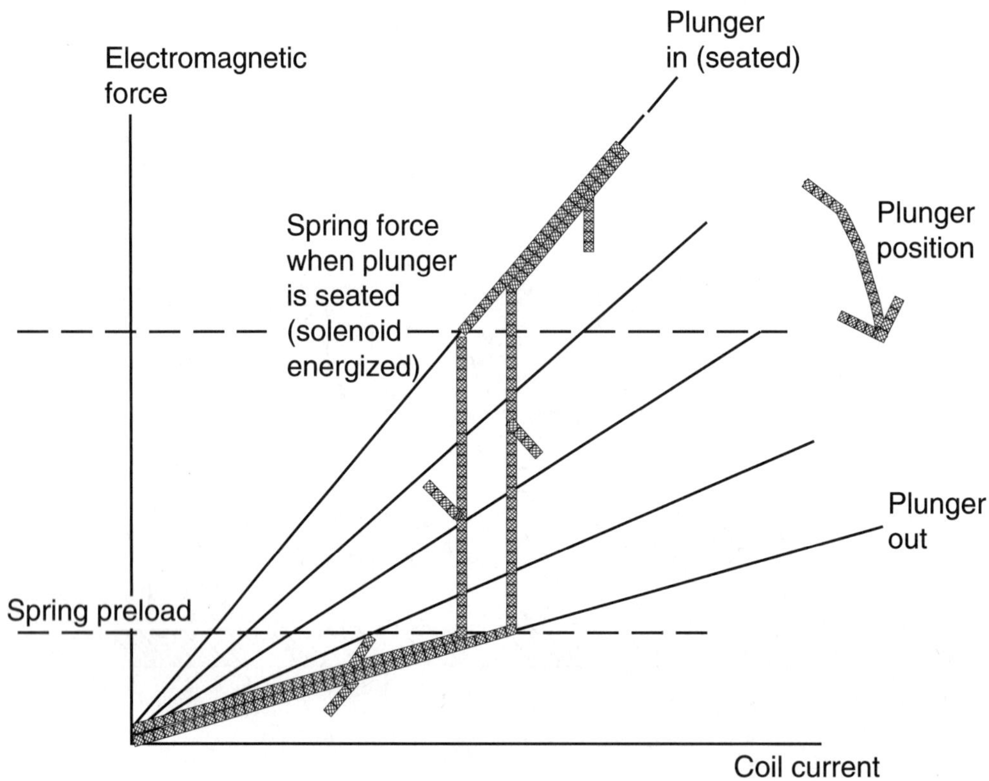

Figure 13–3 Characteristics of DC solenoid with spring

After several unsuccessful attempts, you should conclude that this is impossible. Examine Figure 13–3, which shows the force-position-current characteristics of this solenoid, to understand why. This situation is true when the spring force is much weaker than the force the solenoid can produce.

13.6 Modeling

The next task is to model the system. Assume that you will use a linear potentiometer to sense the position of the system. Initially, you will also use a rotary potentiometer to produce a voltage indicative of the desired position. Figure 13–4 is a

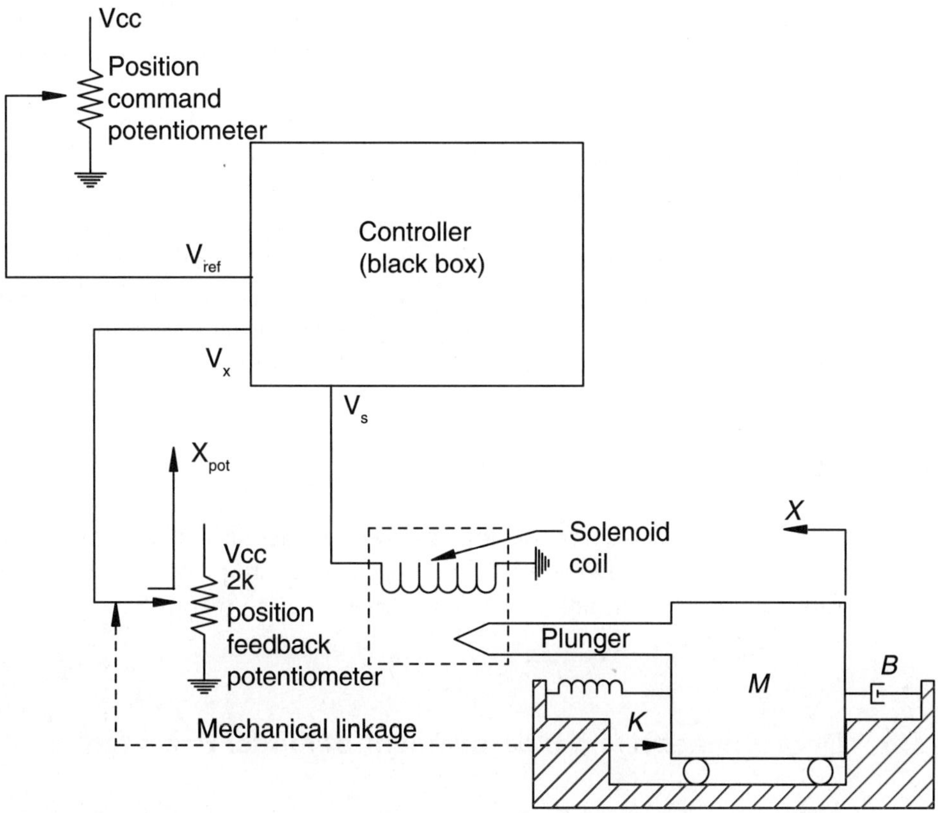

Figure 13–4 Schematic of complete system

good schematic with which to begin the modeling process. One of your assignments is to create a block diagram of all the information, including all the sensors and solenoid characteristics.

You might want to begin this task with an overall block diagram that shows all the relevant components in the system and their interconnections. This representation should not be very specific and might contain blocks with general labels, such as "solenoid dynamics."

Your next task is to obtain model schematics (see Chapter 3) for all relevant components possessing dynamics characteristics. For each component, the result of these efforts should be a specific mathematical representation, such as a transfer function, that can be incorporated in the overall block diagram. At this point, the transfer function will be in terms of symbolic coefficients, where parameter values will be determined later.

Some of the blocks might retain their general definition. An example is the controller, since its definition will not be established until later in the process.

Once all the components have been mathematically defined, they can be incorporated into the working block diagram. This representation incorporates all the transfer functions and blocks that represent the complete system. The phrase "working block diagram" is used since this representation most likely changes as the project progresses. For instance, your experience during testing might motivate you to change the model, or you might select a different driver circuit.

13.7 Solenoid Driver Selection

The next task is to determine how to drive the solenoid. What are the solenoid current requirements? Do you think a 741 op-amp can produce the needed current? Once you have information about the driver, you can incorporate these characteristics in the model.

13.8 Determination and Measurement of Model Parameters

Your next task is to come up with a method for obtaining all the relevant parameter values so that you can put numbers in your model instead of symbolic coefficients.

13.9 Simulation

Now that you have a completed model (model structure and parameter values), you can begin doing some simulations. Get to know the dynamic characteristics of the system. Learn the associated time constants, natural frequencies, range, and other important characteristics. Implement the model in SIMULINK, and confirm that the method is valid.

13.10 Controller Selection (Continuous-Type Device)

Determine which controller you will use (e.g., PLC, analog electronics, LabVIEW). Incorporate any relevant characteristics into the model.

13.11 First Pass at Controller Design: Proportional Feedback

Close the loop using proportional control only. Using the set of specifications previously stated and assuming a maximum steady-state error of 15 percent for a step input, obtain an idea of the required controller gain. Evaluate the system dynamics. Does the system meet the specifications? If necessary, modify any parameters that can be adjusted (e.g., controller gain).

13.12 Second Pass: PD Controller

Assume that you want to improve the speed of response. Which improvements can be achieved using PD or another type of compensator? Design and simulate your new design, and compare it to your earlier design.

13.13 Implementation with a Digital Controller

Assume that you want to use a microcontroller-based controller. What is the bandwidth of your system? What would be the sampling frequency of the microcontroller?

Summary

1. This chapter attempts to give the student an experience similar to that of a control engineer. This experience can be realized as a special problem used outside class or as class lecture material.

Idea Portfolio

1. Write a 1-page article that will be published in the local paper describing your efforts in developing such a system.
2. Using a minimum number of technical terms, write a 1-page article for *Time* magazine describing your dynamic system.
3. Using technical terms, write a 1-page article for *Control Engineering* describing your dynamic system.
4. If time permits, look into what would be required to write a business plan to attract investors.

Appendix A

Derivation of the Relationship between *Tr* and *BW* and τ

Most oscilloscopes provide a means for correlating rise time, *Tr*, and the system's bandwidth. As discussed in Chapter 5, rise time is the time it takes the system to go from 10 to 90 percent of its final value when responding to a step input or a square wave of sufficiently low frequency that allows the transient response to die out before the next pulse of the square wave. The oscilloscope facilitates this correlation by providing dotted, horizontal lines at the 10 and 90 percent of a range comprised of 10 vertical divisions. Many oscilloscopes provide the following formula in their front panels:

$$BW = \frac{0.35}{T_R}$$

This appendix derives the preceding formula.

A first-order system's response to a step input is given by:

$$v(t) = V_F\left(1 - e^{-t/\tau}\right)$$

where V_F is the final value and τ is the time constant.

This equation can also be expressed as:

$$\frac{v(t)}{V_F} = \left(1 - e^{-t/\tau}\right)$$

or

$$e^{-t/\tau} = \frac{V(t)}{V_F} - 1$$

Because this equation is valid for any *t*, it can be evaluated at any two arbitrary values of time (e.g., t_1 and t_2). Evaluating the preceding expression at t_1 and t_2 and taking the ratio of the two resulting equations produces:

$$\frac{e^{-t_2/\tau}}{e^{-t_1/\tau}} = \frac{1 - \dfrac{V(t_2)}{V_F}}{1 - \dfrac{V(t_1)}{V_F}} = e^{-(t_2 - t_1)/\tau}$$

Taking the natural log of the preceding equation:

$$\ln\left[e^{-(t_2 - t_1)/\tau}\right] = \ln\left[\frac{1 - \dfrac{V(t_2)}{V_F}}{1 - \dfrac{V(t_1)}{V_F}}\right]$$

which results in:

$$\frac{-(t_2 - t_1)}{\tau} = \ln\left[\frac{1 - \dfrac{V(t_2)}{V_F}}{1 - \dfrac{V(t_1)}{V_F}}\right]$$

or:

$$t_2 - t_1 = -\ln\left[\frac{1 - \dfrac{V(t_2)}{V_F}}{1 - \dfrac{V(t_1)}{V_F}}\right]\tau$$

Because the rise time, T_r, is defined as the time between the 10 and 90 percent of the final value, then:

$$T_r = t_2 - t_1 = -\ln\left[\frac{1-0.9}{1-0.1}\right]\tau$$

Next, τ is related to the corner frequency, or bandwidth in rad/s, ω_0, because:

$$\tau = \frac{1}{\omega_0} = \frac{1}{2\pi f_0} = \frac{1}{2\pi BW}$$

Note: BW and f_0 are the same.

Then, we can substitute for τ:

$$T_r = -\ln\left[\frac{1-0.9}{1-0.1}\right]\frac{1}{2\pi BW}$$

Evaluating all the constants results in:

$$T_r = \frac{0.35}{BW}$$

Using the preceding equation with the relationship between τ and BW we obtain:

$$T_r = 2.2\tau$$

Notice that this relationship is only exact for first-order systems.

Appendix B

MATLAB Basics

MATLAB, which resemble command prompt originally found in DOS, works very much like a command-line program. However, it has many toolboxes and graphical user interfaces (GUI) that leverage the Microsoft® Windows environment. The command prompt looks like:

```
>>
```

indicating that it is ready for a command.
Entering:

```
>>b=33;
```

enters into the workspace a variable named *b* with a value of 33. The semicolon supresses any immediate output. If the semicolon was not used, the following would result:

```
b =

    33
```

which helps ensure the right data was entered.
Entering:

```
>>A=[1 2;3 4;5 6;7 8]
```

would define a matrix *A* that is 4×2.

The command `who` displays the names of the variables in the workspace. The command `whos` in addition to displaying the variable name, it displays the type and number of elements associated with each variable.

Entering:

```
>>b=5+3j
```

is a complex number with a real part of 5 and an imaginary part of 3. It produces an output of:

```
c =

    5.0000 + 3.0000i
```

Entering:

```
>>r=3*4/2
```

produces:

```
r =

    6
```

Useful Commands

Any DOS command can be invoked from the MATLAB environment by simply preceding the command with a "!". For instance, to display all the text files in the A drive enter:

```
>>!dir a:*.txt
```

The command:

```
pwd
```

shows the working directory.

Consider the following polynomial:

$$s^3 + 7s^2 + 14s + 8$$

This polynomial in s, which we will call *poly1,* is captured in MATLAB by entering:

```
>>poly1=[1 7 14 8]
```

Notice that $poly1(1) = 1$, $poly1(2) = 7$, $poly1(3) = 14$, and $poly1(4) = 8$. To find the roots of the polynomial, we invoke the `roots()` function:

```
>>roots(poly1)
```

which produces:

```
ans =

   -4.0000

   -2.0000

   -1.0000
```

The function `damp()` is similar, except that complex pair roots are shown as pairs, with their respective natural frequencies and damping ratios.

Because transfer functions are such common elements of MATLAB, a function has been created that captures them (i.e., the ratio of two polynomials), for instance:

```
>>G1=tf(2, [1 2])
>>G2=tf(3, [4 5])
```

defines two transfer functions (1) $G_1(s)$ and (2) $G_2(s)$.

Similarly, the following expression combines the two transfer functions:

```
>>G=(G1*G2)
```

and

```
>>CLTF=minreal(G/(1+G))
```

calculates the closed-loop transfer function

Functions

The following functions are available in the MATLAB environment:

`abs`	absolute value
`ans`	variable created to hold the value of a calculation
`acos`	arcosine
`atan`	arctangent (2 quadrant)
`atan2`	arctangent (4 quadrant)
`axis`	specifies the x and y range displayed in a plot
`clear`	clears the MATLAB workspace
`conv`	multiplies two polynomials
`cos`	cosine
`diary`	saves the session (commands entered, anwers)
`eig`	computes the eigen values
`exp`	exponential with base *e*
`format`	controls the display format
`grid`	toggles the grid on the current plot
`help`	help utility
`hold`	freeze current plot and axis
`imag`	extracts imaginary component of a complex number
`linspace`	produces linearly spaced numbers
`load`	loads variable(s) stored in the working directory
`log`	logarithm (base *e*)
`log10`	logarithm (base 10)
`logspace`	produces geometrically spaced numbers

`margin`	computes gain and phase margins and corresponding crossover frequencies
`max`	calculates maximum value of the elements in an array
`min`	calculates minimum value of the elements in an array
`nichols`	Nichols chart
`nyquist`	Nyquist plot
`plot`	generates plot
`poly`	computes polynomial coefficients from roots
`polyval`	computes value of polynomial
`pzmap`	plots the pole-zero map of a linear system
`real`	extracts real component of a complex number
`residu`	partial fraction expansion
`rlocfind`	determines gain associated with a point in the root locus
`roots`	determines roots of a polynomial
`sin`	sine
`sqrt`	square root
`tan`	tangent
`text`	places text on current graph
`title`	places the title on current graph
`who`	shows present variables in the MATLAB workspace
`whos`	shows present variables, size and type in MATLAB workspace
`xlabel`	places x-axis label on current plot
`ylabel`	places y-axis label on current plot

Script Files

Think of script files as batch or macro files. They are text files that can be created or viewed with Notepad or any other word processor. MATLAB also has a utility for creating and debussing script files. The following example generates an array with one cycle of the sine function (1kH) and plots it. Figure AB–1 shows the output.

```
% script file sho_sine.m
% this script file generates an array with one cycle of the sine
% function (1kH). It also plots it.

% generate the input
t=linspace(0,0.001,100);
% calculate the result
sine_out=sin(2*pi*1000*t);
```

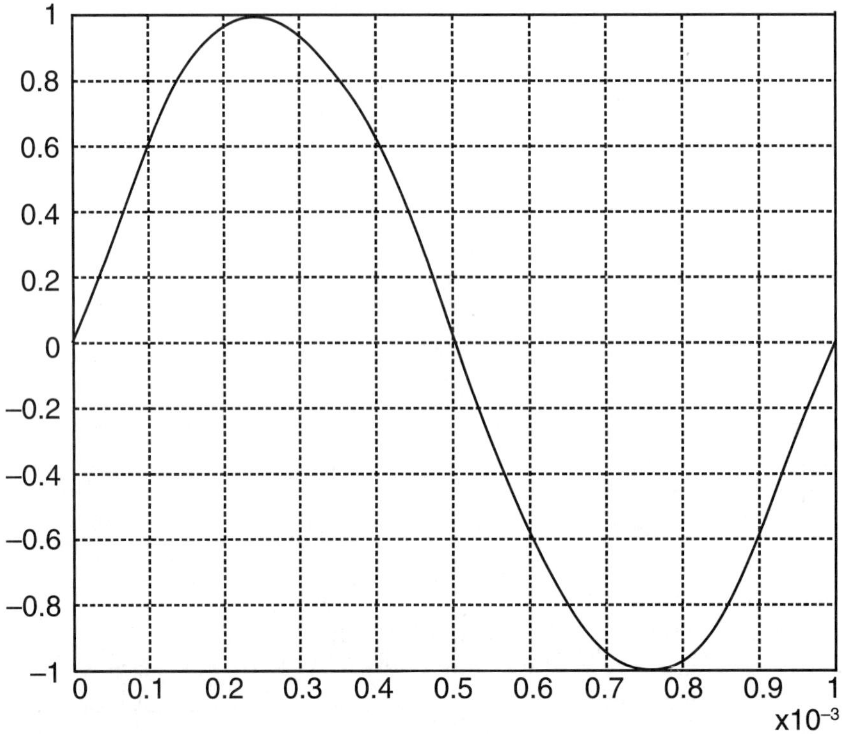

AB–1

```
% plot the result
plot(t,sine_out);
% turn on grid
grid
```

Notice that the preceding file introduces into the workspace two new arrays: (1) *t* and (2) *sine_out*.

Function Files

Function files are similar to subroutines in that they receive a(n) input(s) and produce output(s). The following function is a text file named smooth.m. This example is extremely slow, because it operates one array element at a time. Entering at the MATLAB prompt:

```
>>smooth_data=smooth(raw_data,4)
```

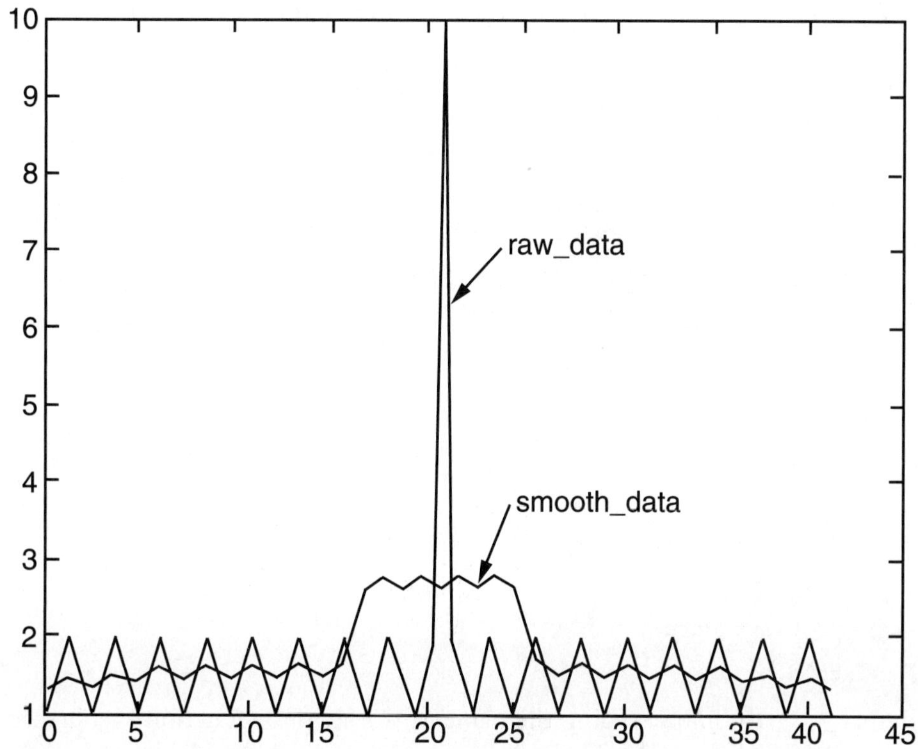

AB–2

produces the plot in Figure AB–2. Notice that it also creates an array, smooth_data, that contains the smoothed data.

```
function [out_data]=smooth(in_data,n1)
% this function smooths the data in the input array by averaging
% each point with the preceding and the following n1 points
% It also plots the input and output
% usage:
% [out_data]=smooth(in_data,n1);
%         or simply
%    smooth(in_data,n1);
% where:
% in_data = this is the array containing the input data
% out_data = this is the array containing the output data
```

```
% n1 = number of data points on each side to smooth
n_ele=length(in_data);
for i=1 : n_ele;
      sum1=0;
      for j=1 : n1;
        index1=max([i-j , 1]);
        index2=min([i+j , n_ele]);
       sum1-sum1+in_data( index1 ) + in_data( index 2 );
    end
     sum1=sum1+in_data(i);
 out_data(i)=sum1/(2*n1+1);
end
plot(in_data);
hold on;
plot(out_data);
hold off;
```

By simply typing the name of the script or function file, preceded by the word 'help,' one obtains the comments prior to the first blank line in the file. This is extremely helpful when remembering the function and usage of a file. For example, typing:

```
>>help sho_sine
```

produce the first three lines of the script file `sho_sine.m`.

Typing:

```
disp(` The pressure at the nozzle is: `)
```

produces a text output, which is useful when documenting intermediate script file results.

Plotting SIMULINK Simulation Results

Simulation data can be captured in a "Save to Workspace" sink block. The data can then be manipulated and plotted in the MATLAB environment using the name used in the "Save to Workspace" block. It can also be retrieved from "Scope" block by using the line command:

```
>>SIMPLOT(DATA)
```

Appendix C

MATLAB Function asymbode.m for Asymptotic Bode Magnitude Plotting

This MATLAB function produces an asymptotic, magnitude Bode Plot that clearly shows the corner frequencies associated with the poles and zeros.

The recommended implementation is to modify this script file (using a text editor or MATLAB's built-in m-file editor) to reflect the poles and zeros of the system under investigation, then rename it (e.g., probl.m). It can also be called as any MATLAB function by uncommenting the first line. The program expects the transfer function to be in Bode format, for example:

$$T(s) = \frac{K_e\left(\frac{s}{z_1} + 1\right)}{\frac{s}{p_1} + 1} \cdots$$

Program Listing

```
% function asymbode=(System_Type,Error_Coef,FreqRange,Zeros,Poles)
% ASYMBODE.M
% Written for Dr. Thaler to do simple Asymptotic Bode
% plotting without using the Bode command for starting
% values.
```

```
% Written by Colin R. Cooper
% Last Mod: 11/6/96
% label for magnitude and frequency axis added 6/21/03 (nfm)
% System parameters must be defined.
% ---------------------------------------------------------
System_Type =             [2];
Error_Coef=             [50];     % Error coefficent
FreqRange =         [.01 1000];   % Define plot range
Zeros =                   [3];
Poles =              [.1 10 10];
Err_Coef_Flag = 'y';    % 'y' displays the error coef.
                        % 'n' does not display.

LType = '-';            % Line type for Asymp curve
EType = '-r';           % Line type for Error coef. curve
                        % '-'  for solid line
                        % '--' for dashed line
                        % ':'  for dotted line
                        % '-.' for dash-dot line
% ---------------------------------------------------------
% Do not modify anything below this line.
% =========================================================
% Make the Corner Freq. vector and sort by frequency
cf = [Zeros, Poles];
sl = [20*ones(size(Zeros)), -20*ones(size(Poles))];
[cf,ind] = sort(cf);
sl = sl(ind);
% Select the correct starting magnitude for the System Type
SL = -20*System_Type;
if System_Type == 0
    cf_dB = 20*log10(Error_Coef);  % Starting magnitude at min freq.
else
```

```
    cf_dB = -
20*system_Type*log10(FreqRange(1)/((Error_Coef)^(1/System_Type)));
    Err_Coef_dB = [cf_dB, 0];        % Values for Err_Coef line
    Err_Coef_fr = [FreqRange(1), Error_Coef^(1/System_Type)];
end

% Generate the points for the asymptotic diagram

sl = [0, sl];     % Insert starting slope
cf = [FreqRange(1), cf, FreqRange(2)];    % Inserts end points
for n = 1:length(cf)-1
  SL = SL + sl(n);
  cf_dB(n+1) = cf_dB(n) + SL*log10(cf(n+1)/cf(n));
end
% Plot the output.
figure (1)

% If the system is type 0, or you don't want the error coef line
if (System_Type == 0) | (Err_Coef_flag == 'n')
   semilogx(cf,cf_dB,LType,...
       cf(find(sl==-20)),cf_dB(find(sl==-20)),'+',...
       cf(find(sl==20)),cf_dB(find(sl==20)),'o')
% If the system is type 1,2 and you want to show the error coef.
else
  semilogx(cf,cf_dB,LType,...
      cf(find(sl==-20)),cf_dB(find(sl==-20)),'+',...
      cf(find(sl==20)),cf_dB(find(sl==20)),'o',...
      Error_Coef_fr,Err_Coef,dB,EType)
end

%additions 6/21/03 nfm
xlabel( 'Frequency (rad/sec)')
ylabel( 'magnitude (dB)')
grid on
```

Usage

Consider the following transfer function:

$$G(s) = \frac{50\left(\frac{s}{3}+1\right)}{s^2\left(\frac{s}{0.1}+1\right)\left(\frac{s}{10}+1\right)\left(\frac{s}{10}+1\right)}$$

This MATLAB file was edited to reflect the characteristics of the preceding transfer function. Notice that it will plot the asymptotic Bode Plot for frequencies in the range of 0.01 to 1000 rad/s. To produce the plot, type the following MATLAB prompt:

```
>>asymbode
```

which produces Figure AC–1. Note that the screen plot is in color. When the transfer function has no zeros, enter nothing inside the brackets.

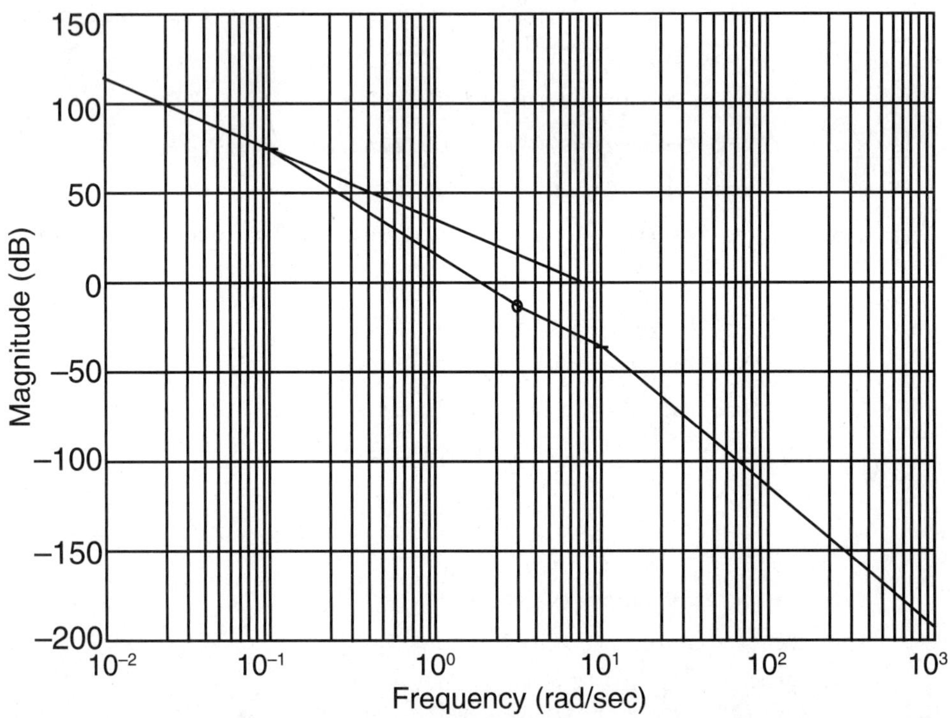

Figure AC–1 Example output of the MATLAB function asymbode.m

Appendix D

Allen-Bradley's PID Instruction

(Courtesy of Allen-Bradley)

Process Control Instruction

This chapter describes the MicroLogix 1200 and MicroLogix 1500 Proportional Integral Derivative (PID) instruction. The PID instruction is an output instruction that controls physical properties such as temperature, pressure, liquid level, or flow rate using process loops.

The PID Concept

The PID instruction normally controls a closed loop using inputs from an analog input module and providing an output to an analog output module. For temperature control, you can convert the analog output to a time proportioning on/off output for driving a heater or cooling unit. An example appears on page 19-17.

The PID instruction can be operated in the timed mode or the Selectable Time Interrupt (STI mode). In the timed mode, the instruction updates its output periodically at a user-selectable rate. In the STI mode, the instruction should be placed in an STI interrupt subroutine. It then updates its output every time the STI subroutine is scanned. The STI time interval and the PID loop update rate must be the same in order for the equation to execute properly. See Using the Selectable Timed Interrupt (STI) Function File on page 18-12 for more information on STI interrupts.

PID closed loop control holds a process variable at a desired set point. A flow rate/fluid level example is shown below.

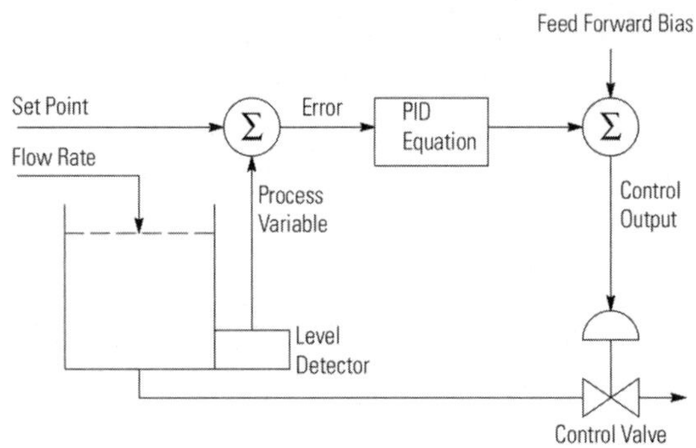

The PID equation controls the process by sending an output signal to the control valve. The greater the error between the setpoint and process variable input, the greater the output signal. Alternately, the smaller the error, the smaller the output signal. An additional value (feed forward or bias) can be added to the control output as an offset. The PID result (control variable) drives the process variable toward the set point.

The PID Equation

The PID instruction uses the following algorithm:

Standard equation with dependent gains:

$$Output = K_C\left[(E) + \frac{1}{T_I}\int(E)dt + T_D \cdot \frac{d(PV)}{dt}\right] + bias$$

Standard Gains constants are:

Term	Range (Low to High)	Reference
Controller Gain K_C	0.01 to 327.67 (dimensionless)[(1)]	Proportional
Reset Term $1/T_I$	327.67 to 0.01 (minutes per repeat)[(1)]	Integral
Rate Term T_D	0.01 to 327.67 (minutes)[(1)]	Derivative

(1) Applies to MicroLogix 1200 and 1500 PID range when Reset and Gain Range (RG) bit is set to 1. For more information on reset and gain, see PLC 5 Gain Range (RG) on pa ge19-13.

The derivative term (rate) provides smoothing by means of a low-pass filter. The cut-off frequency of the filter is 16 times greater than the corner frequency of the derivative term.

PD Data File

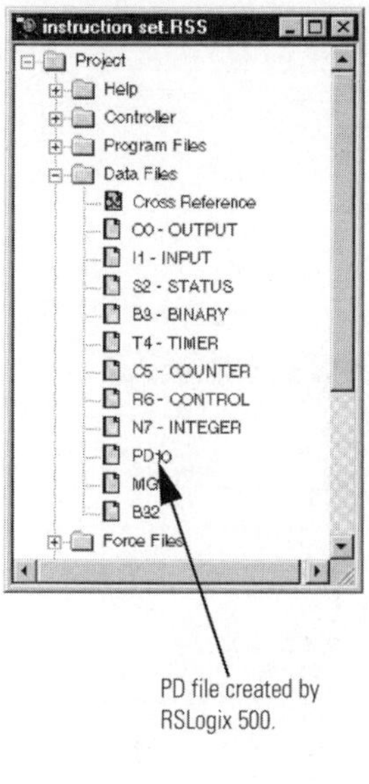

PD file created by RSLogix 500.

The PID instruction implemented by the MicroLogix 1200 and 1500 controllers is virtually identical in function to the PID implementation used by the Allen-Bradley SLC 5/03 and higher processors. Minor differences primarily involve enhancements to terminology. The major difference is that the PID instruction now has its own data file. In the SLC family of processors, the PID instruction operated as a block of registers within an integer file. The Micrologix 1200 and 1500 PID instruction utilizes a PD data file.

You can create a PD data file by creating a new data file and classifying it as a PD file type. RSLogix automatically creates a new PD file or a PD sub-element whenever a PID instruction is programmed on a rung. The PD file then appears in the list of Data Files as shown in the illustration.

Each PD data file has a maximum of 255 elements and each PID instruction requires a unique PD element. Each PD element is composed of 20 sub-elements, which include bit, integer and long integer data. All of the examples in this chapter use PD file 10 sub-element 0.

PID - Proportional Integral Derivative

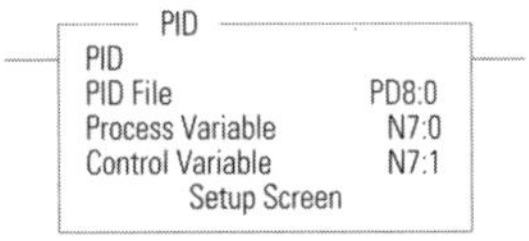

Instruction Type: output

Table 19.1 Execution Time for the PID Instruction

Controller	When Rung Is:	
	True	False
MicroLogix 1200	295.8 μs	11.0 μs
MicroLogix 1500	251.8 μs	8.9 μs

It is recommended that you place the PID instruction on a rung without any conditional logic. If conditional logic exists, the Control Variable output remains at its last value, and the CVP CV% term and integral term are both cleared when the rung is false.

NOTE In order to stop and restart the PID instruction, you need to create a false-to-true rung transition.

The example below shows a PID instruction on a rung with RSLogix 500 programming software.

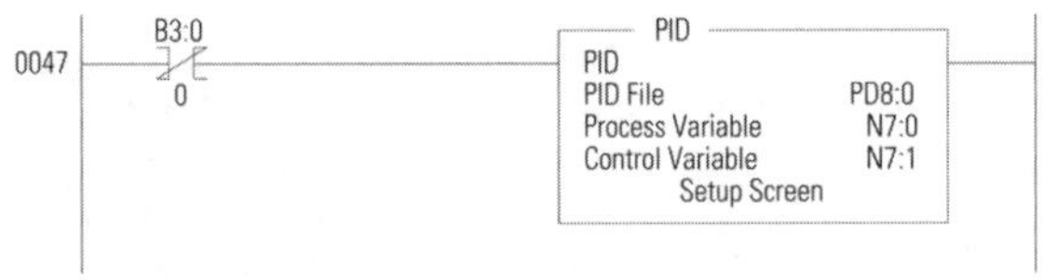

When programming, the setup screen provides access to the PID instruction configuration parameters. The illustration below shows the RSLogix 500 setup screen.

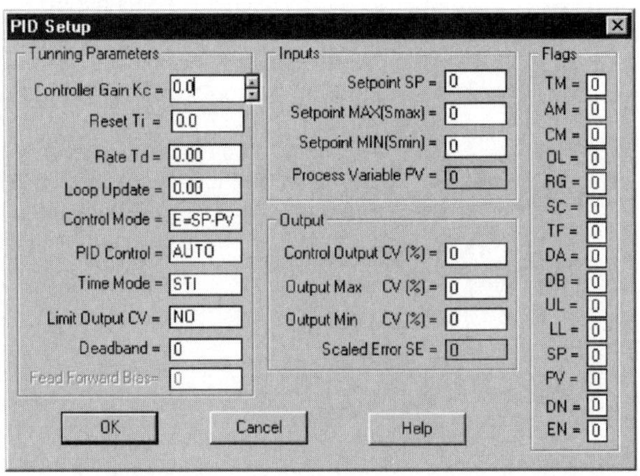

Input Parameters

The table below shows the input parameter addresses, data formats, and types of user program access. See the indicated pages for descriptions of each parameter.

Input Parameter Descriptions	Address	Data Format	Range	Type	User Program Access	For More Information
SPS - Setpoint	PD10:0.SPS	word (INT)	0 to 16383(1)	control	read/write	19-4
PV - Process Variable	user defined	word (INT)	0 to 16383	control	read/write	19-4
MAXS - Setpoint Maximum	PD10:0.MAXS	word (INT)	-32,768 to +32,767	control	read/write	19-5
MINS - Setpoint Minimum	PD10:0.MINS	word (INT)	-32,768 to +32,767	control	read/write	19-5
OSP - Old Setpoint Value	PD10:0.OSP	word (INT)	-32,768 to +32,767	status	read only	19-5
OL - Output Limit	PD10:0/OL	binary	1 = enabled 0 = disabled	control	read/write	19-6
CVH - Control Variable High Limit	PD10:0.CVH	word (INT)	0 to 100%	control	read/write	19-6
CVL - Control Variable Low Limit	PD10:0.CVL	word (INT)	0 to 100%	control	read/write	19-6

(1) The range listed in the table is for when scaling is not enabled. With scaling, the range is from minimum scaled (MINS) to maximum scaled (MAXS).

Setpoint (SPS)

Input Parameter Descriptions	Address	Data Format	Range	Type	User Program Access
SPS - Setpoint	PD10:0.SPS	word (INT)	0 to 16383(1)	control	read/write

(1) The range listed in the table is for when scaling is not enabled. With scaling, the range is from minimum scaled (MINS) to maximum scaled (MAXS).

The SPS (Setpoint) is the desired control point of the process variable.

Process Variable (PV)

Input Parameter Descriptions	Address	Data Format	Range	Type	User Program Access
PV - Process Variable	user defined	word (INT)	0 to 16383	control	read/write

The PV (Process Variable) is the analog input variable.

Setpoint MAX (MAXS)

Input Parameter Descriptions	Address	Data Format	Range	Type	User Program Access
MAXS - Setpoint Maximum	PD10:0.MAXS	word (INT)	-32,768 to +32,767	control	read/write

If the SPV is read in engineering units, then the MAXS (Setpoint Maximum) parameter corresponds to the value of the setpoint in engineering units when the control input is at its maximum value.

Setpoint MIN (MINS)

Input Parameter Descriptions	Address	Data Format	Range	Type	User Program Access
MINS - Setpoint Minimum	PD10:0.MINS	word (INT)	-32,768 to +32,767	control	read/write

If the SPV is read in engineering units, then the MINS (Setpoint Minimum) parameter corresponds to the value of the setpoint in engineering units when the control input is at its minimum value.

NOTE *MinS - MaxS* scaling allows you to work in engineering units. The deadband, error, and SPV are also displayed in engineering units. The process variable, PV, must be within the range of 0 to 16383. Use of *MinS - MaxS* does not minimize PID PV resolution.

Scaled errors greater than +32767 or less than -32768 cannot be represented. If the scaled error is greater than +32767, it is represented as +32767. If the scaled error is less than -32768, it is represented as -32768.

Old Setpoint Value (OSP)

Input Parameter Descriptions	Address	Data Format	Range	Type	User Program Access
OSP - Old Setpoint Value	PD10:0.OSP	word (INT)	-32,768 to +32,767	status	read only

The OSP (Old Setpoint Value) is substituted for the current setpoint, if the current setpoint goes out of range of the setpoint scaling (limiting) parameters.

Output Limit (OL)

Output Parameter Descriptions	Address	Data Format	Range	Type	User Program Access
OL - Output Limit	PD10:0/OL	binary	1 = enabled 0 = disabled	control	read/write

An enabled (1) value enables output limiting to the values defined in PD10:0.CVH (Control Variable High) and PD10.0.CVL (Control Variable Low).

A disabled (0) value disables OL (Output Limiting).

Control Variable High Limit (CVH)

Output Parameter Descriptions	Address	Data Format	Range	Type	User Program Access
CVH - Control Variable High Limit	PD10:0.CVH	word (INT)	0 to 100%	control	read/write

When the output limit bit (PD10:0/OL) is enabled (1), the CVH (Control Value High) you enter is the maximum output (in percent) that the control variable attains. If the calculated CV exceeds the CVH, the CV is set (overridden) to the CVH value you entered and the upper limit alarm bit (UL) is set.

When the output limit bit (PD10:0/OL) is disabled (0), the CVH value you enter determines when the upper limit alarm bit (UL) is set.

If CV exceeds the maximum value, the output is not overridden and the upper limit alarm bit (UL) is set.

Control Variable Low Limit (CVL)

Output Parameter Descriptions	Address	Data Format	Range	Type	User Program Access
CVL - Control Variable Low Limit	PD10:0.CVL	word (INT)	0 to 100%	control	read/write

When the output limit bit (PD10:0/OL) is enabled (1), the CVL (Control Value Low) you enter is the minimum output (in percent) that the Control Variable attains. If the calculated CV is below the minimum value, the CV is set (overridden) to the CVL value you entered and the lower limit alarm bit (LL) is set.

When the output limit bit (PD10:0/OL) is disabled (0), the CVL value you enter determines when the lower limit alarm bit (LL) is set. If CV is below the minimum value, the output is not overridden and the lower limit alarm bit (LL) is set.

Output Parameters

The table below shows the output parameter addresses, data formats, and types of user program access. See the indicated pages for descriptions of each parameter.

Output Parameter Descriptions	Address	Data Format	Range	Type	User Program Access	For More Information
CV - Control Variable	User-defined	word (INT)	0 to 16,383	control	read/write	19-7
CVP - Control Variable Percent	PD10:0.CVP	word (INT)	0 to 100	control	read/write	19-7
SPV - Scaled Process Variable	PD10:0.SPV	word (INT)	0 to 16383	status	read only	19-7

Control Variable (CV)

Output Parameter Descriptions	Address	Data Format	Range	Type	User Program Access
CV - Control Variable	User-defined	word (INT)	0 to 16,383	control	read/write

The CV (Control Variable) is user-defined. See the ladder rung below.

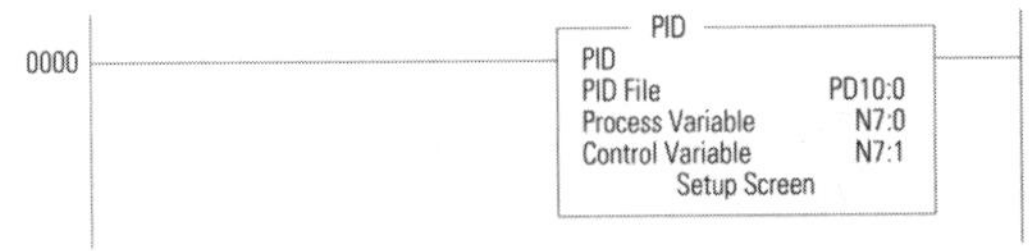

Control Variable Percent (CVP)

Output Parameter Descriptions	Address	Data Format	Range	Type	User Program Access
CVP - Control Variable Percent	PD10:0.CVP	word (INT)	0 to 100	control	read/write

CVP (Control Variable Percent) displays the control variable as a percentage. The range is 0 to 100%. If the PD10:0/AM bit is off (automatic mode), this value tracks the control variable (CV) output. Any value written by the programming software is overwritten. If the PD10:0/AM bit is on (MANUAL mode), this value can be set by the programming software, and the control variable output tracks the control variable percent value.

Scaled Process Variable (SPV)

Input Parameter Descriptions	Address	Data Format	Range	Type	User Program Access
SPV - Scaled Process Variable	PD10:0.SPV	word (INT)	0 to 16383	status	read only

The SPV (Scaled Process Variable) is the analog input variable. If scaling is enabled, the range is the minimum scaled value (MinS) to maximum scaled value (MaxS).

If the SPV is configured to be read in engineering units, then this parameter corresponds to the value of the process variable in engineering units. See Analog I/O Scaling on page 19-17 for more information on scaling.

Tuning Parameters

The table below shows the tuning parameter addresses, data formats, and types of user program access. See the indicated pages for descriptions of each parameter.

Tuning Parameter Descriptions	Address	Data Format	Range	Type	User Program Access	For More Information
KC - Controller Gain - K_c	PD10:0.KC	word (INT)	0 to 32,767	control	read/write	19-9
TI - Reset Term - T_i	PD10:0.Ti	word (INT)	0 to 32,767	control	read/write	19-9
TD - Rate Term - T_d	PD 10:0.TD	word (INT)	0 to 32,767	control	read/write	19-9
TM - Time Mode	PD10:0.TM	binary	0 or 1	control	read/write	19-10
LUT - Loop Update Time	PD10:0.LUT	word (INT)	1 to 1024	control	read/write	19-10
ZCD - Zero Crossing Deadband	PD10:0.ZCD	word (INT)	0 to 32,767	control	read/write	19-11
FF - Feed Forward Bias	PD10:0.FF	word (INT)	-16,383 to +16,383	control	read/write	19-11
SE - Scaled Error	PD10:0.SE	word (INT)	-32,768 to +32,767	status	read only	19-11
AM - Automatic/Manual	PD10:0/AM	binary (bit)	0 or 1	control	read/write	19-12
CM - Control Mode	PD10:0/CM	binary (bit)	0 or 1	control	read/write	19-12
DB - PV in Deadband	PD10:0/DB	binary (bit)	0 or 1	status	read/write	19-12
RG - PLC 5 Gain Range	PD10:0/RG	binary (bit)	0 or 1	control	read/write	19-13
SC - Setpoint Scaling	PD10:0/SC	binary (bit)	0 or 1	control	read/write	19-13
TF - Loop Update Too Fast	PD10:0/TF	binary (bit)	0 or 1	status	read/write	19-13
DA - Derivative Action Bit	PD10:0/DA	binary (bit)	0 or 1	control	read/write	19-14
UL - CV Upper Limit Alarm	PD10:0/UL	binary (bit)	0 or 1	status	read/write	19-14
LL - CV Lower Limit Alarm	PD10:0/LL	binary (bit)	0 or 1	status	read/write	19-14
SP - Setpoint Out of Range	PD10:0/SP	binary (bit)	0 or 1	status	read/write	19-14
PV - PV Out of Range	PD10:0/PV	binary (bit)	0 or 1	status	read/write	19-15
DN - Done	PD10:0/DN	binary (bit)	0 or 1	status	read only	19-15
EN - Enable	PD10:0/EN	binary (bit)	0 or 1	status	read only	19-15
IS - Integral Sum	PD10:0.IS	long word (32-bit INT)	-2,147,483,648 to 2,147,483,647	status	read/write	19-15
AD - Altered Derivative Term	PD10:0.AD	long word (32-bit INT)	-2,147,483,648 to 2,147,483,647	status	read only	19-15

Controller Gain (K_c)

Tuning Parameter Descriptions	Address	Data Format	Range	Type	User Program Access
KC - Controller Gain - K_c	PD10:0.KC	word (INT)	0 to 32,767	control	read/write

Gain K_c (word 3) is the proportional gain, ranging from 0 to 3276.7 (when RG = 0), or 0 to 327.67 (when RG = 1). Set this gain to one-half the value needed to cause the output to oscillate when the reset and rate terms (below) are set to zero.

NOTE Controller gain is affected by the reset and gain range (RG) bit. For information, see PLC 5 Gain Range (RG) on page 19-13.

Reset Term (T_i)

Tuning Parameter Descriptions	Address	Data Format	Range	Type	User Program Access
TI - Reset Term - T_i	PD10:0.Ti	word (INT)	0 to 32,767	control	read/write

Reset T_i (word 4) is the Integral gain, ranging from 0 to 3276.7 (when RG = 0), or 327.67 (when RG = 1) minutes per repeat. Set the reset time equal to the natural period measured in the above gain calibration. A value of 1 adds the maximum integral term into the PID equation.

NOTE Reset term is affected by the reset and gain range (RG) bit. For information, see PLC 5 Gain Range (RG) on page 19-13.

Rate Term (T_d)

Tuning Parameter Descriptions	Address	Data Format	Range	Type	User Program Access
TD - Rate Term - T_d	PD 10:0.TD	word (INT)	0 to 32,767	control	read/write

Rate T_d (word 5) is the Derivative term. The adjustment range is 0 to 327.67 minutes. Set this value to 1/8 of the integral gain T_i.

NOTE This word is not effected by the reset and gain range (RG) bit. For information, see PLC 5 Gain Range (RG) on page 19-13.

Time Mode (TM)

Tuning Parameter Descriptions	Address	Data Format	Range	Type	User Program Access
TM - Time Mode	PD10:0.TM	binary	0 or 1	control	read/write

The time mode bit specifies when the PID is in timed mode (1) or STI mode (0). This bit can be set or cleared by instructions in your ladder program.

When set for timed mode, the PID updates the CV at the rate specified in the loop update parameter (PD10:0.LUT).

When set for STI mode, the PID updates the CV every time the PID instruction is scanned in the control program. When you select STI, program the PID instruction in the STI interrupt subroutine. The STI routine should have a time interval equal to the setting of the PID "loop update" parameter (PD10:0.LUT). Set the STI period in word STI:0.SPM. For example, if the loop update time contains the value 10 (for 100 ms), then the STI time interval must also equal 100 (for 100 ms).

NOTE When using timed mode, your processor scan time should be at least ten times faster than the loop update time to prevent timing inaccuracies or disturbances.

Loop Update Time (LUT)

Tuning Parameter Descriptions	Address	Data Format	Range	Type	User Program Access
LUT - Loop Update Time	PD10:0.LUT	word (INT)	1 to 1024	control	read/write

The loop update time (word 13) is the time interval between PID calculations. The entry is in 0.01 second intervals. Enter a loop update time five to ten times faster than the natural period of the load. The natural period of the load is determined by setting the reset and rate parameters to zero and then increasing the gain until the output begins to oscillate. When in STI mode, this value must equal the STI time interval value loaded in STI:0.SPM. The valid range is 0.01 to 10.24 seconds.

Zero Crossing Deadband (ZCD)

Tuning Parameter Descriptions	Address	Data Format	Range	Type	User Program Access
ZCD - Zero Crossing Deadband	PD10:0.ZCD	word (INT)	0 to 32,767	control	read/write

The deadband extends above and below the setpoint by the value entered. The deadband is entered at the zero crossing of the process variable and the setpoint. This means that the deadband is in effect only after the process variable enters the deadband *and* passes through the setpoint.

The valid range is 0 to the scaled maximum, or 0 to 16,383 when no scaling exists.

Feed Forward Bias (FF)

Tuning Parameter Descriptions	Address	Data Format	Range	Type	User Program Access
FF - Feed Forward Bias	PD10:0.FF	word (INT)	-16,383 to +16,383	control	read/write

The feed forward bias is used to compensate for disturbances that may affect the CV output.

Scaled Error (SE)

Tuning Parameter Descriptions	Address	Data Format	Range	Type	User Program Access
SE - Scaled Error	PD10:0.SE	word (INT)	-32,768 to +32,767	status	read only

Scaled error is the difference between the process variable and the setpoint. The format of the difference (E = SP-PV or E = PV-SP) is determined by the control mode (CM) bit. See Control Mode (CM) on page 19-12.

Automatic / Manual (AM)

Tuning Parameter Descriptions	Address	Data Format	Range	Type	User Program Access
AM - Automatic/Manual	PD10:0/AM	binary (bit)	0 or 1	control	read/write

The auto/manual bit can be set or cleared by instructions in your ladder program. When off (0), it specifies automatic operation. When on (1), it specifies manual operation. In automatic operation, the instruction controls the control variable (CV). In manual operation, the user/control program controls the CV. During tuning, set this bit to manual.

NOTE Output limiting is also applied when in manual.

Control Mode (CM)

Tuning Parameter Descriptions	Address	Data Format	Range	Type	User Program Access
CM - Control Mode	PD10:0/CM	binary (bit)	0 or 1	control	read/write

Control mode, or forward-/reverse-acting, toggles the values E=SP-PV and E=PV-SP.

Forward acting (E=PV-SP) causes the control variable to increase when the process variable is greater than the setpoint.

Reverse acting (E=SP-PV) causes the control variable to decrease when the process variable is greater than the setpoint.

PV in Deadband (DB)

Tuning Parameter Descriptions	Address	Data Format	Range	Type	User Program Access
DB - PV in Deadband	PD10:0/DB	binary (bit)	0 or 1	status	read/write

This bit is set (1) when the process variable is within the zero-crossing deadband range.

PLC 5 Gain Range (RG)

Tuning Parameter Descriptions	Address	Data Format	Range	Type	User Program Access
RG - PLC 5 Gain Range	PD10:0/RG	binary (bit)	0 or 1	control	read/write

When set (1), the reset (TI) and gain range enhancement bit (RG) causes the reset minute/repeat value and the gain multiplier (KC) to be divided by a factor of 10. That means a reset multiplier of 0.01 and a gain multiplier of 0.01.

When clear (0), this bit allows the reset minutes/repeat value and the gain multiplier value to be evaluated with a reset multiplier of 0.1 and a gain multiplier of 0.1.

Example with the RG bit set: The reset term (TI) of 1 indicates that the integral value of 0.01 minutes/repeat (0.6 seconds/repeat) is applied to the PID integral algorithm. The gain value (KC) of 1 indicates that the error is multiplied by 0.01 and applied to the PID algorithm.

Example with the RG bit clear: The reset term (TI) of 1 indicates that the integral value of 0.1 minutes/repeat (6.0 seconds/repeat) is applied to the PID integral algorithm. The gain value (KC) of 1 indicates that the error is multiplied by 0.1 and applied to the PID algorithm.

NOTE The rate multiplier (TD) is not affected by this selection.

Setpoint Scaling (SC)

Tuning Parameter Descriptions	Address	Data Format	Range	Type	User Program Access
SC - Setpoint Scaling	PD10:0/SC	binary (bit)	0 or 1	control	read/write

The SC bit is cleared when setpoint scaling values are specified.

Loop Update Too Fast (TF)

Tuning Parameter Descriptions	Address	Data Format	Range	Type	User Program Access
TF - Loop Update Too Fast	PD10:0/TF	binary (bit)	0 or 1	status	read/write

The TF bit is set by the PID algorithm if the loop update time specified cannot be achieved by the controller due to scan time limitations.

If this bit is set, correct the problem by updating your PID loop at a slower rate or move the PID instruction to an STI interrupt routine. Reset and rate gains will be in error if the instruction operates with this bit set.

Derivative Action Bit (DA)

Tuning Parameter Descriptions	Address	Data Format	Range	Type	User Program Access
DA - Derivative Action Bit	PD10:0/DA	binary (bit)	0 or 1	control	read/write

When set (1), the derivative (rate) action (DA) bit causes the derivative (rate) calculation to be evaluated on the error instead of the process variable (PV). When clear (0), this bit allows the derivative (rate) calculation to be evaluated where the derivative is performed on the PV.

CV Upper Limit Alarm (UL)

Tuning Parameter Descriptions	Address	Data Format	Range	Type	User Program Access
UL - CV Upper Limit Alarm	PD10:0/UL	binary (bit)	0 or 1	status	read/write

The control variable upper limit alarm bit is set when the calculated CV output exceeds the upper CV limit.

CV Lower Limit Alarm (LL)

Tuning Parameter Descriptions	Address	Data Format	Range	Type	User Program Access
LL - CV Lower Limit Alarm	PD10:0/LL	binary (bit)	0 or 1	status	read/write

The control variable lower limit alarm bit is set (1) when the calculated CV output is less than the lower CV limit.

Setpoint Out Of Range (SP)

Tuning Parameter Descriptions	Address	Data Format	Range	Type	User Program Access
SP - Setpoint Out of Range	PD10:0/SP	binary (bit)	0 or 1	status	read/write

This bit is set (1) when the setpoint:

- exceeds the maximum scaled value, or
- is less than the minimum scaled value.

PV Out Of Range (PV)

Tuning Parameter Descriptions	Address	Data Format	Range	Type	User Program Access
PV - PV Out of Range	PD10:0/PV	binary (bit)	0 or 1	status	read/write

The process variable out of range bit is set (1) when the unscaled process variable

- exceeds 16,383, or
- is less than zero.

Done (DN)

Tuning Parameter Descriptions	Address	Data Format	Range	Type	User Program Access
DN - Done	PD10:0/DN	binary (bit)	0 or 1	status	read only

The PID done bit is set (1) for one scan when the PID algorithm is computed. It resets (0) whenever the instruction is scanned and the PID algorithm was not computed (applies to timed mode only).

Enable (EN)

Tuning Parameter Descriptions	Address	Data Format	Range	Type	User Program Access
EN - Enable	PD10:0/EN	binary (bit)	0 or 1	status	read only

The PID enabled bit is set (1) whenever the PID instruction is enabled. It follows the rung state.

Integral Sum (IS)

Tuning Parameter Descriptions	Address	Data Format	Range	Type	User Program Access
IS - Integral Sum	PD10:0.IS	long word (32-bit INT)	-2,147,483,648 to 2,147,483,647	status	read/write

This is the result of the integration $\frac{K_C}{T_I}\int E(dt)$.

Altered Derivative Term (AD)

Tuning Parameter Descriptions	Address	Data Format	Range	Type	User Program Access
AD - Altered Derivative Term	PD10:0.AD	long word (32-bit INT)	-2,147,483,648 to 2,147,483,647	status	read only

This long word is used internally to track the change in the process variable within the loop update time.

Runtime Errors

Error code 0036 appears in the status file when a PID instruction runtime error occurs. Code 0036 covers the following PID error conditions, each of which has been assigned a unique single byte code value that appears in the MSB of the second word of the control block.

Error Code	Description of Error Condition or Conditions	Corrective Action
11H	1. Loop update time $D_t > 1024$	Change loop update time $0 < D_t \leq 1024$
	2. Loop update time $D_t = 0$	
12H	Proportional gain $K_c < 0$	Change proportional gain K_c to $0 < K_c$
13H	Integral gain (reset) $T_i < 0$	Change integral gain (reset) T_i to $0 \leq T_i$
14H	Derivative gain (rate) $T_d < 0$	Change derivative gain (rate) T_d to $0 \leq T_d$
15H	Feed Forward Bias (FF) is out-of-range.	Change FF so it is within the range -16383 to +16383.
23H	Scaled setpoint min MinS > Scaled setpoint max MaxS	Change scaled setpoint min MinS to $-32768 \leq \text{MinS} \leq \text{MaxS} \leq +32767$
31H	If you are using setpoint scaling and MinS > setpoint SP > MaxS, or If you are not using setpoint scaling and 0 > setpoint SP > 16383, then during the initial execution of the PID loop, this error occurs and bit 11 of word 0 of the control block is set. However, during subsequent execution of the PID loop if an invalid loop setpoint is entered, the PID loop continues to execute using the old setpoint, and bit 11 of word 0 of the control block is set.	If you are using setpoint scaling, then change the setpoint SP to $\text{MinS} \leq \text{SP} \leq \text{MaxS}$, or If you are not using setpoint scaling, then change the setpoint SP to $0 \leq \text{SP} \leq 16383$.
41H	**Scaling Selected** 1. Deadband < 0, or 2. Deadband > (MaxS – MinS) **Scaling Deselected** 1. Deadband < 0, or 3. Deadband > 16383	**Scaling Selected** Change deadband to $0 \leq \text{deadband} \leq (\text{MaxS} - \text{MinS}) \leq 16383$ **Scaling Deselected** Change deadband to $0 \leq \text{deadband} \leq 16383$
51H	1. Output high limit < 0, or 2. Output high limit > 100	Change output high limit to $0 \leq \text{output high limit} \leq 100$
52H	1. Output low limit < 0, or 2. Output low limit > 100	Change output low limit to $0 \leq \text{output low limit} \leq \text{output high limit} \leq 100$
53H	Output low limit > output high limit	Change output low limit to $0 \leq \text{output low limit} \leq \text{output high limit} \leq 100$

Analog I/O Scaling

To configure an analog input for use in a PID instruction, the analog data must be scaled to match the PID instruction parameters. In the MicroLogix 1200 and 1500, the process variable (PV) in the PID instruction is designed to work with a data range of 0 to 16,383. The 1769 Compact I/O analog modules (1769-IF4 and 1769-OF2) are capable of on-board scaling. Scaling data is required to match the range of the analog input to the input range of the PID instruction. The ability to perform scaling in the I/O modules reduces the amount of programming required in the system and makes PID setup much easier.

The example shows a 1769-IF4 module. The IF4 has 4 inputs, which are individually configurable. In this example, analog input 0 is configured for 0 to 10V and is scaled in engineering units. Word 0 is not being used in a PID instruction. Input 1 (word 1) is configured for 4 to 20 mA operation with scaling configured for a PID instruction. This configures the analog data for the PID instruction.

Field Device Input Signal	Analog Register Scaled Data
> 20.0 mA	16,384 to 17,406
20.0 mA	16,383
4.0 mA	0
< 4.0 mA	-819 to -1

The analog configuration screen is accessed from within RSLogix 500. Simply double click on the I/O configuration item in the "Controller" folder, and then double click on the specific I/O module.

The configuration for the analog output is virtually identical. Simply address the PID control variable (CV) to the analog output address and configure the analog output to "Scaled for PID" behavior.

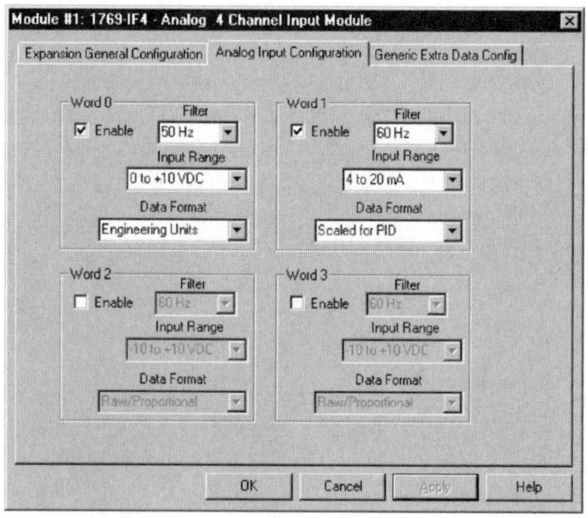

Application Notes

The following paragraphs discuss:

- Input/Output Ranges
- Scaling to Engineering Units
- Zero-crossing Deadband
- Output Alarms
- Output Limiting with Anti-reset Windup
- The Manual Mode
- Feed Forward

ATTENTION Do not alter the state of any PID control block value unless you fully understand its function and how it will affect your process. Unexpected operation could result with possible equipment damage and/or personal injury.

Input/Output Ranges

The input module measuring the process variable (PV) must have a full scale binary range of 0 to 16383. If this value is less than 0 (bit 15 set), then a value of zero is used for PV and the "Process var out of range" bit is set (bit 12 of word 0 in the control block). If the process variable is greater than 16383 (bit 14 set), then a value of 16383 is used for PV and the "Process var out of range" bit is set.

The Control Variable, calculated by the PID instruction, has the same range of 0 to 16383. The Control Output (word 16 of the control block) has the range of 0 to 100%. You can set lower and upper limits for the instruction's calculated output values (where an upper limit of 100% corresponds to a Control Variable limit of 16383).

Scaling to Engineering Units

Scaling lets you enter the setpoint and zero-crossing deadband values in engineering units, and display the process variable and error values in the same engineering units. Remember, the process variable PV must still be within the range 0 to 16383. The PV is displayed in engineering units, however.

Select scaling as follows:

1. Enter the maximum and minimum scaling values MaxS and MinS in the PID control block. The MinS value corresponds to an analog value of zero for the lowest reading of the process variable. MaxS corresponds to an analog value of 16383 for the highest reading. These values reflect the process limits. Setpoint scaling is selected by entering a non-zero value for one or both parameters. If you enter the same value for both parameters, setpoint scaling is disabled.

   For example, if measuring a full scale temperature range of -73°C (PV=0) to +1156°C (PV=16383), enter a value of -73 for MinS and 1156 for MaxS. Remember that inputs to the PID instruction must be 0 to 16383. Signal conversions could be as follows:

Example Values	
Process limits	-73 to +1156°C
Transmitter output (if used)	+4 to +20 mA
Output of analog input module	0 to 16383
PID instruction, MinS to MaxS	-73 to +1156°C

2. Enter the setpoint (word 2) and deadband (word 9) in the same scaled engineering units. Read the scaled process variable and scaled error in these units as well. The control output percentage (word 16) is displayed as a percentage of the 0 to 16383 CV range. The actual value transferred to the CV output is always between 0 and 16383.

When you select scaling, the instruction scales the setpoint, deadband, process variable, and error. You must consider the effect on all these variables when you change scaling.

Zero-Crossing Deadband DB

The adjustable deadband lets you select an error range above and below the setpoint where the output does not change as long as the error remains within this range. This lets you control how closely the process variable matches the setpoint without changing the output.

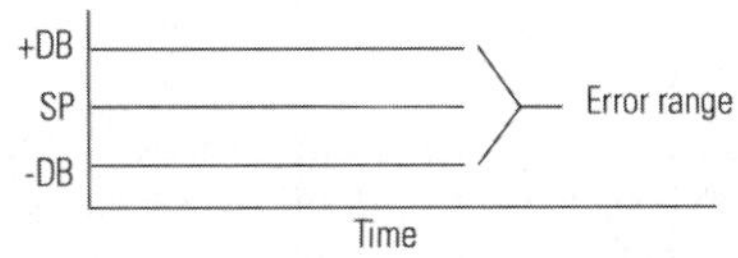

Zero-crossing is deadband control that lets the instruction use the error for computational purposes as the process variable crosses into the deadband until it crosses the setpoint. Once it crosses the setpoint (error crosses zero and changes sign) and as long as it remains in the deadband, the instruction considers the error value zero for computational purposes.

Select deadband by entering a value in the deadband storage word (word 9) in the control block. The deadband extends above and below the setpoint by the value you enter. A value of zero inhibits this feature. The deadband has the same scaled units as the setpoint if you choose scaling.

Output Alarms

You may set an output alarm on the control variable at a selected value above and/or below a selected output percent. When the instruction detects that the control variable has exceeded either value, it sets an alarm bit (bit LL for lower limit, bit UL for upper limit) in the PID instruction. Alarm bits are reset by the instruction when the control variable comes back inside the limits. The instruction does not prevent the control variable from exceeding the alarm values unless you select output limiting.

Select upper and lower output alarms by entering a value for the upper alarm (CVH) and lower alarm (CVL). Alarm values are specified as a percentage of the output. If you do not want alarms, enter zero and 100% respectively for lower and upper alarm values and ignore the alarm bits.

Output Limiting with Anti-Reset Windup

You may set an output limit (percent of output) on the control variable. When the instruction detects that the control variable has exceeded a limit, it sets an alarm bit (bit LL for lower limit, bit UL for upper limit), and prevents the control variable from exceeding either limit value. The instruction limits the control variable to 0 and 100% if you choose not to limit.

Select upper and lower output limits by setting the limit enable bit (bit OL), and entering an upper limit (CVH) and lower limit (CVL). Limit values are a percentage (0 to 100%) of the control variable.

The difference between selecting output alarms and output limits is that you must select output limiting to enable limiting. Limit and alarm values are stored in the same words. Entering these values enables the alarms, but not limiting. Entering these values and setting the limit enable bit enables limiting and alarms.

Anti-reset windup is a feature that prevents the integral term from becoming excessive when the control variable reaches a limit. When the sum of the PID and bias terms in the control variable reaches the limit, the instruction stops calculating the integral sum until the control variable comes back in range. The integral sum is contained in element, IS.

The Manual Mode

In the MANUAL mode, the PID algorithm does not compute the value of the control variable. Rather, it uses the value as an input to adjust the integral sum (IS) so that a smooth transfer takes place upon re-entering the AUTO mode.

In the MANUAL mode, the programmer allows you to enter a new CV value from 0 to 100%. This value is converted into a number from 0 to 16383 and written to the Control Variable address. If your ladder program sets the manual output level, design your ladder program to write to the CV address when in the MANUAL mode. Remember that the new CV value is in the range of 0 to 16383, not 0 to 100. Writing to the CV percent (CVP) with your ladder program has no effect in the MANUAL mode.

PID Rung State

If the PID rung is false, the integral sum (IS) is cleared and CV remains in its last state.

Feed Forward or Bias

Applications involving transport lags may require that a bias be added to the CV output in anticipation of a disturbance. This bias can be accomplished using the processor by writing a value to the Feed Forward Bias element (word FF). (See page 19-11.) The value you write is added to the output, allowing a feed forward action to take place. You may add a bias by writing a value between -16383 and +16383 to word 6 with your programming terminal or ladder program.

Application Examples

PID Tuning

PID tuning requires a knowledge of process control. If you are inexperienced, it will be helpful if you obtain training on the process control theory and methods used by your company.

There are a number of techniques that can be used to tune a PID loop. The following PID tuning method is general and limited in terms of handling load disturbances. When tuning, we recommend that changes be made in the MANUAL mode, followed by a return to AUTO. Output limiting is applied in the MANUAL mode.

NOTE

- This method requires that the PID instruction controls a non-critical application in terms of personal safety and equipment damage.
- The PID tuning procedure may not work for all cases. It is strongly recommended to use a PID Loop tuner package for the best result (i.e. RSTune, Rockwell Software catalog number 9323-1003D).

Procedure

1. Create your ladder program. Make certain that you have properly scaled your analog input to the range of the process variable PV and that you have properly scaled your control variable CV to your analog output.
2. Connect your process control equipment to your analog modules. Download your program to the processor. Leave the processor in the program mode.

ATTENTION Ensure that all possibilities of machine motion have been considered with respect to personal safety and equipment damage. It is possible that your output CV may swing between 0 and 100% while tuning.

NOTE If you want to verify the scaling of your continuous system and/or determine the initial loop update time of your system, go to the procedure on page 19-25.

3. Enter the following values: the initial setpoint SP value, a reset T_i of 0, a rate T_d of 0, a gain K_c of 1, and a loop update of 5.

   Set the PID mode to STI or Timed, per your ladder diagram. If STI is selected, ensure that the loop update time equals the STI time interval.

   Enter the optional settings that apply (output limiting, output alarm, MaxS - MinS scaling, feed forward).
4. Get prepared to chart the CV, PV, analog input, or analog output as it varies with time with respect to the setpoint SP value.
5. Place the PID instruction in the MANUAL mode, then place the processor in the RUN mode.
6. While monitoring the PID display, adjust the process manually by writing to the CO percent value.
7. When you feel that you have the process under control manually, place the PID instruction in the AUTO mode.
8. Adjust the gain while observing the relationship of the output to the setpoint over time.
9. When you notice that the process is oscillating above and below the setpoint in an even manner, record the time of 1 cycle. That is, obtain the natural period of the process.

   Natural Period $\cong$ 4x deadtime

   Record the gain value. Return to the MANUAL mode (stop the process if necessary).

Appendix E

Characteristics of DC Solenoids

The DC solenoid is an electromechanical device that converts an electrical signal to a mechanical movement. Before we describe the characteristics of solenoids, let us review some basic electromagnetic concepts.

Magnetic Principles

The earth is surrounded by a magnetic field that spans from the South Pole to the North Pole. This field can be mathematically and graphically represented as lines of flux, imaginary lines in the direction of the field. The stronger the magnetic field, the more the lines of flux.

A compass is made of a thin strip of magnetized metal that is free to rotate about its center. If we could see the lines of flux produced by the strip near the strip, we would see something like Figure AE–1. When the strip is placed in the earth's magnetic field, it rotates until its north end points in the direction of the earth's lines of flux (i.e., toward the North Pole).

If we were to place a large horseshoe magnet as shown in Figure AE–2, its direction would change, because the horseshoe's magnetic field is much stronger than the earth's magnetic field. Of course, this is only true when the compass is near the horseshoe magnet's magnetic field.

Another important magnetic property of materials is permeability, flux's affinity to go through material. Materials with high permeability are called ferromagnetics. Figure AE–3 shows the horseshoe magnets with two bars of different materials: (1) one with high permeability (cast steel or iron) and (2) low permeability

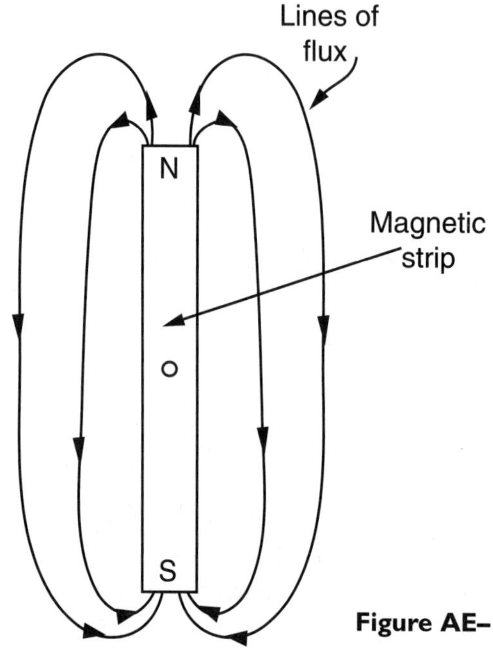

Figure AE–1 Magnetic strip and resulting lines of flux

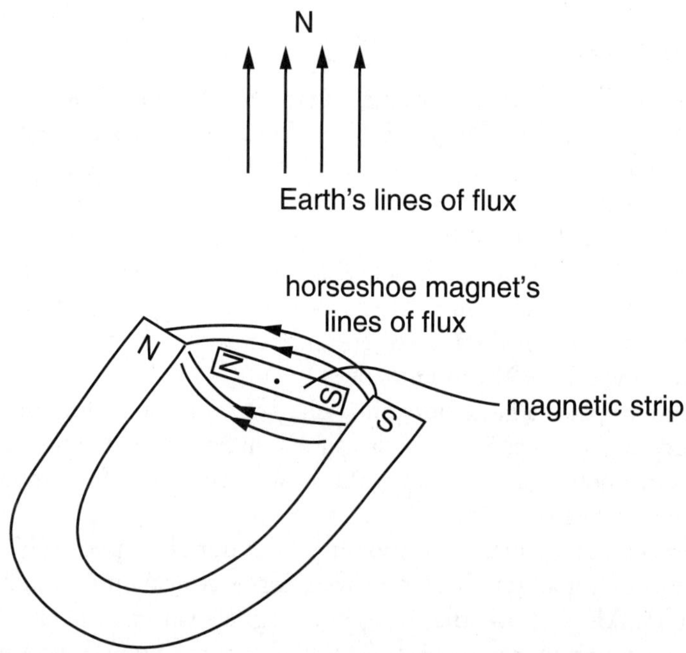

Figure AE–2 Effect of magnetic field on a magnetic strip

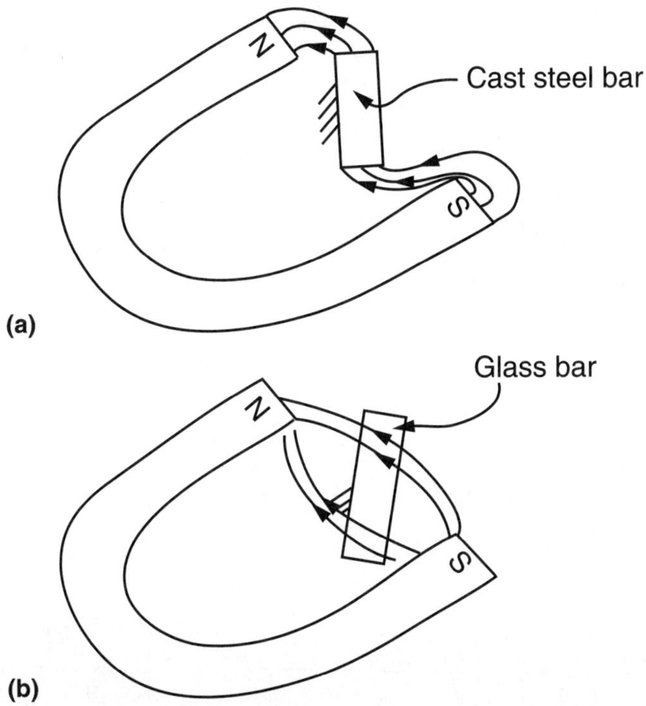

Figure AE–3 Resulting lines of flux in bars of different materials (a) steel, and (b) glass

(glass). Notice how the lines of flux in the first conform to the path established by the iron bar, while in the second the lines of flux ignore the glass bar.

Now consider a common 115 VAC household doorbell, shown in Figure AE–4. When the circuit is energized, the current flowing through the wire creates a magnetic field whose line of flux follows the path formed by the stationary pole and the moving arm (materials with high permeability). An unavoidable portion of the path is the air gap, the space between stationary pole and the moving arm. Air has low permeability, but the lines of flux have to complete their loop and, because they have no other path, they must go through the air gap. The pattern these lines of flux form around the air gap is called fringing.

Another characteristic of lines of flux is that they exert a tension or force that tends to close any air gaps in the loop. The magnitude of the force is proportional to the number of lines of flux crossing the air gap, which is proportional to the instantaneous value of the AC current. For current values above a certain threshold (both negative and positive), the closing force produced by the lines of flux overcomes the

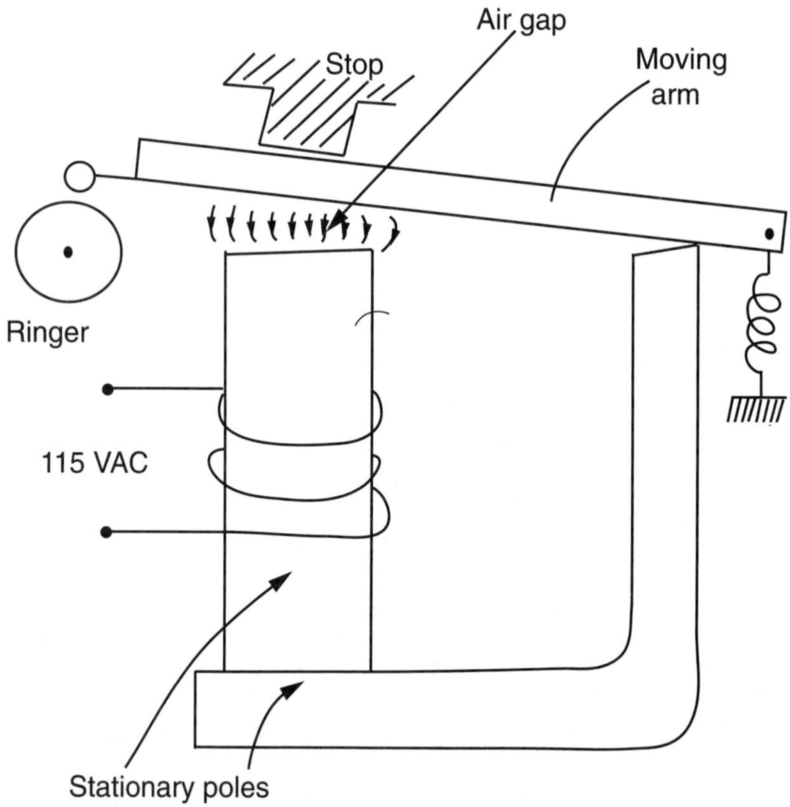

Figure AE–4 Household buzzer

spring force, causing the moving arm to move and strike the ringer. Notice that if the excitation voltage is 115 VAC, 60 Hz, this will take place 2 × 60 times a second. Notice that the magnitude and direction of the lines of flux follow a sinusoidal pattern of 60 Hz. Also notice that the moving arm is always pulled, rather than pushed.

Let us return our focus to the DC solenoid in Figure 13–2. Figure AE–5 shows the main components of the solenoid's assembly: the iron frame, the winding, and the plunger. The section with small circles represents the cross-section of the wires in the winding. When the circuit is energized, a magnetic field is set up. Again, the resulting lines of flux favor paths made of materials with high permeability: the iron frame surrounding the solenoid and the iron plunger. Note that copper has low permeability. Notice also that the magnetic loop encounters an air gap made of the space between the solenoid's frame and the plunger. As observed in

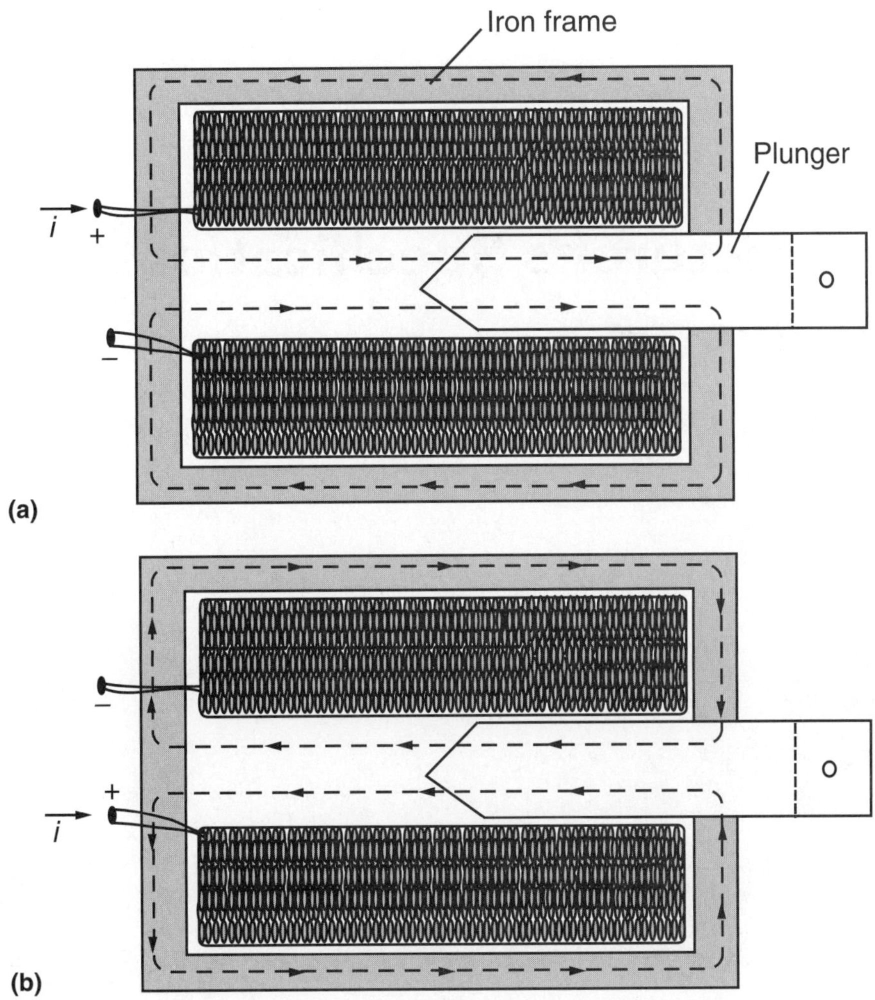

Figure AE–5 Effect of voltage polarity on the solenoid's lines of flux

the doorbell example, the lines of flux attempt to close any air gaps. In this case, this occurs by pulling the plunger inward. Notice that if the polarity of the input DC voltage is reversed, the lines of flux's direction is reversed. However, they still pull the plunger inward. Most commercial solenoids pull rather than push. Notice that to move this plunger outward, we must relay on a spring or another solenoid whose plunger is connected to this one and pulls this one outward.

Appendix F

Manufacturer's Specifications for a DC Motor

(Courtesy of Bodine Electric Company)

42A PM DC Motors 1/10 to 1/3 Hp

ELECTRIC COMPANY

42A motor with optional terminal box Model 0984

Selection Features

- High Starting Torque and Self-Braking
- Skewed Armature for Smooth Low Speed Operation
- Constant force Brush Springs for Long Life and Quietness
- Magnets Stabilized to Common Strength for Consistent Motor to Motor Performance
- Noise-Tested Ball Bearings
- Locked Bearing Minimizes Endplay
- Ceramic Magnets resist Demagnetization
- External Brush Access for Easy Inspection and Replacement
- Models Available with High Speed Shaft Extension for Accessories

Application Information

- Totally Enclosed, Non-Ventilated
- IP-40 Standard or IP-44 with Optional Die Cast Terminal Box Kit Model 0984 (see page 48)
- 3 Conductor Cord including ground, Connection Diagram 074 10101 (see page 6)
- Models with NEMA C Face may be base mounted with Base Kit Model 0991 (see page 48)
- See Page 51 and 54 for Motor Speed Torque Curves

Radial (Fr) and Axial (Fa) Load Capacity
(See page 3)

d= 3/4 in.
Fr = 90 lbs

Fa = 115 lbs

Bodine Data				Output			Input			Armature Data		
Model No.												
Std. Face	Std. Base	Dual Shaft	Type	Rpm	Torque Oz-in.	Hp	V	FF	Amps	Circuit ohms	Inductance mH	Inertia Oz.-in-Sec.^{2}
130V DC, FF=1.0 Continuous Duty												
4029	4028	• 4429	42A3BEPM	2500	50	1/8	130	1.0	1.0	10.8	40	.050
4035	4034	• 4435	42A5BEPM	2500	101	1/4	130	1.0	1.8	5.5	20	.082
4037	–	–	42A7BEPM	2500	134	1/3	130	1.0	2.3	4.2	17	.120
90V DC, FF=1.6 Continuous Duty												
–	4027	–	42A5BEPM	1725	58	1/10	90	1.6	1.1	4.5	3.3	.082
24V DC, FF=1.0 Continuous Duty												
4021	–	–	42A5BEPM	2500	101	1/4	24	1.0	8.8	.27	.64	.082
4022	–	–	42A7BEPM	2500	134	1/3	24	1.0	12	.16	.45	.120

42A Motor shown with NEMA C Face Mount

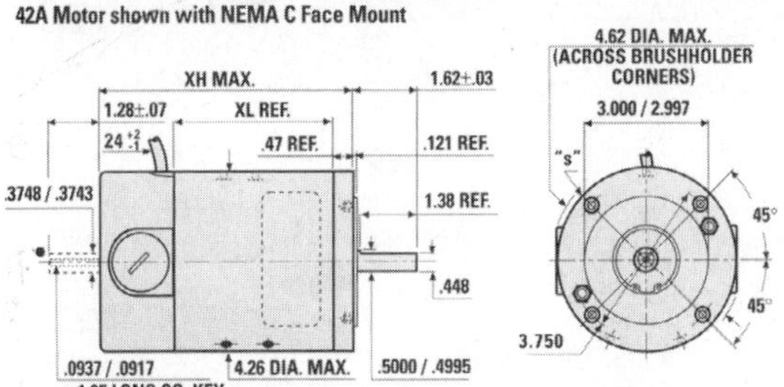

Dimensions (inches)

Type	XH	XL	Lbs.
42A3	5.60	3.30	9.5
42A4	6.16	3.86	10.5
42A5	6.72	4.42	13
42A7	7.87	5.55	16

"S" (4)1/4-20 UNC-2B: .30/.40 in. deep minimum

• Motor has "accessory ready" shaft extension. See page 49.

42A Motor shown with Base Mount

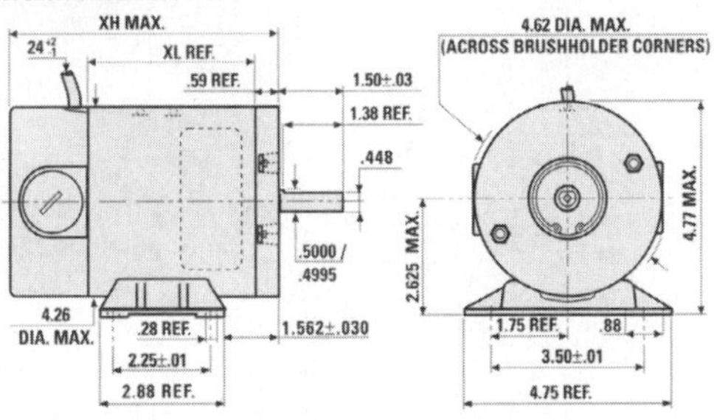

Dimensions (inches)

Type	XH Max	XL Max	Lbs.
42A3	5.72	3.30	9.5
42A4	6.28	3.86	10.5
42A5	6.84	4.42	13
42A7	8.00	5.55	16

24 Volt Motor/Gearmotor Selection Table

Bodine Data						Input	
Model No.	Type	Reference	Output Rpm	Rated Torque Oz-in.	Motor Hp	FF	Amps
24VDC FF=1.0 Continuous Duty							
Permanent Magnet Motors							
0040	24A0BEPM	Page 7	2500	8	1/50	1.0	1.2
0044	24A4BEPM		2500	22	1/18	1.0	2.6
4019	32A5BEPM	Page 8	2500	50	1/8	1.0	5.5
4021	42A5BEPM	Page 9	2500	101	1/4	1.0	8.8
4022	42A7BEPM		2500	135	1/3	1.0	12
Brushless Motors							
3502	22B2BEBL	Page 23	2500	25	1/16	1.0	3.3
3604	22B4FEBL		2500	50	1/8	1.0	5.6
3500	34B3BEBL	Page 23	2500	81	1/5	1.0	9.2
3600	34B3FEBL		2500				

Bodine Data							Input		
Model No.	Type	Reference	Output Rpm	Rated Torque Lb-in.	Ratio	Motor Hp	FF	Amps	Torque Lb-in.
24VDC, FF=1.0 Continuous Duty									
Permanent MagnetParallel Shaft Gearmotors									
4690	24A4BEPM-D3	Page 10	42	40	60	1/18	1.0	2.6	55
4663	24A4BEPM-Z4	Page 11	14	100	180	1/18	1.0	2.6	115
4660	24A4BEPM-Z3		83	30	30				59
4657	24A4BEPM-Z2		417	6.5	6				13
4647	32A5BEPM-W4	Page 13	8	175	312.4	1/8	1.0	5.5	210
4656			26	154	97.5				190
4654	32A5BEPM-W3	Page 13	57	96	43.9				105
4652	32A5BEPM-W2	Page 13	123	46	20.4				52
4649			456	12	5.48				14
4685	42A5BEPM-E3	Page 14	42	270	60	1/4	1.0	8.8	421
4680	42A5BEPM-E2		250	45	10				
4684	42A7BEPM-F3	Page 14	63	220	40	1/3	1.0	12	500
Brushless Parallel Shaft Gearmotors									
3629	22B2BEBL-D3	Page 24	83	29	30	1/16	1.0	3.3	33
3528	22B2FEBL-D3		139	17	18	1/16	1.0	3.3	20
3564	22B2BEBL-Z4	Page 25	21	100	120	1/16	1.0	3.3	115
3661	22B3FEBL-Z3		42	95	60	1/11	1.0	4.4	
3517	34B3BEBL-W3	Page 26	38	143	65.5	1/5	1.0	9.6	115
3611	34B3FEBL-W2		456	12	5.48				14
3585	34B4BEBL-E3	Page 27	42	270	60	1/4	1.0	12	421
Permanent Magnet Right Angle Gearmotors									
4694	24A4BEPM-3F	Page 15	250	9	10	1/18	1.0	2.6	20
4124	32A5BEPM-5F	Page 16	62	55	40	1/8	1.0	5.5	118
4122			250	21	10				44
4636	42A5BEPM-5N	Page 17	125	74	20	1/4	1.0	8.8	159

Bodine offers a broad line of 24 volt products for demanding low voltage applications. Common applications include battery powered devices and medical equipment. Permanent magnet motors and gearmotors, as well as brushless motors, gearmotors and controls are available. 24 volt PMDC motors and gearmotors will run on 12 volt batteries. Rated torque is the same, but speed is halved.

Page numbers are shown for motors and gearmotors. All 24 volt brushless controls are shown on page 29.

Performance Curves

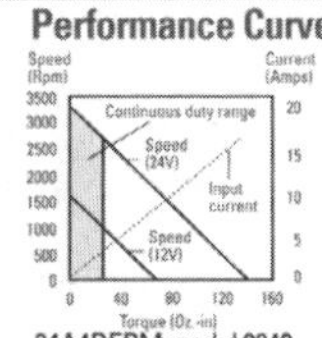

24A4BEPM,model 0040

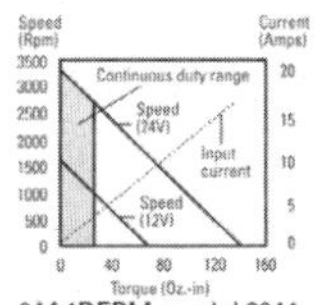

24A4BEPM, model 0044

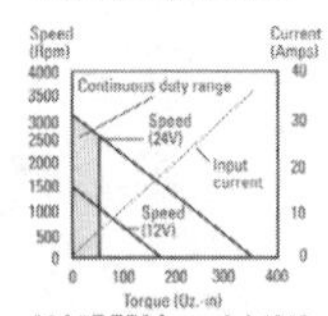

32A5BEPM, model 4019

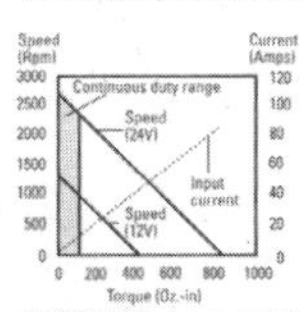

42A5BEPM, model 4021

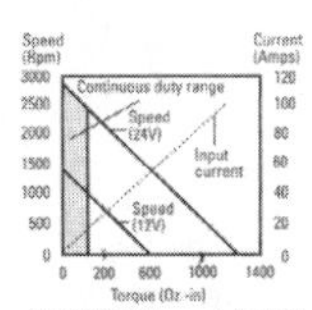

42A7BEPM, model 4022

Appendix G

Guidelines for Developing Your Idea Portfolio

Communication Issues of the Idea Portfolio

Communicate your progress to the instructor every 2 weeks using the accompanying "Progress Report" form. This will allow the instructor to give you feedback and ensure that you keep making progress. Your progress report should contain a Gantt chart showing where you want to be in your project and where you need to be. This tool will also help you improve your project-management skills.

Developing Your Idea into a Commercial Product

Choose an idea with market potential. College is an ideal time to consider becoming an entrepreneur or inventor. Nurture those inclinations. A university can be fertile ground on which to develop your ideas. The United States was forged into an industrial nation by the foresight, hard work, and perseverance of entrepreneurs like Ford, Edison, Packard, Hewlett, and Bell. Remember that you own your ideas. If you want to, have the instructor sign a nondisclosure document so that the instructor cannot reveal your idea without your written permission.

Progress Report

Project ______________________________

Engineer in Charge ____________________

Date ____________________ Week _____ of _____

Address the questions/issues at the end of each chapter.

GANTT CHART

Project selection

Analysis of input and output variables

Basic model development

Preliminary open-loop simulation

Model parameter determination

Controller specification

Preliminary closed-loop simulation

Fine tuning

Final simulation

Implementation suggestions

References

Chapter 1. Introduction

1. Somerville, J.W., and Macia, N.F. "A Feedback Control System for Engineering Technology Laboratory Courses," Proceedings, American Society for Engineering Education, Annual Conference, June 2001, Albuquerque, NM.

Chapter 2. Mathematical Background for Dynamic Systems

1. Spiegel, M.R. (1981). *Applied Differential Equations.* Englewood Cliffs, NJ: Prentice-Hall.
2. Agnew, R.P. (1960). *Differential Equations,* 2nd ed. New York: McGraw-Hill.
3. Betz, H., and Ewing, G. M. (1964). *Differential Equations with Applications,* 2nd ed. New York: Harper and Row.
4. Thaler, G.J. (1989). *Automatic Control Systems.* St. Paul, MN: West Publishing Co.
5. Stanley, W.D. (1997). *Transform Circuit Analysis for Engineering and Technology,* 3rd ed. Upper Saddle River, NJ: Prentice-Hall.
6. Jenness, R.R. (1965). *Analog Computation and Simulation: Laboratory Approach.* Boston: Allyn and Bacon, Inc.

Chapter 3. Modeling of Dynamic Systems

1. Franklin, G.F., Powell, J.D., and Emami-Naeini, A. (1994). *Feedback Control of Dynamic Systems,* 3rd ed. Reading, MA: Addison Wesley.
2. Mayr, O. (1970). *The Origins of Feedback Control.* Cambridge, MA: MIT Press.
3. Anderson, J.D., Jr. (2000). *Introduction to Flight,* 4th ed. Boston: McGraw-Hill.
4. Shearer, J.L., Murphy, A.T., and Richardson, H.H. (1967). *Introduction to System Dynamics.* Reading, MA: Addison Wesley.
5. Beakley, G.C. (1984). *Careers in Engineering and Technology,* 3rd ed. New York: Macmillan.
6. Karnopp, D.C., and Rosenburg, R.C. (1975). *Systems Dynamics: A Unified Approach.* New York: Wiley-Interscience.
7. Lewis, J.W. (1994). *Modeling Engineering Systems.* Solana Beach, CA: High-Text Publishers.
8. Close, C.M., and Frederick, D.K. (1993). *Modeling and Analysis of Dynamic Systems,* 2nd ed. New York: Houghton Mifflin Co.
9. Cannon, R.H. (1967). *Dynamics of Physical Systems.* New York: McGraw-Hill.

10. Cochin, I. (1977). *Analysis and Design of Dynamic Systems.* New York: IEP-A Dun-Donnelley Publisher.
11. Blackwell, C.C., and Giles, R.F. (1974). *An Introduction to Mechanical Engineering System Modeling,* unpublished manuscript. University of Texas at Arlington.
12. Hayt, W.H., Jr., and Hughes, G.W. (1968). *Introduction to Electrical Engineering.* New York: McGraw-Hill Book Co.
13. Nilsson, J.W., and Riedel S.A. (2000). *Electric Circuits,* 6th ed. Upper Saddle River, NJ: Prentice-Hall.

Chapter 4. Feedback Control

1. Gadd, C.J. (1929). "Babylonia Law," *Encyclopedia Britannica,* 14th ed. Vol. 2, 283.
2. Wolf, A. (1939). *A History of Science, Technology, and Philosophy in the XVth Century.* New York: Macmillan.
3. Dorf, R.C., and Bishop, R.H. (1998). *Modern Control Systems,* 8th ed. Reading, MA: Addison Wesley.
4. D.A. Linkens "Adaptive and Intelligent Control in Anesthesis," *IEEE Control Systems,* December 1992, 6–10.

Chapter 5. Stability and Dynamic Response

1. Routh, E.J. (1905). *Dynamics of a System of Rigid Bodies.* London: MacMillan and Co., Ltd.

Chapter 6. Time Domain Performance Characteristics

1. Dorf, R.C., and Bishop, R.H. (1998). *Modern Control Systems,* 8th ed. Reading, MA: Addison Wesley.

Chapter 7. Root Locus Analysis

1. Evans, W.R. (1950). "Control Systems Synthesis by Root Locus Method," *AIEE Trans.*
2. Evans, W.R. (1950). "Graphical Analysis of Control Systems," *AIEE Trans, 67.*
3. Orbit, A. Technical (First Quarter, 2000). Publication for the Adv. of Machinery Production, Management, and Rotor Dynamics. Nevada: Bently.
4. Ogata, K. (1970). *Modern Control Systems.* Upper Saddle River, NJ: Prentice-Hall.

Chapter 8. Frequency Response Analysis

1. Nyquist, H. (January 1932). *Regeneration Theory,* BSTJ.
2. Bode H.W. (1948). *Network Analysis and Feedback Amplifier Design.* New York: Van Nostrand.
3. Bode H.W. (1940). "Relations between Attenuation and Phase in Feedback Amplifier Design." *BSTJ.*
4. Weber, R., While, M., and Manning K. (1959). *College Physics,* 3rd ed. New York: McGraw-Hill.
5. Smith, D. (March 29, 1974). *A Case Study and Analysis of the Tacoma Narrows Bridge Failure.* Engineering Project, Department of Mechanical Engineering, Carleton University, Ottawa, Canada.
6. Farquharson, F.B. (1950). "Aerodynamic Stability of Suspension Bridges with Special Reference to the Tacoma Narrows Bridge," Bulletin 116, Part I, The Engineering Experiment Station, University of Washington.

Chapter 9. Introduction to State Space Methods

1. Thaler, G.J. (1989). *Automatic Control Systems,* St. Paul, MN: West Publishing Company.
2. Macia, N.F. (August 1988). *Noninvasive, Quick Obstruction Estimation Method for the Measurement of Parameters in the Nonlinear Lung Model,* Ph.D. Dissertation, Arizona State University.

Chapter 10. Design of Control Systems

1. Whiteley, A.L. "Theory of Servo Systems with Particular Reference to Stabilization," *Journal IEE,* Vol. 93(2). 553–67.
2. Graham, D., and Lathrop, R.C. "The Synthesis of Optimum Transient Response," *Trans AIEE* (Part 2) Vol. 72, pp. 273–88.
3. Ziergler, J.B., and Nichols, N.B. "Optimum Settings for Automatic Controllers", *ASME.*
4. Corripio, A.B. (2000). *Tuning of Industrial Control Systems,* 2nd ed. Research Triangle Park, NC: ISA.
5. Ibid.
6. Ziergler, J.B., and Nichols, N.B. "Optimum Settings for Automatic Controllers", *ASME.*
7. St. Clair, D.W. (1990). *Controller Tuning in Control Loop Performance.* Newark, DE: Straight-Line Central Company.
8. Franklin, G.F., Powell, J.D., and Emani-Naeni, A. (1994). *Feedback Control of Dynamic Systems,* 3rd ed. Reading, MA: Addison Wesley.

9. Corripio, A.B. (2000). *Tuning of Industrial Control Systems,* 2nd ed. Research Triangle Park, NC: ISA.
10. Astrom, K.J., Wittenmark, B. (1995). *Adaptive Control,* 2nd ed. Reading, MA: Addison-Wesley.
11. Kim, N.K. (2004). "Developing an Inexpensive Device to Train Controller Tuning," Proceedings of ASEE 2004 Annual Conference.

Chapter 11. Implementing the Controls Scheme with Hardware: PLCs

1. Programmable Gain Amplifier, MCP6S2X. Chandler, AZ: Microchip Corp. www.microchip.com.
2. Drzewiecki, T.M., and Macia, N.F. (2003). "Fluidic Technology: Adding Control, Computation, and Sensing Capability to Microfluidics." Microtechnologies for the New Millennium: Smart Sensors, Actuators, and MEMS. Gran Canaria, Spain: SPIE.
3. Norvelle, F.D. (2000). *Electrohydraulic Control Systems.* Upper Saddle River, NJ: Prentice-Hall.
4. *MicroLogix 1200 and MicroLogix 1500 Programmable Controllers, Instruction Set, and Reference Manual,* (2000). Publication 1762-RM001C-EN-P. (Bulletins 1762 and 1764), Madison, WI: Allen-Bradley.

Chapter 12. Introduction to Digital Control Systems

1. Dabney, J.B., and Jarman, T.L. (2001). *Mastering SIMULINK4.* Upper Saddle River, NJ: Prentice-Hall.
2. Ibid.
3. Franklin, G.F., Powell, J.D., and Workman, M. (1998). *Digital Control of Dynamic Systems,* 3rd ed. Reading, MA: Addison Wesley.

Chapter 13. Case Study: A Position Control System Using a DC Solenoid

1. Jackson, H.W. (1986). *Introduction to Electric Circuits,* 6th ed. Upper Saddle River, NJ: Prentice-Hall.
2. Macia, N.F. (1996). "Using a DC Solenoid in a Closed-Loop Position Control System to Teach Control Technology," Washington, DC: Proceedings ASEE 1996 Annual Conference.

Index

D

E

F

G

H

O

P

T

U

V

W

Z